RADIO FREQUENCY POWER IN PLASMAS
TENTH TOPICAL CONFERENCE

AIP CONFERENCE PROCEEDINGS 289

RADIO FREQUENCY POWER IN PLASMAS

TENTH TOPICAL CONFERENCE

BOSTON, MA APRIL 1993

EDITORS:
MIKLOS PORKOLAB
MASSACHUSETTS INSTITUTE OF TECHNOLOGY

JOEL HOSEA
PRINCETON UNIVERSITY

American Institute of Physics   New York

Authorization to photocopy items for internal or personal use, beyond the free copying permitted under the 1978 U.S. Copyright Law (see statement below), is granted by the American Institute of Physics for users registered with the Copyright Clearance Center (CCC) Transactional Reporting Service, provided that the base fee of $2.00 per copy is paid directly to CCC, 27 Congress St., Salem, MA 01970. For those organizations that have been granted a photocopy license by CCC, a separate system of payment has been arranged. The fee code for users of the Transactional Reporting Service is: 0094-243X/87 $2.00.

© 1994 American Institute of Physics.

Individual readers of this volume and nonprofit libraries, acting for them, are permitted to make fair use of the material in it, such as copying an article for use in teaching or research. Permission is granted to quote from this volume in scientific work with the customary acknowledgment of the source. To reprint a figure, table, or other excerpt requires the consent of one of the original authors and notification to AIP. Republication or systematic or multiple reproduction of any material in this volume is permitted only under license from AIP. Address inquiries to Series Editor, AIP Conference Proceedings, AIP Press, American Institute of Physics, 500 Sunnyside Boulevard, Woodbury, NY 11797-2999.

L.C. Catalog Card No. 93-72964
ISBN 1-56396-264-0
DOE CONF-9304112

Printed in the United States of America.

CONTENTS

Preface xv

ION CYCLOTRON RANGE OF FREQUENCIES

Overview of TFTR ICRF Results (Invited) 3
    G. Schilling
**First Results of Ion Cyclotron Resonance Heating
on ASDEX Upgrade (Invited)** 12
    J.-M. Noterdaeme, C. Hoffmann, M. Brambilla, K. Büchl,
    A. Eberhagen, A. Field, C. Fuchs, O. Gehre, J. Gernhardt,
    O. Gruber, G. Haas, A. Hermann, F. Hofmeister, A. Kallenbach,
    G. Lieder, V. Mertens, H. Murmann, S. de Peña Hempel,
    W. Poschenrieder, Th. Richter, F. Ryter, N. Salmon, H. Salzmann,
    W. Schneider, F. Wesner, H.-P. Zehrfeld, H. Zohm, ICRH Team,
    ASDEX Upgrade Team
**High-Power ICRF and LHCD Experiments on
Tore Supra (Invited)** 24
    B. Saoutic, B. Beaumont, A. Bécoulet, J. P. Bizarro,
    D. Fraboulet, X. Garbet, M. Goniche, L. Guiziou,
    G. T. Hoang, T. Hutter, E. Joffrin, H. Kuus, X. Litaudon,
    P. Mollard, D. Moreau, F. Nguyen, A. L. Pecquet,
    Y. Peysson, G. Rey, D. van Houtte, M. Zabiégo
Review of Combined ICRH–NBI Results in TEXTOR 32
    A. M. Messiaen, P. E. Vandenplas, D. Van Eester,
    G. Van Wassenhove, H. Conrads, P. Dumortier,
    F. Durodié, H. Euringer, G. Fuchs, B. Giesen,
    D. Hillis, F. Hoenen, R. Koch, H. R. Koslowski,
    A. Krämer-Flecken, M. Lochter, B. Mullier, J. Ongena,
    T. Oyevaar, H. Soltwisch, H. F. Tammen, G. Telesca,
    R. Uhlemann, R. Van Nieuwenhove, G. Van Oost,
    M. Vervier, G. Waidmann, R. R. Weynants
High-Power ICRF Experiments on TFTR 36
    J. R. Wilson, J. C. Hosea, R. Majeski, C. K. Phillips,
    J. H. Rogers, G. Schilling, J. Stevens, G. Taylor, TFTR Group
    (PPPL), M. Murakami, D. A. Rasmussen
Two Frequency ICRF Operation on TFTR 40
    J. H. Rogers, R. Majeski, J. R. Wilson, J. C. Hosea,
    G. Schilling, J. Stevens, and D. A. Rasmussen
**Modeling of High-Power ICRF Heating Experiments
on TFTR** 44
    C. K. Phillips, J. R. Wilson, M. Bell, E. Fredrickson,
    J. C. Hosea, A. Khudaleev, R. Majeski, M. Murakami,
    M. P. Petrov, A. Ramsey, J. H. Rogers, G. Schilling,
    C. H. Skinner, J. E. Stevens, G. Taylor, K.-L. Wong,
    TFTR Group (PPPL)

Fast Wave Direct Electron Heating in TFTR.................................. 48
    M. Murakami, E. F. Jaeger, F. G. Rimini, D. A. Rasmussen,
    J. E. Stevens, J. R. Wilson, D. B. Batchelor, M. Bell, R. Budny,
    E. Fredrickson, R. C. Goldfinger, G. Hammett, D. J. Hoffman,
    J. C. Hosea, A. Janos, R. Majeski, D. Mansfield, C. K. Phillips,
    J. H. Rogers, G. Schilling, G. Taylor, M. C. Zarnstorff

Sawtooth Stabilization Experiments by ICRF Heating
Alone and its Combination with NBI or
LHCD in JT-60U 52
    H. Kimura, T. Fujii, M. Sato, M. Nemoto, K. Hamamatsu,
    T. Kondoh, M. Matsuoka, S. Moriyama, M. Saigusa, H. Takeuchi,
    K. Ushigusa, D. J. Campbell

ICRF Heating Scenarios in Alcator C-MOD 56
    Y. Takase, P. T. Bonoli, S. N. Golovato,
    M. Porkolab

Ion Cyclotron Heating Experiments and Plans
for the Advanced Toroidal Facility (ATF)............................... 60
    D. A. Rasmussen, D. B. Batchelor, R. H. Goulding,
    D. J. Hoffman, E. F. Jaeger, J. A. Rome, C. E. Thomas

PBX-M Ion Bernstein Wave Heating Overview.......................... 64
    M. Ono, R. Cesario, T. K. Chu, H. Herrmann, B. LeBlanc,
    T. Seki, W. Tighe, N. Asakura, R. Bell, L. Blush, S. Bernabei,
    R. Conn, R. Doerner, J. Dunlap, A. England, A. Grossman,
    J. Harris, R. Hatcher, R. Isler, R. Kaita, S. Kaye, H. Kugel,
    M. Okabayashi, H. Oliver, F. Paoletti, S. Paul, N. Sauthoff,
    L. Schmitz, S. Sesnic, H. Takahashi, G. Tynan

Density Profile Modification During IBW in PBX-M 68
    B. LeBlanc, M. Ono, W. Tighe, J. Dunlap, R. Bell, T. K. Chu,
    A. England, R. Isler, S. Kaye, D. McCune, M. Okabayashi,
    A. Post-Zwicker, H. Takahashi, S. Sesnic

Changes to the Ion Temperature Profile During
IBW Heating in PBX-M 72
    W. Tighe, R. Bell, T. K. Chu, H. Hermann, B. LeBlanc,
    M. Okabayashi, M. Ono, N. Asakura, R. Cesario,
    A. England, R. Isler, R. Kaita, H. Kugel, S. Paul,
    A. Post-Zwicker, H. Takahashi

IBW Generated Ponderomotive Potential Effect
on Edge Plasma in PBX-M... 76
    A. Grossman, L. Schmitz, R. Doerner, M. Ono, H. Kugel,
    M. Okabayashi, R. Bell, G. Tynan, L. Blush, R. W. Conn,
    PBX-M Group

Ion Bernstein Waves in a Toroidal Steady-State Plasma 80
    C. Riccardi, D. Batani, M. Fontanesi, A. Galassi,
    E. Sindoni

LOWER HYBRID RANGE OF FREQUENCIES

Lower-Hybrid Current Drive Experiments, Synergism
with the Fast Wave Near Ion Cyclotron Resonance and
Future Plans on JET (Invited) 87
    C. Gormezano

Lower-Hybrid Current Drive and ICRF Heating Experiments on JT-60U (Invited) 99
    T. Fujii, K. Ushigusa, M. Saigusa, H. Kimura, T. Imai,
    Y. Ikeda, O. Naito, M. Nemoto, S. Moriyama, M. Seki,
    T. Kondoh, K. Hamamatsu, M. Sato, R. Yoshino,
    Y. Kamada, D. J. Campbell

Lower-Hybrid Enhanced Performance in Tore Supra 107
    G. T. Hoang, D. Moreau, E. Joffrin, X. Litaudon,
    Y. Peysson, Tore Supra Team, R. V. Budny, S. Kaye,
    S. A. Sabbagh, V. Fuchs

Measurement of the Parallel Distribution Function During Lower-Hybrid Current Drive in Tore Supra 111
    J. L. Ségui, G. Giruzzi, D. Vézard, W. D. Liu,
    X. Caron, R. L. Meyer

First Results of the 8 GHz LH Experiment in FTU 115
    F. Alladio, M. L. Apicella, G. Apruzzese, E. Barbato,
    R. Bartiromo, F. Bombarda, G. Bracco, G. Buceti,
    P. Buratti, A. Cardinali, R. Cesario, M. Ciotti,
    V. Cocilovo, A. Coletti, I. Condrea, F. Crisanti,
    R. De Angelis, F. De Marco, B. Esposito, T. Fortunato,
    D. Frigione, L. Gabellieri, E. Giovannozzi, G. Granucci,
    M. Grolli, S. Ide, A. Imparato, H. Kroegler, L. Lovisetto,
    G. Maddaluno, M. Marinucci, G. Mazzitelli, D. McNeill,
    P. Micozzi, F. Mirizzi, A. Moleti, F. Orsitto, L. Panaccione,
    M. Panella, V. Pericoli, L. Pieroni, S. Podda, G. B. Righetti,
    M. Roccon, D. Santi, F. Santini, M. Sassi, E. Sternini,
    G. Tonini, A. A. Tuccillo, O. Tudisco, F. Valente, V. Vitale,
    V. Zanza, M. Zerbini

Fast Electron Current Density Profile and Diffusion Studies During LHCD in PBX-M 119
    S. E. Jones, J. Kesner, S. C. Luckhardt, F. Paoletti,
    S. Bernabei, R. Kaita, S. von Goeler,
    PBX-M Group, F. Rimini

Determination of the Energy of Suprathermal Electrons During Lower-Hybrid Current Drive on PBX-M 123
    S. von Goeler, S. Bernabei, W. Davis, D. Ignat,
    S. Jones, R. Kaita, G. Petravich, F. Rimini, P. Roney,
    J. Stevens, A. Post-Zwicker

Comparison of X-Ray Pinhole Camera Images with Calculations Based on Lower-Hybrid Wave Physics 127
    D. W. Ignat, E. J. Valeo, S. von Goeler

Motional Stark Effect Plasma Equilibria During LHCD Experiments on PBX-M 131
    F. Paoletti, S. Batha, S. Bernabei, H. Fishman, R. Hatcher,
    S. Hirshman, D. Ignat, S. Jones, R. Kaita, S. Kaye, J. Kesner,
    C. Kessel, B. LeBlanc, D. Lee, F. Levinton, S. Luckhardt,
    M. Okabayashi, H. Takahashi, S. Sesnic, Y. Sun, S. von Goeler

ELECTRON CYCLOTRON RANGE OF FREQUENCIES

**High-Power 140 GHz ECRH Experiments at the
W7-AS Stellarator (Invited)** 137
  V. Erckmann, R. Burhenn, T. Geist, H. J. Hartfuss, M. Kick,
  M. Maassberg, W7-AS Team, NBI Team, W. Kasparek,
  G. A. Müller, P. G. Schüller, V. I. Il'in, V. I. Kurbatov,
  S. Malygin, V. I. Malygin

Suppression of Disruptions with ECH on JFT-2M (Invited) 149
  K. Hoshino, M. Mori, T. Yamamoto, H. Tamai, T. Shoji,
  Y. Miura, H. Aikawa, S. Kasai, T. Kawakami, H. Kawashima,
  M. Maeno, T. Matsuda, K. Nagashima, K. Oasa, K. Odajima,
  H. Ogawa, T. Ogawa, T. Seike, T. Shiina, K. Uehara,
  T. Yamauchi, N. Suzuki, H. Maeda

**Experiments on Nonlinear Absorption of ECH Waves
in MTX and a Comparison with Theory (Invited)** 157
  B. W. Stallard, S. L. Allen, J. A. Byers, T. A. Casper,
  B. I. Cohen, R. H. Cohen, C. J. Lasnier,
  M. E. Fenstermacher, J. H. Foote, E. B. Hooper,
  M. A. Makowski, W. H. Meyer, J. M. Moller,
  B. W. Rice, T. D. Rognlien, G. R. Smith,
  K. I. Thomassen, R. D. Wood, K. Hoshino,
  K. Oasa, K. Odajima, T. Ogawa, T. Oda, T. Ogo

**Dimensionally Similar Discharges with Central RF Heating
on the DIII-D Tokamak** 165
  C. C. Petty, T. C. Luce, R. I. Pinsker

**Modification of Electrical Conductivity in T-10 by Electron
Cyclotron Heating** 169
  R. W. Harvey, C. B. Forest, O. Sauter, J. Lohr,
  Y. R. Lin-Liu

**ECRH Scenarios at High Magnetic Field and
Electron Density on the FTU Tokamak** 173
  L. Argenti, A. Bruschi, S. Cirant, G. Granucci,
  S. Nowak, A. Simonetto, G. Solari

CURRENT DRIVE AND PROFILE CONTROL

**Review of Tokamak Experiments on Direct Electron
Heating and Current Drive with Fast Waves (Invited)** 179
  R. I. Pinsker

**Status and Comparison of Codes Used for Fast Wave
Current Drive (Invited)** 192
  P. T. Bonoli

**Control of the Current Density Profile with Lower-Hybrid
Current Drive on PBX-M (Invited)** 202
  R. E. Bell, S. Bernabei, L. Blush, T. K. Chu, R. Doerner,
  J. Dunlap, A. England, G. Gettelfinger, N. Greenough,

J. Harris, R. Hatcher, S. Hirshman, D. Ignat, R. Isler,
S. Jardin, R. Kaita, S. Kaye, T. Kozub, H. Kugel,
B. LeBlanc, S. Jones, J. Kesner, D. Lee, F. Levinton,
S. Luckhardt, M. Okabayashi, F. Paoletti, S. Paul,
F. Rimini, N. Sauthoff, S. Sesnic, L. Schmitz, Y. Sun,
H. Takahashi, W. Tighe, G. Tynan, E. Valeo,
S. von Goeler

Current Profile Control and Stability Studies in the Tokamak Physics Experiment (TPX) 210

    M. Porkolab, P. T. Bonoli, J. J. Ramos,
    M. E. Fenstermacher

Advanced Tokamak Operations with ICRF and Lower-Hybrid Power 214

    T. K. Mau, B. J. Lee, D. A. Ehst

Parasitic Effects of Ion Absorption on Fast Wave Current Drive in TPX 218

    P. E. Moroz, D. B. Batchelor, E. F. Jaeger,
    T. K. Mau, D. R. Mikkelsen, M. Porkolab

Lower-Hybrid Counter Current Drive for Edge Current Density Modification in DIII-D 222

    M. E. Fenstermacher, W. M. Nevins, M. Porkolab,
    P. T. Bonoli, R. W. Harvey

Experimental Studies of High-Energy X-Ray Emission and Bootstrap Current Generation in High $\epsilon\beta_p$ Lower-Hybrid Driven Plasmas 226

    J. P. Squire, M. Porkolab, J. A. Colborn, J. Villaseñor

Lower-Hybrid Current Drive, Hard X-Ray Emission, and Fluctuations 230

    L. Vahala, G. Vahala, P. T. Bonoli

Current Drive With the Second ECR Harmonic on T-10 234

    V. V. Alikaev, A. A. Bagdasarov, A. A. Borshegovskij,
    M. M. Dremin, Yu.V. Esipchuk, Yu.A. Gorelov,
    N. V. Ivanov, A. Y. Kislov, L. K. Kuznetsova,
    G. E. Notkin, Yu.D. Pavlov, K. A. Razumova,
    I. N. Roy, N. L. Vasin, V. A. Vershkov, T-10 Team,
    C. B. Forest, J. Lohr, T. C. Luce, R. W. Harvey

Inductive Effects During Second Harmonic Current Drive Experiments on T-10 238

    C. B. Forest, R. W. Harvey, J. Lohr, T. C. Luce,
    Y. R. Lin-Liu, Yu. Esipchuk, G. Notkin,
    K. Razumova, and the T-10 Team

On RF "Helicity Injection" and Alfvén Wave Current Drive 242

    C. Litwin, N. Hershkowitz

Modeling of RF-Assisted Helicity Injection Start-Up and Current Drive 246

    Y. S. Hwang, M. Ono

THEORY

Kinetic Analysis of Minority Gyroresonant Heating:
Conversion Fields in Tokamak Geometry 253
    E. R. Tracy, A. J. Brizard, D. R. Cook,
    A. N. Kaufman

Enhanced Decay Instability and Mode Conversion
to a Strongly-Damped Nonlinear Wave 257
    L. Friedland, A. N. Kaufman, J. J. Morehead

RF Heating of the Ionosphere: An Example of Generic
Linear Mode Conversion 261
    W. G. Flynn, R. G. Littlejohn

A Simple Derivation of Relativistic Full-Wave Equations
at Electron Cyclotron Resonance 265
    D. C. McDonald, R. A. Cairns, C. N. Lashmore-Davies

Electron Cyclotron Absorption and Emission: "Vexatae Questiones" 269
    M. Bornatici, F. Engelmann

Electron Cyclotron Power Absorption in Plasmas with
Non-Maxwellian Electron Velocity Distributions 273
    S. Yue, A. H. Kritz, G. R. Smith

Perpendicular Ion Acceleration by Short
Scale-Length RF Fields 277
    K. J. Reitzel, G. J. Morales, V. K. Decyk

Statistical Approach to LHCD Modeling Using the Wave
Kinetic Equation 281
    K. Kupfer, D. Moreau, X. Litaudon

Full-Wave Theory of a Quasi-Optical Launching System
for Lower-Hybrid Waves: Preliminary Results 285
    G. Cincotti, F. Gori, M. Santarsiero, R. Serrecchia,
    F. Frezza, G. Schettini, F. Santini

The Effect of Poloidal Antenna Width on Lower-Hybrid
Wave Propagation 289
    R. A. Cairns, V. Fuchs

Interaction of ICRF Waves with Lower-Hybrid Driven
Suprathermal Electrons 293
    A. K. Ram, A. Bers, V. Fuchs, R. W. Harvey

Self-Consistent Modeling of RF Heating of Fast Particle
Populations and Beams 297
    D. Van Eester

Global Wave Modeling of Electron Interactions with Fast
Magnetosonic Waves 301
    E. F. Jaeger, D. B. Batchelor, M. Murakami

The Efficiency of Fast Wave Current Drive for a Weakly
Relativistic Plasma 305
    S. C. Chiu, Y. R. Lin-Liu, C. F. F. Karney

Surface Waves in Nonuniform Plasmas and Their Absorption
at the Localization of Surface Plasmons 309
    Yu. M. Aliev, J. Berndt, H. Schlüter, A. Shivarova

ANTENNA DESIGN AND RF TECHNOLOGY

High-Power and Long Pulse Capability of the LH Systems
on Tore Supra and Test Bed Facility.................................... 315
    G. Rey, M. Goniche, G. Berger-By, P. Bibet, J. P. Bizarro,
    J. J. Capitain, J. Carrasco, Y. Demers, D. Guilhem,
    G. T. Hoang, X. Litaudon, R. Magne, D. Moreau,
    Y. Peysson, M. Seki, J. Schlosser, G. Tonon

Characteristics of a Large Multijunction Launcher for
High-Power LHCD Experiments on JT-60U 319
    M. Seki, Y. Ikeda, K. Ushigusa, O. Naito, T. Kondoh,
    S. W. Wolfe, and T. Imai

Combline Antennas for Launching Traveling Fast Waves................... 323
    C. P. Moeller, R. W. Gould, D. A. Phelps, and R. I. Pinsker

Folded Waveguide Designs for Tokamaks................................ 327
    D. J. Hoffman, T. S. Bigelow, C. H. Fogelman, J. J. Yugo,
    J. B. O. Caughman, W. L. Gardner, M. D. Carter,
    P. H. Probert, E. Barbato

Comparison of the Folded Stripline and Stacked Stripline
Concepts to the Folded Waveguide Launcher 331
    W. L. Gardner, J. B. O. Caughman, D. J. Hoffman,
    P. H. Probert

Three-Dimensional Effects for Radio Frequency
Antenna Modeling... 335
    M. D. Carter, D. B. Batchelor, D. C. Stallings

Conversion of the Four-Strap Array in DIII-D
to a Tunable Traveling Wave Antenna................................. 339
    D. A. Phelps, C. P. Moeller, C. C. Petty, R. I. Pinsker,
    P. M. Ryan, R. H. Goulding, D. J. Hoffman

Design of Long-Pulse Fast Wave Current Drive Antennas
for DIII-D 343
    F. W. Baity, D. B. Batchelor, K. C. Bills, C. H. Fogelman,
    E. F. Jaeger, J. L. Ping, B. W. Riemer, P. M. Ryan,
    D. C. Stallings, D. J. Taylor, J. J. Yugo

Effect on Antenna Structure of High-Power RF
During Plasma Operation 347
    G. Haste, C. E. Thomas, A. Fadnek, M. Carter,
    B. Beaumont, A. Becoulet, H. Kuus, B. Saoutic

Power Compensators for Phased Operation of Antenna Arrays
on JET and DIII-D 351
    R. H. Goulding, D. J. Hoffman, P. M. Ryan, G. Bosia,
    M. Bures, D. Start, T. Wade, C. C. Petty, R. I. Pinsker

Electrical Characterization of the JET A_2 Antenna:
Comparison of Model with Measurements 355
    P. M. Ryan, R. H. Goulding, V. Bhatnagar,
    A. Kaye, T. Wade

Self-Consistent 3-D ICRH Antenna Modeling with Plasma 359
    Y. L. Ho, W. Grossmann, A. Drobot, M. D. Carter,
    P. M. Ryan, D. B. Batchelor

A "3-D" ICRF Antenna Coupling Model for Heating
and Current Drive Applications in Tokamaks............................. 363
    M. H. Bettenhausen, J. E. Scharer

Ion Cyclotron Range of Frequencies (ICRF) Heating
of Fast Ions in Fusion Plasmas... 367
    J. E. Scharer, N. T. Lam, R. S. Sund, O. Sauter

Design of the Ion Cyclotron System for TPX............................ 371
    D. W. Swain, S. Shipley, J. Yugo, R. Goulding,
    D. Batchelor, D. Stallings, E. Fredd

Design and Coupling Characteristics of Lower-Hybrid
Launcher for TPX 375
    A. E. Hubbard, M. Porkolab, S. Bernabei,
    N. Greenough, P. Goranson, D. Swain, J. Yugo

Arc Detection and Protection in High-Power Antenna Systems 379
    J. B. O. Caughman, D. J. Hoffman

ICRF Coil for the IDEAL Plasma.. 383
    R. W. Motley, R. Majeski, S. A. Cohen,
    M. Diesso, J. R. Wilson

Antenna Conditioning with Insulating Antenna Tiles
in Phaedrus-T 387
    T. Intrator, P. Probert, M. Doczy, D. Diebold,
    D. Brouchous

A Preliminary Engineering Assessment of the ITER
CDA ECH Launcher 391
    T. S. Bigelow, D. W. Swain, M. Sawan

Megawatt Gyrotrons for ECRH... 395
    M. Blank, T. L. Grimm, W. C. Guss, K. E. Kreischer,
    R. J. Temkin

GENERAL RF PLASMA INTERACTIONS

Regimes for Electron Heating via Mode Conversion
in TFTR.. 401
    R. Majeski, J. C. Hosea, C. K. Phillips, J. H. Rogers,
    G. Schilling, J. E. Stevens, J. R. Wilson

DT Simulation of ICRF Heated Supershots in TFTR
Using TRANSP.. 405
    R. C. Goldfinger, D. B. Batchelor, C. K. Phillips, R. Budny,
    G. W. Hammett, J. C. Hosea, D. M. McCune, J. E. Stevens,
    J. R. Wilson, and the TFTR Team

Behavior of Toroidal Alfvén Eigenmodes During ICRF Heating............. 409
    K. L. Wong, J. R. Wilson, Z. Y. Chang, D. Darrow,
    E. Fredrickson, C. K. Phillips

Computational Path to Second Stability in PBX-M 413
    M. Okabayashi, D. W. Ignat, S. C. Jardin, Y.-C. Sun

Quantitative Radiation Limits on ICRF-Specific Impurities.............. 417
    D. A. D'Ippolito, J. R. Myra

Far Field ICRF Sheath Formation on Walls and Limiters 421
    J. R. Myra, D. A. D'Ippolito
ICRF System and Plasma Performance of the
Ignitor Experiment 425
    F. Carpignano, B. Coppi, P. Detragiache,
    M. Nassi, L. Sugiyama
Phaedrus-T Tokamak Probe Measurements 429
    J. Sorensen, D. Diebold, N. Hershkowitz, T. Intrator,
    R. Majeski, J. Meyer, P. Probert, G. Winz
Anomalous Electron Streaming Due to Waves in
Tokamak Plasmas 433
    S. D. Schultz, A. Bers, A. K. Ram
Microwave Reflectometry for ICRF Coupling Studies
on TFTR 437
    J. B. Wilgen, G. R. Hanson, T. S. Bigelow, D. B. Batchelor,
    I. Collazo, D. J. Hoffman, M. Murakami, D. A. Rasmussen,
    D. C. Stallings, S. Raftopoulos, J. R. Wilson
Ion Cyclotron Resonance Heating for the Plasma
Separation Process 441
    A. Compant La Fontaine
Poloidal Electric Field Due to ICRH and its Effect
on Neoclassical Transport 445
    L. Vacca
Measurement of the Correlation Spectrum of Electrostatic
Potential Fluctuations in a Toroidal ECRH Plasma 449
    E. D. Zimmerman, S. C. Luckhardt, J. C. Rost
Analytical Estimations on the Axial Structure of
Plasma-Waveguide Discharges 453
    Yu. M. Aliev, I. Ghanashev, A. Shivarova, H. Schlüter,
    M. Zethoff
Observation of Edge Electron Heating During 800 MHz Lower
Hybrid Fast Wave Experiments on the Versator II Tokamak 457
    J. Villaseñor, M. Porkolab, G. Gibson, J. Colborn,
    J. Squire
Author Index 461

PREFACE

The Tenth Topical Conference on RF Power in Plasmas was held in the Park Plaza Hotel, Boston, Massachusetts, April 1–3, 1993, under the sponsorship of the Massachusetts Institute of Technology, the Princeton Plasma Physics Laboratory, and the U.S. Department of Energy. The program committee was chaired by Miklos Porkolab (MIT) and co-chaired by Joel Hosea (PPPL); other members included Don Batchelor (ORNL), Claude Gormezano (JET Joint Undertaking), Noah Hershkowitz (Univ. of Wisconsin), T. Imai (JAERI), George Morales (UCLA), Jean-Marie Noterdaeme (Garching), and Ron Prater (General Atomics).

A total of 102 papers were presented, including 13 invited talks. All contributed papers were presented in poster format which stimulated a considerable amount of discussion and exchange of ideas. Of the 137 attendees, 106 were from the U.S. and 31 were from abroad. The meeting was very well attended, partly because of the easy access of Boston by air and partly because the meeting followed immediately after the Sherwood Theory meeting in Providence, Rhode Island. There was also reasonable good student attendance from the local institutions such as MIT. There was a significant participation from Europe, and to some extent, Japan. This ensured a successful conference, for it is recognized that during the past decade the greatest investment in new RF equipment for tokamak experiments was made by the European community and Japan. Consequently, a majority of the new results in the past few years in RF heating were obtained abroad.

Meanwhile, interest in RF heating and current drive has increased in the U.S. due to the recognition that neutral beam heating, the mainstay of the U.S. plasma heating effort, has reached its maximum potential, and extrapolation to the reactor regime is problematic. Also, high-power ICRF heating results on TFTR in the U.S. have contributed to the confidence that this RF heating scheme should extrapolate well to the reactor regime as well as showing that RF produced energetic ions can produce toroidal Alfvén eigenmodes relevant to predicted alpha particle instabilities. The U.S. RF theoretical effort has remained vigorous, as was attested by the good participation by theorists at this conference. Another area of growth has been RF technology, in particular in the area of ICRF antenna design. As a consequence, the U.S. is well positioned to remain competitive in the area of RF heating and current drive in tokamaks. Furthermore, RF experimentation in the U.S. is gaining momentum once more, and the next meeting (to be held in Palm Springs, California, in May, 1995) should see many new results from U.S. tokamaks. Finally, with TPX and ITER on the horizon, interest in heating and current profile control with RF waves is gaining renewed interest and this bodes well to the future of RF related physics research and technology.

Much of the credit for the successful organization and execution of this conference goes to Ms. Carol Arlington of the MIT Plasma Fusion Center, and to Mrs. Phyllis Schwartz of the Princeton Plasma Physics Laboratory. Our sincere thanks are extended to them for their expert and efficient help with all aspects of this conference. The chairman of this conference also wishes to thank Dr. P. Bonoli, Dr. S. Golovato, and Dr. Y. Takase of MIT for their help with sorting papers and organizing the poster sessions.

Miklos Porkolab
Joel Hosea, Co-Chairmen

ION CYCLOTRON RANGE OF FREQUENCIES

Overview of TFTR ICRF Results

G. Schilling

Princeton University Plasma Physics Laboratory

The primary physics objective of the TFTR program has now become the study of alpha-particle effects in D-T plasmas. ICRF heating has the potential to assist this objective in several areas, specifically through alpha-particle behavior simulation by the energetic minority tail ions and by core electron heating of Supershot plasmas. This paper describes the hardware, performance and physics results achieved with the TFTR ICRF system during the D-D plasma optimization campaign in 1992. It also describes the system modifications and experiments planned for TFTR's D-T phase starting in 1993.

1. ICRF System Description.

The ICRF system is shown schematically in Figure 1. It now consists of four antennas, each with two current straps, but with different Faraday screens to allow a comparison to be made in the tokamak environment. Each antenna is movable radially to permit coupling optimization. The antenna characteristics are given in the Table below.[1]

Antenna	P_{rf} (max. achieved)	Faraday screen	$k_{\|\|}$ (m^{-1})
Bay K	2.5 MW	single row, slanted 6°, oval rods	11
Bay L	3.6 MW	double row, straight, circular rods	11
Bay M	5.2 MW	double row, straight, circular rods	9
Bay N	2.7 MW	double row, slanted 6°, oval rods	11

The antennas at Bays K and N were each connected to one 3 MW fixed-frequency source operated at 47 MHz, and the antennas at Bays L and M were each connected to two 2.5 MW variable-frequency sources operated also at 47 MHz. The current straps were usually run with $(0,\pi)$ phasing. Total available rf source power was 16 MW. The maximum power levels achieved into the different plasma conditions during this campaign were:

L-mode, H-minority	11.4 MW
Supershot, ^{3}He-minority	7.6 MW
Pellet-fueled plasma	5.7 MW
Fast wave direct electron heating	3.0 MW

Since a large variety of experiments was performed on TFTR, at different magnetic fields, plasma currents, plasma positions, and with varied amounts of auxiliary heating, it was found necessary to perform intensive antenna conditioning before each series of experimental shots. This was performed on the morning of each day that rf power was required, in parallel with plasma setup/conditioning shots and with neutral beam conditioning. One to three hours of rapid-rate vacuum conditioning up to the 50 kV antenna voltage limit, with a pulse width of 0.5 to 2 s, was followed by one to three hours of antenna conditioning in a He plasma with an H-minority, until the physics experiment was ready to proceed. The exact time spent conditioning depended critically on the past discharge history, and especially on the amount of limiter material deposited onto the antenna surfaces by plasma disruptions.

During the vacuum conditioning phase gas puffs were observed on the torus residual gas analyzer immediately following each rf pulse, suggesting that co-deposited carbon plus gas layers on the antenna surfaces were being outgassed or removed by the conditioning process. Frequently antenna arcs required reconditioning at lower power, indicating that sufficient energy was stored in the transmission lines and flowed into the fault, causing surface modification to high-voltage electrodes. This latter effect was also observed during antenna conditioning in plasma.

2. Alpha-Particle Simulations.

The energetic tail ions produced by ICRF minority heating have been used to simulate alpha-particle behavior in non-D-T plasmas. As is shown on Figure 2, the radial distribution of energetic ions produced by minority heating is similar to the alpha-particle distribution expected from a TRANSP simulation of a D-T Supershot.[2] These simulations have allowed us to both check out alpha-particle diagnostics and to investigate potential alpha-particle physics effects.

2.1. Alpha-Particle Diagnostics.

The Alpha-CHERS diagnostic has been developed to observe the lower energy confined alpha-particles produced during D-T operation in TFTR.[3] CHarge Exchange Recombination Spectroscopy of photons emitted by alpha–particles that have gained an electron through charge exchange with a neutral beam atom will yield information on the alpha energy distribution in the range from thermal to about 0.5-1.0 MeV. These photons are observed as a Doppler-shifted wing on the He^+ 4686 Å (n = 4-3) line. In this experiment ^{3}He-minority heating produced an ion tail with energies up to 400 keV, which was detected and analyzed. The measured time evolution and spatial profile are as expected. The intensity of the observed ^{3}He signal was comparable to the predicted intensity of the D-T alpha-particle signal.

The Alpha Charge Exchange diagnostic has been developed to measure the energy distribution of the fast confined D-T alpha-particles and other energetic ions in the range 0.2 - 3.5 MeV by mass and energy analysis of neutral atoms resulting from ion charge exchange with cold neutrals in the ablation cloud of an injected Li or B pellet.[4] Again ^{3}He-minority heating produced an energetic ion tail which simulates the expected alpha-particle behavior. 1.1 MW of rf power produced a tail temperature of 170 keV, in agreement with modeling predictions.

2.2. Collective Instabilities.

Collective instabilities may result in enhanced alpha-particle losses in D-T plasmas. The energetic tail ions produced during ICRF minority heating have $v_H \sim v_\alpha$ and have been used to investigate some of these instabilities in non-D-T plasmas.

2.2.1. Toroidal Alfvén Eigenmode.

The toroidal Alfvén eigenmode (TAE) has been excited by these trapped particles in H-minority, He-majority L-mode plasmas.[5] The appearance of the mode is evidenced by a number of distinct signatures. Oscillations near 200 kHz are observed on the Mirnov coil signals, Figure 3. Discrete narrow peaks are observed on the frequency spectrum of the microwave reflectometry signal, Figure 4. The different peaks correspond to different toroidal mode numbers, Doppler shifted by plasma rotation.

A plasma density scan shows the $n_e^{-0.5}$ frequency scaling expected of the TAE mode, as is shown on Figure 5.

A series of four probes has been installed on TFTR to detect the loss of fast ions, including fusion alpha-particles. These probes use the machine magnetic fields to function as a magnetic spectrometer, measuring for each ion both the gyroradius (i.e. energy) and the pitch angle for mostly trapped, co-going particles. The detectors see greatly increased energetic tail ion loss at the onset of the TAE mode, Figure 6.[6] This loss appears to be on the order of 10% of the tail population at the highest rf power and mode amplitude. Empirically, the instability appears suddenly when the ICRF power exceeds ~ 3 MW; analysis shows the instability appearing when the energy stored in the tail ions exceeds ~ 75 kJ.

2.2.2. Kinetic Ballooning Mode.

A study of the effect of energetic particle-driven kinetic ballooning modes (KBM) on the β-limit in TFTR discharges has been performed, again using energetic tail ions produced by H-minority heating.[7] Calculations have shown that for TFTR and BPX the alpha-driven KBM modes can reduce the MHD β-limit. Plasmas near the TFTR β-limit were created ($\beta_n \leq 2.2$) with up to 17.5 MW of neutral beam heating, and an energetic ion tail was added via 3-4 MW of ICRF in an attempt to destabilize the KBM's, Figure 7.

A search was then made for the following KBM symptoms: (a) plasma disruptions, β collapse, or decrease in τ_E, (b) increased MHD activity at KBM frequencies, (c) increased minority tail ion loss.

The experimental observations yielded: (a) no signs of plasma degradation due to the ICRF tail ions, (b) no signs of ICRF-induced MHD at KBM frequencies, (c) no signs of KBM-induced tail ion loss. It appears that KBM modes were not generated by the ICRF tail ions in these discharges.

3. Electron Heating in Supershots.

ICRF is an attractive method for providing additional heating power to the TFTR Supershot plasmas. Fast wave direct electron heating has greater coupling difficulties due to weak wave damping, but it avoids the additional non-reactive plasma minority, and it has the potential for current drive. Minority ion heating offers easier coupling at the cost of the additional minority species.

3.1. Fast Wave Direct Electron Heating.

The fast wave direct electron heating transport studies were performed in a high T_e neutral beam-heated Supershot regime.[8] Since no fast ion tails with their attendant buildup and slowing-down times are involved, fast modulation of this heating source is quite possible. The increase of wave damping with $n_e T_e$ leads to strongly centrally-peaked power deposition in the Supershot regime. Application of the rf power and the resulting plasma response are shown in Figure 8. Analysis of the plasma response indicates

    (a) modeling of direct electron heating power deposition by the PICES and
        TRANSP codes yields profiles that are consistent with the measured values,
    (b) heat wave analysis yields that χ_e^{HW} is similar to the χ_e^{PB} obtained from
        power balance analysis within a factor of two,
    (c) density wave analysis yields that D_e^{HW} is within a factor of two of the D_e^{PB}
        obtained from particle balance,
    (d) the total electron transport loss is apparently inconsistent with the predictions
        of the critical temperature gradient model.

This direct electron heating appears to be an attractive alternative to the ^{3}He-minority heating in future D-T supershots, avoiding possible complications due to the presence of minority ions in studying the effects of alphas in D-T plasmas, as well as eliminating the slight plasma performance degradation of the ^{3}He addition.

3.2. ^{3}He-minority Heating.

ICRF ^{3}He-minority heating in D-majority Supershot plasmas has provided efficient core electron heating.[9] Figure 9 shows the results of a TRANSP time-dependent transport analysis, indicating that 5.7 MW of rf power dominates the central electron heating when compared to 22.9 MW of neutral beam power.[2] The central electron temperature obtained by the addition of varying amounts of rf power to Supershots with otherwise fixed parameters is shown on Figure 10; a ~ 40% increase has been obtained with the ~ 6 MW added so far.

The increased central electron temperature leads to an increased alpha-particle slowing down time, increasing $\beta_\alpha(0)$ in D-T plasmas. Figure 11 shows the projected core alpha energy slowing down time for D-T plasmas equivalent to those of Figure 10. The ~ 6 MW of rf power added so far gave a ~ 60% enhancement.

4. ICRF System Modifications for D-T Operation.

The TFTR ICRF system is presently being modified for D-T operation.[10] The base frequency of 47 MHz is being lowered to 43 MHz to provide some toroidal magnetic field headroom at the largest Shafranov shifts, i.e. the highest heating power levels. The ability to drive two of the four antennas either at 43 or 64 MHz is being added. A general upgrade of the transmitters to provide greater reliability at the higher power levels is also in progress.

5. Conclusions.

The TFTR ICRF program has contributed greatly to the information obtained from the 1992 D-D plasma campaign in preparation for D-T. The alpha-particle simulations by energetic minority tail ions have given us a head start on investigating potentially adverse collective instabilities driven by the alphas in the D-T phase. These simulations have also provided alpha diagnostics checkout capability before D-T. Fast wave direct electron heating has demonstrated good efficiency as well as providing transport information through power modulation. ^{3}He-minority heating of D-majority Supershot plasmas has demonstrated good core electron heating efficiency, leading to enhanced β_α on axis in D-T.

Acknowledgment. This work was performed with the help and active participation of the following colleagues:

M. Bell, H. Biglari, M. Bitter, N. L. Bretz, R. Budny, L. Chen, D. Darrow, P. C. Efthimion, E. Fredrickson, G. Y. Fu, B. Grek, L. Grisham, G. Hammett, J. C. Hosea, A. Janos, D. Jassby, F. C. Jobes, D. W. Johnson, L. C. Johnson, R. Majeski, D. K. Mansfield, E. Mazzucato, S. S. Medley, D. Mueller, R. Nazikian, D. K. Owens, S. Paul, H. Park, C. K. Phillips, A. T. Ramsey, J. H. Rogers, J. Schivell, G. L. Schmidt, J. E. Stevens, B. C. Stratton, J. D. Strachan, E. Synakowski, G. Taylor, J. R. Wilson, K. L. Wong, S. J. Zweben *Princeton University Plasma Physics Laboratory*

R. K. Fisher, J. McChesney *General Atomics*

M. P. Petrov, A. V. Khudaleev *Ioffe Institute, St. Petersburg, Russia*

F. Rimini *JET Joint Undertaking*

J. Machuzak *Massachusetts Institute of Technology*

D. B. Batchelor, L. Baylor, C. E. Bush, R. C. Goldfinger, D. J. Hoffman, E. F. Jaeger, M. Murakami, A. L. Qualls, D. A. Rasmussen *Oak Ridge National Laboratory*

Z. Chang, R. J. Fonck, G. McKee, T. Thorson *University of Wisconsin.*

This work was supported by U.S. DoE contract DE-AC02-76-CHO-3073.

References.

[1] J. H. Rogers et al., Bull. Am. Phys. Soc. Vol. 37, No. 6 (1992), paper 5R 11.

[2] R. Goldfinger et al., paper this conference.

[3] G. McKee et al., Bull. Am. Phys. Soc. Vol. 37, No. 6 (1992), paper 5R 23.

[4] S. S. Medley et al., 20th EPS Conference On Controlled Fusion And Plasma Physics, Lisboa, 26-30 July 1993.

[5] J. R. Wilson et al., submitted to Phys. Rev. Lett.
also K. L. Wong et al., paper A12 this conference.

[6] D. S. Darrow et al., "Observation of Fast Ion Losses in TFTR During ICRF-Driven Alfvén Instabilities" presented at the IEA Workshop on Alpha Physics and Tritium Issues in Large Tokamaks, Princeton, NJ, February 18, 1993 (unpublished).

[7] S. J. Zweben et al., "Effect of ICRH-Tail Ion Driven Kinetic Ballooning Modes on the β-Limit" presented at the IEA Workshop on Alpha Physics and Tritium Issues in Large Tokamaks, Princeton, NJ, February 18, 1993 (unpublished).

[8] M. Murakami et al., paper A11 this conference.

[9] G. Taylor et al., 20th EPS Conference On Controlled Fusion And Plasma Physics, Lisboa, 26-30 July 1993.

[10] J. H. Rogers et al., paper A9 this conference.

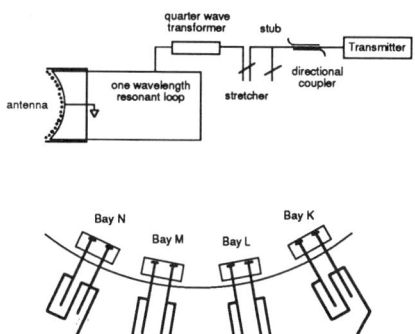

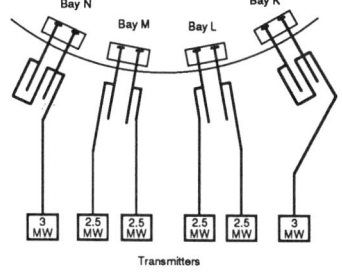

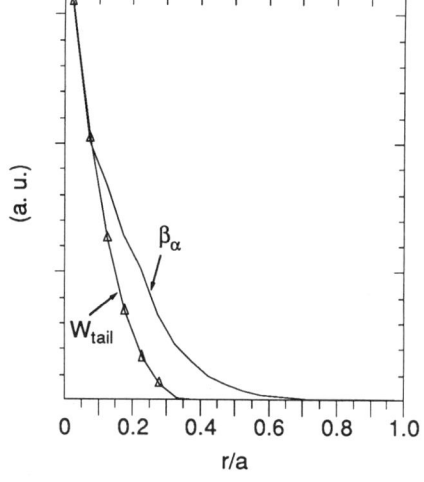

Figure 1. ICRF layout on TFTR.

Figure 2. Radial distribution of energetic minority tail ions and alpha-particles.

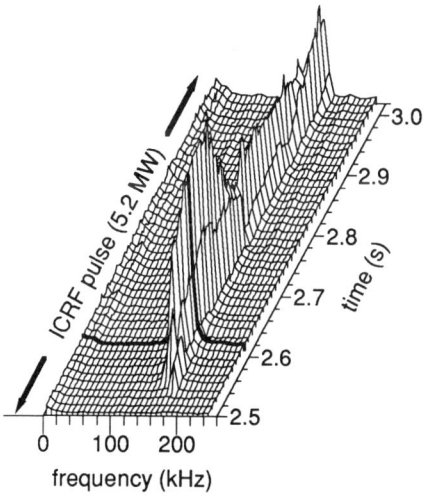

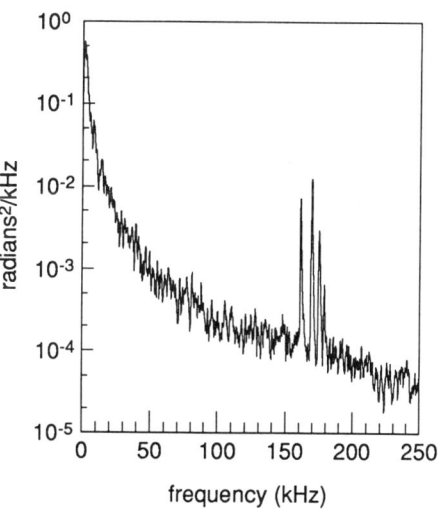

Figure 3. Mirnov coil signals during ICRF H-minority heating in a He-majority L-mode plasma.

Figure 4. Frequency spectrum of microwave reflectometry signal.

10   Overview of TFTR ICRF Results

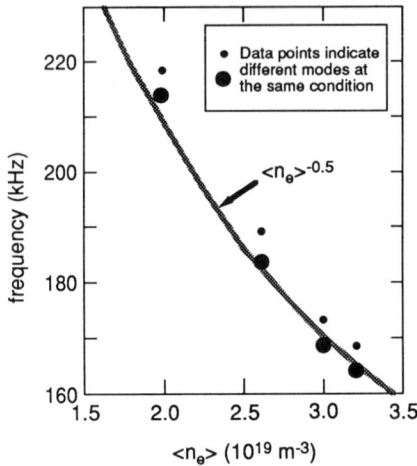

Figure 5. Scaling of the observed oscillation frequency with plasma density.

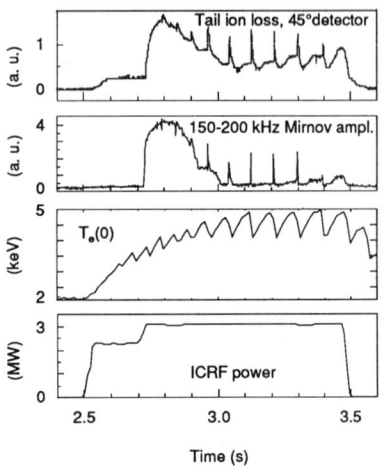

Figure 6. Energetic tail ion loss during TAE mode activity.

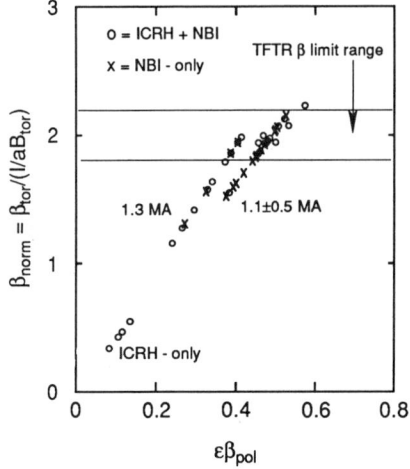

Figure 7. β values for discharges in the KBM search.

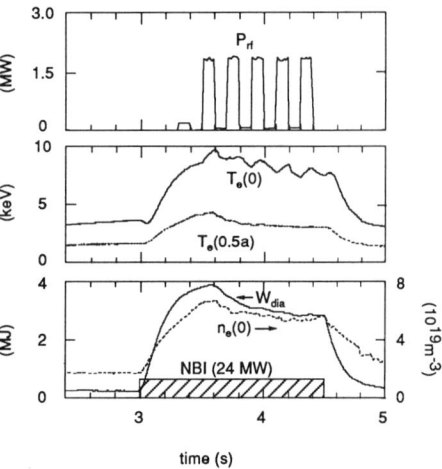

Figure 8. Fast wave direct electron heating of Supershot plasma.

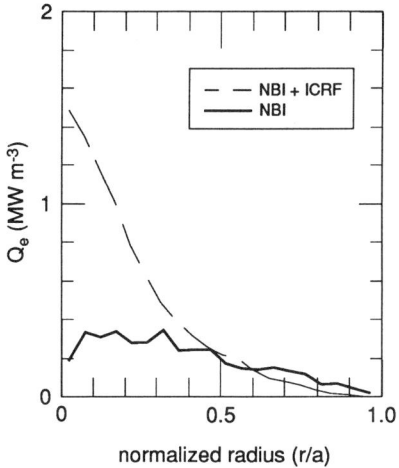

Figure 9. TRANSP indication of strong core electron power deposition during ICRF minority heating of Supershots.

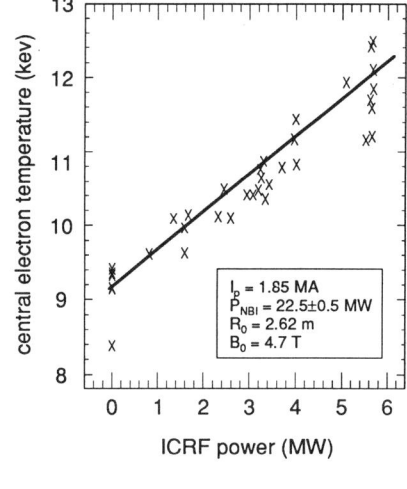

Figure 10. Increase of central electron temperature by ICRF ^{3}He-minority heating of a D-D Supershot plasma.

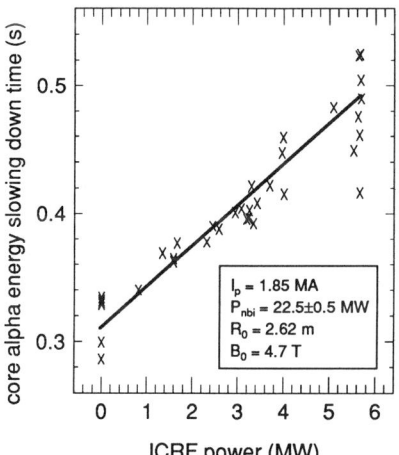

Figure 11. Projected core alpha-particle slowing down time for ^{3}He-minority heating of a D-T Supershot plasma.

First Results of Ion Cyclotron Resonance Heating on ASDEX Upgrade

J.-M. Noterdaeme, C. Hoffmann, M. Brambilla, K. Büchl, A. Eberhagen, A. Field,
C. Fuchs, O. Gehre, J. Gernhardt, O. Gruber, G. Haas, A. Hermann,
F. Hofmeister, A. Kallenbach, G. Lieder, V. Mertens, H. Murmann,
S. de Peña Hempel, W. Poschenrieder, Th. Richter, F. Ryter, N. Salmon,
H. Salzmann, W. Schneider, F. Wesner, H.-P. Zehrfeld, H. Zohm,
ICRH Team, ASDEX Upgrade Team

Max-Planck-Institute for Plasmaphysics, D-8046 Garching, Fed.Rep.Germany

ABSTRACT

ASDEX Upgrade is equipped with an ICRH system consisting of 4 generators of 2 MW power each and 4 double loop antennas. The generators, tuneable in frequency from 30 to 120 MHz, cover several heating scenarios over a wide range of magnetic fields (1 T< B_t <3.9 T): minority heating of H and He_3 and second harmonic heating of H and D.

ICRH-heated discharges in ASDEX Upgrade were so far carried out mainly at 30 MHz and a magnetic field of 2 T (H minority in D and He). Peak powers of 2.4 MW and pulse length up to 2.5 s were achieved (total energy 3.75 MJ).

In L-mode, the density on turn-on of the ICRH stays constant, or even decreases. The ratio of radiated power to total input power is unchanged (60% in an unboronized machine, 30% in a freshly boronized machine) between Ohmic and ICRH phases. The electron temperature increases with 0.9 MW from 1 to 1.25 keV, the loop voltage drops.

Transitions to the H-mode were easily and reliably achieved with ICRH alone (necessary ICRH power as low as 0.9 MW) and the length of the ELMy H-mode phases was limited only by the applied ICRH pulse length (ELMy H-mode phases of up to 2 s were achieved).

The paper presents further results on heating and confinement in L and H-mode, antenna and edge studies and on divertor measurements.

Preliminary experiments, performed with a combination of H minority heating (30 MHz) and H second harmonic (60 MHz) in 600 kA He and D discharges (H minority in the 5 to 20 % range) at 2 T, and with non-resonant heating (30 MHz and 60 MHz at 1.35 T) are briefly discussed.

INTRODUCTION

ASDEX Upgrade is a D-shaped tokamak with R=1.65m, a=0.5m, b=0.8m, dedicated to the study of reactor compatible divertor operation. It went in operation early 1991 [1]. Its capabilities are a toroidal magnetic field up to 3.9 T, and a plasma current up to 2 MA. It can be operated in single null, double null and limiter configurations. The maximum allowable parameters depend on the configuration. For single null operation, the design operating point is 2.7 T and 1.6 MA.

THE ASDEX Upgrade ICRH SYSTEM

The first goal of the ICRH system on ASDEX Upgrade is to provide the necessary heating power for the edge and divertor studies, which constitute the main experimental focus of ASDEX Upgrade [2, 3, 4]

The specific strength of the ICRH system on ASDEX Upgrade is its large frequency range, and the flexibility of the antenna concept making it possible to easily change components of the antenna, and to adapt it to different configurations (single null, double null, limiter).

ASDEX Upgrade is equipped with an ICRH system consisting of 4 generators [5], totalling 8 MW, and four low field side antennas (Fig. 1, 2). The generator frequency is tuneable from 30 to 120 MHz, covering thereby all relevant heating scenarios in the range of $1\ T < B_t < 3.9\ T$: minority heating of H (above 2 T) and He_3 (above 3 T) and second harmonic heating of H (below 3.9 T) and D (above 2 T) (Fig. 3).

Each generator is connected by a 6 inch transmission line, through a double stub matching network[6], and vacuum insulated transmission lines[7] to a double loop antenna [8] (Fig. 2, 4).

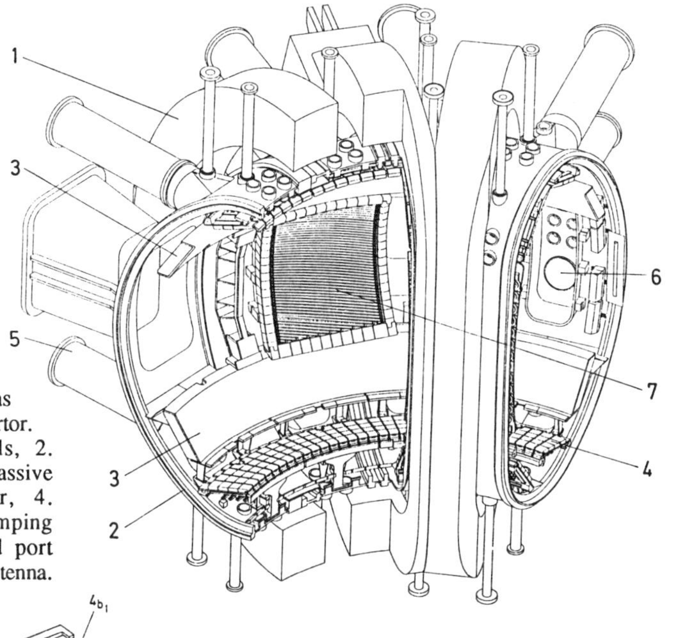

Fig. 1 Cut-away view of ASDEX Upgrade showing one of the ICRH antennas as well as parts of the bottom divertor. 1. toroidal field coils, 2. vacuum chamber, 3. passive stabilizing conductor, 4. divertor plates, 5. pumping ports, 6. antenna feed port (antenna removed), 7. antenna.

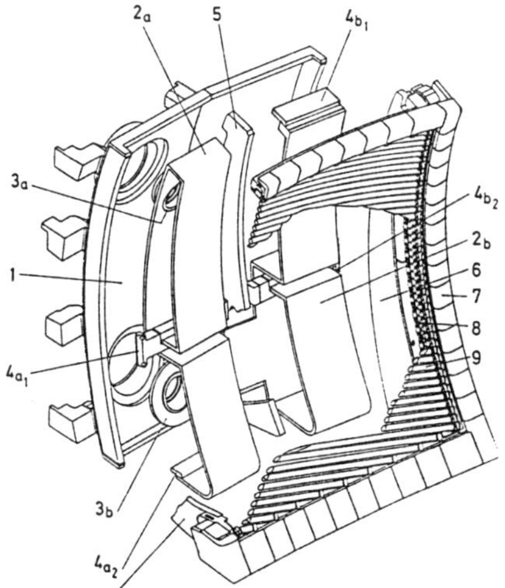

Fig. 2 Detail of one of the ASDEX Upgrade antenna : 1. return conductor, 2a,b. central conductor, 3a,b. antenna feed points, 4a,b. antenna short circuits, 5. septum, 6. distance plate, 7. limiter on cooling frame, 8. springs holding the Faraday screen rods, 9. Faraday screen rods

EXPERIMENTS

ICRH started operation on ASDEX Upgrade at the end of April 1992. During the three experimental periods in 92 and the experimental period starting March 93, about 220 shots were performed with ICRH. All four generators were taken in operation. The maximum power reached per generator is in the range 400 kW to 1400 kW. A total peak power of 2.4 MW was achieved, the pulse length was up to 2.5 s, and the maximum total delivered ICRH energy 3.75 MJ. The antennas were operated with the loops out-of-phase.

ICRH-heated discharges were performed in bottom single null discharges at toroidal fields of 1.35 T and 2 T both with the ion grad B drift towards the X-point (in ASDEX Upgrade, by convention "negative" B_t direction) and away from it, and plasma currents of 350, 600 and 800 kA.

The main operating scenario was hydrogen minority heating at 30 MHz in He or D at 2T.

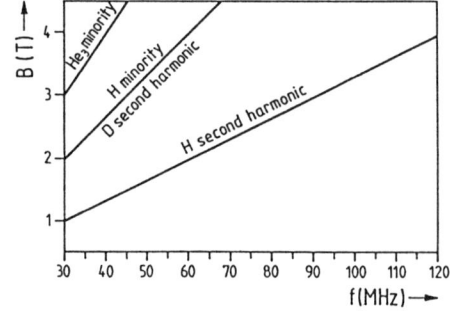

Fig. 3 Frequency range for the ICRH generators

Early in the experimental period (H)He experiments were made in an unboronized machine. After boronization, (H)D and (H)He experiments were performed.

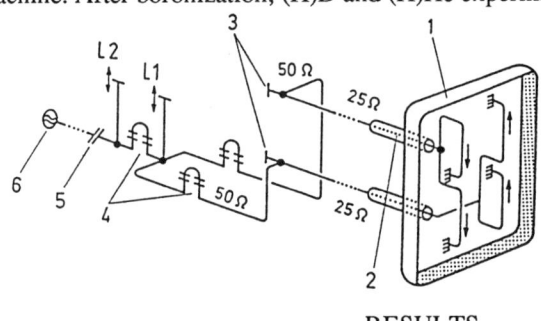

Fig. 4 Schematic layout of the ICRH system. 1. Antenna with the two loops, 2. vacuum insulated transmission lines, 3. pre-matching stub, 4. "U-lines" to adjust the length of different parts, and change the phasing between the loops, 5. DC-break, 6. generator

RESULTS

Heating and Confinement results with H minority heating in He and D

Absorption

To estimate the ratio of power absorbed in the plasma center to the generator power, the change of plasma energy dW/dt, calculated from the equilibrium β at turn-on and turn-off of the ICRH was set equal to $\alpha\, P_{ICRH}$, where P_{ICRH} is the net ICRH power measured at the generator. Losses in the transmission line and in the antenna are about 5 %, so that the ratio of absorbed to *coupled* power is actually α/0.95.

The absorption coefficient α measured at turn-on was typically 0.66 for (H)He and 0.55 for (H)D. Theoretical analysis (Fig. 5) indicates that this difference is in qualitative agreement with the expected dependence of the absorption on the H concentration which is different under both conditions (5% H in He, 15% H in D).

The H concentration is estimated from mass spectrometric measurement of the gas in the divertor exhaust during the discharge. This estimate is consistent in the case of (H)D with the ratio of the hydrogen isotope fluxes from the divertor plates, as measured from the hydrogen α lines.

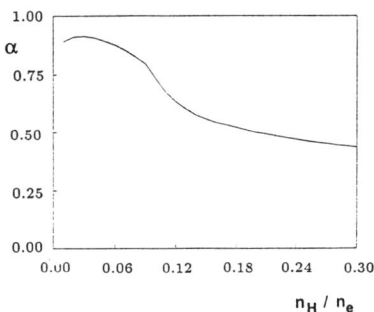

Fig. 5. Calculated multiple-pass absorption coefficient inside the plasma, assuming 15 % losses near the wall at each reflection, as a function of the H concentration.

An increase of the absorption coefficient with power was seen for the turn-off (typically 0.76 for (H)He and 0.60 for (H)D in L-mode). This is consistent with the quasi-linear improvement of the absorption for minority heating with the absorbed power (Fig. 6). Absorption coefficients around 0.9 were measured for H-mode plasmas. In the following, to calculate the total absorbed ICRH power a constant absorption coefficient of 0.75 was used throughout.

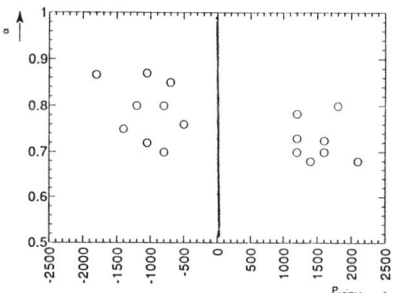 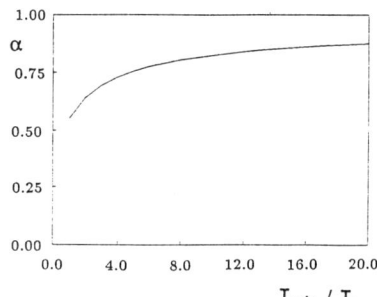

Fig. 6. Measured absorption coefficient on turn on ($P_{ICRH}>0$) and turn off ($P_{ICRH}<0$) of the generators, and comparison with the theoretical dependence of the absorption on the minority temperature for a fixed minority concentration of 10%. The minority tail temperature itself increases linearly with the power absorbed by the minority.

L- Mode

Typical L-mode results of 600 kA ICRF-heated discharges (at $n_e = 3.10^{19}$ m^{-3}) with about 5 % H in He show, at the ICRH turn-on a constant or even decreasing density, despite a gradual increase in the recycling signals (neutral pressure and H$_\alpha$ in the divertor by about a factor 2. This is consistent with a decrease of particle confinement in the L-mode, a common feature in NI heated tokamak discharges, but often masked in RF heated tokamaks by a very strong RF specific increase of recycling at the walls (above the expected one due the increased plasma energy content), due to a direct interaction between the RF and the edge plasma. There is no step-like increase of the recycling at the walls, which would be indicative of such a significant RF-edge interaction.

The ratio of radiated power to total input power is unchanged between Ohmic and ICRH phases (60 % in an unboronized machine, 30 % in a freshly boronized machine) (Fig. 7).

For a boronized machine, Z_{eff} in the center is in the range 2 to 3 . The Abel inverted radiation profile (Fig. 8), measured with 40 bolometer sight lines in a horizontal fan and 24 in a vertical fan, show a hollow radiation profile with a strong peak at the X-point (at the top in the figure).

Total energy content shown in Fig.9 was calculated on the basis of β equilibrium, whose calibration is not yet completed to full satisfaction. With this caution in mind, we can compare confinement time in L- mode discharges, with the ITER 89-P scaling [9] : for a 600 kA, 2 T discharge, and 1 MW of total input power the calculated confinement is of the order of 90 ms (Fig. 9), the ITER 89-P prediction is 86 ms.

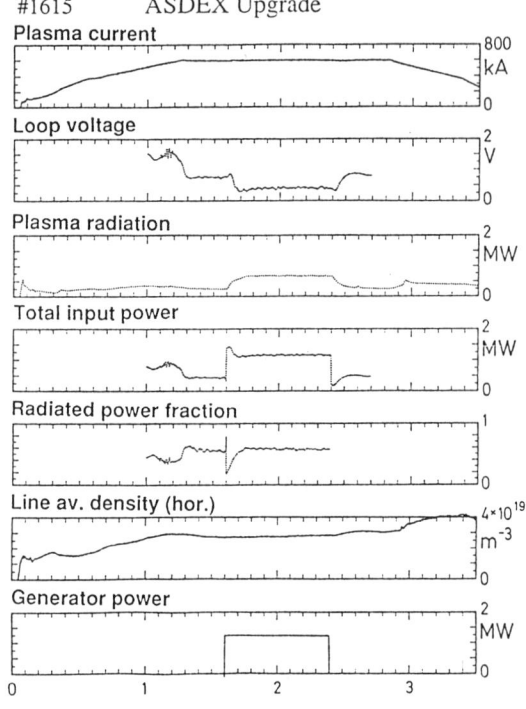

Fig. 7 L-mode at 2 T, 600 kA, unboronized with 1.2 MW ICRH power, (H)He (#1615)

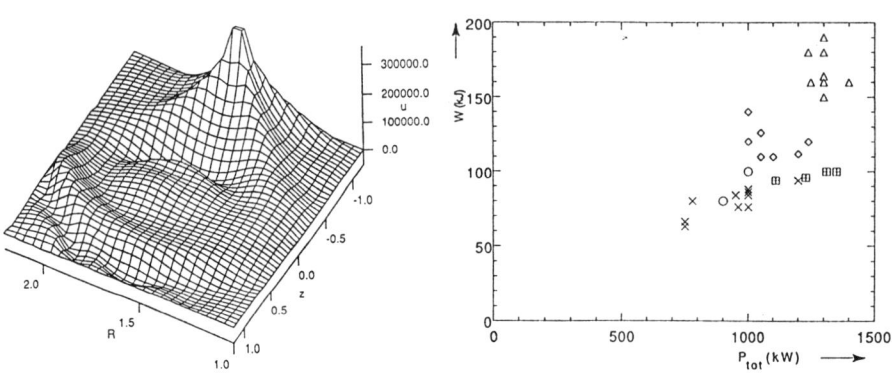

Fig. 8. Radiation profiles in W/m^{-3} for a poloidal cross-section (X-point in the figure at the top, z from top to bottom, R from right to left).

Fig. 9. Energy content versus total power (x) OH, (▣) L at 800 kA (o) L, (◊) grassy H-mode, (Δ) H at 600 kA

H-mode

ELMy H-mode phases up to 2 s long were easily achieved in single null D and He plasmas at low ICRH power levels for the favourable ion grad-B direction of the toroidal field ("negative" B_t) and an antenna limiter to plasma separatrix distance of a few centimeters (Fig. 10). The power threshold P_{thr} for ICRF heated (H minority) deuterium discharges in ASDEX Upgrade at 600 kA is consistent with $P_{thr} = 2\ n_e\ B_t$ (MW, $10^{20} m^{-3}$,T). The magnetic field dependence, also observed on D III-D[10] was established [11] for ASDEX Upgrade on the basis of ohmic H-modes, which were obtained at higher current (800 kA) and lower magnetic fields (≤1.35T).

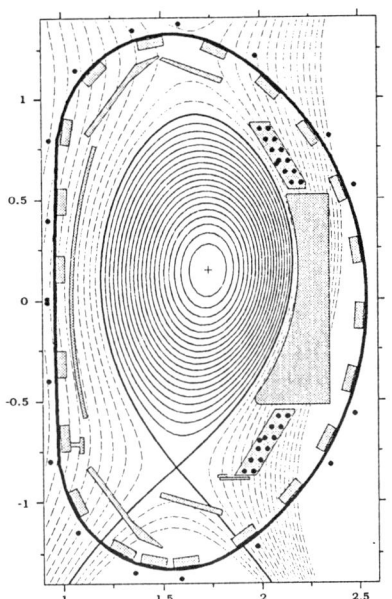

Fig. 10 Flux surfaces for a 600 kA, 2 T, H-mode discharge, boronized, with 1.3 MW ICRH, H minority in D (# 2329 at t= 3.58s).

To compare the H-mode power threshold among divertor experiments with different geometry and size, a normalization with the plasma surface S has been introduced[12, 13]. We obtain for ASDEX Upgrade $P_{thr}/S = 0.05\ n_e\ B_t$ (MW, m^2, 10^{20}m^{-3},T). The value 0.05 for the so-normalized threshold power is similar to JFT-2M [14, 13] but lower than both D III-D [15], with $P_{thr}/S \approx 0.10\ n_e\ B_t$, and JET [13]. This may point to an inverse relationship between normalized threshold power and aspect ratio (R/a) for D-shaped tokamaks : larger aspect ratio machines could have a smaller threshold power for the H-mode transition.

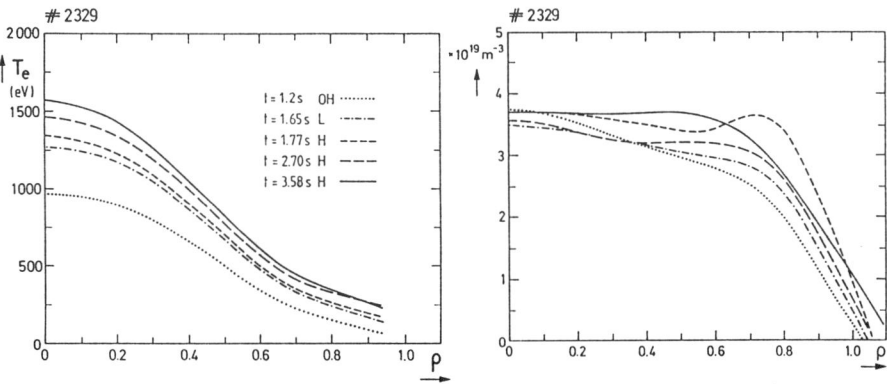

Fig. 11. Temperature and density profiles as a function of normalized flux radius at different times for the discharge shown in fig. 12 (where the times are also marked by ↓).

18 First Results of Ion Cyclotron Resonance Heating

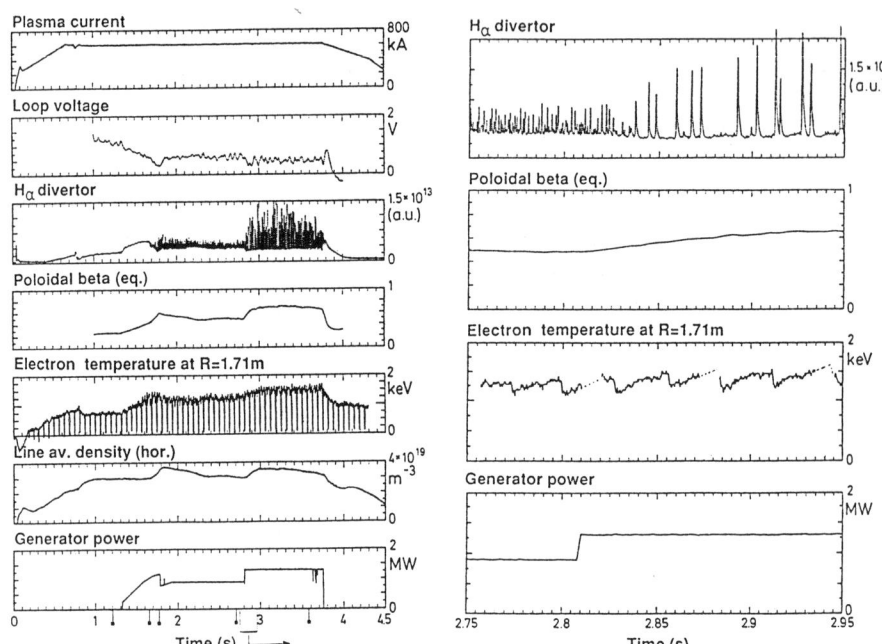

Fig. 12 H- mode discharge (#2329) with two levels of ICRH power (0.9 and 1.3 MW). Plasma current 600 kA, 2T, H minority in D, boronized. At the right, an expanded time scale near the increase in power showing the sudden change of ELM behaviour.

The H-mode could still be achieved (at larger power) for antenna-plasma distances as small as 1 cm. The H-mode threshold for the unfavorable ion grad-B direction is above 2 MW. This high value may have been influenced by unfavorable plasma conditions.

The H-modes achieved with ICRH have, depending on the power level, grassy ELMs (near the threshold power), or singular type-III ELMs[16] (at higher power) with a typical repetition rate of 70 Hz. If the power is increased suddenly during an H-mode (Fig. 12) the change in ELM type, indicating a change in the edge parameters, occurs within 30 ms after the increase in power. This delay is short with respect to the confinement time. The change also does not happen gradually, but seems to indicate a threshold behaviour.

Confinement times for well developed H-mode discharges (with type-III ELMs) are in the range 140 ms for 1.3 MW of total heating power at 600 kA. This is to be compared with the JET/D III-D H-mode scaling[17] (using R as the linear dimension), predicting 116 ms and corresponds to an enhancement factor of 1.85 with respect to ITER-89 P[9]. The confinement time of grassy ELM discharges at lower power (1 MW), is about 25 % lower.

Combination of minority and second harmonic heating

Experiments were performed with a combination of H minority heating (30 MHz) and H second harmonic (60 MHz) in 600 kA He discharges with an H concentration in the 5% range, and D discharges with a H concentration in the 15 % range. Second harmonic heating is usually performed with much higher concentration of the resonant species. With sufficient power at the fundamental frequency (minority heating), the additional power at the second harmonic seems to be as efficient as the minority heating. For example, adding 300 kW at 60 MHz to a plasma preheated with 300 kW at 30 MHz, produced an additional ß increase similar to the initial one obtained with the 30 MHz alone. Interestingly enough, even after the 30 MHz was switched off, the heating efficiency with the 60 MHz (now alone) was similar to that of the 30 MHz. There are indications that the minimum power in minority heating needed for start-up of the second harmonic depends on the minority concentration, as would be theoretically expected if the second harmonic couples to the high energy minority tail produced by the minority preheating. Without minority pre-heating the second harmonic alone may slowly "bootstrap" itself, at high enough power density.

Electron heating

A limited number of discharges with off-resonant heating up to 1.1 MW were done with 30 MHz and 60 MHz at 1.35 T. Stable plasmas were obtained where the ICRH produced a β increase at increased density and constant loop voltage. H_α at the divertor plates in this case increases suddenly on turn-on of the generators, indicating a direct RF-plasma edge interaction. The ICRF power causes an increase of carbon influx from the inner shield as well as of the radiated power. Remarkable is a considerable increase of the carbon ion temperature taken from the Doppler width of a C III line which has its emission slightly outside the separatrix. The heating of the carbon ions is not yet understood, possible explanations are the heating due the high cyclotron harmonics of the carbon ions or parametric effects.

Edge plasma studies

Antenna Resistance

The measured antenna resistance, defined as a lumped resistance at the end of a 25 Ω transmission line for one (of the two) antenna loops, was in the range 1 to 2.5 Ω for distances between plasma separatrix and antenna limiters of 5 to 1 cm. Antenna resistance values at 60 MHz are typically 3 Ω at 3 cm. In both cases, the value dropped only slightly (20 %) at the L to H transition, but decreased further later in time during the H-phase.

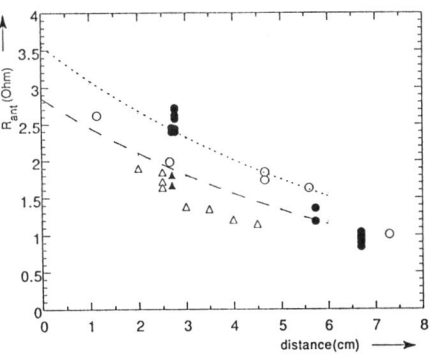

Fig. 13 Values of the antenna resistance as a function of the distance antenna limiter to plasma separatrix. The central conductor is 4 cm behind the antenna limiter. Experimental measurement for L(o), and H-mode discharges (Δ) at 600 kA, $3.5\ 10^{19}$ m^{-3} and (●, ▲) at 800 kA, $4\ 10^{19}$ m^{-3}. Theoretical calculations assume a central density of $3.5\ 10^{19}$ m^{-3}, a density at the separatrix of $1\ 10^{19}$ m^{-3}, and 0 outside the separatrix; the central temperature was taken to be 1.25 keV. The average central conductor to return conductor used for the calculations was 15 cm (...), and 10 cm (- - -).

To compare theoretical and measured values (Fig. 13), we used for the theoretical calculations the same definition of antenna resistance (rather than the usually theoretically defined coupling resistance), but had to assume a constant distance between central conductor and return conductor (even though the actual distance between central conductor and return conductor is 5 cm at the top of the antenna and 20 cm at the bottom). Reasonable agreement is obtained for an average constant distance near 10 cm.

Impurity fluxes from the antenna

The Faraday screen rods are inclined by an angle of 15° to the toroidal direction to align them with the magnetic field for the standard parameters (1.6 MA, 2.7 T). During experiments performed at $q_{\psi 95}=5$ (angle 8°) before the machine was boronized, Ti fluxes were spectroscopically observed from the TiC coated Faraday screen of an operating antenna. The fluxes Γ_{Ti} scale proportional to the line averaged density $\Gamma_{Ti} \propto n_e$ and to the antenna voltage $\Gamma_{Ti} \propto (P_{ICRH})^{1/2}$ (Fig. 14). Ti was however not detected spectroscopically in the plasma interior, nor in the divertor, indicating a good screening.

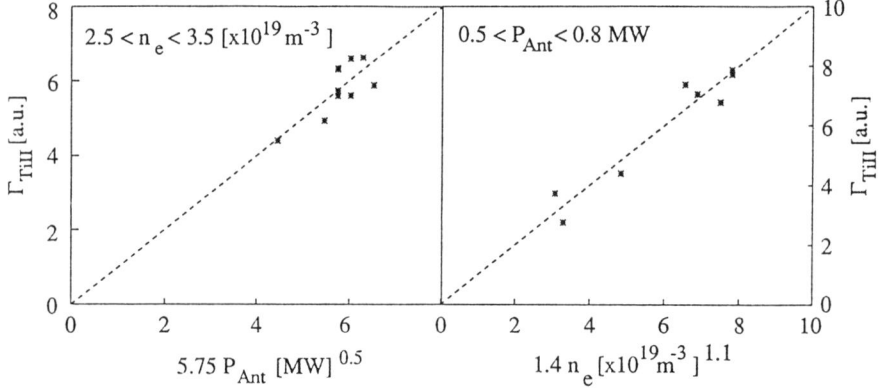

Fig. 14. Scaling of Ti fluxes from the powered antenna with antenna power $\Gamma_{Ti} \propto (P_{ICRH})^{1/2}$ and with line averaged electron density $\Gamma_{Ti} \propto n_e$. The Ti$^+$ flux is assumed proportional to the intensity of a Ti II (334.9 nm) line for which the number of emitted photons per ionized ion has a weak dependence on n_e and T_e.

After boronization, Ti fluxes from the screen were no longer observed, these being substituted by B fluxes. The scaling of the B influx from the screen is given by $\Gamma_B \propto n_e P_{ICRH}$, indicating sputtering by thermal ions (or neutrals). No impurity problems occurred and no unintentional disruptions where produced at the present power levels and pulse length. For a positive toroidal field, the angle between the magnetic field lines and the Faraday screen is 23°. Although the fluxes from the screen do not increase, as compared to the negative B_t direction, the plasma disrupts above a certain power level and pulse length (typically 1 MW, 1.5 s) following a cooling of the boundary and the appearance of a MARFE. The plasma conditions however, seem to be already in the Ohmic phase deteriorated, as apparent from a density limit about a factor 2 lower for the negative B_t direction.

Density limit

Both OH and ICRF heated discharges show a similar physical picture with respect to the density limit [18]. The discharges test the limit by a preprogrammed slow n_e ramp lasting up to the disruption of the plasma current. During ramp-up the boundary temperature lowers, leading to a shrinking of the current density profile and a correlated decrease of the plasma elongation b/a by a few percent. Approximately 0.2 s before the disruption, discharges with $q_{\psi 95}=5$ (2 T, 600 kA) normally develop MARFEs close to the lower active X-point, which move upwards into the main chamber whereas discharges with the lower safety factor $q_{\psi 95}=3.3$ (1.35 T, 600 kA) only show an enhanced bremsstrahlung in the vicinity of the X-point. The ratio of total radiated power to the input power in the high density phase can increase up to 80 %. A few milliseconds before the discharge reaches the density limit, MHD modes develop and lock. The plasma current at first survives, though n_e and the plasma energy quench. The locked mode itself can last a few tens of milliseconds, leading finally to the disruption of I_p. The maximum achievable density increases marginally with ICRH at 2 T (minority heating) but not at 1.35T (non-resonant heating).

Divertor measurements

Asymmetry inner-outer divertor plates

The divertor plates are viewed by a photodiode array from the top of the torus. In OH plasmas and ICRH L-mode, the emission of H_α from the inner plate is the same as that of the outer plate (ratio inner/outer = 1), independently of the direction of the magnetic field. During H-mode (with negative B_t), the ratio is approximately 1.5 during grassy ELMs and between ELMs, but increases to 2-3 during ELMs.

Cooling water calorimetry and wall thermometry indicate an energy load 2.2 to 3.5 times higher on the outer target plates for the negative direction of the magnetic field, and 1.2 to 1.7 for the positive direction. For negative B_t and $q_{\psi 95} \approx 5$, no significant toroidal asymmetries were measured on the divertor target plates by the toroidally resolved cooling water calorimetry. This is in contrast with positive B_t where toroidal asymmetries seem to be apparent.

The higher values for H_α on the *inner* plates during ELMs are not in contradiction to the calorimetric measurements which indicate a larger energy load on the *outer* target plates. Assuming a constant temperature, the increase of H_α means, an increase of density at the plates. On the other hand, the energy loss due to the ELM is much larger than the purely convective loss calculated from the particle loss in the main plasma (appearing as density increase at the divertor plates). Fast spectroscopic measurements in the divertor region (see below) indicate that the ELM energy is deposited in a short (≈ 10 µs) pulse, due to electron heat transport along the field lines at the beginning of the ELM, followed by the density increase on a longer time scale (some 100 µs) [19, 20, 21].

Radiation from the divertor

A multi-chordal spectrometer permits the study of impurity behaviour in the divertor. The system measures the emission (220-850 nm) of various atoms and ions along 16 lines of sight above the lower outer divertor plate with a spatial resolution of about 2 mm. Two detection systems are simultaneously available : high spectral (0.03 nm) with low time resolution (20 ms), and high time resolution (10 µs) for two spectral lines with low spectral resolution (10nm).

The increase of the C III intensity in a single ELM during ICRH, with no noticeable time delay (on the fast time scale of 10 µs) between the cords at various distances of the divertor plate indicate that the ELM causes a very fast increase of T_e due to electron heat conduction along the field lines.

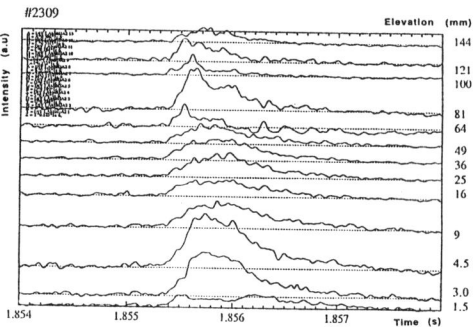

Convective transport, limited to the ion sound velocity ($v_s \approx 10^4$ m/s) would become visible as a time delay. Of further interest is the peaking of the C III intensity in the vicinity of the X-point, which is evident from the profiles shown in Fig. 15. This increase may be the result of an ergodization of the X-point during the ELM, followed by an influx of C II from the private flux region and an excitation during the heat pulse [21].

Fig. 15 C III intensity for cords at different height above the the divertor plate as a function of time, showing the instantaneous increase of the intensity on all cords and the additional peak in the X-point region (cord at 81 mm above the divertor plate).

Energy balance

For ICRH discharges the sum of the total radiation energy from above the X-point, measured by bolometry, and the energy load on the divertor target, detected by the cooling water calorimetry account well for the total input power (between 66 % and 112 %, mean value 84 %). In ELMy H-mode the total power is distributed about evenly between ELM losses, transport losses and radiation losses.

CONCLUSIONS

One year after the first plasma shots on ASDEX Upgrade, the ICRH system came into operation. Good heating results are obtained, with no specific ICRH impurity problems at the present power levels and pulse length. The reliable achievement of the H-mode with ICRH alone, obtained shortly after ICRH start-up was our first highlight. In addition to the large flexibility of the ICRH system allowing the investigation of interesting heating scenarios, with possible information on plasma transport in the center, the extensive diagnostics in the edge and divertor region will provide new insights in the interaction between the RF and the edge. The RF itself may become an important tool to influence the interaction between the plasma and the boundary structures, as the divertor plates.

ACKNOWLEDGEMENTS

It is a pleasure to acknowledge interesting discussions with Prof. Wilhelm and Dr. J. Neuhauser. The excellent and dedicated work of the technical and operational teams of the ASDEX Upgrade machine and power plant, and the ICRH systems merits recognition. Finally, a word of thanks to the people in charge of data acquisition and processing.

REFERENCES

1. Köppendörfer W., et al., "Results of the first operational phase of ASDEX Upgrade", in Plasma physics and controlled nuclear fusion, (14th. Int. Conf., Würzburg, 1992), Paper A-2-3, IAEA (1993) to be published.
2. Köppendörfer W., "ASDEX upgrade. A tokamak experiment with a reactor-compatible poloidal divertor", in Fusion technology, (12 th Symposium, Jülich, 1982), Vol. 1, Pergamon Press (1983) 187.
3. Gruber O., Kaufmann M., Köppendörfer W., Lackner K. Neuhauser J., "Physics background of the ASDEX upgrade project", J. Nucl. Mater. 121 (1984) 407.
4. Kaufmann M., "Edge physics and H-mode studies in ASDEX-Upgrade", Plasma Physics and Controlled Fusion, invited talk at the 1993 EPS Conference, Lissabon (1993) to be published.
5. Wesner F., et al., "The 4x2 MW ICRH system for ASDEX Upgrade", in Fusion Technology, (16th Symposium, London, 1990), Vol. 2, 1181.
6. Hofmeister F., et al., "The RF system and Matching Procedure for ASDEX and ASDEX Upgrade", Fus. Eng. Design (1993) to be published.
7. Wedler H., et al., "Vacuum insulated antenna feeding lines for ICRH at ASDEX Upgrade", Fus. Eng. Des. (1993) to be published.
8. Noterdaeme J.-M., Wesner F., Brambilla M., Fritsch R., Kutsch H.-J. et al., "The ASDEX Upgrade ICRH antenna", Fus. Eng. Des. (1993) to be published.
9. Yushmanov P.N., Takizuka T., Riedel K.S., Kardaun O.J.W.F., Cordey J.G. et al., "Scalings for tokamak energy confinement time", Nucl. Fus. 30 (1990) 1999.
10. Carlstrom T.N., Shimada M., Burrell K.H., DeBoo J., Gohil P. et al., "H-mode transition studies in D III-D", in Controlled Fusion and Plasma Physics, (16th. Conf., Venice, 1989), Europhys. Conf.Abstr. Vol. 1, EPS (1989) 241.
11. Ryter F., et al., "Ohmic H-mode and H-mode threshold in ASDEX Upgrade", in Controlled Fusion and Plasma Physics, (20th. Conf., Lissabon, 1993), EPS (1993) to be published.
12. Goldston R.J., et al., "Burning plasma experiment, confinement", Fus. Techn. 21 (1992) 1076.
13. Kardaun O., et al., "ITER : Confinement analysis based on the H-mode database", in Plasma Physics and Controlled Nuclear Fusion Research, (14th. Int.Conf., Würzburg, 1992), Paper F-1-3, IAEA (1993) to be published.
14. Shoji T., et al., "Divertor bias experiment on JFT-2M", in Plasma Physics and Controlled Nuclear Fusion Research, (14th. Int. Conf., Würzburg, 1992), Paper. A-5-5, IAEA (1993) to be published.
15. Osborne T.H., Brooks N.H., Burrell K.H., Carstrom T.N., Groebner R.J. et al., "Observation of the H-mode in Ohmically heated divertor discharges on D-III-D", Nucl.Fus. 30 (1990) 2023.
16. Zohm H., Osborne T.H., Burrell K.H., Chu M.S. Doyle E.J., "ELM studies on D III-D and a comparison to ASDEX results", in Plasma Physics, (19th. Conf., Innsbruck, 1992), Europhysics Conference Abstracts Vol. 1, EPS (1992) 243.
17. Schissel D.P., DeBoo J.C., Burrell K.H. et al., "H-mode confinement scaling for the D III-D and JET tokamaks", Nucl.Fus. 31.(1991) 73.
18. Mertens V., et al., "Experimental investigation and interpretation of MARFEs and density limit in ASDEX Upgrade", in Controlled Fusion and Plasma Physics, (20th. Conf., Lissabon, 1993), EPS (1993) to be published.
19. Zohm H., et al., "Characterisation of ELMS on ASDEX Upgrade", in Controlled Fusion and Plasma Physics, (20th. Conf., Lissabon, 1993), EPS (1993) to be published.
20. Field A., et al., "Spectroscopic investigations of ELM phenomena in the ASDEX Upgrade divertor with high time resolution", in Controlled Fusion and Plasma Physics, (20th. EPS Conf., Lissabon, 1993), EPS (1993) to be published.
21. Lieder G., et al., "Interpretation of low ionized impurity distributions in the ASDEX Upgrade divertor", in Controlled fusion and plasma physics, (20th Conf., Lissabon, 1993), EPS (1993) to be published.

High-Power ICRF and LHCD Experiments on Tore Supra

B. Saoutic, B. Beaumont, A. Bécoulet, J. P. Bizarro, D. Fraboulet, X. Garbet,
M. Goniche, L. Guiziou, G. T. Hoang, T. Hutter, E. Joffrin, H. Kuus, X. Litaudon,
P. Mollard, D. Moreau, F. Nguyen, A. L. Pecquet, Y. Peysson, G. Rey,
D. van Houtte, M. Zabiégo.

Association Euratom-CEA/DRFC CE-Cadarache/13108 Saint-Paul-lez-Durance
(France)

ABSTRACT

For a given ICRF power, the sawtooth-free period duration increases with plasma density. This trend, correlated with an increasing soft X-ray inversion radius, clearly shows a better stabilization by hot ions at higher density. A tentative explanation is given through a full modelling of the hot ion anisotropy using a Fokker Planck code coupled with a 1/2 D power balance code simulating experimental data. A linear stability analysis of the kink/tearing m=1 mode, in terms of a MHD kinetic functional, shows that the anisotropy of the distribution function may amplify the effect of the hot ion pressure gradient. A two-fold lengthening of the sawtooth-free period is obtained when applying LHCD to ICRF heated plasmas. Analysis of polarimetric data shows that, instead of the continuous decrease of the central safety factor observed when using ICRF alone, the q value "freezes" when LHCD power exceeds 3 MW. This demonstrates a current profile control effect opening the way towards steady-state stabilization. Working with both ICRH and LHCD on the same target plasma allows thorough comparisons of transport with different power deposition profiles and coupling mechanisms. The energy stored in electrons is the same although the electronic temperature gradients are quite different. Nevertheless, due to better ion heating, the energy life time (discarding hot ions contribution) is always higher with ICRH. Local transport analyses allow comparisons between both heating schemes for different confinement situations.

I. SAWTOOTH STABILIZATION BY HOT IONS:

I.1 Introduction: sawtooth stabilization using on axis ion cyclotron resonance frequency (ICRF) minority heating was first observed on JET[1]. This phenomenon was subsequently verified on other machines (TFTR[2], Tore Supra[3]). Since, other mechanisms using the fast wave and leading to stabilization have been discovered: minority ion current drive[4], higher harmonic stabilization[5]... This paper will concentrate on minority heating.

Experimental observations are quite comparable on the various tokamaks[6,8] There is a power threshold under which no stabilization occurs. Stabilization is most readily achieved with on-axis heating and is not possible when the cyclotron layer lies close to or outside the q=1 surface. The threshold power depends on the plasma current, on the monopole/dipole antenna phasing and on the majority ion species of the plasma. It is generally suggested[7,8] that the driving mechanism for this stabilization could be the pressure exerted by the energetic, anisotropic ion population created by ICRF minority heating. These theories are qualitatively consistent with the dependence of the threshold power on plasma current and on ion cyclotron layer position. However, no explanation has yet been given for the dependence on the majority ion species and spectrum shape.

A new experimental trend has been observed on Tore Supra which brings new insight on the mechanism responsible for the stabilization of sawteeth. Figure 1

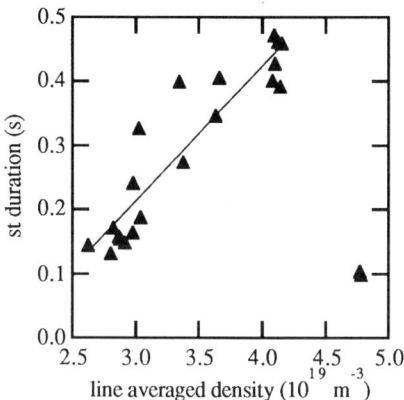

Fig. 1: Sawtooth free period duration vs. averaged central line density.

displays the result of a plasma density scan done with a constant ICRF power of 3.5 MW, just above the threshold for stabilization. Plasma parameters are: plasma current 1.5 MA, major radius 2.28 m, minor radius .72 m, magnetic field 3.7 T, hydrogen minority ($\approx 10\%$) in deuterium, cyclotronic layer on the magnetic axis, dipole configuration, frequency 57.4 MHz. The increase of the duration of the sawtooth-free period with plasma density is well correlated with the increase of the inversion radius observed by soft X-rays. On the other hand, the span of the density scan is small enough, so that the "initial" inversion radius of the last relaxation before the sawtooth free period does not change significantly. Moreover, dramatic changes in the power and hot ion pressure profile are not expected (this will be confirmed later in the paper). This suggests that the position of the q=1 surface along the hot ion pressure profile, if it does explain the dependence on plasma current, does not play the key role for the density scan.

One of the major effects of a density scan could be a change in the velocity space anisotropy of the hot ion distribution. A very simple heuristic argument is that, for a given fast wave electric field, the higher the density, the higher the collisionality, and the lower the anisotropy. We will first quantify this phenomenon. Then, we will see how it fits in the theory of sawtooth stabilization by hot ions. Last, we will envisage some consequences.

I.2 Quantification of the tail anisotropy: the main tool we have been using to quantify the anisotropy is the code VICHTOR[9]. This code is a Fokker-Planck code dedicated to ICRF minority heating. In a toroidal geometry, it incorporates a 2D Hamiltonian phase space description leading to a local quasi linear operator for heating. Its formulation is variational and the test functions it uses, are:

$$f = K \left(e^{-h/t_0} + A \, p \, e^{-p/\Delta p} e^{-h/\Delta t} \right) \quad (1)$$

where p is the parallel energy of a test minority ion when it crosses the cyclotronic layer and h its kinetic energy. For a given fast wave electric field (E), electron and bulk ion temperatures and densities, VICHTOR computes: A, a quantity related to α the proportion of minority ions in the hot tail; Δt, the "effective temperature" of the tail; and Δp, the characteristic anisotropy width of the tail in the p-direction. K is determined through particle conservation.

Once the distribution function is calculated, it is easy to compute the energy going to the electrons, to the bulk ions and the energy stored in the hot ions. However the direct experimental determination of E is not possible. We have then been led to carry out a time dependent power balance in the center of the plasma. This allows us to determine the power going from hot ions to electrons. Knowing the experimental bulk temperatures and densities, we scan the value of the electric field until the hot ion distribution provides the right amount of power to electrons. Then we can check that the amount of power going to the bulk ions is compatible with the power balance request, and, taking into account the total amount of ICRH power launched in the plasma, we can determine an effective radius for the wave power deposition profile (assumed to be gaussian). To gain computing time, we have interfaced the power balance 1/2 D code with VICHTOR through a neural network. This allows us, starting from the experimental plasma density and temperature, to determine either the "effective power deposition radius" and the electric field, or, in the inverse direction, to simulate the electronic and ionic temperatures of the plasma.

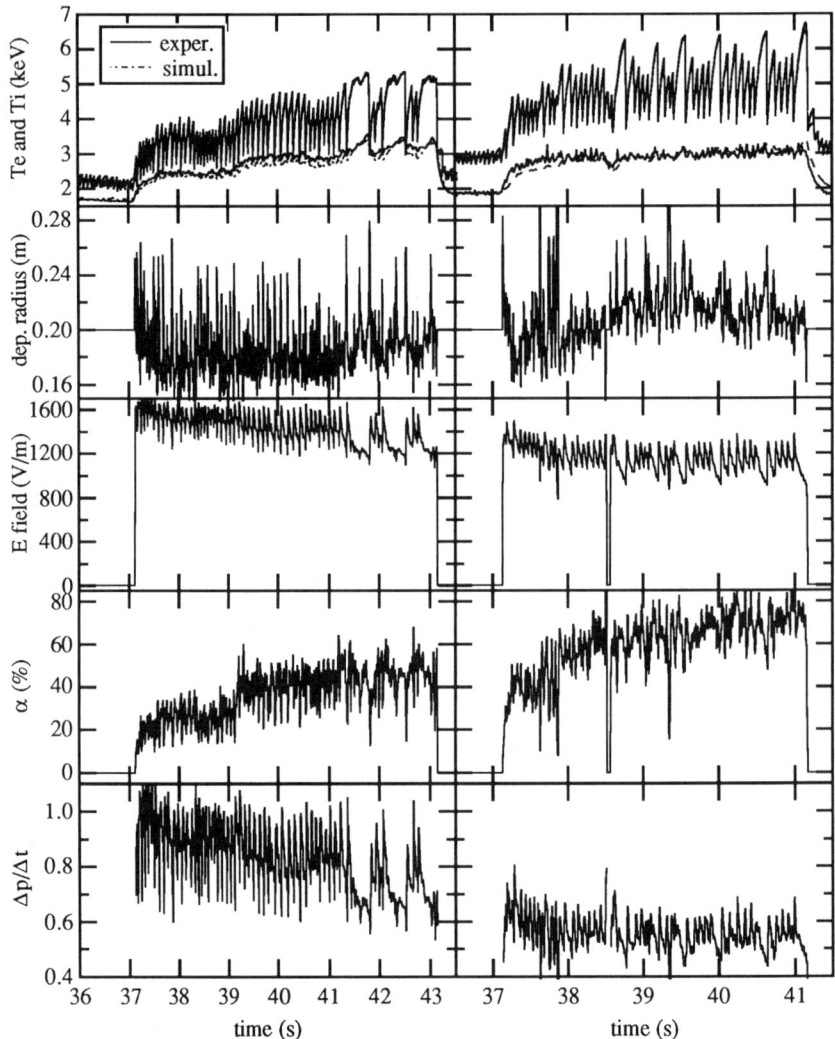

Figure 2 displays the results of simulations for two shots with different electronic densities: #8853 (right column) with an averaged line density of $2.7 \; 10^{19} \; m^{-3}$ and #8871 (left column) with an averaged line density of $4 \; 10^{19} \; m^{-3}$. This figure also displays the electric field and effective radius used in the simulation. It is worth noticing that the power balance code, which does not include transport of fast ions during the sawtooth crash, simulates it as an increase of the effective radius indicative of a widening of the hot ion profile. This can be correlated with fast ion bursts observed on the ripple diagnostic of Tore Supra[10]. Furthermore, we have run our full-

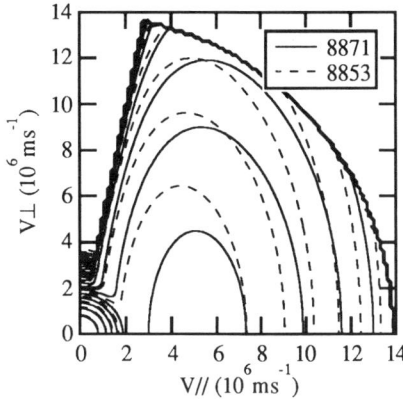

Fig. 3: ICRH distribution function.

wave code ALCYON[11] and compared the electric field and power deposition profile it predicts with the one previously determined: for the top of a monster sawtooth (shot #8871), ALCYON gives 1000 Vm^{-1} to compare with the 1200 Vm^{-1} needed by the simulation and the 1200 Vm^{-1} provided by a simple cold plasma calculation .The deposition radius calculated by ALCYON and the balance power code also agree well. As mentioned before, the "effective power deposition profile" and hot ion pressure are quite comparable in the two shots. A contrario, the ratio $\Delta p/\Delta t$ differs significantly (.55 for shot #8853, .8 for shot #8871). This effect clearly shows on the velocity space anisotropy (figure 3).

I.3 Comparison to MHD stabilization theory: the theoretical model[12] is based on a variational principle equivalent to the energy principle. This principle applied to the (m=1, n=1) mode shows the kink/ tearing destabilizing term as well as the resonant and interchange globally stabilizing ones. Making an asymptotic development of the stabilizing terms in ε (ratio of the radius of the q=1 surface to the major radius) and 1- q_0 (where q_0 is the central safety factor), one sees that, for an isotropic distribution function, the leading term of the interchange contribution from trapped particles can cancel either with the leading term of the interchange contribution from passing particles, or with the leading term of the resonant contribution from trapped particles (the resonant contribution from passing particles being always negligible). However, for trapped particles, it is mainly the deeply trapped ones who contribute to the interchange and resonant terms [8]. As distribution functions, such as the one shown in figure 3, do not account for many of these particles, we can conclude that for an ICRH type of distribution function, the stabilizing term is mainly driven by the interchange contribution from passing particles and reads as:

$$\delta W_{stab} = \int_0^a dr\, R(r) \int_0^{\lambda_c} d\lambda\, g(\lambda,r) \frac{\partial}{\partial r}\left(\int_0^\infty dH\, H^{3/2} F(H,\lambda,r) \right) \quad (2)$$

where F is the hot ion distribution function, H the kinetic energy, λ a pitch angle variable $(2/\pi) \cos^{-1}(V_{//}/ V)$ (λ_c being the value for the limit passing/trapped particles), and g is the weighting function for the interchange term. R is a function introduced to cope either with tearing or kink instability: the more resistive the solution is (tearing), the more the q = 1 surface contributes. Using for F the test function (1) and knowing that g is strongly weighted by the barely passing particles, integration leads to:

$$\delta W_{stab} = \int_0^a dr\, R(r)\, (K A) \left(1 - \frac{\partial (K A)}{(K A)\, \partial r} - C r \frac{1}{\Delta p / \Delta t + \varepsilon} \right) \quad (3)$$

Assuming a tearing solution, this formula leads to the conclusion that, when the value of the ratio $\Delta p/\Delta t$ evolves from .55 to .8, the stabilizing term increases by 30%. To explain this trend, it has to be observed that passing particles (especially the barely passing ones) spend more time in the good curvature region. For increasing $\Delta p/\Delta t$, the number of barely passing particles increases and the distribution function builds some features similar to the one of a sloshing particle distribution[13] and becomes more and more stabilizing. Considering that the ICRH power level is just above the threshold, this effect is sufficient to explain that sawteeth are not suppressed for shot #8853.

I.4 Conclusion: from new experimental trends observed on Tore Supra, it has been possible to point out the significance of the velocity space anisotropy of hot ion tails on sawteeth stabilization. As the collisionality on ions is quite different in helium compared to deuterium, such phenomenon could explain the higher power threshold observed when working with a helium majority. One can also notice that the distribution function can dramatically change when passing from dipole to monopole configuration (through the value of the adiabatic barrier for example). Such effects will be investigated. An important conclusion is that, because distribution functions expected for α particles are much more isotropic than for ICRH, one has to be very cautious when extrapolating sawtooth stabilization to ignited plasmas.

II. EFFECT OF LHCD ON SAWTOOTH STABILIZATION BY HOT IONS:

The duration of a sawtooth-free period due to hot ion stabilization, increases when applying lower hybrid current drive (LHCD) to the plasma. This has already been observed on JET[14]. Figure 4a shows the result of a scan of the LHCD power on Tore Supra, with the following parameters: plasma current 1.5 MA, major radius 2.28 m, minor radius .72m, magnetic field 3.7 T, electronic averaged line density $4 \; 10^{19}$ m^{-3}, hydrogen minority ($\approx$ 10%) in deuterium, cyclotronic layer on the magnetic axis, ICRH power 4 MW, dipole configuration, LHCD launcher phase 0°. The sawtooth stabilized phase period lengthens from 0.45 s (ICRH alone) to almost 1 s (ICRH + 3.4 MW of LHCD power). This lengthening is not correlated with an increase of the inversion radius of the soft X-ray emission. As a matter of fact, the inversion radius at the crash of the 1s stabilized phase is even slightly smaller that the one at the crash of a .45 s stabilized phase with no LHCD applied. This suggests that the lengthening is not due to an increase of the hot ions stabilizing efficiency. This statement is also supported by the analysis of the fast neutral spectra. Figure 4b displays the flux of fast neutrals whose energy ranges from 25 to 100 keV. This flux is averaged during sawtooth-free periods in order to avoid hot ion bursts occurring at sawtooth crashes. For zero LHCD launcher phase, the fast neutral spectra do not change regardless the LHCD power level. It is still worthwhile to notice on fig 4b that, for non-zero LHCD launcher phase, some part of the LHCD power couples to the ICRH created hot ions: the number of fast neutrals almost doubles. Very significant too is the black triangle which corresponds to a stabilized phase occurring at an ICRH power level of 2.8 MW, far below the usual ICRH alone threshold level (3.5 MW): in this case, the neutral particle number is the same as that in a 3.5 MW ICRH only experiment.

Another possible explanation for the lengthening could be an increase of the central electronic temperature, leading to longer characteristic resistive times, and consequently to longer periods before the expanding q=1 surface reaches the zone where hot ions cannot stabilize the MHD activity anymore. However, when working at high density, the LHCD power deposition profile is very wide and peaks off-axis. Therefore, the central temperature does not increase much (ranging from 5 to 5.5 keV with LHCD power from 0 to 3.4 MW). Such an explanation must then be discarded.

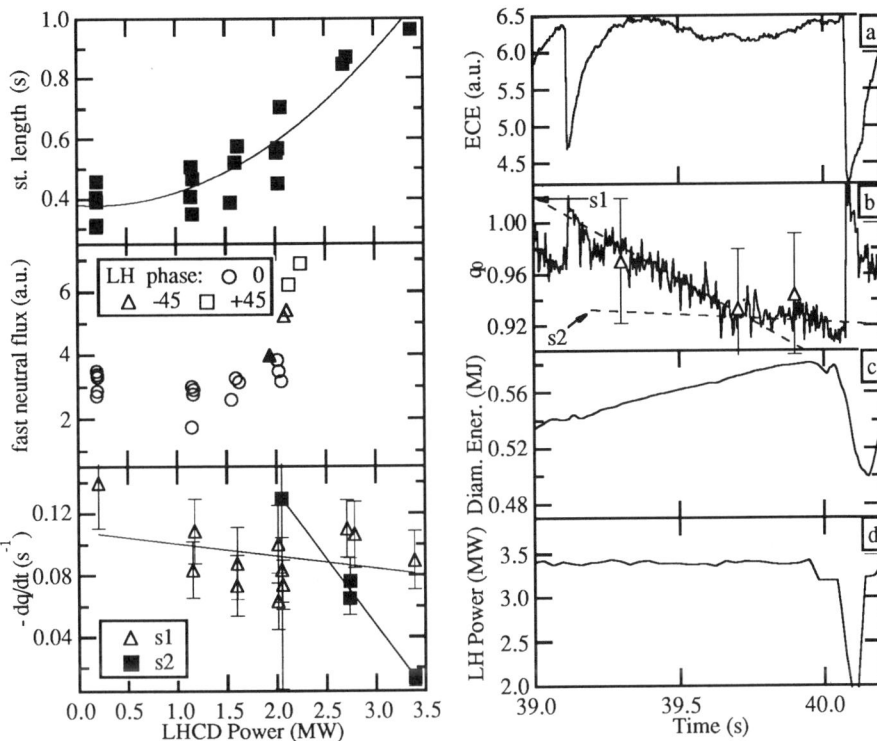

Fig. 4: (a) Sawtooth free period duration, (b) fast neutrals flux and (c) dq/dt vs. LHCD power.

Fig. 5: (a) ECE signal, (b) central safety factor, (c) diamagnetic energy and (d) LHCD power during 1s stabilized phase.

Fig. 5b shows the evolution of the central safety factor (q_0) during the 1 second sawtooth-free period. The solid line corresponds to q_0 computed by Abel inverting the data of a 5 channel polarimeter. Triangles correspond to q_0 computed through the IDENT-D equilibrium code[15] which determines the current profile, in accordance with the Grad-Shafranov equation, by fitting various experimental data (magnetic, interferometric, polarimetric...). After an initial decrease quite similar to the one observed when using ICRH alone, the q_0 value "freezes" and barely evolves anymore until the crash. In order to quantify this phenomenon, we have determined by a least square fit, the slope s1 of the time evolution of q_0 during the first 0.45 s of the sawtooth-free period, and the slope s2 for the sawtooth-free period extending from 0.45 s to the crash (dashed lines in fig. 5b). Fig. 4c illustrates such an analysis carried out for all the shots of the LHCD power scan. The value of s1 (triangles) does not change much as could be expected from the very small range of variation of the central electronic temperature. More than 2 MW of applied LHCD power are necessary to get an effect on the s2 slope (squares). As a matter of fact, the freezing of the q_0 value requires a LHCD power of the order of 3.5 MW. The fact that the LHCD power was marginally at this level during the 1 s sawtooth-free period, supports the statement that the crash is due to loss of power because of arcing in the LHCD launcher (see fig. 5d), and that, by avoiding arcing and/or increasing the LHCD power, the sawtooth free period should be maintained much longer. However, the stored diamagnetic energy continuously rises during the sawtooth-free phase (see fig.

5c), indicating that the temperature profiles have not yet reached full equilibrium. Some resistive evolution of the q-profile must then take place although its central value does not evolve any more (only 20% of the total plasma current is driven by LH waves). Nevertheless, there is a clear evidence that by controlling (even partially) the current profile, it is possible to extend the duration of the sawtooth-free period. Further experiments are planned on Tore Supra to check whether a steady-state stabilization can be achieved this way.

III. LOCAL TRANSPORT ANALYSIS:

Using ICRH then LHCD on the same shot has allowed us to make a thorough comparison of transport between both schemes. Parameters are: plasma current 1.7 MA, major radius 2.30 m, minor radius .74m, magnetic field 3.9 T, electronic averaged line density $4 \; 10^{19}$ m^{-3}, hydrogen minority ($\approx 10\%$) in deuterium, cyclotronic layer on the magnetic axis, ICRH and LHCD power 3.8 MW, ICRH dipole configuration, LHCD launcher phase 0°. Fig. 6 shows the electronic and ionic temperature profiles measured during ICRH and LHCD, and the power deposition profiles determined by ALCYON for the ICRH case and a modified version of the Bonoli-Englade LHCD code[16] for the LHCD case. As quoted above, the power deposition profiles are drastically different in the two cases. The coupling mechanisms also differ: fast ions for ICRH; fast electrons for LHCD. Due to its central and narrow power deposition profile, ICRH temperature profiles are much more peaked and exhibit stronger gradients than the LHCD ones. Nevertheless, as shown on the bar chart of fig. 7, the kinetic energy stored in the electrons is the same in both cases and is in good agreement with the offset-linear Rebut-Lallia scaling law[17].

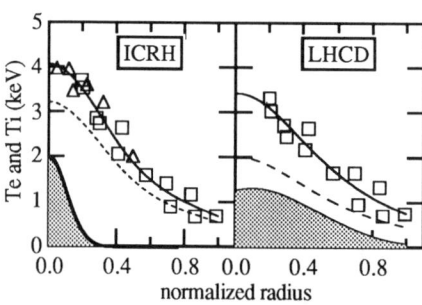

Fig. 6: Temperature and power deposition profiles. Solid line: electrons. Dashed line: ions. Shaded area: power deposition (a. u.). Squares: Thoms. scatt.. exp. points. Triangles: ECE exp. points.

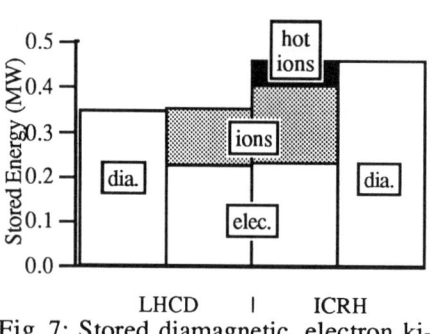

Fig. 7: Stored diamagnetic, electron kinetic and ion kinetic energy.

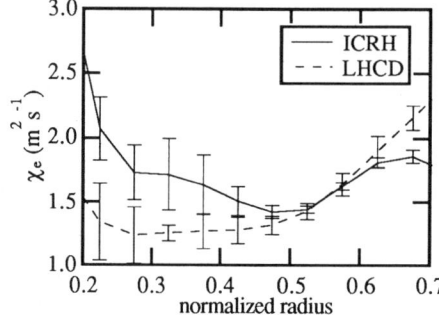

Fig. 8: Electron heat diffusion coefficient profiles.

Figure 8 illustrates the results of local transport analysis carried out using the LOCO code[18]. The electron heat diffusion coefficient χ_e is larger in the confinement zone for ICRH. This clearly follows the trends of critical gradient models. The general picture given by this analysis is: even if the local confinement is more deteriorated in the ICRH case, the fact that the power deposition is central makes up for it, and the global "electron energy confinement" is the same. However, something is gained when using ICRH. Because the low energy part of the hot ions tail collides with ions, some ICRH power is coupled to them. This shows up in the ion kinetic energy which is larger than in the LHCD case where power goes to ions only by equipartition (see fig. 7). This leads to a better global energy confinement for ICRH, even when the hot ion contribution is discarded.

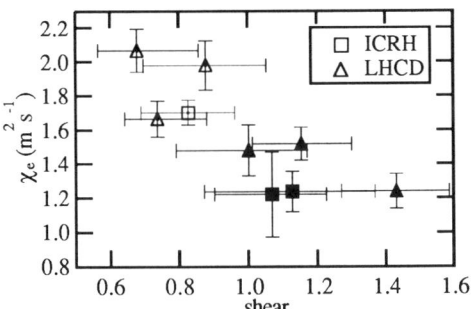

Fig 9: Electron heat diffusion coefficient vs. shear. Open symbols: L mode. Dark symbols: enhanced confinement.

Figure 9 shows a comparison of the electron thermal diffusivity in the confinement region (.5<r/a<.7) for different confinement modes: ICRH and LHCD L-modes, LHCD high internal inductance (li) mode, lower hybrid enhanced performance[19] (LHEP), and ICRH sawtooth stabilized phase. All improved confinement schemes exhibit a lower χ_e correlated with the value of the magnetic shear. A striking point is that the enhancement of confinement during both monster sawtooth and high li seems to be due to the same mechanism: namely, an increase of the magnetic shear.

1   D.J. Campbell, D. H. F. Start, J. A. Wesson et al., Phys. Rev. Lett., **60** (1988) 2148.
2   J.C. Hosea, C.K. Phillips, J.E. Stevens et al., Proc. of the Joint Varenna-Lausanne Int. Workshop on Theory of Fusion Plasmas, Varenna, Italy, (1990) 223.
3   A. Agarici, A. Bécoulet, B. Beaumont et al., EUR-CEA-FC-1457 (1992) 61.
4   D.F.H. Start, V.P. Bhatnagar, G. Bosia et al., Proc. of the IAEA Technical Committee Meeting on Fast Wave Current Drive in Reactor scale Tokamaks, Arles, France, (1991) 227.
5   T. Watari, Plasma Phys. Cont. Fusion, **35A** (1993) A181.
6   C.K. Phillips, J. Hosea, E. Marmar et al., Phys. Fluids B, Vol. 4, **7** (1992) 2155.
7   R.B. White, M.N. Bussac and F. Romanelli, Phys. Rev. Lett., **62** (1989) 539.
8   F. Porcelli, Plasma Phys. Contr. Fusion, 33 (1991) 1601.
9   A. Bécoulet, D.J. Gambier and A. Samain, Phys. Fluids B, 3 (1991) 137.
10   V. Basiuk, J.P. Roubin, A. Bécoulet et al., Proc. of 19th EPS Conf. on Controlled Fusion and Plasma Physics, Innsbruck, Austria, **16C** (1992) 175.
11   D.J. Gambier, A. Samain, Nuclear Fusion, **25** (1985) 283.
12   D. Edery, X. Garbet, J.P. Roubin et al., Plasma Phys. Contr. Fusion, **34** (1992) 1089.
13   R.J. Hastie, Y.P. Chen, F.J. Ke et al., Chinese Phys. Lett., **4** (1987) 561.
14   D. Moreau et al., Plasma Phys. Contr. Fusion, 33 (1991) 1621.
15   J. Blum et al., Nucl. Fus., **30** (1990) 1475.
16   J.P. Bizarro and D. Moreau, Phys. Fluids B, **5** (1993) No 4 (in press).
17   P.H. Rebut, P.P. Lallia and M.L. Watkins, Proc. 12th Int. Conf. on Plasma Physics and Controlled Nuclear Fusion Research, Nice, France, (1988) IAEA-CN-50/D-4-1.
18   G.R. Harris, H. Capes, X. Garbet, Nuc. Fus., **32** (1992) 1967.
19   D. Moreau, B. Saoutic et al, Proc. 14th Int. Conf. on Plasma Physics and Controlled Nuclear Fusion Research, Würzburg, Germany, (1992) IAEA-CN-56/E-2-1(C).

REVIEW OF COMBINED ICRH-NBI RESULTS IN TEXTOR

A.M. Messiaen[1]*, P.E. Vandenplas[1], D. Van Eester[1], G. Van Wassenhove[1],
H. Conrads[2], P. Dumortier[1], F. Durodié[1], H. Euringer[2], G. Fuchs[2],
B. Giesen,[2] D. Hillis[3], F. Hoenen[2], R. Koch[1], H.R. Koslowski[2],
A. Krämer-Flecken[2], M. Lochter[2], B. Mullier[1], J. Ongena[1]*,
T. Oyevaar[4], H. Soltwisch[2], H.F. Tammen[4], G. Telesca[2], R. Uhlemann[2],
R. Van Nieuwenhove[1]*, G. Van Oost[1], M. Vervier[1], G. Waidmann[2],
R.R. Weynants[1]

(1) Laboratoire de Physique des Plasmas/Laboratorium voor Plasmafysica-Association "Euratom-Etat belge"/Associatie "Euratom-Belgische Staat" Ecole Royale Militaire - B-1040 Brussels - Koninklijke Militaire School
(2) Institut für Plasmaphysik, Forschungszentrum Jülich, GmbH Association "Euratom-KFA", D-5170 Jülich, FRG
(3) Oak Ridge National Laboratory, P.O. Box 2009, Oak Ridge, Tennessee 37831-8071, USA.
(4) Associatie Euratom-FOM, FOM-Instituut voor Plasmafysica "Rijnhuizen", Postbus 1207, NL-3430 BE Nieuwegein (The Netherlands).

ABSTRACT

The synergism observed between NBI and ICRH is theoretically interpreted for the interaction at the second and third ion cyclotron harmonic. It is also shown that the performances of supershot-like discharges obtained with balanced injection can be substantially improved by beam-RF interaction both at $2\,\omega_{CD}$ and $3\,\omega_{CD}$.

INTRODUCTION

The synergy of the combined heating by neutral beam injection and by ICRH is now being theoretically interpreted by the use of a coupled wave/Fokker-Planck equation solver [1]. The wave equation fully accounts for the nonmaxwellian distribution function of the beam species and retains finite temperature corrections to all orders in slab geometry. The beam distribution function is found integrating the relevant Fokker-Planck equation. On the other hand to describe the interaction of the beam with the target plasma in toroidal geometry simulations by the code system TRANSP of Princeton are used more in particular for the interpretation of the generation of non-inductively driven currents. The synergy occurs through modifications of the target plasma parameters both for NBI and ICRH and through the direct interaction of the RF with the beam. The interpretation of the heating results by combination of neutral beam injection (either by deuterium co-injection or balanced injection with maximum injection energy E_b of 55 keV) and ICRH on a D(H) plasma has been undertaken for two scenarii : (i) at $\omega = 2\,\omega_{CD} = \omega_{CH}$ ($\omega/2\pi$ = 32.5 MHz for B_t = 2.25 T) and (ii) at $\omega = 3\,\omega_{CD}$ ($\omega/2\pi$ = 38 MHz for B_t = 1.7 T). The ICRH is coupled by means of two antenna pairs, on one of which the electrostatic shield is removed. The behaviour

---
* NFSR Researcher

of the two antenna pairs (shielded or not) is quite similar provided that the unshielded antenna feeder area is covered [2]. TEXTOR has been operated with boronized wall and siliconized wall conditions. ICRH is as easily operated with either of the two wall conditions and analogous results are obtained. A detailed account of the effect of siliconization on the improvement of the confinement time will be given later [3].

Figure 1a depicts the theoretically predicted power deposition profile into the D-co-injected beam and plasma species (H minority, D-bulk and electrons) for the two scenarii considered and their experimental parameters, assuming full absorption of the RF waves by the plasma after a sufficiently large number of transits. For the $\omega = 2\, \omega_{CD}$ scenario most of the RF power (~ 80 %) is directly absorbed by the H minority (which redistributes its power mostly to the electrons) and less than 10 % of the wave power is coupled to the D-beam. In the $\omega = 3\, \omega_{CD}$ case most of the power is coupled to the D-beam in the central part of the plasma. In Fig. 1b, the beam perpendicular energy density in the $v_\perp$, $v_\parallel$ plane is depicted when $\omega = 2\, \omega_{CD}$. It shows the appearance of a highly energetic tail due to the absorbed RF power.

RESULTS

(i) <u>At $\omega = \omega_{CH} = 2\omega_{CD}$</u>, it has been shown that there is a synergistic effect on confinement and heating when ICRH is added to co-injection leading to a larger incremental confinement time of ICRH and particularly to a three times larger heating efficiency of the ions by the RF [4,5]. As a result, the hot ion mode together with the improved confinement regime (I-mode) [6] remain in presence of combined heating. With balanced injection I-mode discharges with supershot characteristics are observed. Their performances can even be enhanced by the addition of ICRH. In Fig. 2 the corresponding energy and temperature increases together with the peaked density profile and the hollow radiation profile are shown. No sign of high Z pollution (siliconized wall conditions) is observed. For comparison the corresponding energy and temperature levels obtained for pure balanced injection (at a slightly lower density) are shown between the vertical dotted lines. It is essential to take the density into account since T_{io} is a strongly decreasing function of density. The larger central ion temperature increase (for same density) with respect to the electronic one is therefore to be noted. The neutron yield (referred to the same density), which results from beam-target and beam-beam reactions, increases by 50 % due to the addition of ICRH and reaches 10^{14} neutrons/s.

With 1.8 MW of co-injected NB power it has been possible to drive half of the plasma current non-ohmically for stationary conditions (> 2 s) and for any value of the total plasma current. This value is further increased (up to 70 %) by the addition of ICRH and the corresponding total current drive efficiency rises above 0.5×10^{19} MAm^{-2}/MW [4].

The synergistic effect on heating, current drive and neutron yield can be explained in the following way : (i) For a large part it is due to the electron heating of the target plasma resulting from the application of ICRH which leads to an increase of the beam slowing down times and a rise of its critical energy. The electron temperature increase is mainly due to minority hydrogen heating causing an energetic minority tail to be formed which relaxes collisionally on the electrons. This produces an increase of the beam driven current and of the beam power fraction coupled to the ions. This last effect explains the apparent increase of the ion heating efficiency due to ICRH which mainly heats the electrons through the minority tail and compensates the decrease of the beam power coupled to the electrons. (ii) It also results, to lesser extent, from the direct RF power coupling to the beam. As shown in Figure 1b this leads to the deformation of the beam energy density distribution the contribution of which towards large $v_\perp$ is strongly enhanced. This explains the further reduction in beam collisionality and shows that the beam contributes to the development of a highly energetic perpendicular tail. (iii) Let us finally note the influence of the observed decrease of plasma toroidal rotation [4] (for co-injection + ICRH) which can increase the relative velocity between the beam and the plasma and increase the neutron reactivity.

(ii) <u>At $\omega=3\omega_{CD}$</u>, a large amount of RF power is coupled to the D-beam (see Fig. 1a) and a highly energetic tail is produced which strongly enhances the beam target neutron yield due to co-injection in proportion to the total NBI+RF power [7]. When applied to balanced injection a substantial increase of neutron reactivity and stored energy is observed as shown in Fig. 3 where 2 levels of RF power have been added to the balanced injection (boronized wall conditions). The energy and the neutron yield are increased roughly in proportion to the total additional power, the central density reaching 9×10^{13} cm^{-3}. Note that the density peaking is increased by the addition of ICRH and that the neutron yield increase with respect to balanced injection due to the additional ICRH would at constant density be nearly 4. At maximum power the energy confinement time has still an improvement of 1.5 with respect to the ITER L89-P scaling. The central electron and ion temperature are also shown in Fig. 3. Let us remark that the hot ion mode character remains also during ICRH and that notwithstanding the large density increase, the temperatures do not appreciably decrease when ICRH is coming.

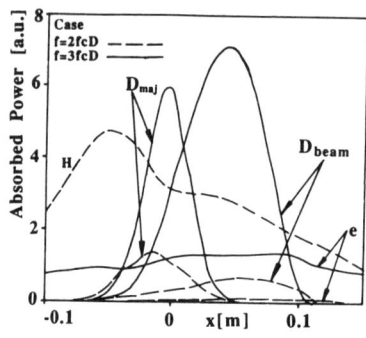

Fig. 1a - Relative RF power absorption by the D-co-injected beam and by plasma species (D_{maj}, H, e) for $\omega = \omega_{CH} = 2 \omega_{CD}$ and $\omega = 3 \omega_{CD}$ cases ($N_H/N_D \approx 5\%$).

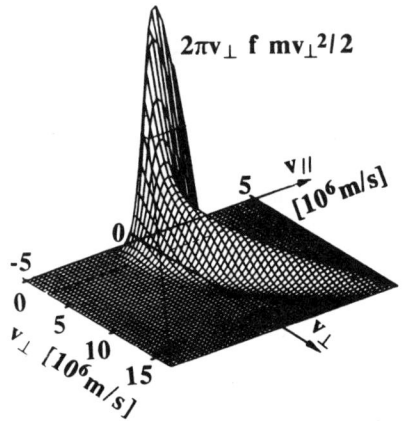

Fig. 1b - Perpendicular energy density $2\pi v_\perp f (v_\|, v_\perp)\ 1/2\ mv_\perp^2$ of the beam in the cartesian $v_\perp$, $v_\|$ plane (case $\omega = 2 \omega_{CD}$, RF power density absorbed by the beam : 0.2 MW/m^3).

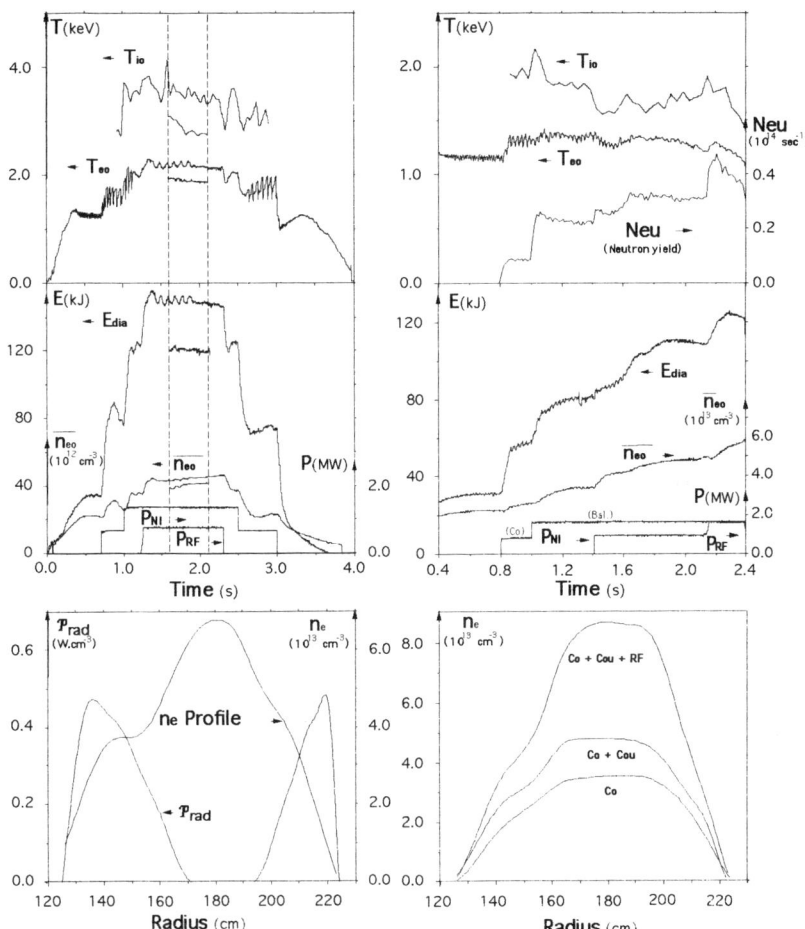

Fig. 2 - Balanced injection (with P_{RF} at $\omega = 2\omega_{CD}$, I = 0.4 MA, B_t = 2.25 T, $N_H/N_D \cong 8\%$, E_b = 50 keV, siliconized wall).
(a) Display of the diamagnetic energy E_{dia}, central temperature T_{eo} (ECE), T_{io} (CXES), central average chord density $\bar{n}_{eo}$ and additional power versus time; between the dotted lines are indicated the corresponding traces when no RF is present.
(b) Radial profiles of density and radiated power.

Fig. 3 - (a) Display of E_{dia}, T_{eo}, T_{io}, $\bar{n}_{eo}$, the neutron yield and the additional power versus time.(b) Radial density profiles from the co-injection, balanced-injection and combined heated cases ($\omega = 3\omega_{CD}$, I_p = 0.35 MA, B_t = 1.7 T, E_b = 41 keV, boronized wall).

REFERENCES

1. D. Van Eester, this conference.
2. R. Van Nieuwenhove et al., Nuclear Fusion, 32, 1913 (1992).
3. J. Winter et al., sub. for publication; J. Ongena et al. EPS Conf. Lisboa, July 1993.
4. A.M. Messiaen et al., Plasma Phys. Contr. Fusion, 35, A15 (1993).
5. J. Ongena et al., Proc 14th Int.Conf. Plasma Phys. Contr. Fusion, Würzburg, IAEA-CN-56/E-3-4 (1993).
6. J. Ongena et al., Nuclear Fusion, 33, 283 (1993).
7. G. Van Wassenhove et al., Europhysics Conf. Abstract, Vol. 16E, 141 (1992).

HIGH POWER ICRF EXPERIMENTS ON TFTR*

J. R. Wilson, J. C. Hosea, R. Majeski, C. K. Phillips, J.H. Rogers,
G. Schilling, J. Stevens, G. Taylor and the TFTR Group, Princeton University,
M. Murakami, D. A. Rasmussen, ORNL

ABSTRACT

ICRF heating experiments have been conducted in a variety of conditions on the TFTR tokamak. Power levels up to 11.4MW have been applied. During NBI driven supershot discharges the central electron temperature has been increased from 9 kev to 13 kev via ^3He minority heating with 6 MW of RF power. This temperature increase leads to a 70% increase in the projected alpha energy slowing down time. In gas fueled L-mode discharges the energetic hydrogen minority tail is observed to strongly influence the MHD stability of the discharges. Besides the stabilization of the sawtooth instability previously reported, the destabilization of both the m=1 fishbone and the TAE (toroidal Alfven eigenmode) instabilities have been observed. The TAE instability is accompanied with significant (~10%) loss of high energy ions and degradation in global confinement time.

INTRODUCTION

The TFTR research program is focused on the goal of exploring the physics of the burning D-T plasma. To this end the application of rf power in the ICRF (ion cyclotron range of frequencies) regime has several points of applicability. During D-T operation ICRF heating will be applied to supershot plasmas to heat electrons with the aim of increasing the alpha particle slowing down time and hence the population of energetic alphas. Preliminary experiments in DD plasmas have yielded significant central electron temperature increases (from 9 to 13 kev) with 6 MW of ICRF power added to 24 MW of NBI power. In addition ICRF heating will be used to directly accelerate both tritium ions ($2\Omega_T$ at 43 MHz) and alphas ($2\Omega_\alpha$ at 64 MHz).

In preparation for D-T experiments ICRF heating has proven to be a valuable tool for simulation of alpha particle physics and for check-out of alpha diagnostics. The energetic hydrogen and helium-3 ion tails produced during minority fundamental cyclotron heating have stored energies and particle velocities that are comparable to those of the expected alpha population in DT plasmas on TFTR. Energetic hydrogen tails have been used to investigate the TAE instability and have also been observed to drive the fishbone instability. Energetic ^3He tails have been used in proof of principle experiments to verify that the α-CHERS (charge exchange recombination spectroscopy) and α-charge exchange diagnostics can observe an energetic helium population.

L-MODE DISCHARGES WITH TAE MODES

Experiments in L-mode plasmas have been performed at B_T=3.4T, Ip= 1.4 and 1.85 MA with hydrogen minority heating at a frequency of 47 MHz. Majority gasses have included both deuterium and helium. The amount of hydrogen minority is determined by the intrinsic background recycling. This regime was chosen for its technical ease both for the rf and the machine operation. Power levels of up to 11.4 MW have been applied (Fig. 1).

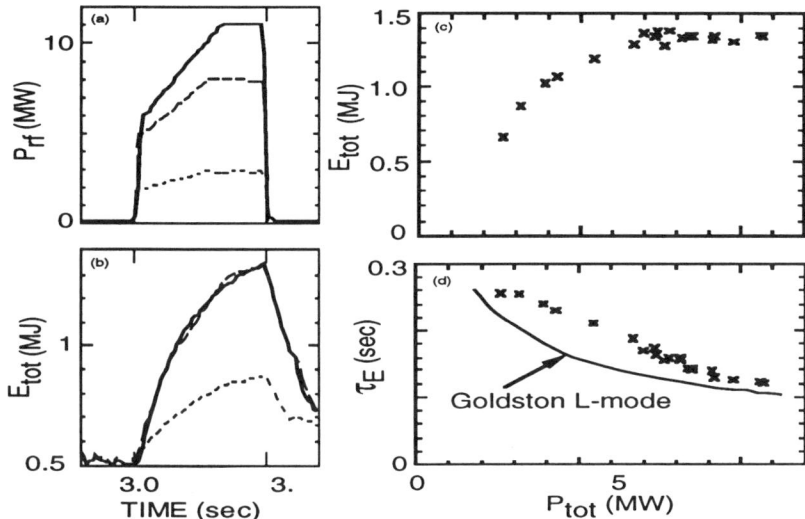

Figure 1 L-mode discharges with H minority ICRF heating
(a) Time evolution of ICRF power for three discharges (b) Time evolution of total stored energy for the same discharges (c) Peak stored energy for a sequence of discharges as a function of ICRF power (d) Global energy confinement time for the same discharges

At low to moderate power levels global energy confinement is seen to be as much as 1.4 times better than L-mode. This value of confinement enhancement is not unusual for high-recycling plasmas on TFTR which have a significant fast ion population from either NBI or ICRF. As the power is increased further the stored energy saturates faster than $P^{-0.5}$ and confinement approaches L-mode. Coincident with this saturation is the appearance of high levels of TAE mode activity[1] and enhanced loss of fast ions as observed by particle detectors external to the plasma.

From the inferred missing stored energy an estimate can be made of the maximum amount of energy that could be lost as fast ions. This is found to be a maximum of ~ 30% of the input rf power. A best estimate of the total from the amount seen directly on the particle probes is ~10% although the uncertainly in extrapolating to a fully toroidal loss leads to a factor of 2 uncertainty in this number. An alternative explanation of the saturation has been given by Cottrell et al[2]. They point out that extremely energetic minority ions can have such large orbit excursions into the cold outer plasma that they feel on average a drag that is greater than would naively be expected. This leads to a less energetic minority population and hence less of an enhanced confinement. Simulations with the SNAP and FPP codes have been performed to estimate the magnitude of this effect. The results of these simulations lead us to believe that, for these experiments, the effect is not large enough to explain the entire loss. In the future separation of the two mechanisms would best be accomplished by scanning the plasma current.

High-Power ICRF Experiments on TFTR

ICRF ADDED TO SUPERSHOTS

Up to 7.5MW of ICRF in the ^{3}He minority mode has been added to NBI driven supershot plasmas at B=4.8T and Ip= 1.85 and 2.2 MA.[3] A total power of 37.5 MW has been injected into TFTR in this regime. Figure 2 shows the time evolution of the electron temperature and the estimated central alpha slowing down time for a 23 MW NBI driven supershot with and without 5.7 MW of ICRF.

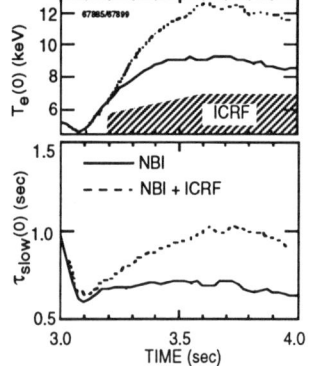

Figure 2 Time evolution of central electron temperature and the derived alpha slowing down time for 23 MW NBI w/wo 5.7 MW ICRF

The significant increase in central election temperature from 9 kev to 12.5 kev is seen to lead to an increase in the alpha velocity slowing down time τ_{slow} (0) of 40% to greater than 1 sec. To take full advantage of this increase during DT (i.e. to reach a saturated β_α value) the length of the auxiliary heating must be lengthened from 1 to ~1.5 sec. (TRANSP simulation)

In figure 3 the temperature profile and the calculated deposition profile of power flowing to the electrons is shown.

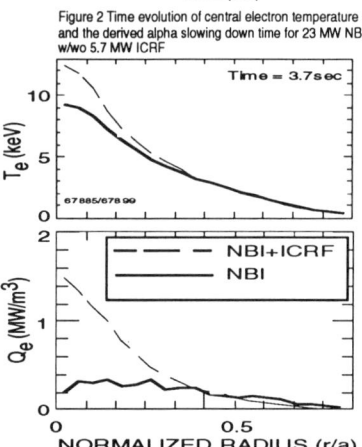

Figure 3 Electron temperature profile and power deposition profile w/wo ICRF

The strong central deposition of power into electrons from ICRF heating can be seen. Even though the ICRF power level is only 25% of the NBI power level it dominates the electron heating inside of r/a = 0.25.

In figure 4 the incremental; changes in alpha energy slowing down time, central election temperature, radiated power fraction and confinement enhancement are shown for an ICRF power range of 0-6MW added to a 23 MW NBI driven supershot.

Both the electron temperature and hence the alpha slowing down time increase linearly with ICRF heating power. The amount of power radiated from the plasma increases slowly enough with rf power level so that the fraction of power radiated falls with increasing rf power level. The global energy confinement time, as with NBI only supershots, is virtually independent of power and hence the enhancement factor H $\equiv \tau_E/\tau_L$ increases with rf power (figure 4).

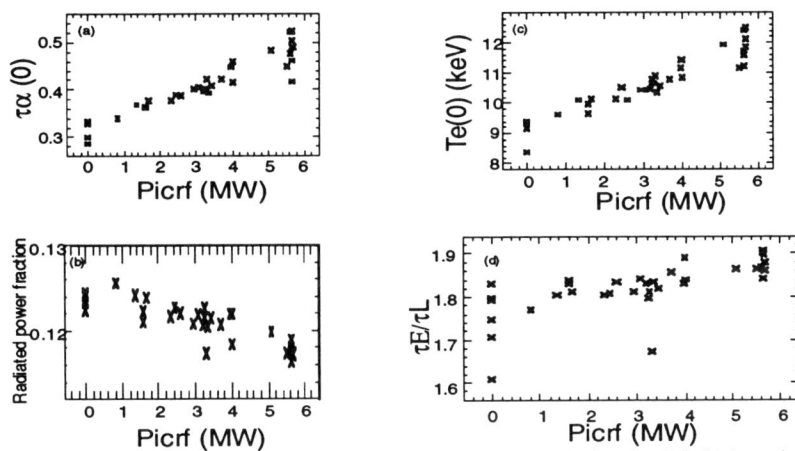

Figure 4 Addition of electron heating to Supershot discharges (a) Derived central alpha slowing down time (b) Radiated power fraction (c) Central electron temperature (d) Ratio of energy confinement time to L-mode

A study of the performance of these discharges has shown that the Shafranov shifts are large enough that the present fixed operating frequency of 47 MHz is too high, for the maximum value of toroidal field obtainable on TFTR, to allow the resonant absorption layer to be placed on axis. Therefore for future work, this frequency has been lowered to 43 MHz which should allow some headroom in toroidal field strength. In addition two of the four ICRF antennas have been configured to work at both 43 and 64 MHz[4]. This will allow hydrogen minority heating at full toroidal field for TAE studies and a test of direct alpha heating at its second harmonic resonance during DT operations.

ACKNOWLEDGMENT

This work would not have been possible without the full support of the TFTR group which we gratefully acknowledge. The research was sponsored by the Office of Fusion Energy, US Department of Energy, under contract DE-AC02-76-3073.

REFERENCES:

1. J. R. Wilson et al, (1992) IAEA Conference Wursburg, Germany, IAEA-CN-56/E-2-2.
2. G. A. Cottrell and D. F. H Start Nucl. Fus. 31, 61 (1991)
3. G. Taylor et al, (1992) to be published Phys Fluids B
4. J. H. Rogers et al, (1993) this conference.

TWO FREQUENCY ICRF OPERATION ON TFTR

J. H. Rogers, R. Majeski, J. R. Wilson, J. C. Hosea, G. Schilling, and J. Stevens
Princeton Plasma Physics Laboratory, Princeton, NJ 08543

and D. A. Rasmussen
Oak Ridge National Laboratory, Oak Ridge, TN 37831

ABSTRACT

Modifications have been made to allow two of the ICRF antennas (bays L and M) on TFTR to operate at either of two frequencies, 43 MHz or 64 MHz. The two frequency operation will allow a combination of ^{3}He-minority and H-minority heating at near full field on TFTR. Distributing the RF power between different ion species is expected to result in lower energy minority ions which have better confinement and better energy coupling to the bulk plasma. The higher frequency, 64 MHz, may also be useful in direct electron heating and current drive experiments at lower toroidal fields. The two frequency capability was accomplished by lengthening the resonant loops (2λ at 43 MHz, 3λ at 64 MHz) and replacing the conventional quarter wave impedance transformers with a tapered impedance design. The other two antennas (bays K and N) will operate at a fixed frequency, 43 MHz. Models of the antenna, resonant loops and impedance matching system are presented.

INTRODUCTION

Two frequency operation may improve operational reliability as well as provide an opportunity to investigate some interesting physics. Previous experience has shown that significant reflected power signals are difficult to avoid during injection of ICRF power into low absorption plasmas, such as experienced during direct electron heating or current drive experiments. The reflected signal is modulated in time, presumably due to variations in the mode structure (which change the antenna loading as well as the coupling between antennas). This is a significant operational difficulty because the RF transmitters can be damaged by the large reflected signals. On TFTR, a large component of this signal is comprised of cross talk from other antennas. However, any signal received by an antenna at another frequency will be reflected at the mismatch at the resonant loop (in the case of bay K or N) or at the impedance matching elements.

The combination of frequencies chosen will provide an opportunity for a number of interesting experiments. For maximum toroidal magnetic field on TFTR, 43MHz can be used for on axis ^{3}He minority heating (or tritium second harmonic heating), and 64MHz can be used for on axis H minority heating (and deuterium second harmonic heating). Splitting up a given injected power between different ion species will result in the minority ions having lower energy, enhancing the confinement. During the upcoming D-T operations on TFTR, the capability of distributing the RF energy between the different species will give an additional control with which to optimize the nuclear reaction rate. Also, the two frequency operation may have a significant impact on direct electron heating and current drive experiments. The coupling to the electrons has a strong temperature dependence. One frequency can be used for H minority heating (which couples energy to the electrons well) to heat the electrons, while the other frequency directly heats the electrons.

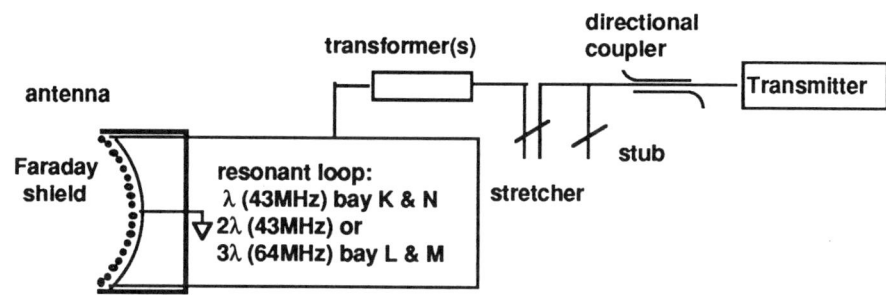

Fig. 1 Schematic of the TFTR antenna power feed system.

RF DESIGN

The modifications which will allow two frequency operation are simple. First, the resonant loops which feed the antenna were lengthened so that they are two wavelengths long at 43 MHz and three wavelengths long at 64 MHz (see fig. 1). Second, instead of using quarter wave transformers to bring the input impedance closer to 50Ω, a tapered transformer design was used. Only bays M and L were modified in this way since they are fed on each strap by FMIT transmitters which are tunable from 40-80 MHz. Bays K and N will continue with the previous configuration and operate at a single frequency, 43 MHz.

A broad band impedance transformer was designed which is optimized for the two operating frequencies.[1] The input impedance of the resonant loop is ~ 3kΩ for vacuum conditioning and ~ 240Ω for plasma operation. The transformer is optimized for coupling to a 240Ω load and is composed of two sections of length λ/4 at 53.5 MHz. The first section has impedance 74.5Ω, the second 149.1Ω. Ideally, the 149.1Ω section would be attached directly to the resonant loop; however, due to physical constraints, a 7.0 m long (~λ at 43 MHz, ~ 1.5λ at 64 MHz), 50Ω line connects the loop to the high impedance side of the transformer.

In order to model the RF design, a lumped element model of the antenna was developed (figure 2). This model incorporates a three component transmission line approximation for the vacuum feed through.[2] The current strap is approximated as a transmission line along the distance which it runs along the antenna box wall (separation << strap width). The radiating part of the current strap is modeled by two transformers (coupling the two adjacent straps) with resistors tied to an inductive ground in the center. In addition, stray capacitance to ground is represented by a single capacitor for each strap at the end of the radiating element. The S-parameters of the actual antenna were measured, and the component values of the model were determined by optimizing the match between the model S-parameters and the measured S-parameters. When the appropriate length resonant loop is included in the model to give a resonant frequency of 43MHz, the second resonant frequency, 63.85 MHz agrees with that measured on bay L.

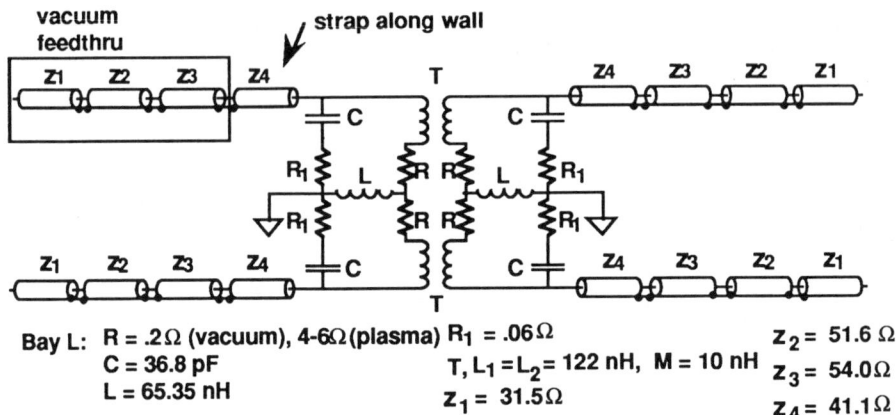

Bay L: R = .2Ω (vacuum), 4-6Ω (plasma) R_1 = .06Ω
C = 36.8 pF
L = 65.35 nH
T, $L_1 = L_2$ = 122 nH, M = 10 nH
Z_1 = 31.5Ω
Z_2 = 51.6 Ω
Z_3 = 54.0Ω
Z_4 = 41.1Ω

Fig. 2 The Lumped element model of the TFTR antenna.

PSpice circuit simulation software (© 1992 MicroSim Corp.) was used to estimate the input impedance of the resonant loops with 0-π phasing, including the 7.0 m length of 50Ω transmission line and tapered transition. The VSWR of this system is shown in figure 3 as a function of resistance, R. Note that this resistance is not the same as the integrated radiation resistance, since the later is distributed over a finite length, however the two are proportional. The effective total series resistance in the loop, measured in the 50Ω transmission line, is .95R. For comparison, figure 4 shows the VSWR for the same antenna with the old configuration of quarter wave transformers optimized for 43 MHz. In this arrangement, a 3λ/4 length of 50Ω is attached to the resonant loop, followed by a 35Ω, λ/4 length, and then a 90Ω, λ/4 length. This combination is optimized for a resonant loop impedance of ~300Ω, which is somewhat higher than the new transformers (i.e. lower loading resistance, R).

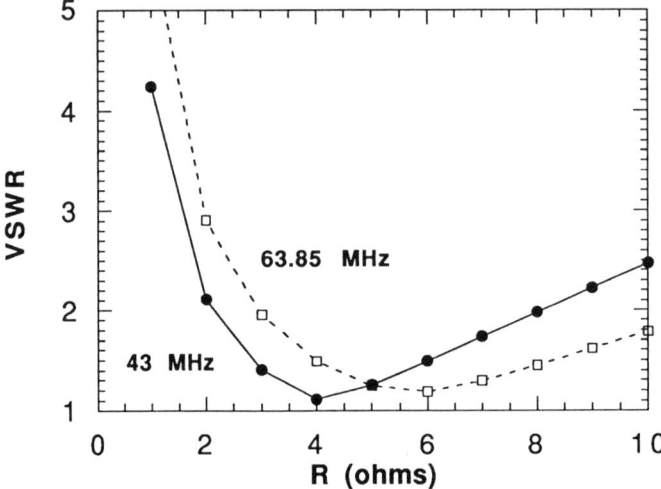

Fig. 3 The VSWR from the 50Ω transmission line into the input of the new transformers for bays M and L.

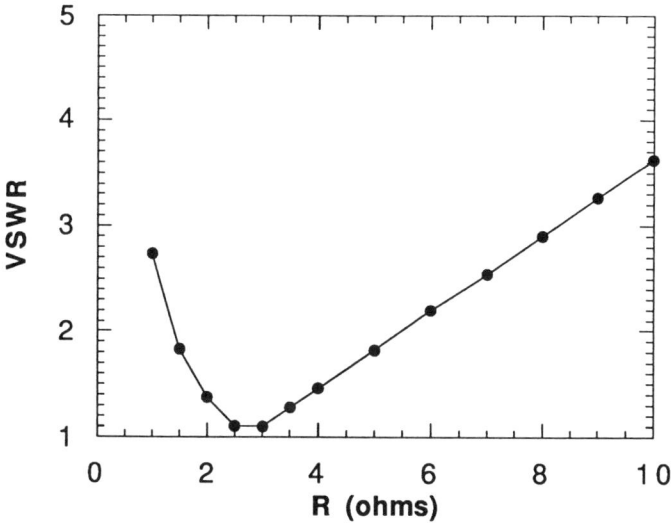

Fig. 4 The VSWR for the same antenna as figure 3, but with the old configuration of quarter wave transformers.

Several other technical improvements will also be implemented during the upcoming run. A feedback circuit will be used to vary the plasma major radius to maintain constant antenna loading, which will help compensate for the large density variations during supershots. An analog calculation of P/V^2, which is proportional to the plasma loading, is performed using an operational amplifier circuit. P is the net transmitted power and V is a voltage near a high voltage point in the resonant loop. During the next run, the transmitters will also have the capability of locking the phase difference between straps. This will be a great aid to electron heating and current drive experiments. Also, the tuning element positions (stub and stretcher) will be decoupled from the relative phase between loops, which may simplify impedance matching.

ACKNOWLEDGEMENTS

This work was supported by the U.S. Department of Energy contract # DE-AC02-76-CHO-3073.

REFERENCES

1. G. Matthaei, L. Young, and E. M. T. Jones, Microwave Filters, Impedance Matching Networks, and Coupling Structures, (Artech House Books, Dedham, 1980).

2. G. J. Greene, P. L. Colestock, J. C. Hosea, C. K. Phillips, D. N. Smithe, J. E. Stevens, J. R. Wilson, W. Gardner, and D. Hoffman, Proc. 8th Topical Conf. on RF Power in Plasmas (AIP, New York, 1989) p. 254.

MODELING OF HIGH-POWER ICRF HEATING EXPERIMENTS ON TFTR*

C.K. Phillips, J.R. Wilson, M. Bell, E. Fredrickson, J.C. Hosea, A. Khudaleev#,
R. Majeski, M. Murakami†, M.P. Petrov#, A. Ramsey, J.H. Rogers, G. Schilling, C.H. Skinner, J.E. Stevens, G. Taylor, K.-L. Wong,
and the TFTR Group
Princeton Plasma Physics Laboratory, Princeton, NJ 08540

INTRODUCTION

Over the past two years, ICRF heating experiments have been performed on TFTR in the hydrogen minority heating regime with power levels reaching 11.2 MW in helium-4 majority plasmas and 8.4 MW in deuterium majority plasmas[1]. For these power levels, the minority hydrogen ions, which comprise typically less than 10% of the total electron density, evolve into a very energetic, anisotropic non-Maxwellian distribution. Indeed, the excess perpendicular stored energy in these plasmas associated with the energetic minority tail ions is often as high as 25% of the total stored energy, as inferred from magnetic measurements. Enhanced losses of 0.5 MeV protons consistent with the presence of an energetic hydrogen component have also been observed [2]. In ICRF heating experiments on JET at comparable and higher power levels and with similar parameters, it has been suggested [3,4] that finite banana width effects have a noticeable effect on the ICRF power deposition. In particular, models indicate that finite orbit width effects lead to a reduction in both the total stored energy and the tail energy in the center of the plasma, relative to that predicted by the zero banana width models. In this paper, detailed comparisons between the calculated ICRF power deposition profiles and experimentally measured quantities will be presented which indicate that significant deviations from the zero banana width models occur even for modest power levels (P_{rf} ~ 6 MW) in the TFTR experiments.

EXPERIMENTAL OBSERVATIONS AND ANALYSIS

ICRF heating experiments on TFTR were performed in the deuterium majority/hydrogen minority regime at power levels up to 8.4 MW with (0-π) phasing. A plasma current of 1.8 MA was used and the toroidal field was set at about 3.2 T to insure on-axis heating for a source frequency of 47 MHz. Central densities ranged from 3.8-5.5 X 10^{13} cm^{-3} with Zeff typically about 1.5-1.8 for these discharges. According to spectroscopic measurements, the H/(H+D) ratio at the edge of these plasmas was in the range of 5%-8%, leading to an inferred central hydrogen density of about 6%. At ICRF power levels above 3 MW, the sawteeth were either stabilized or significantly lengthened during the RF pulse. The time evolution of the central electron temperature and density for discharges with 3.2 and 8.4 MW are displayed in Fig. 1.

From the magnetics measurements for the total stored energy, W_{tot}, and the diamagnetic stored energy, W_{dia}, the excess perpendicular stored energy, $W_\perp$, in each discharge can be estimated as $W_\perp = 3\ W_{dia} - 2\ W_{tot}$. This measurement is compared in Fig. 2 with estimates obtained from an equilibrium energy balance analysis, W_{kin}, and from calculations of the integrated tail stored energy, W_{code}, obtained using the SNAP RF model[5]. Since the ion temperature was not measured directly in these discharges, the ion temperature profile was calculated assuming that $\chi_i = \alpha * \chi_e$, with α determined so that the measured and calculated neutron production rates were in agreement. For these discharges, α was typically in the range of 1-2. By subtracting the measured electron stored energy and the inferred ion stored energy components from W_{tot}, the kinetic estimate of the tail stored energy was obtained. Note that $W_\perp$ and W_{kin} are consistent to within the experimental accuracy of the data. However, W_{code} is consistently higher than both $W_\perp$ and W_{kin}, the discrepancy becoming more pronounced at the higher RF power levels.

Passive charge exchange measurements of the minority hydrogen distribution as a function of energy, obtained using the alpha charge exchange diagnostic, are shown in Fig. 3 for different RF power levels. Estimates for the hydrogen tail temperature for each discharge, obtained from straight line fits to the curves, are compared against the model predictions in Fig. 4. Despite the large discrepancies between the measured and predicted values for the volume-integrated fast ion energy content, as evidenced in Fig. 2, the model predictions for T_{tail} agree reasonably well with the charge exchange measurements.

DISCUSSION

Comparisons between the predicted ICRF power deposition profiles and the measured total fast ion stored energy and effective temperature for the high RF power TFTR experiments lead to the paradoxical conclusion that while the computed central fast ion temperatures are in rough agreement with the data, the computed volume integrated fast ion stored energy is a factor of 2-3 too high, particularly at the higher RF power levels. A similar disagreement between the computed and measured total stored energy of the fast ion has been noted in the JET experiments and ascribed to finite orbit width effects[3,4]. According to the JET model[3], finite orbit width effects become significant when the radial width of the fast ion orbit is comparable to the width of the focal spot of the RF power deposition. This condition is satisfied for tail temperatures above a critical value, which for both TFTR and JET is on the order of 0.875-1 MeV. In the TFTR discharges studied here, the tail temperature, as measured and as computed with a zero banana width model, is significantly lower than this critical value in most cases. Furthermore, the finite orbit width model also predicts a significant decrease in the peak tail temperature, a result which is inconsistent with the TFTR data. While the predicted tail temperatures typically exceed the measured temperatures by about 10%, the overall implication is that finite orbit width effects can not account for the discrepancy between the experimental results and the model.

An alternate cause for the differences between the measurements and the model may be the presence of TAE mode activity in these discharges[1,6]. The relative amplitude of the magnetic perturbations associated with the TAE modes is plotted in Fig. 5, showing a strong increase with applied RF power. The TAE modes excited in these experiments tend to peak off-axis. It is possible that enhanced hot ion loss associated with these modes could lead to a decrease in the fast ion stored energy at radii where the TAE mode peaks. Such an enhanced hot ion loss has been observed in the ^{4}He(H) experiments[1,6]. This could lead to an overall decrease in the volume integrated fast ion stored energy while having a minimal effect on the central tail temperature. Self-consistent transport studies of the power deposition in these discharges will need to be completed before the relative importance of the TAE mode activity and the finite orbit width effects on the fast ions can be quantified.

*Work supported by U.S.D.O.E. Contract # DE-AC02-76-CHO-3073
†Oak Ridge National Laboratory, Oak Ridge TN 37831-8072
#Ioffe Institute, St. Petersburg, Russia

REFERENCES

1. G. Taylor, M. Bell, H. Biglari, et al., to be published in Phys. of Fluids B.
2. S.J. Zweben, R. Boivin, D.S. Darrow, et al., to be published in Proc. of 14th Int. Conf. on Plasma Physics and Controlled Nuclear Fusion Research, September 1992, Wurzburg, Germany, paper IAEA -CN-56/A-6-3.
3. G.A. Cottrell and D.F.H. Start, Nuclear Fusion **31**,61 (1991).
4. M.A. Kovanen, W.G.F. Core and T. Hellsten, Nuclear Fusion **32**,787 (1992).
5. D.N. Smithe, C.K. Phillips, G.W. Hammett, and P.L. Colestock, Radio-Frequency Power in Plasmas, (AIP, N.Y. , 1989) p. 338.
6. J.R. Wilson, M. Bell, H. Biglari, et al., to be published in Proc. of 14th Int. Conf. on Plasma Physics and Controlled Nuclear Fusion Research, September 1992, Wurzburg, Germany, paper IAEA-CN-56/E22.

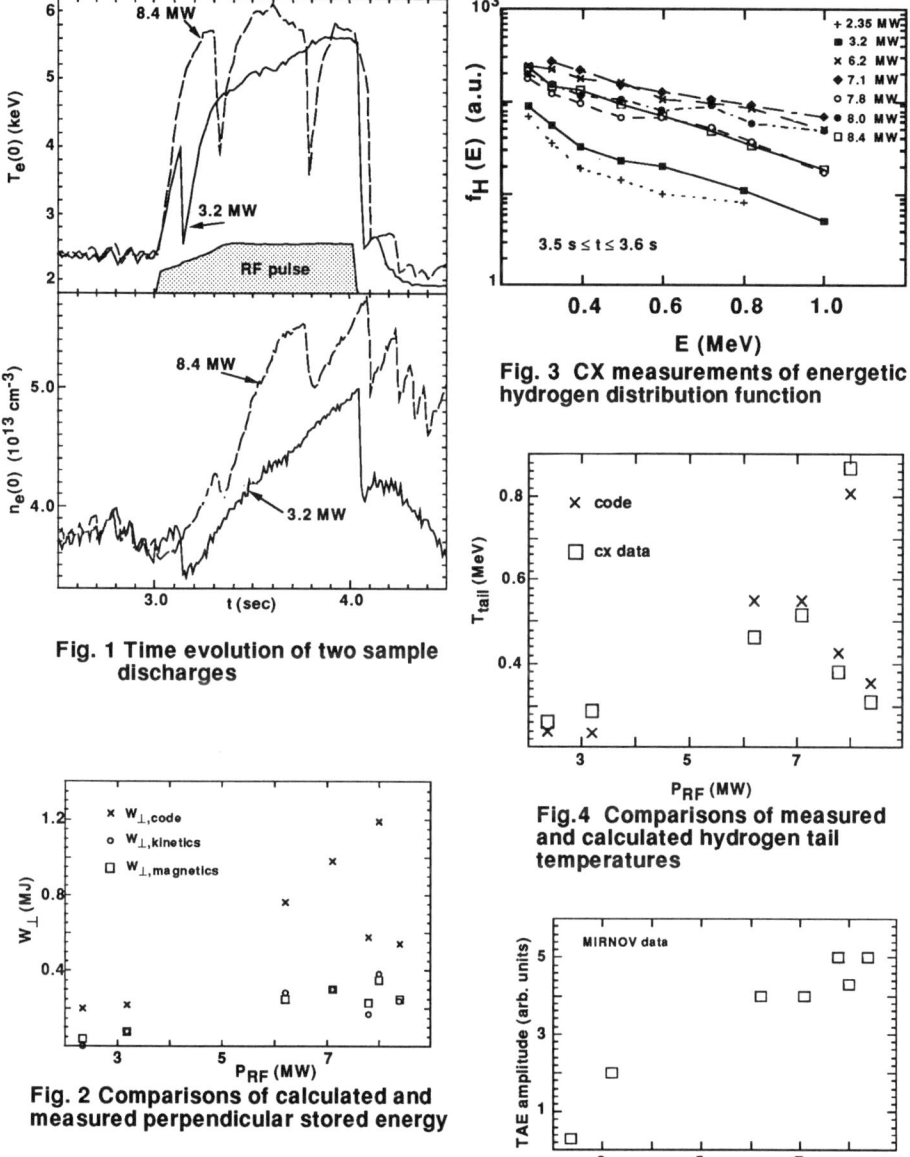

Fig. 1 Time evolution of two sample discharges

Fig. 2 Comparisons of calculated and measured perpendicular stored energy

Fig. 3 CX measurements of energetic hydrogen distribution function

Fig. 4 Comparisons of measured and calculated hydrogen tail temperatures

Fig. 5 Amplitude of magnetic perturbation of TAE mode

FAST WAVE DIRECT ELECTRON HEATING IN TFTR

M. Murakami[a], E. F. Jaeger[a], F. G. Rimini[b], D. A. Rasmussen[a], J. E. Stevens, J. R. Wilson,
D. B. Batchelor[a], M. Bell, R. Budny, E. Fredrickson, R. C. Goldfinger[a], G. Hammett,
D. J. Hoffman[a], J. C. Hosea, A. Janos, R. Majeski, D. Mansfield, C. K. Phillips,
J. H. Rogers, G. Schilling, G. Taylor, and M. C. Zarnstorff

Plasma Physics Laboratory, Princeton University, Princeton, NJ 08543-0451
[a] Permanent Address: Oak Ridge National Laboratory, Oak Ridge, TN 37831-8072
[b] Permanent Address: JET Joint Undertaking, Abingdon, OXON, OX14 3EA, UK

ABSTRACT

Direct electron heating experiments were carried out in two regimes: $B_T = 4.6$ T with D^+ supershots; and $B_T = 2.3$ T with ^{3}He majority. The electron power deposition profiles measured with modulation of RF power are found to be strongly peaked in the core with the total volume-integrated power of up to 80% of the modulated power. The magnitude and profile shape agree well with those predicted by a full-wave code.

INTRODUCTION

Direct electron heating with fast waves is a prerequisite to a viable current drive scenario for future steady-state tokamaks; it is also an attractive alternative to the normal ion resonance method in TFTR DT operation, since it does not require the addition of non-reactive ion species and avoids the complication of minority ion tails in studying the effects of alphas. Since the direct electron heating is relatively weak, it is essential to minimize competing ion damping. Direct electron heating experiments have been carried out in two regimes: high field ($B_T = 4.6$ T) and low field (2.3 T). The results obtained are similar to those of earlier work.[1-3] Modulation of the RF power facilitated measurements of heating power deposition profiles. The centrally peaked electron power deposition represents a large perturbation in electron transport in NBI-heated plasmas even with modest RF power.

EXPERIMENTS IN THE HIGH FIELD REGIME

The high field experiments ($B_T = 4.6$ T at $R_0 = 2.62$ m) were conducted in the ^{3}He minority regime but with no ^{3}He present. The only ion resonance is the deuterium fundamental resonance on the high field side. Figure 1 shows the RF power waveform and time evolution of several plasma parameters. Strong D^0 NBI of 24 MW into a low edge recycling D^+ plasma creates a supershot[4] target plasma with high central electron temperature $T_e(0) = 8$ keV and high central density $n_e(0) = 7 \times 10^{19}$ m^{-3}, which are necessary for reasonable single-pass absorption. The 47-MHz RF power was $\approx 100\%$ modulated ($\Delta P_{FW} = 1.65$ MW) with a 5-Hz square wave. Modulation of $T_e(0)$ observed in response to the RF modulation demonstrates efficient direct electron heating. However, as is usual in supershots with NBI alone, the supershot performance later degrades, owing either to MHD instability or to limiter recycling. The electron temperature profile was measured by a 20-channel ECE grating polychromator with 0.2-ms time resolution, and the electron density was derived from Abel-inversion of 10-channel FIR interferometer signals with 1-ms time resolution.

HEATING POWER PROFILES

The electron heating is analyzed by two methods. First, the step in the power per unit volume absorbed by electrons is inferred from the change in T_e slope through $\Delta q_e = (3/2)n_e\Delta(\partial T_e/\partial t)$ at the time when the RF power is turned on or off, provided the heat transport remains constant during the RF power transition and there is no discontinuity in electron density. The typical period to determine the slope is 10 to 15 ms before and after the RF transition. A second method involves a Fourier transform of T_e data whose fundamental frequency (ω_m) component can be related to the square wave amplitude of the power deposition through $\tilde{q}_e = (2/\pi)\omega_m(3/2)n_e\tilde{T}_e$ when the modulation is fast enough ($\omega_m\tau_E \approx 5$).

Figure 2 shows the power deposition profiles measured for the first RF pulse (3.5 to 3.6 s). The electron absorbed power profiles derived from $\Delta(\partial T_e/\partial t)$ analysis are strongly peaked in the core, but become small at the outer half of the plasma radius. The total volume-integrated electron absorbed power $P_e(a)$ reaches 80($\pm$15)% of ΔP_{FW}. The Fourier transform method also gives similar results, although it could only be used to determine core deposition owing to low signal at outer radii. Since the single-pass absorption is less than 10% under the present experimental conditions, these results indicate that multiple-pass absorption is taking place.

The measured power deposition profiles are consistent with those predicted by a 3-D, full-wave code, PICES (Poloidal Ion Cyclotron Expansion Solution) with multiple (100) modes.[5] The predicted power deposition profiles (Fig. 3) based on the experimental profiles show the centrally peaked $q_e(r)$ with integrated power of 75($\pm$10)% of the input power. The fundamental deuterium ion resonance at the normalized radius $\rho \approx 3/4$ is predicted to absorb 20($\pm$10)% of the input power, as shown in Fig. 3. The calculations of the electron and ion power absorption are subject to uncertainties due to eigenmode structure and shear Alfvén resonance near the edge, both of which are sensitive to details of the profiles assumed.[6]

The evolution of plasma conditions in the discharge influences the electron heating. Supershot performance degradation (not unique to discharges with RF) starts soon ($\approx$30 ms) after the first RF pulse ends. The central electron temperature starts decreasing. The peaked density profile, characteristic of supershots, becomes broad, reducing the central density. Density modulation synchronized with RF power modulation begins shortly after the degradation starts and is probably caused by limiter recycling. The electron density modulation (only present in the latter part of the discharge) must be taken into account in the power deposition analysis. Figure 4 shows the electron power density calculated from the change in the slope of the electron stored energy density ($\Delta(\partial U_e/\partial t)$) based on the output of the time-dependent profile analysis code TRANSP (with 5-ms time resolution). Comparing this with the power deposition profile solely based on $\Delta(\partial T_e/\partial t)$, we observe that the density modulation has little effect on the profiles within $\rho = 0.3$, but has substantial effects on the volume-integrated power at outer radii. The total electron absorbed power $P_e(a)$ is 60($\pm$10)%, of which about 30% is the density-related loss at outer radii. The total (electron and ion) power deposited in the plasma can be calculated from the modulated components of the plasma stored energy measured from the diamagnetism and equilibrium position. The modulated components of both magnetic signals (using "boxcar" averaging to remove slow evolution of the plasma energies) have the same peak-to-peak amplitudes. This is consistent with the absence of the velocity space anisotropy without minority ions. These measurements show that the total absorbed power $P_{tot}(a)$ is 80($\pm$10)% of the input power. The difference ($\approx$20%) between this value and the total electron absorbed power could be attributed to the power absorbed by D$^+$ ions in the fundamental resonance, although time resolution and statistics in the ion temperature measurements are insufficient to detect the modulation in the presence of large heat input from NBI.

EXPERIMENTS IN THE LOW FIELD REGIME

Low field experiments were conducted at $B_T = 2.3$ T with ^{3}He as the majority ion. The competing ion resonances include the fundamental H resonance at the high field edge, the second harmonic ^{3}He resonance at the center, and the third harmonic D resonance toward the low field side. Direct electron heating is clearly seen in the modulation of central electron temperature in response to the modulated RF power (100% modulation of 1.5 MW RF power with a 4 Hz square wave) The ion temperature measurements showed that the second harmonic ^{3}He is weak.[7] Figure 5 shows results of the electron heating power deposition analyses for one of the cases. The measured volume-integrated power absorbed by the electrons is 30 to 50% of the modulated power.

DISCUSSION

Direct electron heating has been observed in the low and high field regimes in TFTR. Although the competing ion resonances in these regimes are rather different, the observed electron power deposition profiles are similar, and are strongly centrally peaked. In both regimes the power deposition in the core (within $\rho = 1/4$) increases with increasing central electron pressure (product of $n_e(0)$ and $T_e(0)$). Although the magnetic field dependence is not certain, the core power depositions at two different fields are roughly proportional to corresponding electron damping rates calculated at the center, as shown in Fig. 6. In the high field regime, as much as 90% of the RF power can be damped directly on the electrons. The measured deposition profiles agree well with those predicted by the PICES code. Comparative experiments showed that the overall heating efficiency with direct electron heating is similar to that with either NBI alone or ^{3}He minority heating[8] with the same power input. The direct electron heating offers an attractive alternative in DT experiments on TFTR for studying the effects of alpha particles without the complication of minority ion tails. In addition, modulation of strongly centrally peaked power to the electron channel is a profitable way to study electron transport in beam-heated TFTR plasmas.

ACKNOWLEDGMENTS

We wish to acknowledge contributions of many members of the TFTR Group, and useful discussions with Drs. K. McGuire, C. C. Petty, R. I. Pinsker, and S. Scott. The research was sponsored by the Office of Fusion Energy, U.S. Department of Energy, under contract DE-AC02-76-CHO-3073.

REFERENCES

[1] D.F.H. Start et al., Nucl. Fusion **30**, 2170 (1990).
[2] T. Yamamoto et al., Phys. Rev. Lett. **63**, 1148 (1989).
[3] C. C. Petty et al., Phys. Rev. Lett. **69**, 289 (1992).
[4] J. D. Strachan et al., Phys. Rev. Lett. **58**, 1004 (1987).
[5] E. F. Jaeger and D. B. Batchelor, in Ninth Topical Conference on Radio Frequency Power in Plasmas, Charleston, AIP Conf. Proc. **244**, 197 (1990).
[6] E. F. Jaeger et al., this conference.
[7] J. R. Wilson et al., in Proc. 14th Int. Conf. on Plasma Phys. Controlled Nuclear Fusion Research (Würzburg, Germany, 1992) IAEA-CN-56/E-2-2.
[8] G. Taylor et al., to be published in Phys. Fluids.

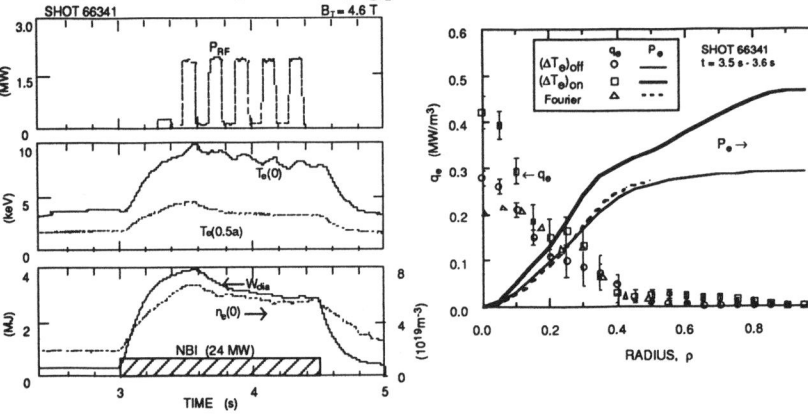

Fig. 1 Time evolution of modulated RF power and plasma parameters in the high field regime ($B_T = 4.6$ T with D^+ supershot)

Fig. 2 Electron power deposition profiles measured for the first RF pulse shown in Fig. 1.

M. Murakami et al. 51

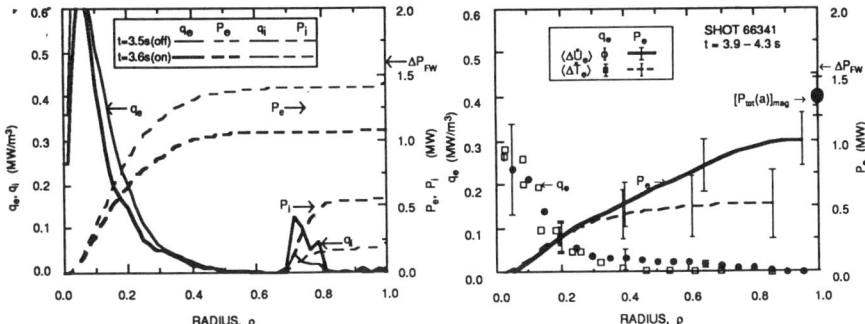

Fig. 3 Electron and ion power deposition profiles (power density and volume-integrated power) predicted by PICES full-wave code for the experimental conditions at the turn-on and turn-off of the first RF pulse shown in Fig. 1.

Fig. 4 Electron power deposition profiles (power density and volume-integrated power) measured for the last three RF pulses (t = 3.9 – 4.3 s) shown in Fig. 1. Dots represent averaged values of the change of the time-derivatives of the electron stored energy at the RF power transitions, and squres represent corresponding values for the time-derivative of the electron temperature alone. The total absorbed power magnetically measured is shown by the large dot..

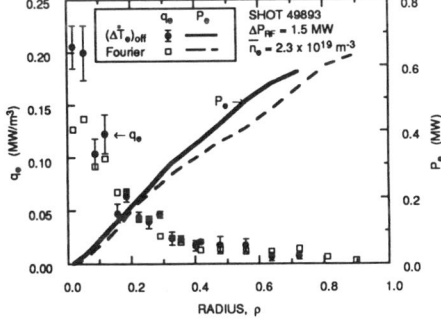

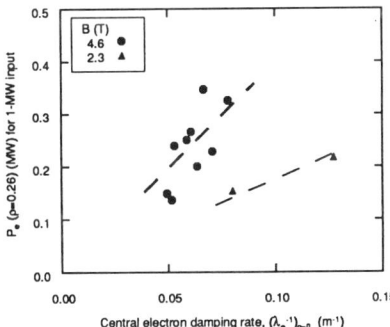

Fig. 5 Electron power deposition profiles (power density and volume-integrated power) measured in the low field experiment (B_T = 2.3 T with ^{3}He majority ions).

Fig. 6 Volume-integrated power absorbed by electrons in the core ($\rho < 1/4$) normalized to input power versus single-pass electron damping rates calculated at the center.

SAWTOOTH STABILIZATION EXPERIMENTS BY ICRF HEATING ALONE AND ITS COMBINATION WITH NBI OR LHCD IN JT-60U

H. Kimura, T. Fujii, M. Sato, M. Nemoto, K. Hamamatsu, T. Kondoh,
M. Matsuoka, S. Moriyama, M. Saigusa, H. Takeuchi and K. Ushigusa.
Japan Atomic Energy Research Institute, Naka, Naka, Ibaraki, Japan, 311-01

D.J. Campbell
JET Joint Undertaking, Arbingdon, Oxfordshire OX14 3EA, U.K.

ABSTRACT

Extension of an operation range for the sawtooth stabilization by a second harmonic minority ion ICRF heating is being pursued in JT-60U. The sawtooth stabilization is obtained at relatively high density ($\bar{n}_e \sim 3.8 \times 10^{19} m^{-3}$) and low-q ($q_{eff}=4$ at $I_p=2.4MA$) for OH target plasmas with only 2.8MW of ICRF power. The sawtooth stabilization is thus obtained at high $\langle n_e \rangle / P_{tot}$ value of $0.8 \times 10^{19} m^{-3} MW^{-1}$, which is notably larger than the value achieved on JET, where fundamental resonance minority ion heating is employed. Another peculiar effect of the second harmonic heating is that stabilization occurs only after the inversion radius expands sufficiently, i.e., $r_{inv}/a \geq 0.26$. The sawtooth stabilization by ICRF heating becomes easier with NBI-heated target plasmas (longest stable period up to 1.7sec), while it tends to be difficult with the LHCD target plasmas.

INTRODUCTION

Stabilization of sawtooth oscillations is important for optimization in high performance discharges in present-day experiments (e.g., to get high internal inductance (l_i) for low-q discharges). Eventually control of sawtooth activities will be one of the key techniques for burn control and helium ash exhaust in a fusion reactor. Therefore, reliable methods for them should be developed in reactor-grade plasmas. In the former JT-60 ICRF experiments, we obtained clear symptoms of the sawtooth stabilization in the high density and low-q regimes, utilizing the second harmonic minority ion heating scheme[1]. In current JT-60U ICRF experiments, we are demonstrating more firmly the sawtooth stabilization in relatively high density and low-q discharges with the same heating scheme. Performance on the sawtooth stabilization and conditions for the sawtooth stabilization by the second harmonic minority ion ICRF heating are discussed for Ohmic, NBI and LHCD target plasmas.

ICRF HEATING ALONE

The ICRF heating alone experiments were done mainly with helium discharges containing residual minority hydrogen. The frequency is 116MHz, which gives the hydrogen second harmonic cyclotron resonance at 3.82T. We found that the electron temperature profile shape is strongly dependent on the resonance position. Figures 1(a) and 1(b) compare the T_e profile for B_{T0} (toroidal field at the machine center (R=3.32m) in the vacuum) of 3.97T and B_{T0} of 4.05T, respectively. Including the internal magnetic field[2], the resonance position (R_{res}) is estimated to be ~3.53m (($R_{res}-R_{ax}^{eq}$)/a~0.0) for B_{T0} of 3.97T and ~3.60m (($R_{res}-R_{ax}^{eq}$)/a~0.06) for B_{T0} of 4.05T, where R_{ax}^{eq} is the magnetic axis position derived from equilibrium analysis. Discharge conditions of these two examples are quite similar except small difference in B_{T0}. ICRF power was injected at an early phase of the discharges in order to avoid affect of sawtoothing in comparison. It is clearly shown that the T_e profile shape changes considerably with such small difference in the resonance positions and that the T_e profile is indeed more peaked with closely on-axis ICRF resonance position (Fig. 1(a)).

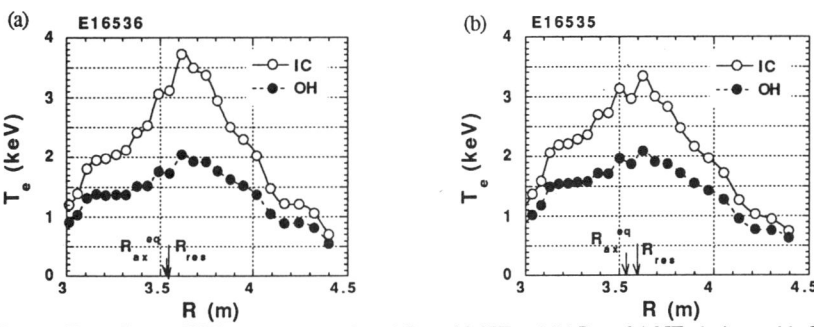

Fig. 1 Te profiles by ECE measurements for (a) B_{T0} of 3.97T and (b) B_{T0} of 4.05T, during and before ICRF heating. $I_P \sim 2.5MA$, $\bar{n}_e \sim 3.8 \times 10^{19} m^{-3}$, $P_{IC} \sim 3MW$.

With the optimum resonance position so far (i.e., B_{T0}=3.97T), we could obtain the sawtooth stabilization at relatively high electron density and low-q regime ($\bar{n}_e$ =3.8×10^{19}m^{-3}, q_{eff}=4.1 at I_p=2.4MA) with only 2.8MW of ICRF power. The example is shown in Fig. 2, where time evolution of the diamagnetic plasma stored energy, $\bar{n}_e$, the central electron temperature by an ECE polychromator $T_e(0)$ and the ICRF power is indicated. The longest sawtooth period was 0.61sec, which is larger than the energy confinement time of this shot (τ_E=0.46s).

The sawtooth inversion radius, which is deduced from an Abel inversion of signals of the soft X-ray detector array in the poloidal cross-section, expanded to $r_{inv}/a \sim 0.32$ at the crash of the stable period. In this particular case, the sawtooth stabilization is obtained at $\langle n_e \rangle / P_{tot}$ of $0.8 \times 10^{19} m^{-3} MW^{-1}$, which is about 60% larger than the upper boundary value in JET ICRF experiments[3], where fundamental resonance minority ion heating is employed. This is considered to be due to the fact that JT-60U employs second harmonic minority ion heating, which is favourable for producing energetic ions in high density regimes[1].

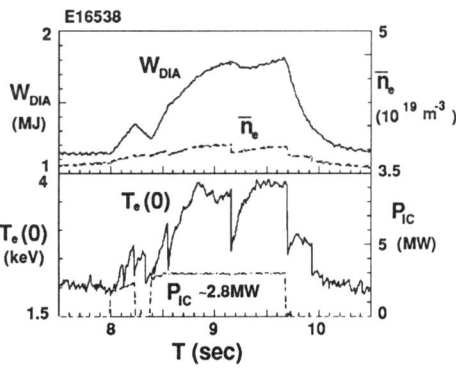

Fig.2 Time evolution of ICRF heating alone, showing sawtooth stabilization at largest $\langle n_e \rangle / P_{tot}$ value.

COMBINED ICRF AND NBI

Typical example of the combined ICRF and NBI heating experiment is shown in Fig. 3 (shot E16376), where time evolution of the diamagnetic plasma stored energy, $\bar{n}_e$, $T_e(0)$ by the ECE polychromator, the ICRF power, the NBI power and a channel number of the soft X-ray detector array corresponding to the sawtooth inversion is indicated. The present experiments were performed with deuterium discharges containing residual minority hydrogen. B_{T0} was 4.05T. So, the resonance position was somewhat off-axis (($R_{res}-R_{ax}^{eq}$)/a~0.08). Total neutral beam power was ~9.6MW. We obtained stable period of 1.74sec by 3.1MW of ICRF power, as shown in the electron temperature trace in Fig. 3. Figure 4 indicates maximum sawtooth period during the combined IC and NB heating (open square) as a function of the ICRF power together with the data of the ICRF heating alone. Considering B_{T0} of 4.05T for the present combined IC+NB experiment, sawtooth stabilization by ICRF heating is realized more easily with NBI-heated target plasmas than with ohmic target plasmas.

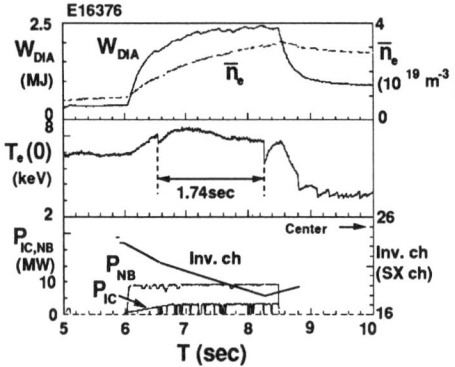
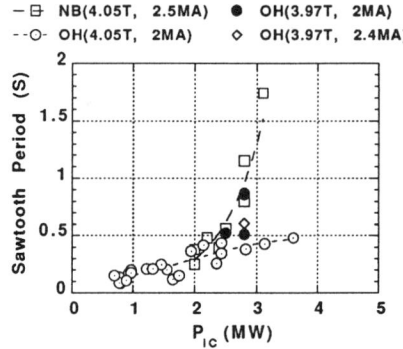

Fig. 3 Time evolution of combined ICRF and NBI heating. I_P=2.5MA, B_{T0}=4.05T, q_{eff}=4, $\bar{n}_e$~3×10^{19}m^{-3}, P_{IC}=3.1MW, P_{NB}=9.6MW.

Fig. 4 Maximum sawtooth periods during combined IC and NB heating and IC heating alone versus ICRF heating power.

COMBINED ICRF AND LHCD

The combined ICRF and LHCD experiments were conducted with helium discharges with minority hydrogen. B_{T0} was 3.97T, so the ICRF resonance position was almost on-axis. Figure 5 shows time evolution of the combined ICRF and LHCD experiment. Peak refractive index of LH waves ($N_\parallel^{peak}$) is 1.6 and injected LH power (P_{LH}) was 1.5MW. Reduction rate of the loop voltage ($\Delta V_L/V_L$) by LHCD was -0.39 and it was further reduced to -0.55 by ICRF heating. As shown in Fig. 5, sawtooth stabilization could not be obtained with the LHCD target plasma in the present experiment. Similar result was obtained with higher value of $N_\parallel^{peak}$ (=2.3) of LH waves.

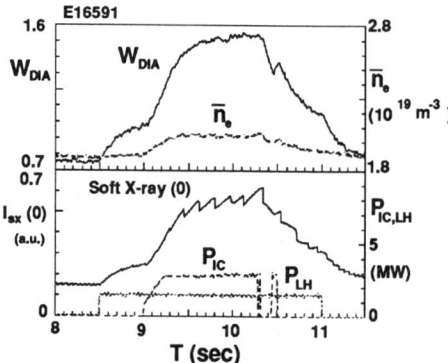

Fig. 5 Time evolution of the combined ICRF and LHCD experiment. I_P=2MA, B_{T0}=3.97T, q_{eff}=5.3, $\bar{n}_e$~2×10^{19}m^{-3}, P_{IC}=2.8MW, P_{LH}=1.5MW.

We found that the sawtooth inversion radius became shorter in application of the LH waves than without them. In the LHCD target plasma case, r_{inv}/a is reduced to 0.22, while it remains at a value of 0.26 in the Ohmic target plasma case, before ICRF injection. Relation between the sawtooth stabilization and the inversion radius will be discussed in the next section. Measurement of hard X-ray emission shows that hard X-ray intensity decreases after injection of ICRF waves only in the plasma core, suggesting absorption of LH waves by fast protons produced by ICRF waves in the plasma core.

DISCUSSION AND CONCLUSIONS

Figure 6 shows sawtooth periods as a function of the inversion radius at the crash, using a channel number difference between two soft X-ray detectors viewing the sawtooth inversion and the plasma center as a measure (~4.2cm/ch), for the three heating schemes presented in this paper. Arrows indicate direction of time passage during stable periods. In the ICRF heating alone with the nearly on-axis resonance condition, we found that the sawtooth stabilization can be obtained only after a size of the inversion radius exceeds a critical value. The result is in contrast with the JET results, where the stabilization tends to be easier with decreasing inversion radius[4].

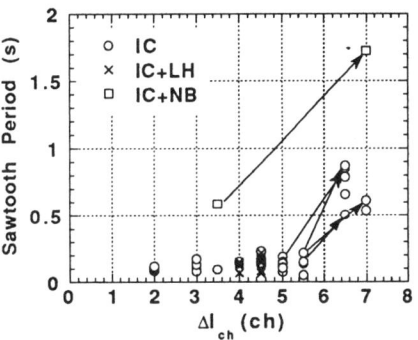

Fig.6 Sawtooth periods versus soft X-ray channel number difference between the sawtooth inversion and the plasma center for various heating schemes.

For efficient stabilization, fast protons should be confined inside q=1 surface. Therefore, the JT-60U results suggest that for the second harmonic heating case, the orbit effect is much more important for confining fast protons inside q=1 surface in comparison with the fundamental resonance heating. The reason can be understood by the Fokker-Planck calculation using present experimental parameters, i.e., the distribution of minority fast ions by the second harmonic heating is well below the one by the fundamental heating in the lower energy range (≤ 400keV) but well over that in the upper range (≥ 400keV). The reason why the sawtooth stabilization could not be obtained with LHCD target plasmas can be explained by smaller inversion radii than those with Ohmic target plasmas. However, Fig. 6 also indicates that for IC+NB case the stabilization is obtained from much smaller inversion radius ($r_{inv}/a \sim 0.2$) and hence variation range of r_{inv}/a is quite large in comparison with IC alone case. The reason why the stabilization can be easily obtained in IC+NBI case is not well understood so far. In conclusion, the second harmonic minority ion heating provides efficient sawtooth stabilization at relatively high $\bar{n}_e$ and low q regime but the initial inversion radius for the stabilization should be larger than for the fundamental resonance heating, suggesting that the orbit effect is more pronounced for confining fast ions inside q=1 surface for the second harmonic heating.

REFERENCES

1. H. Kimura et al., "Higher Harmonic ICRF Heating Experiments in JT-60", to be published in Plasma Phys. Contr. Fusion.
2. M. Sato et al., Japan Atomic Energy Research Institute Report, JAERI-M 92-073, 342 (1992).
3. D. J. Campbell et al., Proc. 15th Eur. Conf. on Contr. Fusion and Plasma Heating (Dubrovnik, 1988), EPS, Vol. 12B, Part I, 377.
4. F. Porcelli et al., Plasma Phys. Contr. Fusion **33**, 1601 (1991).

ICRF HEATING SCENARIOS IN ALCATOR C-MOD

Y. Takase, P. T. Bonoli, S. N. Golovato, M. Porkolab
MIT Plasma Fusion Center, Cambridge, MA 02139

ABSTRACT

Alcator C-MOD tokamak ($R = 0.67$ m, $a = 0.21$ m, $\kappa = 1.8$, $B \leq 9$ T, $I \leq 3$ MA) will start operating in April 1993. Initially 2 MW of rf power at 80 MHz will be available, which will be upgraded to 4 MW by early 1994. Additional 2–4 MW of 40–80 MHz tunable power may be installed in 1996 in collaboration with PPPL. With the 80 MHz transmitters, He^3 minority heating at $B = 7.9$ T and H minority heating at 5.3 T are the main heating scenarios, with possibilities of He^3 and H second harmonic minority heating (at 3.9 T and 2.6 T, respectively) and direct electron heating by ELD/TTMP. The tunable transmitter will enable fundamental He^3 and H minority heating at lower fields, and Alfvén wave heating below the ion cyclotron frequency at fields above 4 T in H plasmas. Advanced tokamak scenarios with high bootstrap current fraction and high normalized beta can also be studied in combination with pellet fueling and 4.6 GHz LHCD under TPX-like conditions ($B \sim 4$ T, $\bar{n}_e \lesssim 1 \times 10^{20}$ m^{-3}) for pulse lengths exceeding the L/R time.

INTRODUCTION

Alcator C-Mod will operate with typical parameters $R = 0.67$ m, $a = 0.21$ m, $\kappa = 1.8$, $B \leq 9$ T, and $I \leq 3$ MA. The auxiliary heating is provided by the ICRF fast magnetosonic wave. RF operation will begin this summer with a movable single-strap ("monopole") antenna with 2 MW of rf power at 80 MHz available from the transmitter. After half a year of operation the single-strap antenna will be replaced by two pairs of two-strap ("dipole") antennas, with 4 MW available from the transmitter. Additional 2–4 MW of 40–80 MHz tunable power may be installed in 1996 in collaboration with PPPL. Two new antennas are being considered for the 1996–1997 time period, one being a folded waveguide launcher in collaboration with ORNL and the other being an antenna aimed at electron heating and current drive in collaboration with PPPL. These antennas are planned to be installed in a third port and tested sequentially. Addition of a 2 MW LHCD system at 4.6 GHz (a long-pulse version of the Alcator C LH system) in 1997 is also planned, with a future upgrade to 4 MW. In this paper, different ICRF heating scenarios as well as LHCD + ICRF advanced tokamak scenarios with high bootstrap current fraction and high normalized beta are discussed. Details of the planned ICRF experiments and hardware is described in the companion paper.[1]

ICRF HEATING SCENARIOS

In this paper we refer to different minority heating scenarios with the minority species in parentheses, e.g., D(He^3) indicates He^3 minority in D majority and D(2H) indicates second harmonic H minority in D majority, etc. Different heating scenarios currently being considered for C-Mod are summarized in Table I. With the 80 MHz transmitter available initially, He^3 minority heating in D majority plasma at $B = 7.9$ T and H minority heating in either D or He^3 majority plasma at 5.3 T will be the main

	B(T)	f = 80 MHz	f = 40 MHz
Minority Heating	7.9	D(He^3)	
	5.3	D(H),He^3(H)	
	3.9	D(2He^3),H(2He^3)	D(He^3)
	2.6	D(2H),He^3(2H)	D(H),He^3(H)
Electron Heating	$\gtrsim 4.0$		Alfvén H plasma ($\omega < \Omega_H$)
	$\lesssim 3.5$	high harmonic D plasma($\omega > 2\Omega_D$)	

Table I. ICRF heating scenarios being considered for C-Mod.

heating scenarios. Ultra-high density ($n_{e0} \gtrsim 10^{21}$ m^{-3}) scenarios prototypical of Ignitor can be studied with D(He^3) heating and pellet fueling. In this scenario, the shear Alfvén wave mode conversion layer exists near the plasma edge on the high field side of the fundamental D resonance, which may cause deleterious effects such as edge heating and impurity generation. At reduced fields He^3 and H second harmonic minority heating at 3.9 T and 2.6 T, respectively, would be possible. However, there are a number of possible parasitic resonances due to the more tightly packed resonance layer structure at higher harmonics. In particular, the second harmonic D (fundamental H) resonance on the high field side edge and the third harmonic D resonance at $r/a \simeq 1/3$ on the low field side in the D(2He^3) scenario, and the shear Alfvén resonance on the high field side edge in the H(2He^3) scenario are potentially deleterious. Note that the He^3(H) and H(2He^3) scenarios are topologically equivalent to T(D) and D(2T) scenarios in a D-T plasma. There is also a possibility of direct electron heating by ELD/TTMP with an antenna properly designed for electron heating at $B \lesssim 3.5$ T in D plasmas. At $B = 3.5$ T both second harmonic D and fourth harmonic D (second harmonic H) resonances are just outside the plasma, but the third harmonic D resonance is at the plasma center. At $B = 3.0$ T the only resonances are the third and fourth harmonic D resonances located at $r/a \simeq 1/2$.

The 40–80 MHz tunable transmitter would provide operating flexibility by enabling fundamental He^3 and H minority heating at lower fields, and Alfvén wave heating below the ion cyclotron frequency in H plasmas at $B \gtrsim 4$ T. It will also enable a variety of multiple frequency ICRF heating scenarios.

Other possibilities include electron heating with the mode-converted IBW or shear Alfvén wave. These mode-converted short-wavelength waves may be detected and followed by the CO$_2$ laser scattering diagnostic, currently in preparation. Data from this diagnostic will provide important information on propagation and absorption of these mode-converted waves.

Some results of modeling for D(H) and D(He^3) scenarios have been reported earlier.[2,3] Further analyses of different minority heating scenarios were performed using the slab geometry full-wave code FELICE.[4] The magnitude of "single-pass" transmis-

58 ICRF Heating Scenarios in Alcator C-MOD

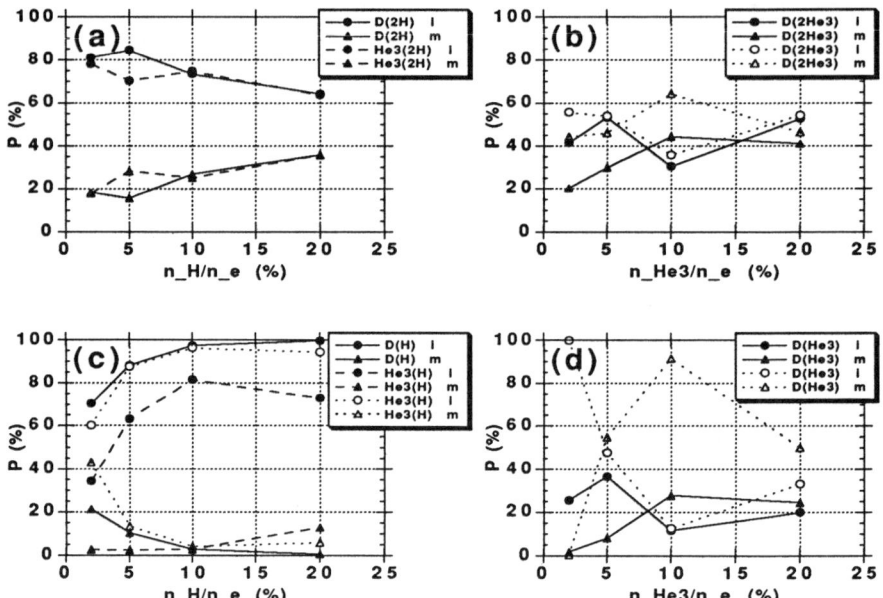

Fig. 1. Power partition for different minority heating scenarios: (a) $B = 2.6\,\text{T}$, D(2H) and He^3(2H), $n_{e0} = 2 \times 10^{20}\,\text{m}^{-3}$, $T_H = 100\,\text{keV}$; (b) $B = 3.9\,\text{T}$, D(2He^3), $n_{e0} = 2 \times 10^{20}\,\text{m}^{-3}$, $T_{He^3} = 100\,\text{keV}$; (c) $B = 5.3\,\text{T}$, D(H) and He^3(H), $n_{e0} = 7 \times 10^{20}\,\text{m}^{-3}$, $T_H = 10\,\text{keV}$; and (d) $B = 7.9\,\text{T}$, D(2He^3), $n_{e0} = 7 \times 10^{20}\,\text{m}^{-3}$, $T_{He^3} = 10\,\text{keV}$. Solid symbols: radiative b.c.; open symbols: reflective b.c.

sion across the resonance / mode conversion layer can be estimated using the radiative boundary condition, while the complete power balance between different power absorption mechanisms and mode conversion can be determined using the reflective boundary condition. The results are summarized in Figs. 1(a)–(d).

In minority heating scenarios where the minority ion-cyclotron resonance layer (fundamental or second harmonic) coincides with a majority ion-cyclotron harmonic layer, i.e., D(H), D(2H), and He^3(2H) scenarios, transmission across the resonance layer is negligibly small and the radiative and reflective boundary conditions give the same results. The D(H) scenario is most desirable with almost complete absorption by the minority ions for $n_H/n_e \gtrsim 10\%$. For D(2H), He^3(2H), and He^3(H) scenarios, absorption by minority ions is still respectable, greater than 50% over a wide range of minority concentration.

In D(He^3), D(2He^3), and He^3(H) scenarios, transmission across the resonance layer is not negligible, and reflections must be taken into account. The D(He^3) scenario has a particularly low "single-pass absorption". Both D(He^3) and D(2He^3) scenarios favor low minority concentration ($n_{He^3}/n_e \lesssim 5\%$) to maximize absorption by minority ions. It should be noted, however, that for these scenarios mode conversion may not be treated correctly by the FELICE code which uses the small $k_\perp \rho_i$ expansion.

ADVANCED TOKAMAK SCENARIOS

Addition of the 4.6 GHz LHCD system in C-Mod offers a possibility to investigate "long pulse" advanced tokamak scenarios, such as those being proposed for TPX. At toroidal fields of $B \lesssim 5\,\text{T}$ Alcator C-Mod can operate for pulse lengths of $\gtrsim 7\,\text{sec}$, which is twice the L/R time and an order of magnitude longer than the skin time. Fully non-inductive current drive under ITER-like conditions can be demonstrated at $n_{e0} \simeq 10^{20}\,\text{m}^{-3}$, $B \simeq 5\,\text{T}$, $q_\psi \simeq 3$ (corresponding to $I_p \simeq 1.5\,\text{MA}$) with 3 MW of absorbed LHCD power (4 MW source power), and confinement can be studied in such non-inductively driven plasmas with strong H minority ICRF heating.

LHCD can also be used for current profile control through off-axis current drive, providing access to advanced tokamak regimes with high bootstrap current fraction, high β_p, high β_N, and reversed shear equilibria for times much longer than the current profile relaxation time scale. A high internal inductance regime may be accessed with FWCD after installing a suitably designed electron heating / current drive antenna. A reversed shear scenario with parameters $B = 4.0\,\text{T}$, $I_p = 0.62\,\text{MA}$, $\langle n_e \rangle = 0.86 \times 10^{20}\,\text{m}^{-3}$ ($n_{e0}/\langle n_e \rangle = 3.7$), $T_{e0} = T_{i0} = 7.5\,\text{keV}$, $\beta_p = 5.4$, $\beta_N = 3.1$, $f_{BS} = 0.81$, $P_{LH} = 0.62\,\text{MW}$, $P_{tot} = 8\,\text{MW}$. obtained using off-axis LHCD combined with pellet fueling and on-axis ICRF heating appears to be particularly promising. The bootstrap current is not localized in the extreme edge region because of the peaked pressure profile, and the q profile has a negative shear core with $q_0 = 2.9$, $q_{min} = 2.0$ at $r/a \simeq 0.5$, and $q_{95} = 7.8$.

CONCLUSIONS

Alcator C-Mod is a versatile and cost-effective facility to explore a wide variety of ITER-relevant and Advanced Tokamak issues. The high-field capability provides access to the ultra-high density regime prototypical of Ignitor, and ensures that tokamak operation is not limited by β limits. Initial emphasis of the C-Mod program will be on divertor improvements such as the radiative divertor and the gaseous divertor. The present divertor hardware is suitable for testing an ITER-like divertor concept without any modification. During this period a comparison of different ICRF heating regimes will be made and access to improved confinement regimes will be explored. Advanced tokamak regimes with high bootstrap current fraction, high β_N, and high confinement will be explored. Initially the current density profile will be controlled transiently using current ramping and shape ramping, but with installation of the LHCD system in 1997 "quasi-steady-state" (i.e., much longer than the current profile relaxation time scale) current profile control will become possible. A promising advanced tokamak scenario has been identified with central particle fueling by pellet injection, on-axis ICRF heating, and off-axis LHCD.

ACKNOWLEDGMENTS

This work was supported by U.S. Department of Energy Contract No. DE-AC02-78ET51013.

REFERENCES

[1] S. N. Golovato, et al., this meeting.
[2] Y. Takase, et al., in Radio-Frequency Power in Plasmas (Proc. 8th Top. Conf., Irvine, 1989) p. 346.
[3] Y. Takase, et al., in Radio Frequency Power in Plasmas (Proc. 9th Top. Conf., Charleston, 1991) p. 189.
[4] M. Brambilla, Nucl. Fusion **28**, 549 (1988).

ION CYCLOTRON HEATING EXPERIMENTS AND PLANS FOR THE ADVANCED TOROIDAL FACILITY (ATF)*

D.A. Rasmussen, D.B. Batchelor, R.H. Goulding, D.J. Hoffman, E.F. Jaeger,
J.A. Rome, C. E. Thomas
Oak Ridge National Laboratory, Oak Ridge, TN 37831

ABSTRACT

In Advanced Toroidal Facility (ATF) Ion Cyclotron Resonant Frequency (ICRF) experiments to date, a single-strap tunable antenna with carbon limiters has been used at power densities up to 13.6 MW/m^2 (900 kW). Hydrogen minority heating experiments at 14.4 MHz were performed in deuterium plasmas formed initially with both second-harmonic electron cyclotron heating (ECH) and helium neutral beam injection (NBI). No ICRF heating was observed in the low-density ECH target plasmas. Two distinct NBI plus ICRF operating modes are described. The moderate heating of these discharges is probably limited by impurities sputtered from the vessel walls and/or poorly confined orbits for the heated particles. A two-strap, phased antenna, designed to reduce impurities and capable of both minority and direct electron heating is described. A second antenna design for optimizing the coupling to well confined ion orbits, additional ICRF heating schemes, and methods for reducing the RF edge interactions in a stellarator geometry are also being considered.

INTRODUCTION

The ATF is an $\ell = 2$ torsatron that has 12 field periods and was designed for studies of plasma stability and confinement in a high shear, current free toroidal device[1]. In experiments to date the ICRF power has been launched from a single-strap tunable asymmetric resonant double loop (ARDL) antenna[2]. The ARDL antenna circuit consists of a current strap with a fixed feed point and grounded tuning capacitors on each end. The antenna was tunable from 9 to 30 MHz. A single-tier Faraday shield was covered on the plasma side with carbon tiles and the antenna was surrounded with carbon tile bumper limiters. The Faraday shield and current strap were tilted 10° in order to match to the stellarator field-line pitch. The antenna design includes the capability for 15 cm of remotely controlled radial motion. For the 1991 experiments described here a modified BBC transmitter (1 MW) tuned to 14.4 MHz was used. Details of the antenna design and results of earlier experiments are described in references 3-7.

EXPERIMENTS WITH A SINGLE-STRAP ANTENNA

In the ATF experiment, the plasma startup and initial heating are generally obtained by first- or second-harmonic ECH. For second harmonic heating, the electron cyclotron cutoff limits the central density of this "target" plasma, at $B_0 \cong 1T$, to about 1×10^{13} cm^{-3}. Typical edge density near the antenna Faraday shield is about 2×10^{12} cm^{-3}. This density does not provide adequate plasma loading ($\leq 0.2 \Omega$) for high power ICRF[3]. In order to enhance the plasma loading, substantial gas puffing in the afterglow of ECH discharges was used to transiently increase the plasma density. The result of turning off the ECH and increasing the gas puff, is a plasma with a central density of $\cong 1.2\times10^{13}$ cm^{-3} and a central T_e of < 50 eV. For this brief series of ECH plus ICRF experiments the antenna loading improved, but no evidence of ICRF heating of the bulk plasma was observed.

The target plasma density and stored energy can be further increased with NBI. For the series of NBI/ICRF experiments described below the maximum central density during NBI was 5×10^{13} cm^{-3} and typical edge densities were 1×10^{13} cm^{-3} providing a maximum plasma

---

*Research sponsored by the Office of Fusion Energy, U.S. Department of Energy, under Contract No. DE-AC05-84OR21400 with Martin Marietta Energy Systems, Inc.

loading of 1.5 Ω. This increased loading allowed up to 900 kW of power to be launched, which is a power density of 13.6 MW/m^2. The launched power is near the design limits of the antenna and the transmitter (1 MW). The power density is close to the record achieved on Tore-Supra (15 MW/m^2) under better loading conditions.

A sequence of nearly identical (B_0 = 0.95 T) NBI heated discharges, for the standard ATF vacuum flux surface configuration, was run in order to allow antenna tuning and high power operation. The standard vacuum flux surface configuration in the ϕ = 15° plane for ATF is shown (dashed lines) in figure 1. Overlayed with the vacuum flux surfaces are the mod-B contours. The minority hydrogen ion cyclotron resonance locations for 14.4 MHz with B_0 = 0.95 T are indicated. The hydrogen concentration was typically 5%. Two distinct NBI operating modes were observed. In the first mode, the combination of relatively low input NBI power ($\cong$ 500 kW for He injection) and inadequate machine conditioning resulted in ATF discharges that suffered a "collapse" of the stored energy when beam heating was initiated. Thus, the plasmas had a central electron temperature of 50 - 100 eV at the time the ICRF was applied. This "target" plasma would remain in a collapsed state throughout the discharge unless ICRF heating was applied. When ICRF power greater than $\cong$ 200 kW was applied, the plasma edge would heat, and as a consequence more efficient coupling of the NBI power would eventually result in a global plasma "reheat". The discharge parameters as a function of time for a typical discharge (shot #20094) of this type are shown in figure 2. A discharge (shot #20087) with a shortened ICRF pulse is also shown for comparison. For the discharge (#20094) with a full ICRF pulse the peak diamagnetic stored energy was $\cong$ 7 kJ. The T_e profile is peaked with a central value of $\cong$ 400 eV. The central T_i is $\cong$ 200 eV. The central n_e is $\cong$ 4 x10^{13} cm^{-3} with the profile essentially flat out to a normalized radius, ρ, $\cong$ 0.85.

With further beam and vacuum vessel conditioning (titanium gettering), a limited number of the second type of "noncollapsing" target discharges was obtained. The discharge parameters as a function of time for noncollapsing discharges with and without ICRF are shown in figure 3. The addition of ICRF appears to result in a modest stored energy increase for this discharge, compared to the non ICRF case, but it also seems to accelerate the thermal collapse process. An energetic ion tail was observed, with a charge exchange neutral particle analyzer, but no significant increase in the ion bulk temperature was observed. The premature thermal collapse is probably the result of impurities (an increase in titanium radiation is measured) entering the plasma after being sputtered from the vessel walls due to the RF heating of the edge plasma. Possible benefits from high power RF plasma conditioning of the antenna and vacuum vessel could not be explored because of limited experimental time.

DISCUSSION OF RESULTS

One candidate for explaining the lack of significant ion heating in these experiments is the effect of the direct ion orbit losses. One set of representative orbits computed by calculating the contours of the adiabatic invariant, J*, for low collisionality conditions[8], and E_R=0, is shown in figure 4. The co-, counter and trapped orbits of 1 Kev ions which mirror at 1 T are all indicated. As can be seen, trapped ions at large major radii have orbits which drift directly out of the plasma. These orbits are strongly modified when realistic plasma potential profiles are included in the calculations. Figure 5a shows the reduction in direct loss orbits when the potential profile[9] shown in figure 5b is used. The main effect of the potential is to move the location of the trapped/passing boundary for thermal ions, while orbits for ion energies more than twice the potential variation are generally unaffected.

FUTURE PLANS

The modification of the existing antenna to a double-strap antenna with 180° phasing is underway. Figure 6 shows the strap arrangement and corresponding poloidal and toroidal cross sections. This phased antenna should reduce the influx of impurities and have improved coupling for both minority ion and direct electron heating. The direct electron heating efficiency at 23 MHz, $B_0 = 1$ T, $T_e = 1$ KeV and $n_e = 2\times10^{13}$ cm^{-3} is greater than 5%. Heating of the electrons and/or the bulk ion population should result in improved ion confinement. Candidate vessel locations and heating scenarios for a second antenna are currently under study. Heating with an antenna located on the top of the plasma may allow selective coupling to ions that are on well confined orbits which will subsequently thermalize with the bulk species. High field launch locations and antenna designs for IBW heating are also under consideration. In addition, extensive antenna conditioning along with improved wall conditioning techniques (i.e. boronization) are expected to reduce the impurity influx.

SUMMARY

The ARDL antenna was used to launch up to 900 kW of 14.4 MHz ICRF into NBI target plasmas in ATF. This power density is quite high for the modest plasma loading. The observed heating of the plasma was probably limited by an influx of impurities released when the edge plasma was heated. Future plans include the modification of the existing antenna to one with a double strap. This should help control impurity influx and also allow the possibility of direct electron heating. A planned second antenna will be optimized to avoid possible effects of poorly confined ion orbits.

ACKNOWLEDGEMENTS

The authors acknowledge the members of the ATF operations and physics groups for their efforts and patience. We especially acknowledge the efforts of G.C. Barber and D.O. Sparks for the installation and operation of the transmitter. We thank the Engineering Division for their considerable efforts during the design and construction of the antenna. We also thank S.L. Milora, H.H. Haselton, C.C. Baker, J.L. Dunlap, J.F. Lyon, M.J. Saltmarsh and J.Sheffield for their encouragement and support.

REFERENCES

1 J. F. Lyon et al., Fusion Technol. **10**, 179 (1986) and R. J. Colchin, et al., Phys. Fluids B **2**, 1347 (1990).
2 D. J. Hoffman, F. W. Baity, W. R. Becraft, J. B. O. Caughman, T. L. Owens, "Experimental Measurements of the Ion Cyclotron Antennas' Coupling and RF Characteristics," (6th Topical Mtg. on the Tech. of Fusion Energy, San Francisco, CA, March 3-7, 1985).
3 M. Kwon et al., Nuclear Fusion, **32**, (1992) 1225.
4 M. Kwon, T. D. Shepard, R. H. Goulding, C. E. Thomas, R. J. Colchin, D. J. Hoffman, and M. R. Wade, in Proc. 9th Topical Conf. Applications of RF Power to Plasmas, AIP Conf. Proc. **244**, 146 (1991).
5 T. D. Shepard, M. D. Carter, R. H. Goulding, M. Kwon, in Proc. 8th Topical Conf. Applications of RF Power to Plasmas, AIP Conf. Proc. 190, 334 (1989).
6 F. W. Baity, in Proc. 7th Topical Conf. Applications of RF Power to Plasmas, AIP Conf. Proc. **159**, 378 (1987).
7 F. W. Baity, D. J. Hoffman, and T. L. Owens, in Proc. 6th Topical Conf. Applications of RF Power to Plasmas, AIP Conf. Proc. **129**, 32 (1985).
8 J.A. Rome, et. al., Sherwood Theory Conf., Newport, RI 1993.
9 S. C. Aceto, et. al., in Proc. 19th EPS Conf. on Controlled Fusion and Plasma Physics, Innsbrook, Austria, June 29-July 3, 1992.

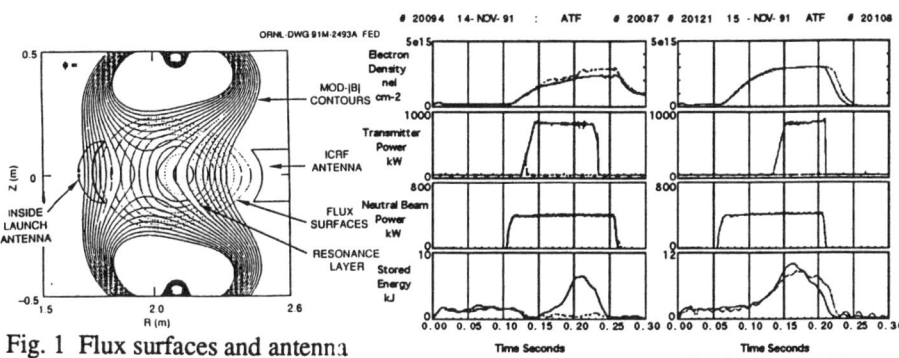

Fig. 1 Flux surfaces and antenna location for standard operating configuration.

Fig. 2 A typical "reheat" discharge with ICRH. A "collapsed" discharge without ICRH is shown for comparison.

Fig. 3 A typical "non-collapsing" discharge with ICRH.

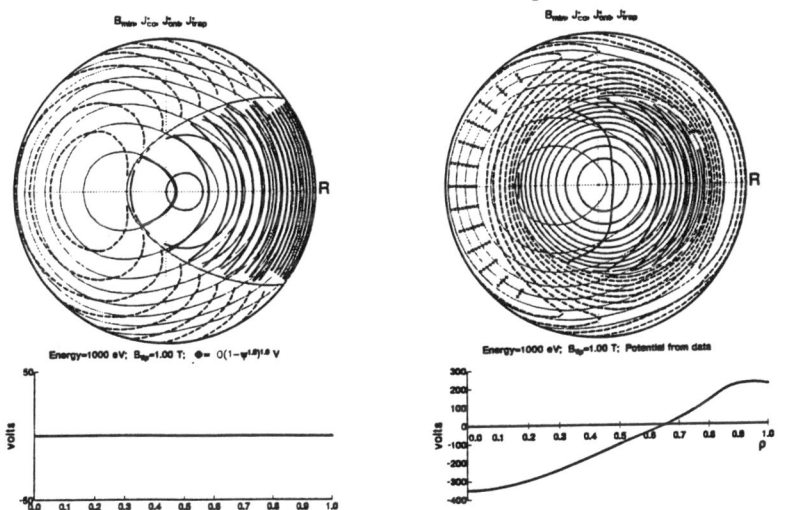

Fig. 4 a) An example of ion orbits calculated with b) zero radial electric field. Direct loss occurs for 1keV trapped ions at large major radius.

Fig. 5 a) An example of ion orbits calculated with b) a non zero radial electric field. Direct loss orbits are strongly modified.

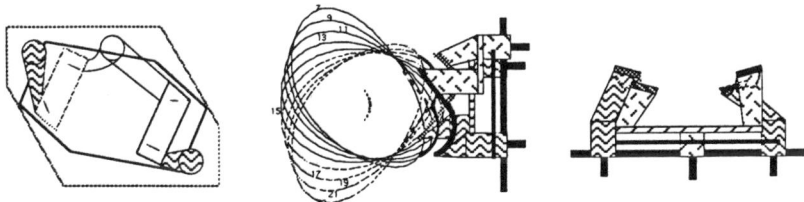

Fig. 6 Strap arrangement and poloidal and toroidal cross sections of the double-strap antenna.

PBX-M ION BERNSTEIN WAVE HEATING OVERVIEW

M. Ono, R. Cesario[1], T.K. Chu, H. Herrmann, B. LeBlanc, T. Seki[2], W. Tighe, N. Asakura[3], R. Bell, L. Blush[4], S. Bernabei, R. Conn[4], R. Doerner[4], J. Dunlap,[5] A. England[5], A. Grossman[4], J. Harris[5], R. Hatcher, R. Isler[5], R. Kaita, S. Kaye, H. Kugel, M. Okabayashi, H. Oliver, F. Paoletti[6], S. Paul, N. Sauthoff, L. Schmitz[4], S. Sesnic, H. Takahashi, and G. Tynan[4], *Princeton University, Princeton, NJ, [1]ENEA, Frascati, Italy, [2]NIFS Japan, [3]JAERI, Japan, [4]UCLA, Calif, [5]Oak Ridge National Laboratory, TN, [5]MIT, MA.*

ABSTRACT

A high power ion Bernstein wave heating system has been introduced on PBX-M for heating and for controlling the plasma pressure profile in an effort to achieve the stable high beta "second stability" regime. The pressure profile can be controlled through local bulk ion heating as well as density profile control. In bean-shaped plasmas with plasma currents range from 180 kA to 250 kA, good ion heating up to the highest applied rf power, ($\approx$ 700 kW,) has been observed. The observed broadening of the ion temperature profile is consistent with localized off-axis bulk ion heating as predicted by IBW ray tracing calculations. Application of IBW also resulted in a greatly modified density profile. The ability for IBW to change the density profile appears to be particularly attractive for controlling the bootstrap current profile for advanced tokamaks. Many important IBWH-related edge physics results were also obtained, including ponderomotive edge plasma modification and parametric instability onset conditions. The experimental plan for the next IBW run includes investigation of synergy with LHCD, attainment of high bootstrap current fraction discharges utilizing the IBW density profile control, and exploration of high beta plasma regimes.

INTRODUCTION

The ion Bernstein wave (IBW) heating system (2 MW at 40-80 MHz) has been installed on PBX-M as a means of controlling the pressure profile.[1] The PBX-M IBW pressure profile control strategy is centered on localized bulk ion heating[2] and density profile control.[3] The IBW system presently consists of two antennas, each connected to the 2 MW, 40-80 MHz FMIT transmitter. The IBW antenna elements are phased (0-π) to reduce the low-$n_{\parallel}$ related edge losses,[4] and the antennas are placed in the outer mid-plane region to optimize accessibility of IBW to the plasma core.[5] The RF frequencies are 47 and 54 MHz with B_T = 1.2 T and 1.4 T, respectively, which correspond to the $5\Omega_D$ resonance near the plasma center.

SUMMARY OF 1992 PBX-M IBW RUN RESULTS

Heating Results - During the present run, IBW power of up to 700 kW was applied to plasmas with a mixture of hydrogen and deuterium. Good comparison shots with and without IBW were taken for various IBW power levels. The stored energy showed a general increase comparable to NBI at similar power levels. The temporal rise and fall of the stored energy were relatively rapid compared to the plasma density behavior. Moreover, for the experimental parameter regimes used in PBX-M, IBWH was in the saturated confinement regime, for which the stored energy is only weakly dependent on the plasma density. These observations indicate that the IBW was indeed depositing power in the core of the plasma.

To investigate the feasibility of pressure profile control by localized ion heating, it is important to check if the predicted IBW power deposition profile is consistent with the observed heating. To facilitate this comparison, the eighteen-channel charge-exchange recombination spectroscopy (CHERS) diagnostic, which uses an impurity oxygen (O^{+7}) line excited by a deuterium neutral beam (with a deposited NBI power of $\approx$ 700 kW) to obtain time-resolved ion temperature profiles. The CHERS system therefore measures the bulk ion temperature, instead of the non-thermal rf heated ion species component.

Three important ion heating results are obtained: 1) the rise in the ion stored energy, ($\Delta[<n_e>T_i(0)]$), is linear with the applied IBW power up to the highest powers; 2) the measured ion temperature profile shows a general broadening with a steepened gradient, $[dT_i(r)/dr]/T_i(r)$, near the predicted IBW power deposition layer as expected; and 3) a detailed examination of the earlier time shows that the ion temperature starts to rise near the power deposition layer. The TRANSP analysis of the IBW heating shows reasonable consistency with the heating result, with the usual radially-increasing χ_i profile. This type of localized ion heating may be utilized for a direct measurement of the local ion thermal diffusivity, χ_i. More detail is given by W. Tighe et al.[2]

Particle confinement improvement and IBW-induced peaked profiles - In PBX-M, for some of the discharges, IBW has yielded very peaked density profiles. This is a clear indication of particle transport changes in the plasma core region, perhaps related to similar peaked profiles observed in JIPPT-II-U.[6] The profile changes in PBX-M took place relatively slowly over 150-200 msec. This density peaking was observed in a circular ohmic as well as in NBI-heated bean-shaped plasmas. The application of IBW into a strongly NBI-heated H-mode discharge causes the profile to steepen from the very flat, H-mode-like profile in the early phase to a supershot-like peaked profile in the later phase as shown in Fig. 1. The central density reached 8×10^{13} cm^{-3} which, without IBW, is not possible even with intense gas puffing. The density gradient reached 5×10^{12} cm^{-4} (comparable to the H-mode edge gradient) in the plasma core region (r ≈ 10-15 cm). This steepened density region may be related to the poloidal velocity shear stabilization of turbulence by IBW.[7] The model predicts that due a non-linear plasma response to the IBW wave field, the poloidal velocity shear layer (very much like the H-mode case) can be created near the power absorption region. Since the wave absorption layer position can be varied, this type of study might lead to a reactor-relevant tool for active plasma transport control. With the density peaking, a complete elimination of sawteeth was observed. This suppression of sawteeth is consistent with the raising of q(0) above one due to the generation of off-axis bootstrap current, as indicated by the TRANSP analysis. The details of IBW-induced transport change is discussed in the companion paper by B. LeBlanc at this conference.[3] A CHERS diagnostic for measuring poloidal rotation during IBW is being prepared.

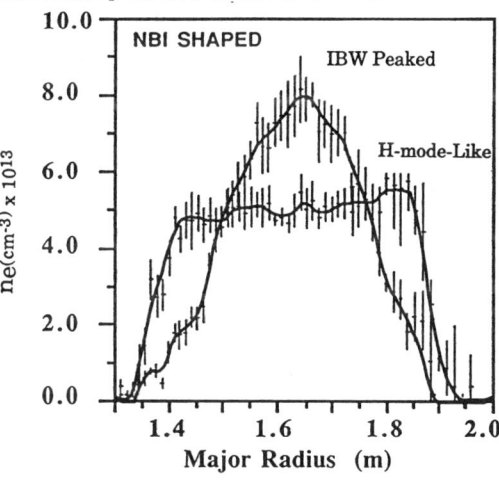

Fig. 1. Peaking of density profile during IBW. Bean-shaped, Ip≈ 250 kA, P_{IBW} ≈ 500 kW and P_{NBI} ≈ 2 MW, deuterium, f = 47 MHz, and B_T= 14.5 kG.

Edge plasma modification - To insure good heating efficiency and to reduce possible impurity generation, it is quite important to understand the edge physics occurring during IBW. This topic has been previously addressed by DIII-D experiments.[8] In the PBX-M IBW experiment, considerable progress has been made in understanding the IBW-edge plasma interactions, including plasma edge modification during IBW, the IBW antenna loading, and the conditions for parametric instability activity. The measured edge density during IBW by a fast reciprocating probe shows a strong reduction of the edge scrape-off density. This reduction of the density confirms the validity of the observed antenna loading based on the electron plasma wave excitation. A model using the ponderomotive force has shown good agreement with experimental observation.[11] The strong edge modification by IBW may be used, for example, in controlling the heat flux into the divertor plates. This reduced edge density during IBW also minimizes the parasitic antenna-plasma sheath effects during IBW.

Parametric instabilities - The topic of parametric instabilities during IBW was first addressed by the DIII-D IBW experiment.[8] Associated with high power IBW, strong parametric instability activity was often observed during the DIII-D experiment, which correlated well with the edge-produced high energy ion tail and electron heating. On PBX-M, this problem was investigated in detail both experimentally and theoretically.[10] Theoretical work, which includes the convective and gradient effects, shows that the parametric instability growth rate depends very strongly on the edge density profile. According to the theory, the parametric activity should increase if the plasma is moved away from the antenna (creating a low density gap region). To test this hypothesis on PBX-M, the plasma position was deliberately varied while monitoring the parametric activity. Under the normal IBW operating conditions, very little parametric activity was observed ≤ 50 dB below the pump ($\omega \approx \omega_{rf}$). However, when the plasma edge was moved away from the antenna by about 10 cm, the parametric instability activity increased to within 20 dB of the pump, as shown in Fig. 2.[10] This result shows that the parametric instability can be controlled during IBW. The probe measurements thus far showed no sign of edge electron heating nor a significant change in the floating potential during IBW.

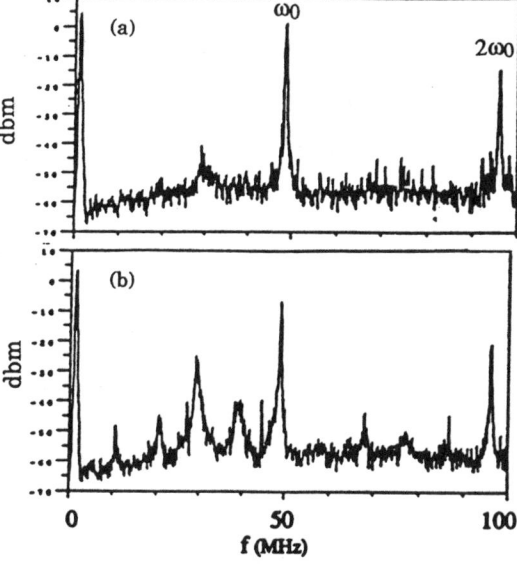

Fig. 2. Parametric instability behavior as a function of the plasma position. (a) R = 166 cm. (b) R = 162 cm. $\bar{n}_e \approx 1 \times 10^{13} cm^{-3}$, $P_{IBW} \approx 100$ kW, $I_p \approx 100$ kA, Circular, deuterium, 47 MHz and B_T= 14.5 kG.

PLANNED IBW EXPERIMENTS ON PBX-M

In addition to extending the present results to higher power levels with improved diagnostics, the future PBX-M IBW experimental plan has four new experimental topics: boron-nitride-clad Faraday shield, high fraction bootstrap current regime, synergy with LHCD, and ponderomotive kink stabilization.

Boron Nitride Faraday shield - In order to test the role of plasma sheaths, one of the IBW antennas was fitted with a boron-nitride-clad Faraday shield. The new shield should effectively insulate the antenna from the plasma, and thus, eliminating possible sheath related problems.[9] This performance of the BN Faraday shield antenna will be compared with that of the antenna with a metallic Faraday shield.

High Fraction Bootstrap Current Regime - The density peaking during IBW produced a peak bootstrap current density contribution of 25% of the total current density (17% of the total current) [$T_e(0) \approx 1.1$ keV]. These fractions can be significantly increased with the additional available heating power (4 -6 MW NBI and 2 MW IBW). This high bootstrap current also raises q(0), which is attractive for access to second stability regime. This bootstrap current generated in the plasma core region may be more favorable for MHD stability (particularly against kinks) compared to the H-mode-like case which generates significant edge currents. The ability to produce peaked density profiles without requiring central fueling may be particularly useful in future devices such as TPX and DEMO reactors.

IBW Synergy with LHCD - The analysis of IBW on PBX-M shows that directly-launched IBW may be utilized to enhance the quality and efficiency of the LHCD synergy. Although electron Landau damping of ion Bernstein waves is relatively weak under normal circumstances, as the wave approaches a major resonance, the electron damping increases strongly.[5] Under appropriate conditions, the localized electron Landau absorption can be quite significant. In view of the synergy with LHCD, the creation of a hot electron target at a desired location (by choosing the resonance layer position), should result in improved localization for LHCD. Moreover, since the IBW electron Landau heating essentially fills the so-called spectral gap, the LHCD efficiency is also expected to increase. If successful, this could lead to a better understanding of the JET results and an improvement of the LHCD performance in PBX-M and TPX.

IBW Ponderomotive Stabilization of External Kinks - It was recognized that the strong ponderomotive force of a low frequency slow wave launcher, such as an IBW antenna, could be used to stabilize external MHD modes such as kinks.[12] The required RF power, unfortunately, is predicted to be relatively high, making this scheme only marginally suitable for the PBX-M-type parameters. However, it was recently proposed that the use of a feedback system which senses the growth of the MHD mode (at low amplitude, $\delta B \leq 10$ G) and modulates the antenna power in response to the mode amplitude and phase could reduce the required RF field by an order of magnitude, and the RF power by two orders of magnitude.[13] This reduction in the RF field and power requirements would make the concept reactor relevant. From a technological standpoint, the high frequency RF system is well-suited for the rapid modulation. The existing transmitters connected to the PBX-M IBW antennas are capable of RF modulation with a frequency of up to 10 kHz, permitting the test of this concept on PBX-M with only small hardware modifications.

*This work was supported by DoE Contract No. DE-AC02-76-CHO-3073.

REFERENCES

[1] M. Ono, S. Bernabei, N. Asakura, et al., in the Proceedings of the Europhysics Topical Conference on Radiofrequency Heating and Current Drive of Fusion Devices, European Physical Society, 16E, 213 (1992); S. Bernabei , et al., in Bull. Am. Phys. Soc. 37, 1437 (1992), to be published in Phys. Fluids B.
[2] W. Tighe et. al., paper B 18, this conference.
[3] B. LeBlanc et al., paper B17, this conference.
[4] M. Ono, in proceedings of Ninth Topical Conference on Radio-Frequency Power in Plasmas (Charleston, 1991), AIP Conference Proceedings 244 (New York, 1992), p223.
[5] M. Ono, Phys. Fluids B 5, 241 (1993).
[6] T. Seki, R. Kumazawa, T. Watari, et al., Nuclear Fusion, 32, 2189 (1992).
[7] H. Biglari, M. Ono, P. H. Diamond, G. G. Craddock, Ref. 4, p376.
[8] R.I. Pinsker, M.J. Mayberry, M. Porkolab, S. C. Chiu, and R. Prater, Ref. 4, p169.
[9] A. Grossman, et al., paper B 19, this conference.
[10] R. Cesario, et al., in Bull. Am. Phys. Soc. 37, 1572 (1992).
[11] R. Majeski, P. Robert, T. Tanaka, et al., Bull. Am. Phys. Soc. 36, 2419 (1991).
[12] D.A. D'Ippolito, Phys. Fluids 31, 340 (1988); J.P. Goedbloed and D.A. D'Ippolito, Phys. Fluids B2, 2366 (1990).
[13] M. Okabayashi, M. Ono, D. D'Ippolito et al., the *IAEA Technical Committee Meeting on the Avoidance and Control of Tokamak Disruptions*, Culham, U.K., Sept. , 1991.

DENSITY PROFILE MODIFICATION DURING IBW IN PBX-M

B. LeBlanc, M. Ono, W. Tighe, J. Dunlap[#], R. Bell, T. K. Chu, A. England[#], R. Isler[#],
S. Kaye, D. McCune, M. Okabayashi, A. Post-Zwicker[#], H. Takahashi, and S. Sesnic
PPPL, Princeton University, Princeton, NJ 08540
[#]ORNL, Oak Ridge, TN 37831

ABSTRACT

Application of IBW power to neutral beam heated PBX-M discharges was observed to progressively peak the electron density profile, inducing gradients in excess of 5×10^{12} cm^{-4} and central density up to 8×10^{13} cm^{-3}. Abated MHD activity occurred during the peaking phase. Core ion thermal diffusivity dropped during the same period to values neighboring 1 m^2/s. Induced large n_e gradients were spatially correlated with the IBW deposition profile. Bootstrap current induced in mid minor radius region broadened the total current profile. Work supported by DOE contract no. DE-AC02-76-CH0-3073.

INTRODUCTION

Ion Bernstein Wave (IBW) power was successively applied to PBX-M plasmas. By adjusting the density in front of the antenna, operational techniques to surmount the parametric instability were established.[1] Bulk ion heating was observed.[2]

The absorption of IBW power in PBX-M plasmas routinely induced density increase during the earlier times of the RF pulse; the augmentation emerged over the whole electron density profile with little change to its shape. Subsequently, the electron density profile underwent progressively substantial changes resulting ultimately in a peaked shape. This phenomenon, reported previously for ohmic discharges,[3] is also observed in NBI (neutral beam injection) plasmas where it occurred over intervals (100 to 200 ms) long compared to the IBW rise time (25 ms) and resulted in peaked density profiles with parameter $n_{e0}/\langle n_e \rangle$ values up to 2.7, local density gradients reaching 5×10^{12} cm^{-4}, and central density attaining 8×10^{13} cm^{-3}. We will refer to "peaking phase" as the latter part of the density profile modification sequence.

This paper deals with the results obtained in NBI plasmas. First we review IBW assisted NBI discharges, then analyze the core region confinement and ascertain the bootstrap current. Discussion and conclusion follow.

IBW ASSISTED NBI HEATED PLASMA

Figure 1 synopsizes the temporal behavior of an IBW assisted NBI heated discharge. Values of parameters labeled with physical units can be read on the vertical scale, others have arbitrary units. The Mirnov trace was zero offset for clarity. The nominal plasma current and toroidal field were 250 kA and 1.45 T respectively. Two injectors (NBI) were turned on successively at 0.35 s and 0.375 s. Two IBW antennas were energized simultaneously at 0.45 s. The RF frequency was 47 MHz. The total power delivered is shown for each heating system: NBI power reached 2.0 MW and IBW 0.5 MW. Sawtoothing activity visible on the soft x-ray, Mirnov coil and line integrated ($n_e l$) traces began after the beam turn-on. The D_α signal

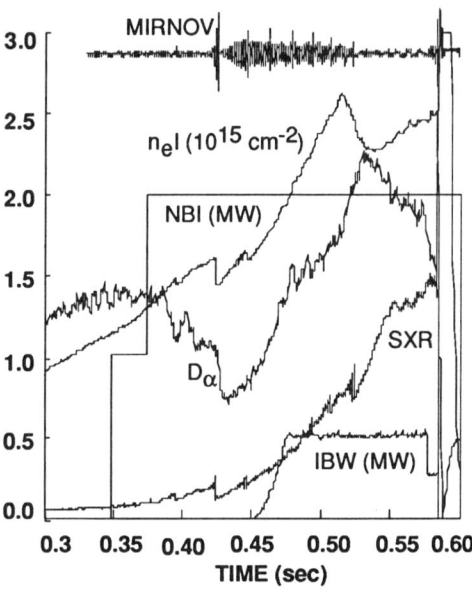

Fig 1. Temporal evolution of key discharge diagnostics and auxiliary heating systems.

shows an H-mode transition which occurred at 0.425 s, following a sawtooth. Further observation of the (rising) D_α trace shows the discharge leaving the H-mode before the application of IBW power. During the early RF phase, $n_e l$ increased steadily at a rate of 1.56×10^{16} cm^{-2} s^{-1}; sawteeth are no longer seen on the $n_e l$ trace, but remain visible on the soft x-ray and Mirnov coil signals. The line integrated density increase stopped at 0.51 s and a descent followed accompanied by an increase in the rate of rise of the D_α signal. The $n_e l$ later recovered and increased at a rate of 3.3×10^{15} cm^{-2}s^{-1}. The amplitude of the Mirnov oscillation decreased throughout the IBW duration. An MHD quiescent period began at 0.52 s as can be seen on the SXR and Mirnov traces. The discharge termination appears induced by a growing oscillation which began at the first antenna turn-off.

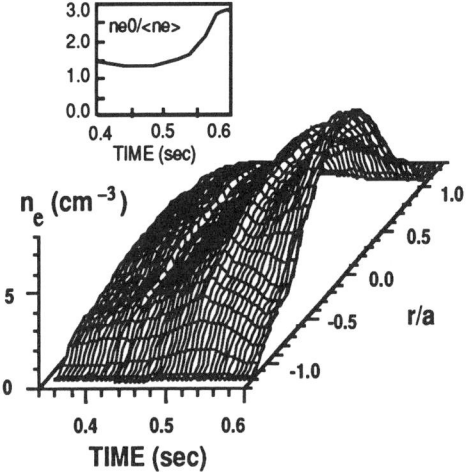

Fig 2. Time history of ne profile against normalized minor radius, a/r. In medallion: $n_{eo}/\langle n_e \rangle$ against time.

Figure 2 illustrates the modifications undergone by the electron density profile during the sequence just described. This figure is a reconstruction of the time evolution of the electron density profile obtained by concatenating high spatial resolution Thomson scattering profiles[4] of equivalent discharges (time points: 0.4, 0.44, 0.48, 0.54, and 0.575 sec) and 2-mm interferometer line integrated density data (time resolution of 0.5 msec) shown in Fig. 1. The sawtooth associated with the H-mode transition (0.425 sec) is visible and also the hump corresponding to the $n_e l$ rollover at 0.51 sec. We can also see the density profile peaking after the $n_e l$ recovery. The shape parameter $n_{eo}/\langle n_e \rangle$ (shown in medallion) reached 2.7. The peaking phase starts at ≈ 0.52 s.

Figure 3 gives plots of the temporal evolution of kinetic parameters measured at the plasma center. The density n_{eo} is extracted from the data shown in Fig. 2. The temperatures T_{eo} and T_{io} were obtained from Thomson scattering and charge exchange recombination spectroscopy respectively. It can be seen that while n_{eo} increased by a factor of 2.5, T_{eo} lowered by only 20%. On the other hand, T_i, after an initial augmentation following the onset of the neutral beam injection, rolled over and underwent a decline during the MHD active phase of the discharge, but eventually recovered and increased to its highest level during the peaking phase. The central electron density (n_{eo}), after the sawtooth at 0.425 s and the hump at 0.51 s, reached 8×10^{13}cm^{-3}.

There is experimental indication that central Z_{eff} increased during the RF pulse, from lower than 4 to values in the vicinity of 5. The fact that T_{eo} does not collapse is suggestive of fully stripped low-Z impurities like oxygen or carbon. The experimental documentation of plasma composition for these discharges is incomplete and will be addressed in the future. Preliminary spectroscopic analysis does not show evidence of metallic impurity accumulation. The neutron rate increased during the peaking phase, which is also suggestive of sustained

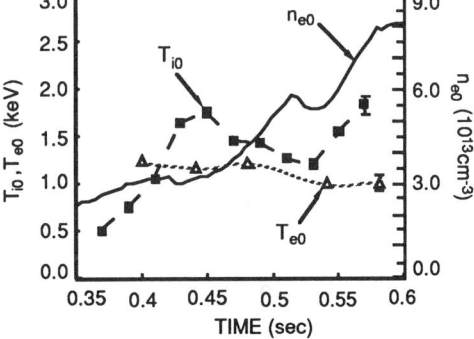

Fig. 3: Evolution of kinetic parameters at plasma center.

T_e since beam-target reactions generate the largest fraction of the neutrons.

CORE CONFINEMENT IMPROVEMENT

The experimental results are consistent with an improvement of the confinement in the core region. In order to permit the transport analysis of these plasmas, the TRANSP[5] code was recently upgraded to read-in the IBW power deposition profile; the latter is computed with a ray tracing code which folds in the measured $n_e(R)$ and $T_e(R)$. In Fig. 4, we show the temporal evolution of the ion effective thermal diffusivity, χ_i, at distances, r, of 5, 10, and 15 cm from the magnetic axis; this corresponds radially to the inner half of the plasma. Note that the TRANSP analysis "omits" the plasma termination at 0.585 s. The vertical error bar was extracted from an inaccuracy analysis done previously on a similar PBX-M set of

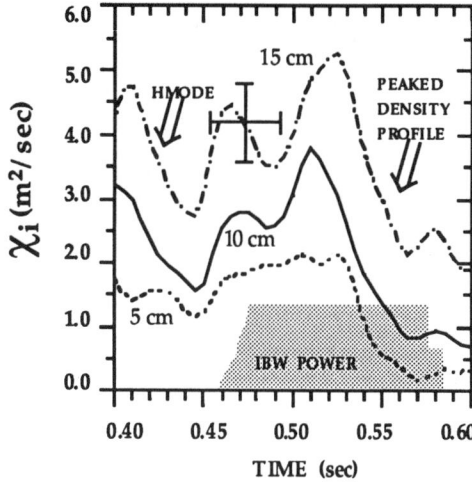

Fig. 4. Core χ_i against time at 5, 10, and 15 cm from the plasma center.

data involving the same diagnostic array.[6] The horizontal error bar is determined by the time resolution of relevant diagnostics. In the present case, it is set to 20 ms, which is the integration time of the ion temperature measurement. The vertical error bar is likely to increase as we get closer to plasma center because of the difficulty of measuring small temperature gradients. We can see χ_i abating as the discharge entered the H-mode and then rising up as the plasma left the H-regime before the application of RF power. The subsequent meandering of the ion diffusivity could be the result of increased MHD activity. A spectacular drop of χ_i began at ≈ 0.52 s and marked the onset of the peaking phase. The ion diffusivity dropped over the three minor radii shown. At 10 cm from the magnetic axis, χ_i fell from 3.5 m^2/sec to values around 1 m^2/sec. The limited time resolution of the analysis does not permit us to establish a temporal correspondence between the onset of the χ_i reduction and the $n_e l$ rollover nor the MHD quieting.

BOOTSTRAP CURRENT

Associated with the peaked profile observed during the later times of the IBW pulse are large values of the local spatial gradient, ∇n_e, of the electron density. The ∇n_e profile is plotted in Fig. 5; the spatial resolution is ≈ 1.2 cm. Also overlaid in this figure is a mapping of the RF power coupled to the bulk ions, P_{IBWI}. There is spatial correlation between the highest gradients (in excess of 5×10^{12} cm^{-4}) and largest values of P_{IBWI}. The maximum ∇n_e is slightly outside of the P_{IBWI} maximum location, $r_{IBW} = 15$ cm. Fig. 6 shows an overlay of TRANSP calculation of the total and the bootstrap current profiles at two time points: just before the application of IBW power (0.45 s) and at the end of the RF pulse (0.575 s). It can be seen that the (total) current density profile, j_{TOT}, broadened noticeably during

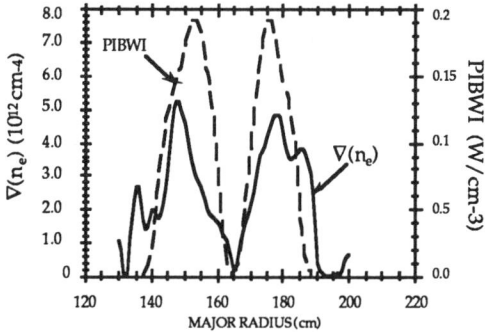

Fig. 5. Overlay of ∇n_e and of IBW power deposition profile mapping.

the IBW coupling. This current profile modification resulted from the generation of sizable and localized bootstrap current in the mid minor radius region. The current density appears to have flattened sufficiently to rise q_0 above one as predicted by TRANSP and supported by the sawtooth activity suppression.

DISCUSSION

It has been argued that IBW power can be used to suppress inner plasma turbulence by inducing a sheared poloidal (ion) flow.[7] There is indirect evidence that this might be the case with these IBW assisted NBI plasmas. As we have presented, the confinement barrier (region of large ∇n_e) coincided with the region of absorption of IBW power, and χ_i fell at r_{IBW} (and also at smaller r) as the density profile was peaking. Preliminary studies of the particle confinement, not presented here, show a decrease of the diffusion coefficient in the core region during the peaking phase.

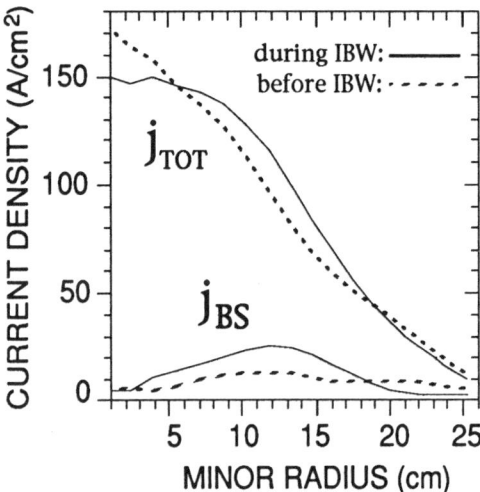

Fig. 6. Total and bootstrap current profiles: before IBW (dotted) and with IBW (solid).

The bootstrap current fraction shown in Fig. 6 is relatively modest because of the moderate temperature ($T_{eo} \approx 1$ keV) and high collisionality ($\nu_{*e} \approx 2$) of these discharges where P_{NBI} is limited to 2 MW. In practice, with higher beam power ($P_{NBI} \approx 4$ MW), T_{eo} values of ≥ 2 keV are readily obtained, At this temperature, $\nu_{*e} \approx 0.4$, and initial calculations predict the peak bootstrap current fraction to be $\approx 70\%$ of the local current density, and over 50% of the total current.

CONCLUSION

We have obtained IBW induced, peaked density profiles with n_{e0} up to 8×10^{13} cm^{-3} and large density gradients (in excess of 5×10^{12} cm^{-4}) located in the mid minor radius region. The peakedness parameter $n_{e0}/<n_e>$ reached 2.7 which is comparable to TFTR supershots for which $n_{e0}/<n_e> \leq 3$.[8] The core ion thermal diffusivity is reduced during the density peaking phase of the discharge to values ≈ 1 m^{2}s^{-1}. There is good spatial correlation between large values of ∇n_e and mapping of the IBW power deposition profile. Radially localized bootstrap current (a result of the IBW induced ∇n_e profile) broadens the current profile sufficiently to raise q_0 above one. IBW provides a promising reactor relevant technique which can peak the pressure and broaden the current profile.

REFERENCES

[1] M. Ono, at this conference.
[2] W. Tighe, at this conference.
[3] M. Ono, Proceedings of Ninth Topical Conference on Radio-Frequency Power in Plasmas (Charleston, 1991), AIP Conference Proceedings 244 (New York, 1992), p. 223.
[4] B. LeBlanc, et al., Rev. Sci. Instrum. 61, 3566 (1990).
[5] R.J. Goldston, et al., J. Comp. Phys. 43, 61 (1981)
[6] B. LeBlanc, et al., PPPL 2840, 1992.
[7] H. Biglari, M. Ono, P.H. Diamond, G.G. Graddock, Proceedings of Ninth Topical Conference on Radio-Frequency Power in Plasmas (Charleston, 1991), AIP Conference Proceedings 244 (New York, 1992), p. 376.
[8] D. Meade, et al. in Plasma Physics and Controlled Nuclear Fusion Research 1990, volume 1, p. 9, IAEA, Vienna, 1991.

Changes to the Ion Temperature Profile during IBW Heating in PBX-M*

W. Tighe, R. Bell, T.K. Chu, H. Hermann, B. LeBlanc, M. Okabayashi, M. Ono, N. Asakura[1], R. Cesario[2], A. England[3], R. Isler[3], R. Kaita, H. Kugel, S. Paul, A. Post-Zwicker[3], and H. Takahashi

Princeton University Plasma Physics Laboratory, NJ, USA, [1]Jaeri, Japan, [2]ENEA, Frascati, Italy, [3]Oak Ridge National Laboratory, TN, USA

ABSTRACT

The introduction of Ion Bernstein Wave Heating (IBWH) experiments in PBX-M have demonstrated significant bulk ion heating and ion temperature (T_i) profile modification. T_i profiles were obtained using a multi-channel, charge-exchange recombination spectroscopy system (the CHERS diagnostic). With IBWH, there are density increases; so to produce comparative discharges when no IBWH is present, additional gas must be injected. For $\bar{n}_e$, of ~3 x 10^{13} cm^{-3}, the central ion temperature, T_{i0}, was observed to increase by 350 eV when 0.6 MW of IBWH was applied for ~300 msec. The product $\bar{n}_e T_{i0}$, a measure of the ion stored energy, increases linearly with IBWH power. The T_i profile broadened with IBWH and the ion temperature gradient steepened near the predicted power deposition layer, providing evidence of localized heating. Examination of ΔT_i profile evolution indicated that the IBWH power was, in fact, deposited in a localized, off-axis region.

INTRODUCTION

This paper deals with observed changes to the T_i profile during IBWH. A primary role of the IBWH program on PBX-M is to provide pressure profile modification through localized bulk ion heating. Ion heating is considered here since the ion energy transport is thought to be less anomalous compared to that of electrons. With IBWH, local ion heating is achieved by matching the frequency of the ion Bernstein wave to the ion-cyclotron harmonic resonance for a selected ion within the plasma. This resonance allows effective coupling and is selective by adjustment of the IBW frequency and the ion species within the plasma. A detailed review of experimental and theoretical aspects of IBWH is available[1].

In the data presented here, we have used deuterium as the selected heated ion and matched to the $5\Omega_D$ frequency, located near the center of the tokamak. Ray-tracing calculations show that for these conditions the IBWH power deposition is off-axis, centered ~8 cm from the mid-point of the plasma. Furthermore, these calculations show that the power is deposited primarily to the ions (< 10% is deposited to the electrons).

EXPERIMENT

The experimental arrangements for the application of IBWH in PBX-M are described by M. Ono[2] (this conference). For the data considered in this paper, the rf pulse duration was 200 msec for which the highest power level was 0.7 MW (shorter pulses achieved power levels >1 MW). The ion temperature was measured by CHERS (CHarge Exchange Recombination Spectroscopy). This is a multichannel diagnostic[3] that provides radial profiles of the bulk ion temperature as a function

---

* Supported by USDOE Contract No.DE-AC02-76-CHO-3073

of time. Up to 50 frames are available during each discharge. To improve the signal-to-noise ratio, integration times of 10 or 20 msec were used (the minimum integration time between frames is ~3 msec). CHERS uses a heating neutral beam (NW NBI) to provide this data. The neutral beam power was 1 MW injected with ~0.7 MW absorbed.

With the application of IBWH power there is a significant increase in density. The line averaged density shows a marked rise as the IBWH power reaches its operational level. Both the rate of density rise and the maximum value attained both increase with IBWH power. Some discussion of the effects of IBWH on density is provided by B. LeBlanc[4] (this conference). For the purpose of investigating ion heating, it is important that comparable discharges, with and without IBWH, be used. To compensate for the density increases during IBWH, gas was injected into the tokamak when no IBWH was applied. A near overlap of the density conditions (both temporally and radially) was achieved during the period of interest. This was verified by Thomson scattering and microwave interferometer data. This overlap allows a direct comparison of the temperature profiles for the discharges of interest.

Comparisons of T_i profiles are shown in Fig. 1. In each plot the upper (solid circle) curve represents the temperature profile when IBWH was applied and the lower (open circle) curve represents the temperature profile for a matched density condition and no IBWH applied. Three cases of IBWH power are shown, .22 MW, .42 MW, .6 MW. The line averaged densities are 2.2, 2.9, and 3.1 x 10^{13} cm^{-3}, respectively. In every case the central ion temperature is higher when IBWH power was applied. Moreover, as the IBWH power is increased, the temperature difference increases from ~150 eV at .22 MW to more than 350 eV at .6 MW. The profile shape also changes. With the application of IBWH the profile broadens and the slope in the outer region steepens. The degree of profile modification will be discussed later.

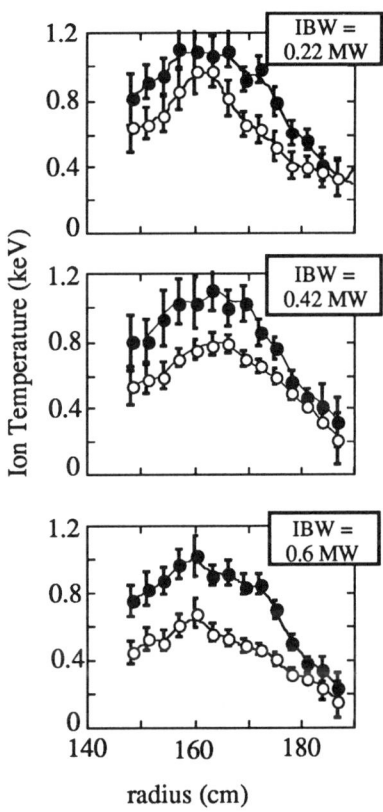

fig. 1 $T_i(r)$ profiles taken with (closed circles) and without (open circles) IBWH for varying IBWH power.

While the assumption here is that IBWH is directly heating the ion population, it is possible that other mechanisms are in place. Examination of Thomson Scattering data shows that the electron temperature also rises when IBWH is applied and heat may flow from the electrons to the ions. For the case of 0.6 MW of IBWH power, the electron temperature rises from ~550 eV to ~750 eV. The ion temperature, however, is significantly higher in this case, reaching a level of ~1 keV so that heat flows from the ions to the electrons.

Aside from direct changes to the ion temperature profile, we can also fold in the changes to the density and examine the behavior of the product of temperature and density as a function of IBWH power. This should reflect the stored energy in the ion population. For this purpose, $T_i(0)$ was averaged over a 10 cm central region of the plasma and a 20 msec time window. The line averaged density was averaged over the same 20 msec time window. These values were then used to evaluate $<T_i(0)*n_el>$ which is plotted as a function of IBWH power in Fig.2. Some selection of the data has been made. Primarily, discharges that had significantly different initial conditions (densities and temperatures before the application of IBWH) have been ignored. Two plots are shown, one taken prior to any observed changes and the other at the peak of the change in T_i and n_el. Most significantly it appears that the stored energy of the ions increases virtually linearly with IBWH power in the range of power used in these experiments.

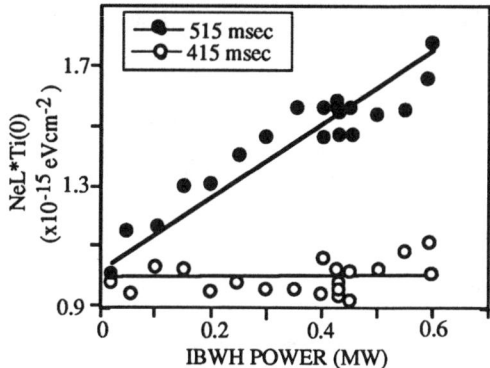

fig. 2. $<T_i(0)*n_el>$, a measure of the ion stored energy, rises linearly with IBWH power. Data taken prior to (open circles) and at the peak (closed circles) of IBWH.

LOCALIZED HEATING

As mentioned earlier, there is some evidence in the ion temperature profile of localized heating and profile modification. Taken late in the application of IBWH these effects are not so obvious since a steady state situation has been reached. Still changes are apparent. Fig. 1 shows a broadening of the profile and a steepening of the gradient near the half-radius region. There is little or no change

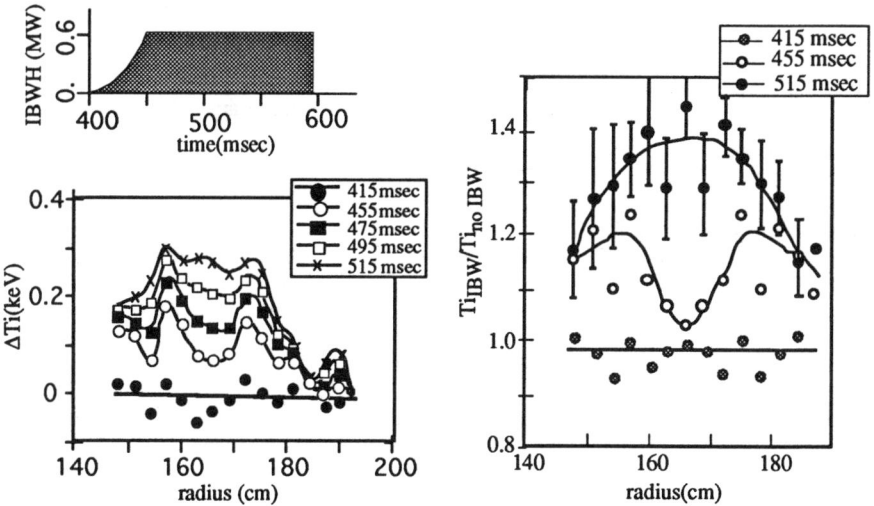

fig. 3. Evolution of ΔT_i profiles during early period of IBWH. Initial heating is localized and off-axis.

fig. 4. Ratios of T_i profiles displaying significant profile changes at the start of IBWH.

in the profile shape near the center. The steepened gradient (dT_i/dr) provides an indication as to the location of heat deposition. However, looking early in the application of IBWH a clearer indication of localized heating and changes to the T_i profile should be possible. This analysis involves taking comparable discharges and subtracting the temperature profile with IBWH from that without IBWH and examining the time evolution of this difference. To reduce the error two sets of such discharges, with the same density and IBWH power, have been averaged.

Fig. 3 shows that while IBWH power is near zero at t=415 msec, no heating has occurred. Just as the IBWH power reaches its operational level at t=455 msec, the central ions remain unheated while there is a rise in ion temperature in the off-axis region (~8 cm from the axis -- in good agreement with the ray-tracing calculations). In time, the central ion temperature begins to rise and eventually the profile fills in.

The ratio of the temperature profiles (T_i[with IBW]/T_i[without IBW]) is a more sensitive parameter to changes to the profile shape. This is shown in Fig. 4. The density and temperature prior to the application of IBWH are very nearly the same so that the ratio of ion temperatures is ~1 at this time. At t=455 msec, there is a significant rise in the ratio in the outer regions but little change in the central regions. At later time the ratio starts to flatten indicating smaller changes to the profile. While the error bars here are large, and further experimentation and analysis are needed, the results are compelling and qualitatively as one would expect.

There are two additional comments with respect to this analysis. First, there is sawtooth activity during this period. For the data shown, the data are averaged through the sawtooth prior to the generation of the temperature profiles. If one interpolates data selected just before the sawtooth crash, the effects of localization are enhanced. On the other hand if one interpolates data taken immediately after the sawtooth crash, the effects of localization are reduced. This is consistent with the idea of sawtooth smoothing but these results are very preliminary and require further study.

Second, if ΔT_i profiles are generated for comparable discharges but for which no IBWH power is applied in either case, the result, as expected, fluctuates about zero but well within the error bars. This may indicate that the error in the present analysis is slightly overestimated. This aspect of the analysis is presently being examined.

CONCLUSIONS AND DISCUSSIONS

With the application of IBWH on PBX-M, we have observed a significant increase in the ion temperature. With a power level of 0.6 MW, IBWH increases the ion temperature by ~350 eV. In this case the ion temperature exceeds the electron temperature indicating that the higher T_e observed may be due to higher T_i. A measure of the stored ion energy ($<T_i(0)*n_e l>$) has been seen to increase linearly with IBWH power. By examining the evolution of the ΔT_i profiles, there is evidence for off-axis, localized ion heating and ion temperature profile modification with IBWH.

REFERENCES

1. M. Ono, Phys. of Fluids B **5** (2), p. 241
2. N. Asakura et al., submitted to Nucl. Fusion
3. M. Ono et al., this conference.
4. B. LeBlanc et al., this conference.

IBW Generated Ponderomotive Potential Effect on Edge Plasma in PBX-M

A. Grossman, L. Schmitz, R. Doerner, M. Ono*, H. Kugel*,
M. Okabayashi*, R. Bell*, G.Tynan, L.Blush,
R.W.Conn, and the PBX-M group†

*Institute of Plasma and Fusion Research
University of California, Los Angeles, CA 90024*

* *Princeton University, Princeton NJ 08544*

† *PPPL, Princeton University, NJ
ORNL, Oak Ridge, TN
MIT, Cambridge, MA*

ABSTRACT

Edge plasma modifications have been observed during IBW(f=54MHz) on PBX-M at very low power (below 100 kW). Reductions of up to a factor of 2 are seen in the density profile over a radial extent of several cm, with little change to the floating potential or electron temperature, as measured by the fast reciprocating probe. At higher powers, the density reduction can be as much as an order of magnitude, with no significant edge electron heating. A 1-D analytical model and a 2-D fluid code are used to model these experimental results, which appear to be due to the generation of a ponderomotive potential barrier in the edge plasma. The experimental results are compared with the analytic and numerical model[1] using an RF generated ponderomotive potential. These results are significant for edge plasma modification concepts such as the rf-limiter and divertor[2], and encouraging for IBWH, where power was observed not to be deposited in the plasma edge.

INTRODUCTION

The divertor design is a key unresolved issue in the planning for future tokamak reactors. The most promising techniques under consideration at present are the high recycling divertor with a cold dense plasma at a divertor plate designed to have a high sputtering threshold, and the gas target divertor. These techniques require a high density, low temperature edge plasma, which directly conflicts the low density, high temperature edge plasma requirements for noninductive current drive techniques. As an alternative to this technique, it has been proposed[1,2] to use the ponderomotive forces to control the flow of edge plasma. This force is the nonlinear time-averaged force of a strong rf field whose energy density is inhomogeneous. The use of ponderomotive forces to divert edge plasma and control its flow to divertor plates in the low density regime has been demonstrated[2] in the PISCES linear edge simulation facility at UCLA, where the ponderomotive force was used to actively control the flow of edge plasma and protect plasma facing components. Here we report on preliminary experiments to test these ideas in the PBX-M tokamak. As will be shown, the existing antennas used for the ion Bernstein wave experiment in that tokamak provide an RF field which is close to what is needed to set up a ponderomotive barrier along field lines to alter the flow to divertor plates. The barrier can modify edge plasma profiles substantially, and the fast scanning probe is the key diagnostic for detecting global changes in the edge density and potential profile which accompany the creation of a ponderomotive barrier.

Important features of the use of RF ponderomotive forces to stabilize and divert plasmas are that (1) only comparatively modest RF power levels are required, (2) heating is not needed and the processes used are nondissipative, (3) nonresonant frequencies are readily achieved in tokamak geometry and (4) only surface penetration of the rf waves need occur.

EXPERIMENT

A fast reciprocating probe has been installed at the outboard midplane to measure the evolution of the edge plasma parameters, electric field, and fluctuation-induced transport during the application of IBW. In these measurements, only the IBW antenna (system 3) far away from the probe is energized, at a toroidal distance of 14m from the probe. Moreover, IBW system 3 is not connected to the probe along field lines. The length along field lines from the probe to the lower/upper divertor plates is about 30m. The probe is synchronized with the IBW so that it enters the plasma before the IBW is turned on, reaches the dwell point at the IBW turn on, dwells for 10-20 msec, and is retracted during the IBW pulse. Thus on every shot, the probe provides a SOL density and potential profile before the IBW turns on, and a profile after the IBW is turned on. The experimental profiles of the density and floating potential before and during the applied rf at 40kW, 60kW and 90kW (shots 296473, 296476, and 296483 respectively) are shown in Fig. 1. A fit to this experimental data was obtained using the 2D numerical model of the edge plasma with ponderomotive potentials described in the next section. The model assumed Bohm diffusion and classical mobilities, and the two radial boundary conditions on the density and potential were obtained from the experimental data. Since the temperature profile in the SOL is observed not to change with the application of IBW, the same experimentally measured temperature profile (from shot 291984) was used as input to the code for all three cases.

THEORETICAL ANALYSIS

Theoretical understanding of the ponderomotive effects used in the RF ponderomotive divertor experiments are based on the potential:

$$\Psi_{rf} = \frac{1}{4m}\frac{e^2 E_\perp^2}{\omega^2 - \omega_c^2} + \frac{1}{4m}\frac{e^2 E_\parallel^2}{\omega^2} \tag{1}$$

The negative perpendicular gradient of this potential is the ponderomotive force, which for frequencies away from cyclotron resonance acts mainly on electrons because of the $\frac{1}{m}$ dependence in Ψ_{rf}. This force is particularly strong for IBW because the antenna electric field is predominantly parallel to the magnetic field[6]. Ions follow the electron motion to preserve quasineutrality and there is a net plasma flow away from the ponderomotive barrier. Note that even though only the parallel RF electric field appears, there can be both a parallel and perpendicular gradient of this potential. The parallel component can create a barrier to plasma flow, while the component perpendicular to a magnetic field produces cross-field drifts of plasma particles, which can act as an RF divertor in itself.

To model the changes in the density and potential profiles under the influence of a ponderomotive potential, let $D_\perp^{Cl}$ be the classical plasma diffusion coefficient due to electron-ion collisions, $D_\perp^i$ be the diffusion coefficient arising from ion-neutral collisions and $\mu_\perp^i$ and $\mu_\perp^e$ be classical mobilities obtained from these diffusion coefficients by the Einstein relations, $D_\perp^i = \mu_\perp^i T_i$ and $D_\perp^e = \mu_\perp^e T_e$. Then we write the flux of electrons and the flux of ions as the sum of diffusion/mobility terms and drift terms:

$$\vec{\Gamma}_\perp^e = -(D_\perp^{Cl}+D_\perp^e)\vec{\nabla}_\perp n + \mu_\perp^e n\vec{\nabla}_\perp \phi + \mu_\perp^e n\vec{\nabla}_\perp \Phi_e + n\frac{\vec{B}\times\vec{\nabla}_\perp \phi}{B^2} + n\frac{\vec{B}\times\vec{\nabla}_\perp \Phi_e}{B^2} - \frac{\vec{B}\times\vec{\nabla}_\perp P_e}{qB^2} \tag{2}$$

$$\vec{\Gamma}_\perp^i = -(D_\perp^{Cl}+D_\perp^i)\vec{\nabla}_\perp n - \mu_\perp^i n\vec{\nabla}_\perp \phi + \mu_\perp^i n\vec{\nabla}_\perp \Phi_i + n\frac{\vec{B}\times\vec{\nabla}_\perp \phi}{B^2} - n\frac{\vec{B}\times\vec{\nabla}_\perp \Phi_i}{B^2} + \frac{\vec{B}\times\vec{\nabla}_\perp P_i}{qB^2} \tag{3}$$

where $E_\perp = -\vec{\nabla}_\perp \phi$, ϕ is the electrostatic potential and Φ_e is the ponderomotive potential on electrons and Φ_i is the ponderomotive potential on ions. These are substituted into the continuity equations:

$$\vec{\nabla}\cdot\vec{\Gamma}_\perp + \vec{\nabla}\cdot\vec{\Gamma}_\parallel = S \tag{4}$$

$$\vec{\nabla}\cdot\vec{J}_\perp + \vec{\nabla}\cdot\vec{J}_\parallel = 0 \tag{5}$$

We are effectively averaging n(z) along z in what follows, we may replace n_w with $\gamma_w n = \frac{2}{3}n$. Assuming sound speed flow, we may write:

$$\Gamma_\parallel^e = \gamma_w n c_s e^{\frac{\phi_w - \phi_f}{T_e}} e^{\frac{\Phi^e(0) - \Phi^e(L)}{T_e}} \tag{6}$$

while

$$\Gamma_\parallel^i = \gamma_w n c_s \tag{7}$$

so

$$\frac{J_\parallel}{q} = \gamma_w n c_s (1 - e^{\frac{\phi_w - \phi_f}{T_e}} e^{\frac{\Phi^e(0) - \Phi^e(L)}{T_e}}) \tag{8}$$

In the limit of no ponderomotive potential or no gradient in the ponderomotive potential, this reduces to the usual current used for Langmuir probe characteristics. With a ponderomotive potential mainly at L (the wall) the electron density in the SOL is increased because this potential acts as a barrier to flow to the wall. With a ponderomotive potential mainly at z=0, electrons are driven towards the wall and electron density in the SOL is decreased.

An expression for the radial density profile in the presence of a ponderomotive barrier and an electrostatic barrier is readily derivable in in slab geometry. (Noting that $\Gamma_\perp = \Gamma_x$, and $\Gamma_\parallel = \Gamma_z$ Averaging along the magnetic field, with connection length L, and taking the source to be ionization of neutral atoms with density n_n, and taking a linear combination of the continuity equations so as to eliminate electrostatic potential:

$$D_\perp^A \frac{\partial^2 n}{\partial x^2} - \frac{\mu_\perp^i \mu_\perp^e}{\mu_\perp^i + \mu_\perp^e} \frac{\partial}{\partial x}[n \frac{\partial(\Phi_e + \Phi_i)}{\partial x}] = \frac{\Gamma_\parallel^w}{L} - [\frac{\mu_\perp^i}{\mu_\perp^i + \mu_\perp^e}] \frac{J_\parallel^w}{qL} - n n_n \langle \sigma v \rangle_{ion} \quad (9)$$

where the ambipolar diffusion coefficient has been defined as $D_\perp^A = \frac{D_\perp^e \mu_\perp^i + D_\perp^i \mu_\perp^e}{\mu_\perp^i + \mu_\perp^e}$. In the limit where ponderomotive forces and biased surfaces control the parallel fluxes, the equation for the density profile becomes:

$$\frac{1}{n}\frac{\partial^2 n}{\partial x^2} = \frac{\gamma_w c_s - n_n \langle \sigma v \rangle L}{L(D_\perp^e \mu_\perp^i + D_\perp^i \mu_\perp^e)}[(\mu_\perp^i + \mu_\perp^e) - \mu_\perp^i(1 - e^{\frac{\phi_w - \phi_f}{T_e}} e^{\frac{\Phi^e(0) - \Phi^e(L)}{T_e}})] \quad (10)$$

In the limit where $\frac{\phi_w - \phi_f}{T_e}$ and $\frac{\Phi_e(0) - \Phi_e(L)}{T_e}$ are independent of x, the general form of the density profile is:

$$n(x) = n_A e^{\frac{-x}{\lambda_n}} + n_B e^{\frac{x}{\lambda_n}} \quad (11)$$

In the limit of no ponderomotive potentials and no nonambipolar current fluxes to the walls, and replacing the classical ambipolar diffusion coefficient with the anomalous coefficient, this expression reduces to the familiar result for the scrape-off layer length[4]. Physically, this new result suggests that the ponderomotive potential barrier can be used to change the scrape-off layer length. With only a very small amount of cross-field mobility, it is possible to substantially modify the the scrape-off layer length with ponderomotive forces and/or electrically biased walls, provided that $\mu_\perp^e < \mu_\perp^i$, which is clearly fulfilled for classical mobility from neutral collisions. Writing the expression for the scrape off layer length in terms of the ambipolar diffusion coefficient, which is then replaced with an anomalous diffusion coefficient (Bohm diffusion), and taking the classical mobility limit, $\mu_\perp^e \ll \mu_\perp^i$,

$$\lambda_n = \sqrt{\frac{D_\perp^A L}{\gamma_w c_s - n_n \langle \sigma v \rangle L}} e^{\frac{\phi_f - \phi_w}{2T_e}} e^{\frac{\Phi^e(L) - \Phi^e(0)}{2T_e}} \quad (12)$$

Without a ponderomotive potential, or a gradient in the ponderomotive potential, this equation predicts that for a negative wall bias, in which $\phi_w < \phi_f$, the density e-folding length will increase and decrease in the positive wall bias case in which $\phi_w > \phi_f$, an effect previously derived by LaBombard[5] in the absence of an RF potential. With a ponderomotive potential that is largest at the divertor plate at z=L and weakest at the stagnation point where z=0, so that $\Phi^e(L) > \Phi^e(0)$ then its effect is to make the SOL wider. If most of the ponderomotive potential is at the stagnation point, it can act to make the SOL narrower. For it to have this effect, the ponderomotive potential should be larger than the electron temperature, in agreement with the reflection condition on electrons.

ACKNOWLEDGEMENTS

This work supported by DoE contract No. DE-FG-89ER51121 and DE-AC02-76-CHO-3073.

REFERENCES

1. A. Grossman, L. Schmitz, et al., J. Nucl. Mater. **196 & 198**, 775 (1992).

2. T. Shoji, A. Grossman, et al. J. Nucl. Mater. **176 & 177**, 830 (1990).

3. B. LaBombard, A. Grossman, R.W. Conn, J. Nucl. Mater. **176 & 177** 548 (1991).

4. R. W. Conn, J. Nucl. Mater. **128 & 129** 407 (1984).

5. B. LaBombard, R. W. Conn, G. Tynan, Plasma Physics and Controlled Fusion **32**(7) 483 (1990).

6. M. Ono, Phys. Fluids, **B 5** 241 (1993).

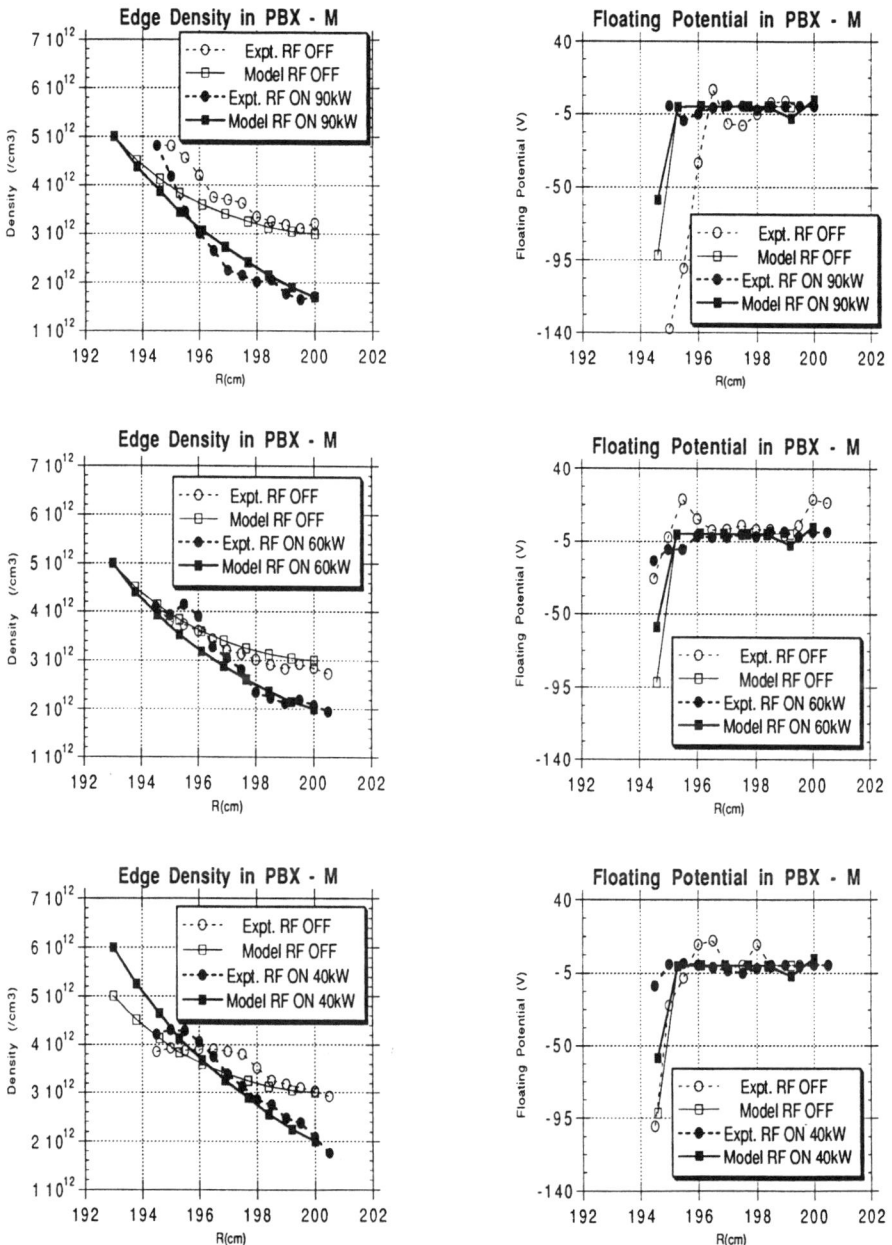

Fig. 1. Effect of rf on the radial profiles of density and floating potential.

ION BERNSTEIN WAVES IN A TOROIDAL STEADY-STATE PLASMA

C.Riccardi, D.Batani, M.Fontanesi, A.Galassi, E.Sindoni
Dipartimento di Fisica, Universita' di Milano, Via Celoria 16, 20133 Milano, Italy.

ABSTRACT

IBW propagation is experimentally investigated in a steady-state toroidal plasma. The waves are externally launched into the plasma by mode conversion of the Electron Plasma waves. The conversion efficiency is analysed as a function of the relevant parametres, by Fourier transforming the signal detected by an interferometric system.

INTRODUCTION

In recent years ion Bernstein waves Heating (IBWH) has been subject of several studies, as it allows direct heating of the tokamak plasmas bulk ions. IBWs can be indirectly excited via Electron Plasma Waves (EPW) mode transformation or directly launched into the plasma. In this paper we present some experimental results on IBW propagation in a steady-state magnetized plasma generated in the Thorello toroidal device. Preliminary results on the IBW propagation obtained in the ACT1 device[1,2], showed that the IBW can be indirectly excited in the plasma by means of a mode transformation. Subsequently IBW heating physics have been investigated through experiments performed on tokamak devices, showing an unexpected electron heating near the plasma edge[3]. In these experiments both linear and non-linear heating process were observed. The aim of the present analysis was the study of the basic physics phenomena involved in the EPW-IBW mode conversion in linear conditions, in a well controlled environment. The mode conversion process is characterized by a strong variation of the wave refractive index, and therefore a reflection phenomena is very likely, due to the criticity of the WKB theory. In this case a large part of the wave energy could be localized in the plasma edge as stationary wave and does not propagate in the centre of the plasma. The EPW-IBW conversion efficiency is analysed as function of the critical parametres, by means of a Fourier analysis of the signal detected by an interferometric system. The experimental results were also compared with the numerical full electrostatic dispersion relation, obtained in the WKB ipotesis, in the range of the ion cyclotron frequency.

EQUIPMENT SET-UP

The steady-state magnetised plasma was produced in hydrogen neutral gas by means of hot-filament electron emission and voltage-induced acceleration[4]. The electron density and temperature were measured with electrostatic Langmuir probes. Typical parametres of the system are: R=40 cm, r=8 cm, maximum axial toroidal magnetic field B_t=2.3kG, n_e=(10^8-10^{10})cm^{-3}, T_e=(1-10)eV, T_i=0.3 eV, neutral gas pressure P=(10^{-5}-10^{-4})mbar. The capability of operating in different discharge regimes makes the weakly ionised plasma of Thorello particularly suitable to study plasma-waves interaction phenomena. IBWs are analysed at the second harmonic of the ion cyclotron frequency ($\omega \approx 2\Omega_i$); typically f=(3-6)MHz. IBWs are externally excited by means of a launching electrostatic slow-wave antenna located at the plasma edge[5]. The antenna consists of four thin metallic blades lying in the vertical plane, along the direction of B; the feeding system allows to apply to the blades a phase shifted signal with continuity between 0° and 90°. Hence we can both change the wavevector component along B, $k_\parallel$, and launch EPWs in a preferential direction along the B lines in the toroidal machine. The antenna system excites waves with $k_\parallel \leq 0.1$ cm^{-1}. The wave vector component perpendicular to B, $k_\perp$, was measured by means of an interferometric system: the signal propagating in the plasma is detected by means of a double r.f. probe moving within the plasma in the radial direction (perpendicular to B); it is then compared to the input signal through a mixer and the interferometric trace, obtained by filtering the mixer output to suppress time-dependent terms, is sent to the Y-channel of a digital oscilloscope.

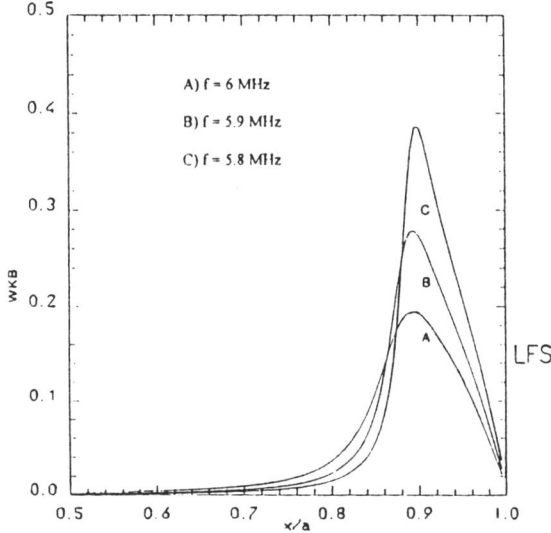

Fig.1 - w parameter, $w = k_\perp (dk_\perp /dx)$, vs x/a, for a) f=1.96fi,b) f=1.9fi.

RESULTS

The results concerning with the EPW-IBW conversion efficiency as a function of the signal frequency. Near to the conversion layer, the density and the refraction index profiles are very strong and the dishomogeneus dispersion relation obtained in WKB hypotesis is no longer correct. Fig.1 shows the WKB parameter, defined as $w = k_\perp (dk_\perp /dx)$, as function of the radial coordinate x/a (a minor radius): the parameter should be w<<1 in the WKB hipotesis, but near to the conversion layer is close to the unitary value. In this condition part of the wave energy could be refracted and trapped as stationary wave in the plasma edge. This energy is lost and does not propagate as

IBW into the plasma core. This process has recently been called Axial Convective Loss[6]. The phenomenum is analysed comparing the experimental data with the WKB theory and considering the wave energy spectrum given by the signal Fourier transform of the signal. Fig.2 shows the experimental dispersion relation compared with the theoretical one, for two frequency values. The theoretical dispersion curves are obtained with the typical density and temperature profiles of the device, in the WKB hypotesis. The EPW-IBW conversion layer defined by the Lower Hybrid frequency resonance: $\omega = \omega_{lh} = \omega_{pi}$ is marked on both the plots; the $k_\perp$ values in the mode conversion layer correspond about 15 cm^{-1} and 18 cm^{-1}.

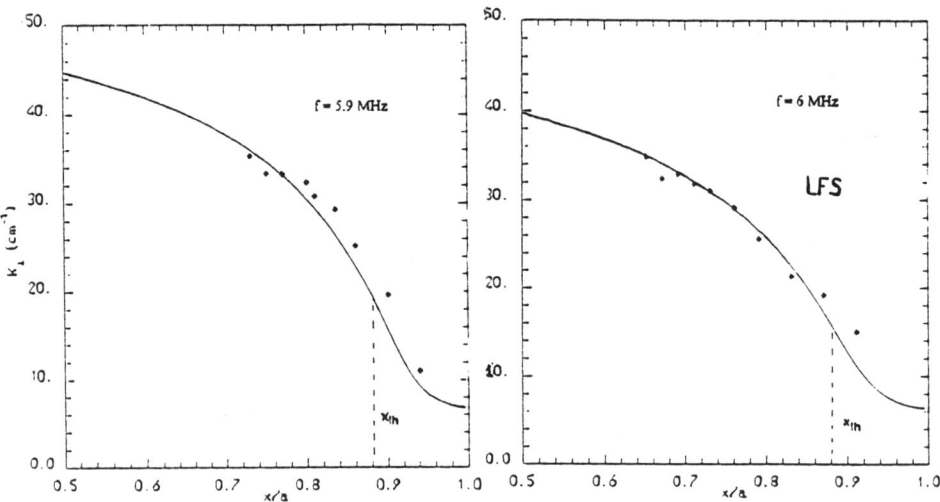

Fig.2 - Comparison of the experimental and theoretical dispersion curves for the parametres: ne(0)=10^{10}cm^{-3},Te(0)=2eV, Ti(0)=0.9eV, B=2.3kG, parabolic density and temperature profiles, a) f=6MHz=1.96fi, b) f=5.9MHz=1.88fi.

The best-fitting between the experimental and theoretical curves is good, even near to the conversion layer, but this analysis does not permit to obtain information about the percentage of EPW energy transformed into IBW one. We tried to analyse and stimate the conversion efficiency by Fourier transforming the interferometric trace. In fig.3 the energy spectrum is showed as a function of the perpendicular wavevector $k_\perp$ for a set of frequencies: only the energy radial component is considered. In case a) the EPW energy spectrum is in the $k_\perp$ range of (0-15) cm^{-1}, while the IBW spectrum corresponds to the part where $k_\perp > 15$cm^{-1}; in case b) the separation between the two regimes is at the $k_\perp$ value of 18cm^{-1}, in agreement with the WKB theory. To evaluate the conversion efficiency, the ratio between the IBW energy and the EPW energies each, obtained by integrating the energy spectrum over its own $k_\perp$ range is calculated. The efficiency turns out to be higher near to the second harmonic of the ion cyclotron frequency where the $k_\perp$ gradient near to the conversion layer is less strongs than in the case of lower frequency. The evidence of a partial EPW-IBW conversion has been analysed in a large frequencies range. The EPW energy which is lost by Axial Convective Loss is partially absorbed at the plasma edge by the electrons.

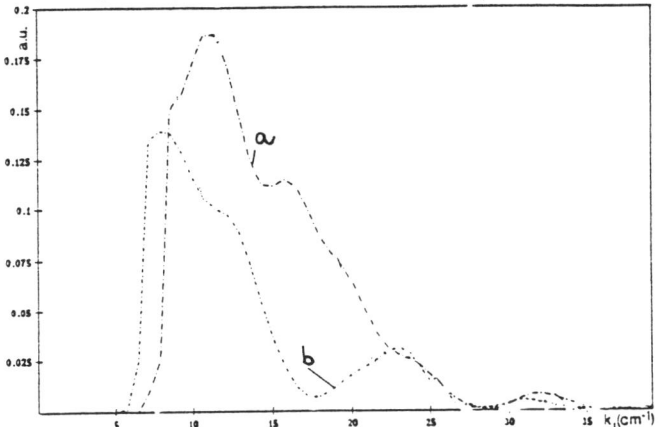

Fig.3- Energy spectrum of the e.s. wave obtained from the Fourier transformed of the signal at two frequencies: a) $f=1.96 fi$ and b) $f=1.90 fi$.

CONCLUSIONS

The analysis of the EPW-IBW conversion in linear conditions has been experimentally carried out by studying the wave energy in the plasma. The EPW - IBW conversion efficiency ratio has been evaluated through a Fourier analysis of the e.s. signal. The results confirm the presence of only a partial energy conversion, which increase with the frequency in the range of the second harmonic of the ion cyclotron frequency. Most part of the EPW energy is absorbed by the electrons at the plasma edge. Further studies will be carried out on the phenomena involved with the energy loss at the plasma edge.

AKWNOLEDGEMENT

We would like to thank Prof.G.Lampis, Dr.S.Bernabei for the very helpful discussion and Mr.G.Braga for the techincal assistance.

REFERENCES

1- M.Ono, L.Wong and G.A.Wurden, Phys.Fluids, 26, 1(1983).
2- M.Ono and L.Wong, Phys.Rev.Lett., 45, 13(1980).
3- R.I.Pinsker, C.C.Petty, M.J.Mayberry, M.Porkolab, W.W. Heibrink, Conf.Proc., Innsbruk, EPS(1992).
4- S.Alba, G.Carbone, M.Fontanesi, A.Galassi, C.Riccardi, E.Sindoni, Plasma Physics, 34 (1992).
5- D.Batani, M.Fontanesi, A.Galassi, C.Riccardi, E.Sindoni, Conf.Proc., Innsbruk, EPS (1992).
6- M.Ono, Phys.Fluids B, 5(2), 1993.

LOWER HYBRID RANGE
OF FREQUENCIES

LOWER-HYBRID CURRENT DRIVE EXPERIMENTS, SYNERGISM WITH THE FAST WAVE NEAR ION CYCLOTRON RESONANCE AND FUTURE PLANS ON JET

C. Gormezano
JET Joint Undertaking, Abingdon, Oxfordshire, OX14 3EA, UK

ABSTRACT

Acceleration of electrons up to energies in the MeV range has been observed in JET during combined application of Lower Hybrid and Fast Waves in fully non-inductively driven discharges. Without Fast Wave, the energy of the fast electrons remains below 200 keV. Up to 18% of the Fast Wave power has been estimated to be coupled to the fast electron population. This additional acceleration of electrons is defined as synergy between LH and Fast Wave. Specific experiments have indicated that the most likely responsible mechanism is the damping on fast electrons of an Ion Berstein Wave generated after mode conversion of the Fast Wave. The new JET RF systems, which are being prepared, (10 MW of Lower Hybrid and up to 20 MW of ICRF power at arbitrary phasing) will allow to assess if this synergism can be used to achieve the current drive efficiencies required in a reactor.

INTRODUCTION

Fast electrons whose velocities exceed significantly the range corresponding to the damping of Lower Hybrid waves have been observed in JET during combined application of Lower Hybrid and Ion Cyclotron Fast Waves, in conditions where the full plasma current was non-inductively driven, i.e. when the electric field was zero. As a result, large current drive efficiencies have been obtained. This synergism might lead to a significant improvement in the non-inductive current drive efficiencies in reactor type plasmas.

A brief review of data obtained during the last JET LHCD campaign will be presented in the first section. In section 2, results from specific experiments including an ICRF power scan, application of non-resonant ion cyclotron power and variation of the ICRF wave spectrum (monopole, dipole) will be presented. In section 3, the results will be discussed in the perspective of the several physics mechanisms which can be invoked to explain such synergistic effects. New hardware for both the LHCD and ICRH systems are being prepared and will be available for the next JET campaign. Their characteristics will be presented in section 4. The potential use of this synergism on JET and application to a reactor will also be discussed.

I. LHCD DATA AND SYNERGISTIC EFFECTS

The maximum power injected with the prototype JET LHCD system, which operates at 3.7 GHz, was 2.4 MW at $N_{//} = 1.8 \pm 0.2$ /1/ /2/. The plasma could be preheated by ICRH either in the dipole (peaked at $k_{//} = 7$ m^{-1}) or in the monopole (peaked at $k_{//} = 0$) phasing. Full replacement of the inductively driven plasma current has been achieved at 0.4 MA with 2 MW of LHCD power only and at 2 MA for 1.5 sec with the combined application of 2.3 MW of LHCD power and 3 MW of ICRH power /3/. In H-mode conditions, 1 MA full current drive was maintained for 4 sec with up to 40% bootstrap current /4/.

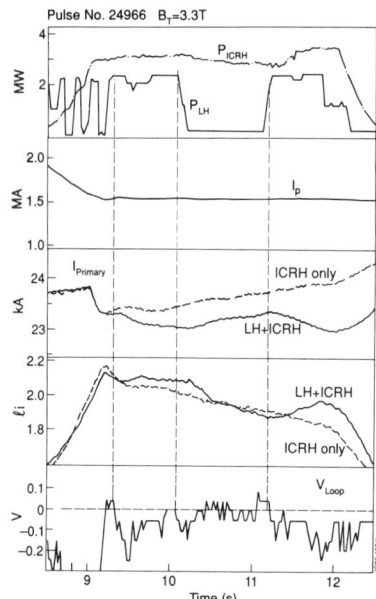

Fig. 1 Full current drive at $I_p = 1.5$ MA. Plasma inductance and primary current for an ICRF heated only discharge are indicated for comparison.

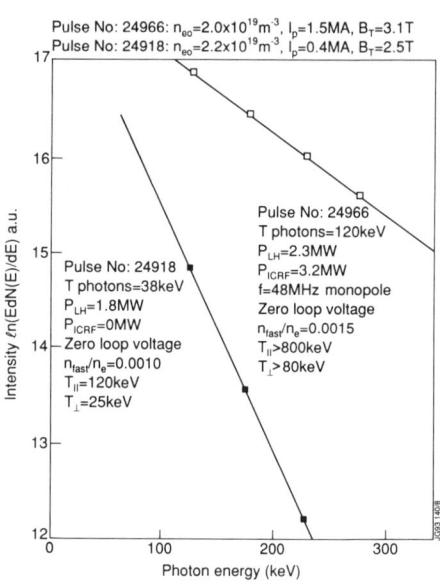

Fig. 2 Photon temperature at zero loop voltage with and without ICRF. The fast electron energy distribution (n_{fast}, $T_{\parallel}$, $T_{\perp}$) are estimated using the FEB and ECE data.

In the pulse shown in Fig. 1, the modulated LHCD power is applied after ramping down the plasma current from 3 MA to 1.5 MA in order to achieve a target plasma with higher internal inductance and more peaked electron temperature profile. During LHCD application, the surface loop voltage becomes negative. The resistive loop voltage, which takes into account correction from the respective time derivative of plasma inductance and plasma current, is zero which is confirmed by the fact that the primary current, which in this case is the only source of poloidal flux, remains constant, or even slightly decreases. The non-inductive current profile as measured from the FEB camera /5/ is similar to the ohmic current profile. Further, no rapid transients in loop voltage are observed.

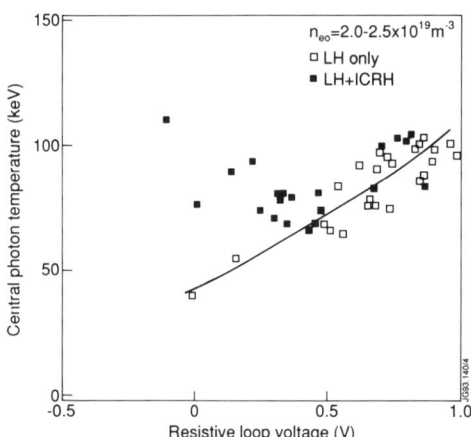

Fig. 3 Central photon temperature versus resistive loop voltage. Time derivative of plasma inductance and plasma current are taken into account.

The FEB camera which is a 19 channel X-ray camera is the main tool to estimate the radial electron distribution. As shown in Fig. 2, photon energy distribution are given for two shots with full current drive with and without ICRF. The electron energy distribution are estimated using also the non-thermal ECE emission /6/. This acceleration of electrons, in the presence of ICRF, beyond the domain of velocities corresponding to the LHCD parallel wave spectrum (100 keV), in the absence of electric field is the "signature" of the synergism which is described in this paper. As shown in Fig. 3, high photon temperatures can be achieved when the loop voltage is not zero, as anticipated from electric field acceleration.

This synergism is further confirmed by calculations carried out with an LHCD numerical simulation code /7/. The physics of Lower Hybrid waves is modelled via a ray-trajectory method, with the dispersion relation including both the electromagnetic part and thermal corrections. The electron distribution function is calculated by the two-dimensional relativistic Fokker-Planck equation, taking into account effects due to spatial fast electron diffusion and the residual toroidal electric field. In this model, the diffusion of fast electrons is due to the stochasticity of the magnetic field. Typical values of the diffusion coefficient are 0.5 m^2/s for thermal electrons and 1.5 m^2/s for 100 keV electrons. The output of the code is then used to simulate the hard X-ray spectrum emitted by the suprathermal population and compared with the output of the FEB camera. In Fig. 4 the results are shown for 4 different pulses: LHCD only at zero loop voltage, LHCD only at moderate power with residual electric field, LHCD only at high electron temperature (8 keV) and residual electric field, and finally (Fig. 4b) a combined LHCD-ICRF pulse at zero loop voltage. All cases with LHCD only are very well simulated, with only a small discrepancy at the plasma centre in the high electron temperature case; but this might be explained by a slightly higher electric field at the centre than assumed in the computation. The large discrepancy between computation and experimental data in the LHCD-ICRF combined pulse can be attributed to the synergism.

This synergism can also be substantiated from the estimation of the power coupled to the bulk electrons from electron power balance during LH power modulation with and without ICRF /8/. About 80% of the LH power is estimated to be coupled to the bulk electrons in pulses with LHCD alone. The power coupled to the electrons can slightly exceed the launched LH power in ICRF heated discharges in some specific conditions, as discussed later. Although error bars are large (±20%), there is a systematic difference between coupled power when synergism is present.

When synergism is observed, current drive efficiencies can be high. When the plasma current is fully non-inductively driven, the non-inductive current drive efficiency has been calculated in two ways /9/:

- an engineering efficiency $\gamma_{eng} = \dfrac{\bar{n}_e R I_p}{P_{LHCD} + P_{ICRF}}$

- a physics efficiency $\gamma_{phy} = \dfrac{\bar{n}_e R I_{RF}}{P_{LHCD} + P_{syn}}$ where $I_{RF} = I_p - I_{BS}$ (I_{BS} being the estimated bootstrap current) and P_{syn} being the part of the ICRF power which is transferred by synergy to the fast electrons created by LHCD. P_{syn} is determined from electron power balance as discussed above.

As discussed in /4/, these efficiencies present a large scatter for full current drive when plotted against $<T_e>/(5 + Z_{eff})$, as in JT-60. This scattering disappears when data are plotted against $n_e <T_e>/(5 + Z_{eff})$ as in Fig. 5 and 6. No apparent saturation is observed in the domain of operation, i.e. for central densities up to 3.6 10^{19} m^{-3}. It is not clear if these features: (i) absence of saturation (ii) linear dependence with plasma pressure, are also a signature of the LHCD - ICRF synergism.

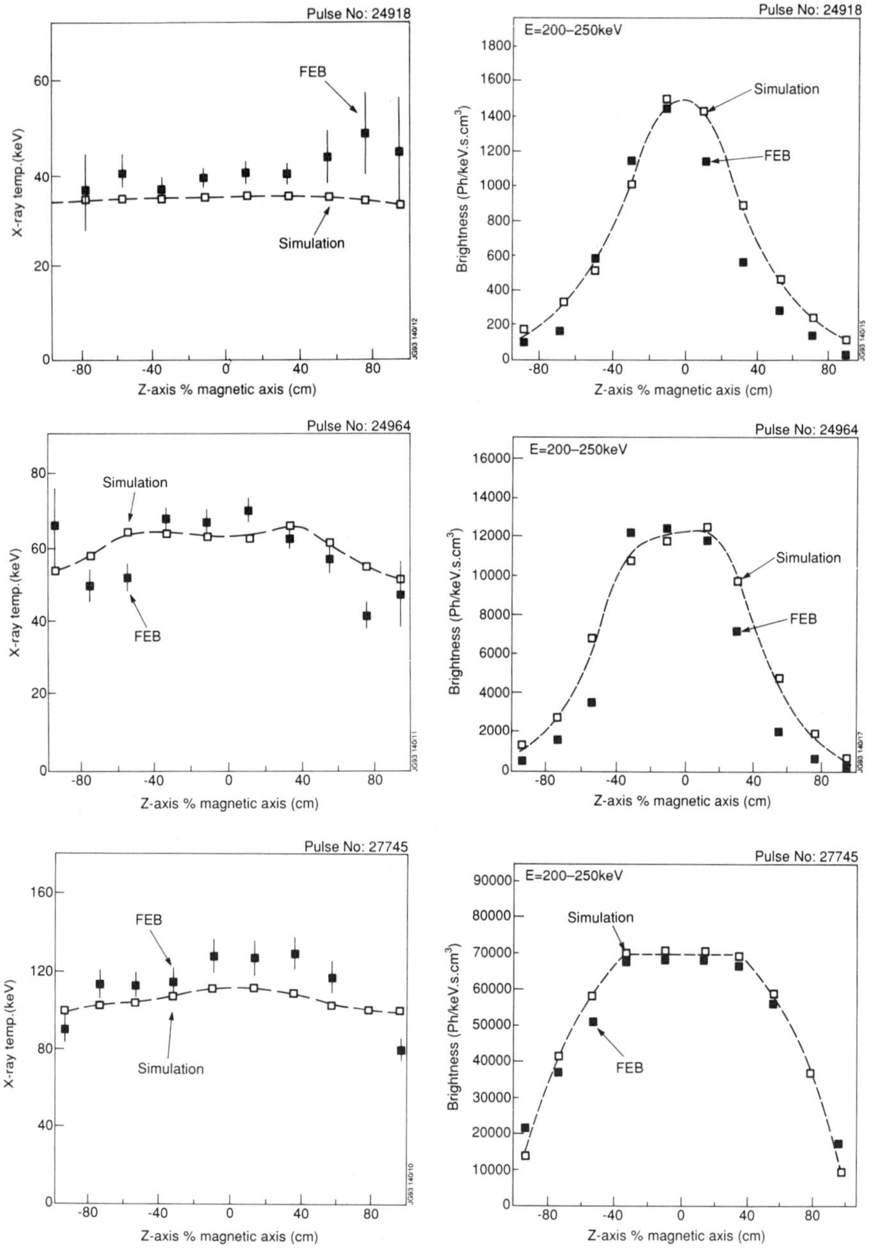

Fig. 4a *Comparison between FEB chord integrated signals and ray tracing simulation code*

	I_p(MA)	B_T(T)	$n_{eo}(10^{19}m^{-3})$	T_{eo}(keV)	P_{LH}(MW)	P_{ICRF}(MW)	V_{Res}(V)
#24918	0.4	2.5	2.2	1	1.8	0	0
#24964	1.5	3.3	2.2	3	1.1	0	0.1
#27745	3.1	3.3	1.3	8	1.3	0	0.3

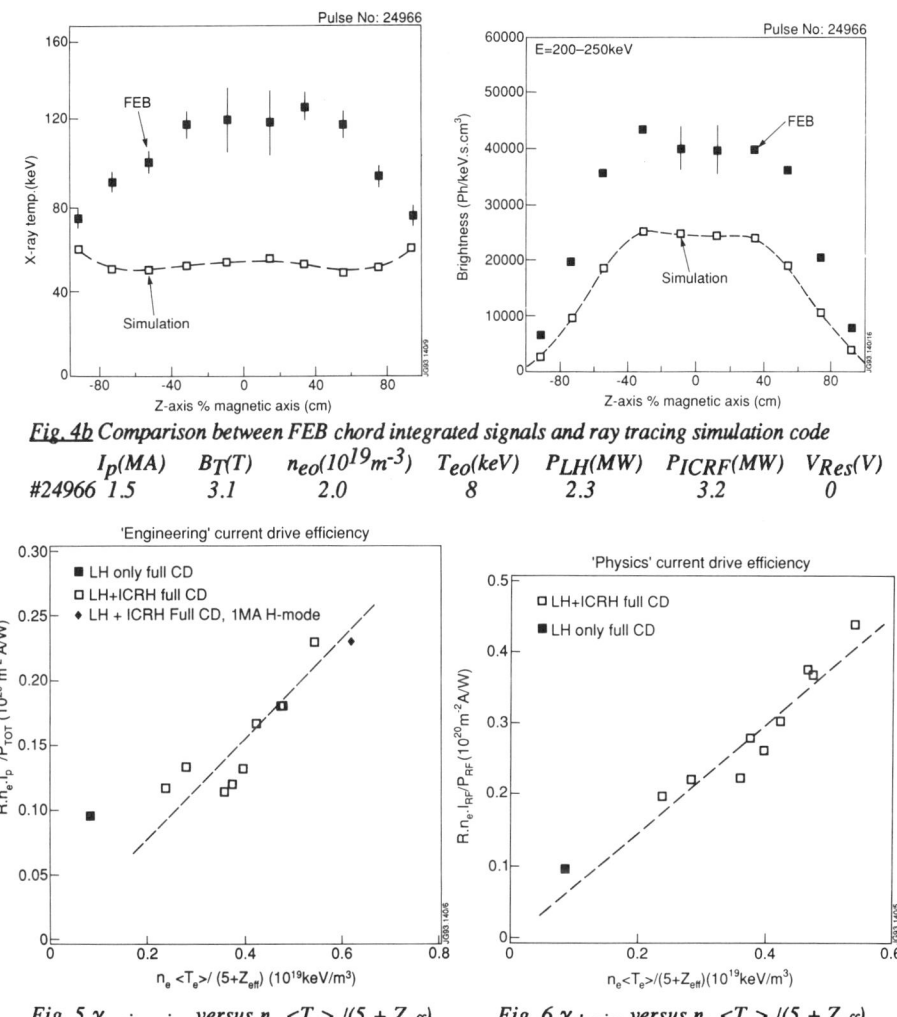

Fig. 4b Comparison between FEB chord integrated signals and ray tracing simulation code

	I_p(MA)	B_T(T)	$n_{eo}(10^{19}m^{-3})$	T_{eo}(keV)	P_{LH}(MW)	P_{ICRF}(MW)	V_{Res}(V)
#24966	1.5	3.1	2.0	8	2.3	3.2	0

Fig. 5 $\gamma_{engineering}$ versus $n_e \langle T_e \rangle / (5 + Z_{eff})$ *Fig. 6* $\gamma_{physics}$ versus $n_e \langle T_e \rangle / (5 + Z_{eff})$

2. SPECIFIC EXPERIMENTS ON LHCD - ICRF SYNERGY

The physics mechanisms which can be invoked in order to explain this synergism are the following:

<u>TTMP</u>: direct absorption of the ICRF Fast Wave on the fast electron population. There is no need for a directional wave since the damping can take place on the LHCD induced fast electrons which are travelling in one toroidal direction. Theoretical simulations indicate that a substantial fast electron population, in excess of 2%, is required to obtain a proper damping, as shown by Cox /10/ and Bhatnagar /11/.

<u>Mode conversion to an Ion Bernstein wave at the ion-ion hybrid resonance layer</u>. This wave (IBW) has a low $N_{//}$ at birth and can then be damped on a pre-existing fast electron population before being upshifted along its ray trajectory, as indicated in

Fig. 7. This scheme has been discussed by Jacquinot in /9/ indicating that up to 25% of the ICRF power can be converted in an IBW in optimum conditions: (i) low $N_{//}$ values ($N_{//} < 2$) for which damping at the resonance layer is not too high, (ii) low minority concentration ($\leq 10\%$) so that the distance between cut-off and ion-ion hybrid layer remains small compared with the wavelength to allow tunnelling through the cut-off region, (iii) low fast ion tail energy to avoid blurring the ion-ion hybrid layer by the resonance layer. A detailed numerical analysis of the IBW damping on a pre-existing fast electron population is yet to be done.

<u>Injection of wave helicity</u>, as suggested by Chan /12/. A fast ICRF wave can dissipate wave helicity, as a result of which a non-resonant, time average force, can be set up which pushes all particles which respond to the wave. This effect is expected to be weak since the parallel electric field of the ICRF wave is small.

<u>Channelling fast ion free energy to Lower Hybrid waves</u>. For a reactor, Fisch and Rax /13/ have proposed to make use of the energy stored in the alpha particle population to improve current drive efficiency. The proposed mechanism utilises the poloidal component of the LH wave to channel free particle energy to the LH wave by using radial transport of alphas induced by the LH wave itself. As a consequence, the LH wave will be amplified. In JET, alphas can be replaced by the ICRF produced fast ion tails.

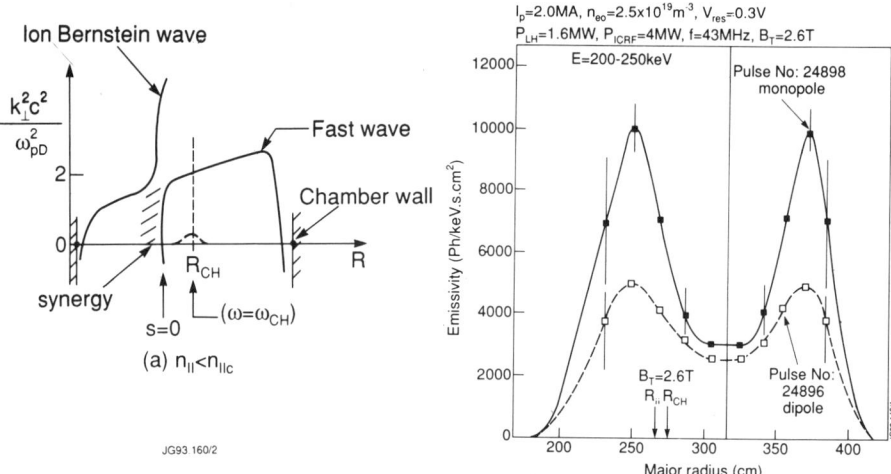

Fig. 7 Mode conversion of a Fast Wave into Ion Bernstein Wave for $n_{//} < 2$.

Fig. 8 Comparison monopole dipole. R_{CH} and R_{ii} correspond respectively to the cyclotron resonance layer and to the ion-ion hybrid layer.

Few specific experimental scans aiming at demonstrating the supporting physics behind the observed LHCD Fast Wave synergism were performed during the last JET campaign. Many of the data presented in this section are taken from scattered experiments.

2a. Comparison monopole dipole: Large changes in the low $k_{//}$ part of the wave spectrum are anticipated when the ICRF antenna configuration is changed from monopole to dipole. Corresponding radial X-ray emission and photon temperature profiles are shown in Fig. 8 for two LHCD-ICRF pulses. The fast electron profile corresponding to a dipole configuration is slightly hollow, as anticipated from simple LH wave propagation considerations in this range of density. The photon emission increases significantly and becomes more hollow when a monopole configuration is used. The photon temperature increases from 60 to 70 keV. This increase of about 15% in photon temperature, which corresponds to an increase by a factor of 2 in parallel electron temperature, has been observed in several monopole dipole comparisons.

It is important to note that the peak in the fast electron distribution is close to the location of the ion-ion hybrid layer which is computed using the best estimate for the minority concentration (5%). The corresponding increase in the photon temperature is also located close to the ion-ion hybrid layer.

The localisation of the synergy is also observed by comparing pulses with and without ICRF. The increase in photon temperature is well localised close to the ion-ion hybrid layer. The resolution of the FEB camera does not allow to give a good estimate of the width of the synergy zone, which is of the order of 20 to 30 cm.

2b. Dependence on ICRF power: The LHCD Fast Wave synergism is found to depend strongly upon the ICRF power. The power coupled to electrons during combined LHCD-ICRH pulses, as measured from power balance during full LH modulation, is compared with the 80% of the LH power which is coupled to the electrons in LHCD only pulses. This is supported by a series of shots which show, as indicated in Fig. 9, that the maximum "synergistic" power reaches 18% of the incident power, at about 4 MW of ICRH power. Note that during these experiments the minority concentration was roughly constant, below 10%.

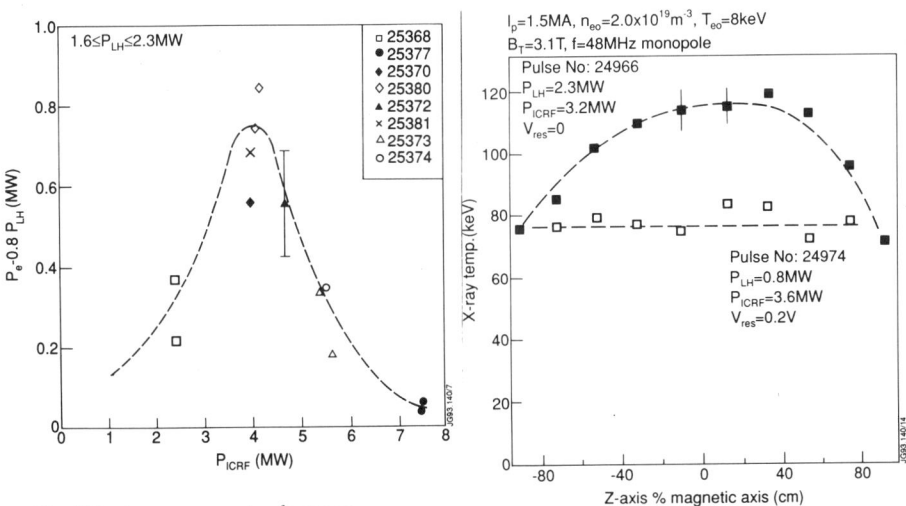

Fig. 9 Fast wave power to electrons versus P_{ICRF} ($1.5 < I_p < 2$ MA, $B_T = 3.45$ T, $n_{eo} \sim 2\ 10^{19}$ m^{-3})

Fig. 10 Influence of LH power on the "photon" temperature.

2c. Dependence on LH power: As shown in Fig. 10, the synergism almost disappears when the LH power is not high enough (≤ 1.5 MW). Acceleration of electrons via synergistic effects, as in pulse 24966 where the loop voltage is zero, is more effective than acceleration by the remaining electric field (V_{res} = 0.2 V), as in pulse 24974, where the plasma current is not fully driven. It is likely that a minimum number of fast electrons, therefore a minimum LH power, is necessary to start the synergy effect in a given plasma.

2d. Application of non resonant ICRF power: In an experiment aimed at studying the effect of non-resonant ICRF on fast electrons, 1 MW of ICRF power at 25 MHz was superimposed on 3 MW of ICRF power at 48 MHz in (H) ^{4}He plasma. 48 MHz corresponds to the resonance of hydrogen ions at the plasma centre. For the corresponding parameter of the discharge (B_T = 3.3 T), 25 MHz corresponds to the ^{4}He resonance at 3.0m. Then, without ^{3}He injection, there is no minority cyclotron resonant absorption at 25 MHz. The time dependence of various signals corresponding to this case is shown in Fig. 11. The level of the X-ray emission depends strongly upon the level of LH power but is insensitive to the application of non-resonant ICRF. Although not optimised, this experiment shows no interaction between non-resonant ICRF and fast electrons.

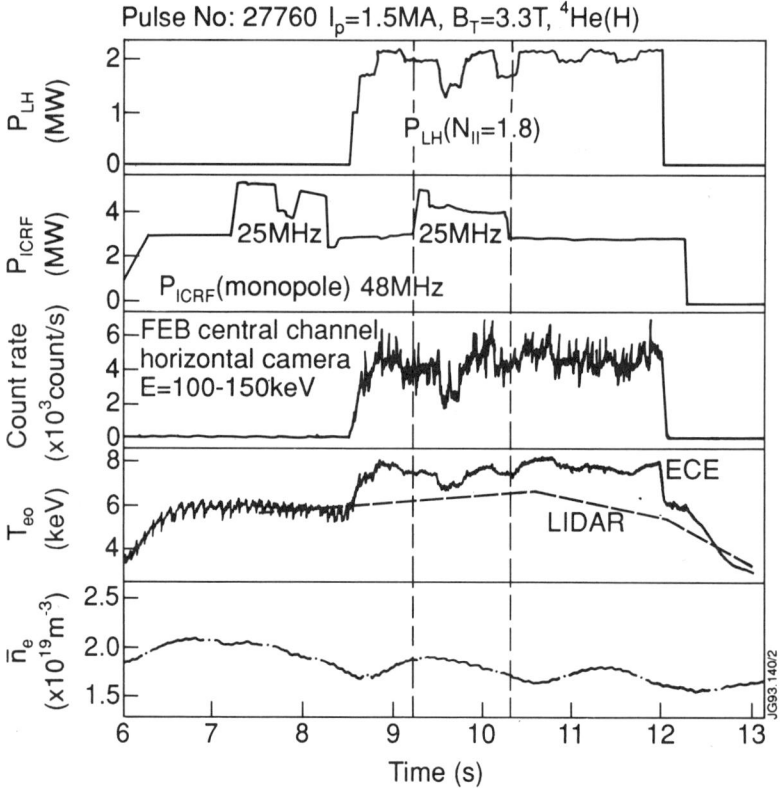

Fig. 11 Influence of ICRF frequency. H_2 minority in ^{4}He plasma - 48 MHz resonance on H_2 located at R = 3.1m and 25 MHz resonance on ^{4}He located at R = 3.0m. There was no ^{3}He in the discharge.

3. DISCUSSION

A decisive proof of principle experiment supporting a physics scheme behind the LHCD-ICRF synergism as observed in JET, is missing. Nevertheless, indications can be inferred by comparing the main physics characteristics as listed in Table 1 with the data presented in section 2.

	Table I
TTMP	• localised at max $\beta_{fast}/r^2 \rightarrow$ plasma centre • strong competition from damping on minority species • theory predicts an effective damping if the fast electron population is very large (> 2%) • damping very sensitive to $k_{//}$ (low $k_{//}$ needed) • proportional to ICRF power
Mode conversion to Ion Bernstein wave	• localised near the ion-ion hybrid resonance layer • depends upon minority concentration • depends upon temperature of fast ions $\rightarrow$ non-monotonic dependence on ICRF power • very sensitive to $k_{//}$ (mode conversion if $N_{//} < 2$)
Injection of wave helicity	• localised near the cyclotron resonance layer • proportional to ICRF power • non-resonant damping but accelerates fast electrons • small effect predicted by theory

Channelling fast ion free energy to Lower Hybrid waves	• poloidal damping of LH waves needs to be close to plasma centre
	  $\rightarrow$ density dependence
	• dependence on fast ion tail energy and radial gradients
	• amplification of LH wave
	  $\rightarrow$ no acceleration of fast electrons
	• dependence on RF power?
	• theory remains to be done

Effective TTMP damping is unlikely to be the cause of synergism. Detailed analysis of the non-thermal emission during LHCD indicates the fast electron population amounts to about 0.2% of the bulk electron population /6/ corresponding to a central β_{fast} of 0.016%. This excludes TTMP since theory predicts that significant damping occurs only if the fast electron beta exceeds 0.2%. Injection of wave helicity is unlikely because expected effects are much lower than observation. Tapping the free energy from the fast ion population is supposed to boost the LH wave power which is inconsistent with the observed acceleration of the LHCD produced fast electrons. But the supporting theory of this mechanism is not yet developed and conclusions cannot really be drawn.

In the available set of experimental data, there are no indications that mode conversion of the Fast Wave does not take place. This cautious statement is motivated by the fact that critical experiments, such as a variation of the ion-ion hybrid resonance by changing the frequency or a variation of minority concentration, especially at high ICRF power, are not available. Data such as the acceleration of LH fast electrons, which takes place at the ion-ion hybrid layer if the fast electron population is high enough (large LH power or peaked profiles) and if the Döppler broadening of the minority resonance is not too large, (which occurs at high ICRF power with a low minority concentration) support the hypothesis of mode conversion to IBW. In addition, the maximum amount of power coupled by the Fast Wave to the electrons is within the range predicted by theoretical considerations. The observed localisation of synergistic effects is a strong experimental fact supporting mode conversion to IBW as the mechanism creating the synergy.

4. PROSPECTS

A route to increase non-inductive current drive efficiency is to maintain the current with very high energy electrons, to avoid collisions even at the density required in a reactor. Fast electrons induced by LHCD have their energy limited to about 100 keV due to accessibility conditions. Acceleration of these electrons by a Fast Wave is a very attractive scheme.

In JET, it might be possible to use TTMP damping on fast electrons as a tool for increasing current drive efficiency if the density of fast electrons can be raised to values so that β_{fast} exceeds 0.2%, compared to the 0.016% already achieved, by operating with higher LHCD power at higher densities. On a reactor, TTMP damping on fast electrons remains a possibility because it is predicted /11/ that: (i) damping by TTMP can reach 100% even for β_{fast} less than 0.2% if no other competing damping mechanisms are present, (ii) the damping in ITER type plasma will take place on a layer greater than the one in JET.

Mode conversion to IBW appears to be a likely mechanism to explain the LHCD-ICRF synergism observed in JET. In particular, current profile control could benefit from synergism via mode conversion, both from enhanced current drive and from enhanced localisation of the maximum non-inductive current drive at the ion-ion hybrid resonance.

One of the aims in the next JET experimental campaign is to assess if this synergism can be used to achieve the current drive efficiencies required in an ITER type reactor, in plasma conditions which are reactor relevant. In addition to experiments aimed at a better identification of the physics mechanisms of the synergy, it will be attempted to extend the operational range for which this synergy is effective to higher densities ($n_{eo} > 3 \; 10^{19}$ m^{-3}) and to identify ways of coupling a larger amount of ICRF power to the fast electron population.

In JET, both RF systems are being upgraded during the present shutdown /14/. The LHCD system will have 3 times more power capability with a narrower wave spectrum: 32 waveguides in an horizontal row instead of 16. Eight new ICRF antennae will be grouped in modules of 4 adjacent current straps allowing a better directivity and a narrower $k_{//}$ spectrum to be achieved. External compensation of ICRF circulating power will be added in order to optimise the maximum available power at arbitrary phasing in poorer coupling conditions for frequencies between 42-48 MHz. It will be possible to superimpose both wave spectra, as shown in Fig. 12,

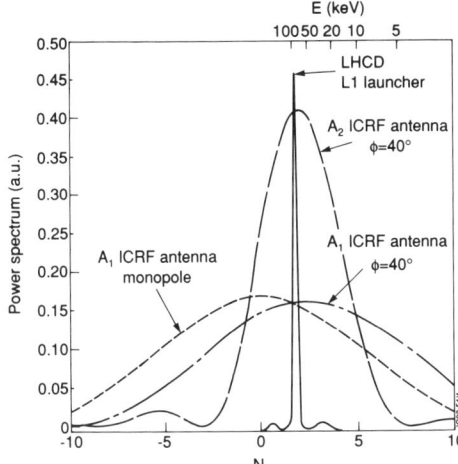

Fig. 12 Wave spectrum of the L1 launcher and of the A_1 and A_2 antennae respectively.

by setting a 40° phasing between straps, indicating that a gain by a factor of 2 in the available "synergistic" power appears possible.

Note that optimum synergism has been obtained with $P_{ICRF} \approx 1.5 - 2\, P_{LHCD}$, but synergistic effects disappeared for ICRF power above 4 MW in the last JET experiments. If mode conversion is the mechanism responsible for synergy, some scenarios will have to be developed in order to use the respective maximum power of 20 MW and 10 MW for ICRF and LHCD systems. With Hydrogen minority, the concentration shall remain low enough to allow tunnelling of the Fast Wave through the cut-off zone. Therefore, some spreading of the power deposition by operating the 4 ICRF modules at slightly different frequencies could be used together with operating at higher density to reduce the energy of the minority ions in order to avoid blurring the ion-ion hybrid layer. Another possibility is to use ^3He as a minority since the damping of the Fast Wave at the resonance is lower. Also, higher minority concentration can be used.

It has been shown that the non-inductive current drive efficiency which is required to maintain an ignited steady-state reactor of the ITER EDA type, with a low aspect ratio and high plasma current, shall exceed $\gamma = 1.2\, 10^{20}$ m^{-2} A/W for densities in the range required for divertor operation /15/ /9/. The anticipated current drive efficiencies do not exceed about $\gamma = 0.5$ m^{-2} A/W for LHCD and for any other conventional method. Therefore, full current drive in a reactor can only be expected for high bootstrap current devices with improved confinement, i.e. low plasma current, unless much higher current drive efficiencies can be achieved.

But, if synergy can be used in a reactor to increase current drive efficiency, it is the "engineering" current drive efficiency which matters since the main heating power is provided by the alphas particles. Therefore, in order to have a significantly increased current drive efficiency, it is necessary that:

the electrons are accelerated to very large energies to approach the maximum theoretical efficiency which is around $2\, 10^{20}$ m^{-2} A/W /16/

most of the ICRF power is coupled to the fast electrons.

As suggested by Jacquinot, nearly 100% mode conversion /9/ can be reached by launching the Fast Wave from the high field side using the D/T ion-ion hybrid layer in a reactor, without the constraint of optimum concentration.

LHCD-ICRF synergy via mode conversion is clearly a route towards effective current drive in a reactor, provided that new experiments confirm the physics ideas developed in this paper.

ACKNOWLEDGEMENTS

Discussions and contributions from Y Baranov, M Brusati, A Ekedahl, L Eriksson, P Froissard, J Jacquinot, C Lashmore-Davies (AEA), F Rimini, D Start are gratefully acknowledged, especially from A Ekedahl for gathering many of the data presented here.

REFERENCES

/1/ M Pain, H Brinkschulte, M Brusati, J A Dobbing, A Ekedahl et al, 13th Symposium on Fusion Engineering, Knoxville (1989)

/2/ A Ekedahl, H brinkschulte, M Brusati, J A Dobbing, C Gormezano et al, in Proc. Europhysics topical Conf. on RF Heating and Current Drive of Fusion devices (Brussels, July 1992) 221

/3/ F Rimini, M Brusati, C D Challis, P Froissard, C Gormezano et al, in Proc. Europhysics topical Conf. on RF Heating and Current Drive of Fusion devices (Brussels, July 1992) 229

/4/ The JET Team, Non Inductive Current Drive in JET, 14th Int. Conf. on Plasma Phys. and Cont. Nucl. Fusion, Wurzburg, Germany, 1992 (CN-56/E-1-1)

/5/ P Froissard, X-ray emission from suprathermal electrons created by LHCD in JET, Thesis (1992)

/6/ M Brusati, A Airoldi, D Bartlett, P Froissard, C Gormezano et al, in Proc. Europhysics topical Conf. on RF Heating and Current Drive of Fusion devices (Brussels, July 1992) 225

/7/ Y Baranov, M Brusati, A Ekedahl, P Froissard, C Gormezano et al, to be presented at the 20th Eur. Conf. Cont. Fusion and Plasma Phy., Lisbon (1993)

/8/ C Gormezano, M Brusati, A Ekedahl, P Froissard, J Jacquinot, F Rimini in Proc. IAEA Tech. Comm. Meeting on Fast Wave Current Drive in Reactor Scale Tokamaks, Arles 1991, 224

/9/ J Jacquinot, V P Bhatnagar, C Gormezano and the JET Team, in Proc. Europhysics topical Conf. on RF Heating and Current Drive of Fusion devices (Brussels, July 1992) to be published in Plasma Phys. and Cont. Fusion

/10/ M Cox, M R O'Brien, C D Warrick, IAEA Tech. Comm. Meeting on FWCD in Reactor Scale Tokamaks, Arles, 1991, (p 122)

/11/ V P Bhatnagar, J Jacquinot, C Gormezano, D F H Start, IAEA Tech. Comm. Meeting on FWCD in Reactor Scale Tokamaks, Arles, 1991, p 110

/12/ V S Chan, R L Miller, T Ohawa, Phys. Fluids B 2, 944 (1990)

/13/ N Fisch and J M Rax, Phys. Rev. Lett. 69, 612

/14/ A S Kaye, in Plasma Phys. and Cont. Fusion 35 (1993) A71

/15/ P H Rebut, D Boucher, C Gormezano, B E Keen, M L Watkins, Plasma Phys. and Cont. Fusion, 35 (1993) A1

/16/ N Fisch, Review of Modern Physics, Vol 59 (1987) 175

LOWER-HYBRID CURRENT DRIVE AND ICRF HEATING EXPERIMENTS ON JT-60U

T. Fujii, K. Ushigusa, M. Saigusa, H. Kimura, T. Imai, Y. Ikeda, O. Naito, M. Nemoto,
S. Moriyama, M. Seki, T. Kondoh, K. Hamamatsu, M. Sato, R. Yoshino, Y. Kamada
Japan Atomic Energy Research Institute, Naka, Ibaraki 311-01, Japan

D. J. Campbell
JET joint Undertaking, Abingdon, Oxfordshire OX14 3EA, U. K.

ABSTRACT

A plasma current of 2 MA is driven by lower hybrid current drive (LHCD) and the maximum CD efficiency of $\eta_{CD}= 2.5 \times 10^{19}$ m^{-2}A/W is achieved, which agrees with the previously reported scaling : $\eta_{CD} \sim 12<T_e>/(5+Z_{eff})$. The direct loss of launched wave power through energetic electrons is identified by measuring the x-ray signal from the divertor plates. The direct loss power is fairly low (typically < 5 %) and depends on the slowing down time of energetic electrons accelerated by launched LH waves. This indicates that the slowing down process is dominant compared with loss mechanisms of fast electrons. The radial diffusion coefficient of fast electrons is less than 1 m^2/s from analyses on change in HX signal on pellet injection and direct lost power by fast electrons. Good coupling characteristics are confirmed for a new large (48 × 4) multijunction launcher. Significantly large loading resistances are obtained with a 2 × 2 phased ICRF antenna, especially ~ 5 ohms with in-phase current mode even at a big separatrix-wall gap of 29 cm. Efficient sawtooth stabilization by 2 ω_{cH} heating is observed in relatively high-density and low-q region, $<n_e>/P_{tot} \approx 0.8 \times 10^{19}$ m^{-3}/MW at $q_{eff} = 4$ and $I_p = 2.4$ MA. The maximum sawtooth period is 0.9 sec for ohmic plasmas and 1.7 sec for NBI-heated plasmas at power levels of $P_{IC} \approx 3$ MW. Combined 2 ω_{CH} ICRF (2.5 MW) and NBI (D beam, ~ 6 MW) heating is carried out for pellet-fueled plasmas. Preliminary results show that the pellet enhanced phase is maintained by 1.4 sec up to sawtooth crash and the energy confinement normalized by heating power is improved by 20 % compared with pellet and NBI heating alone.

INTRODUCTION

Fast electron behaviors produced by lower hybrid current drive (LHCD) is important for understanding LHCD physics. Relatively low power LHCD experiments in JT-60U are performed mainly to estimate the radial diffusion coefficient of fast electrons.
Higher harmonic ICRF heating is investigated in JT-60U, especially second harmonic heating which is one of very attractive heating methods in a reactor grade tokamak . The JT-60U ICRF antenna consists of 2 x 2 loop current straps and an open Faraday shield same as the previous JT-60 antenna, but it has a poloidal solid septum to suppress the mutual coupling between the straps and then to make easy matching in toroidal phasing of strap currents[1]. So far the maximum coupled power is 3.6 MW with two antennas. Sawtooth stabilization is important for optimization in high performance discharges and then is investigated intensively by second harmonic

minority ion ICRF heating in JT-60U. Extension of the operation range toward high density and low-q regime is being pursued. Moreover Combined ICRF and NBI heating of pellet injected plasmas has been performed aiming at sawtooth stabilization at high density and high l_i regime as well as high plasma performance.

LOWER HYBRID CURRENT DRIVE EFFICIENCY

Lower hybrid current drive (LHCD) experiments in JT-60U (R_p~3.45 m, a_p~1 m, κ~1.4) are performed by using a 24x4 multijunction launcher with $P_{LH} < 2.6$ MW. The plasma current of 2 MA is driven by LHCD alone and the maximum current drive efficiency of 2.5×10^{19} m^{-2}A/W is achieved[2]. Figure 1 shows the observed CD efficiency against the empirical scaling for current drive efficiency derived in 1989[3] from JT-60 results; $12<T_e(keV)>/(5+Z_{eff})$ where $<T_e>$ is the volume averaged electron temperature. Open and solid circles show results in JT-60 and JT-60U, respectively. The observed CD efficiency in JT-60U is consistent with the empirical scaling. Recent results from ASDEX[4] and JET[5] are also not inconsistent with this scaling. The power dissipation at the spectrum gap is one of possible explanations for the temperature dependence of CD efficiency[6].

A high power LHCD experiments will be started from April 1993 by adding a new large multijunction launcher. A good coupling characteristics in the new launcher is observed in preliminary experiment[7]. Maximum LH power of ~ 10 MW will be expected in these experiments. Study of current drive efficiency at higher electron temperature regime and the demonstration of large driven current will be performed in these experiments.

BEHAVIORS OF FAST ELECTRONS IN LHCD PLASMAS

Confinement and transport of fast electrons in LHCD plasmas are important for understanding the physics of LHCD. In order to estimate the radial transport coefficient of fast electrons in LHCD plasmas, several experiments and theoretical analysis were performed. All these results suggest that the slowing down is dominant compared with the radial diffusion of fast electrons in JT-60U.

A. Directly lost power through fast electrons

In divertor configuration, the lost fast electrons during LHCD discharges may hit the divertor plates and radiate the thick target bremsstrahlung emission from the plates. The thick target x-ray flux is directly proportional to the fast electron power loss, with little dependence on the electron velocity distribution, when the measured photon energy is much less than the fast electron energy[8]. By measuring the soft x-ray flux from the strike point of the separatrix, the direct loss power through fast electrons was identified in JT-60U LHCD discharges[9]. In Fig. 2, the direct loss power is divided by the injection power and plotted against the slowing down time of fast electrons with a velocity corresponding to the launched peak $N_{\parallel}$. The measured direct loss is nearly proportional to the slowing down time of fast electrons. Observed direct loss power is fairly low and suggesting that the slowing down is dominant compared with the radially loss process.

In order to estimate radial diffusion coefficient of fast electrons from the measurement of direct loss power, a theoretical model which solves one dimensional quasi-linear Fokker-Planck equation with radial diffusion of fast electrons is examined;

$$\frac{\partial f}{\partial t} = v_{eo}\frac{\partial}{\partial u}\left[\widetilde{D}_c\frac{\partial f}{\partial u}\right] + v_{eo}\frac{\partial}{\partial u}\left[\widetilde{D}_w\frac{\partial f}{\partial u}\right] + \frac{1}{\rho}\frac{\partial}{\partial \rho}\left[\rho\widetilde{D}_s|u|\frac{\partial f}{\partial \rho}\right] + S \quad (1)$$

where $\widetilde{D}_w = D_w/(v_{eo}v_e^2)$, $\widetilde{D}_s = D_s/(v_{eo}a^2)$, $\widetilde{D}_c = D_c/(v_{eo}v_e^2)$, v_{eo} is the electron collision frequency at the plasma center, v_e the electron thermal velocity at the plasma center, S the source term to keep the density constant, a the minor radius and $u = v_\parallel/v_e$. D_w, D_c and D_s/u are the quasi-linear diffusion coefficient, the linearlized collision operator and the radial diffusion coefficient of fast electrons, respectively. Eq.(1) is solved numerically for given $T_e(r)$, $n_e(r)$, $D_s(r,v_\parallel)$, $D_w(r,v_\parallel)$ and the lost power through the radial diffusion is calculated. The numerical calculation is performed with typical JT-60U parameters; $T_{eo} = 5$ keV, $n_{eo} = 2\times10^{19} m^{-3}$ and $N_\parallel^{peak} = 1.44$. An assumption of $N_\parallel$ up-shift up to $v_{min} = 3v_e$ is employed. Quasi-linear diffusion coefficient D_w is adjusted so that the driven current and the absorption power have a similar value as the experiments. The electron temperature profile is adjusted so that the absorption power has a similar profile as the Abel inverted hard x-ray signal. Figure 3 shows the numerically calculated direct loss rate against $\tau_{SD}(v_p)/(a^2/D_s(v_p))$ where $\tau_{SD}(v_p)$ is the slowing down time of electrons with the velocity $v_p = c/N_\parallel^{peak}$ and $D_s(v_p)$ is the radial diffusion coefficient of fast electrons at v_p. Calculated direct loss power has a similar dependence on the slowing down time of fast electrons with the experimental results shown in Fig. 2. Since $P_{Loss}/P_{LH} < 0.1$ in the experiments, Fig. 3 suggests that $\tau_{SD}/(a^2/D_s) < 0.05$. This corresponds to $D_s < 0.5$ m^2/s at $N_\parallel^{peak} = 1.44$ where maximum CD efficiency is observed in the experiments.

B. Response on hard x-ray signal after the pellet injection

Pellet injection is one of good method to estimate radial diffusion of fast electrons in LHCD plasmas as long as the discharge is not so strongly disturbed by the pellet injection. Since significant increase in the density at a finite radius due to an abrasion of pellet decreases the fast electron population locally, a time evolution of fast electron population after pellet injection near the pellet abrasion radius may be strongly effected by the radial diffusion of fast electrons inside the abrasion radius. Time evolution of hard x-ray signal are measured after a small pellet injection into the LHCD discharge in JT-60U. To simulate this experiment, diffusion equations for both bulk and non-thermal electrons are solved. It is solved for bulk electrons;

$$\frac{\partial n_b}{\partial t} = -\nabla\Gamma + S \quad (2)$$

with $G = -D_b\partial n_b/\partial r$ and the source S is calculated from the pellet abrasion model;

$$\frac{dL_p}{dr} = -\alpha\frac{n_b^{1/3}T_e^{5/3}}{L_p v_p} - \beta\frac{n_h E_h^{1.549}}{L_b v_p} \quad (3)$$

where L_p is the pellet radius, v_p the pellet velocity, n_h the fast electron density, E_h the energy of fast electron, and α the constant and β adjusted from the measured pellet abrasion depth. The radial diffusion coefficient of bulk electrons D_b is chosen to reproduce the measured time evolution of the line average density after pellet injection. Fast electron population is estimated from

$$\frac{\partial n_h}{\partial t} = S_h - \frac{n_h}{\tau_{SD}} + \frac{1}{r}\frac{\partial}{\partial r}(rD_h\frac{\partial n_h}{\partial r}) \quad (4)$$

where it is assumed that the wave absorption profile S_h does not changed before and after the pellet injection and is approximately described by the Abel inverted hard x-ray profile. From Eqs.(2) - (4) with an appropriate radial diffusion coefficient of fast electrons D_h, the chord integrated x-ray signal is calculated and compared with the experimental results. Figure 4 (a) and (b) show time evolution of measured and calculated hard x-ray signal at the radius of r/a ~ 0.4. To explain the experimentally observed decrease in hard x-ray signal after pellet injection, the radial diffusion coefficient for fast electrons $D_h < 1$ m^2/s is required in this model. This value is consistent with the previously mentioned estimation.

Measurement of the direct loss power of fast electrons and the response of the x-ray signal after the pellet injection indicates that the radial diffusion of fast electrons is not so large compared with that of bulk electrons in JT-60U LHCD plasmas. This conclusion is also supported by other experimental results; for example, the responses of hard x-ray signal and the directly lost power in modulated LH power injection.

COUPLING PROPERTIES OF JT-60U ICRF ANTENNA

Figure 5 shows the loading resistance as a function of separatrix-wall distance for two phasing, $(\pi,0)$ phasing (out-of-phase of strap currents) and $(0,0)$ phasing (in-phase). A density range is $\bar{n}_e = 0.4 - 4.5 \times 10^{19}$m^{-3} for $(\pi,0)$ phasing and $\bar{n}_e = 1 - 2 \times 10^{19}$m^{-3} for $(0,0)$ phasing. The loading resistance decreases with the separatrix-wall distance, but its values are fairly large, especially about 5 Ω even at a big gap of 29 cm for $(0,0)$ phasing. These data for gaps larger than 20 cm is obtained for short ICRF pulses (200 ms) by shifting the plasma position. Residual hydrogen ion tails are observed when the difference between the 2 ω_{CH} resonance position and the plasma center is less than about 0.1a, where a is a minor radius. These tails are generated by second harmonic minority ion heating so that the central plasma heating occurs with $(0,0)$ phasing even at such a big gap.

We have two codes for calculation of the loading resistance, the three-dimensional multi-strap code[10] and the variational method multi-strap code taking account of solid septa and side walls[11]. The calculated results by both codes can be fitted to the experimental data for $(\pi,0)$ phasing although only the result from the variational method multi-strap code is indicated in Fig. 5. But calculated results by both codes can not be fitted to the $(0,0)$ phasing data. The reason for this discrepancy is not well understood. Probably modeling and treatment in the codes for septum effect, self-consistent current distribution, and feeder effect should be improved.

SAWTOOTH STABILIZATION BY 2 ω_{CH} HEATING[12]

Figure 6 shows a typical results of sawtooth stabilization by 2 ω_{CH} minority ion heating alone at $P_{IC} = 2.8$ MW for the plasma parameters of $I_p = 2$ MA, $B_T = 3.97$ T, $q_{eff} = 5.3$, $\bar{n}_e \sim 2 \times 10^{19}$ m^{-3} and helium discharge containing residual hydrogen. The central electron temperature increases from 3.1 keV to 5.5 keV and the profile peaking factor $T_e(0)/<T_e>$ is 2.8. Saw tooth period is extended up to 0.87 sec which is larger than the energy confinement time by a factor of 2.1. The increase in plasma stored energy by diamagnetic measurement is about 0.71 MJ. Then the incremental energy confinement time ($\Delta W/(P_{IC}+I_p\Delta V_L)$) is about 0.3 sec. This value is comparable to the results in JET at a similar plasma current[13]. This suggests that ICRF central heating does not suffer from the large toroidal field ripple of JT-60U.

Figure 7 shows operation range for sawtooth stabilization by 2 ω_{CH} minority ion heating alone in $1/q_{eff}$ - $<n_e>/P_{tot}$ plane, where the data of sawtooth period > 0.6 sec are plotted, comparing JET results[14]. Efficient sawtooth stabilization for the shot of q_{eff} = 4 and I_p = 2.4 MA is obtained at $<n_e>/P_{tot} \approx 0.8 \times 10^{19}$ m^{-3}/MW, which is about 60 % larger than the upper boundary value for the JET results by fundamental resonance minority ion heating. This is considered to be due to the fact that the second harmonic minority ion heating, which is favourable for producing energetic ions at high densities, is used in JT-60U.

Sawtooth stabilization by ICRF heating is realized more easily for NBI-heated target plasmas than for ohmic target plasmas. Then, the longest period of 1.7 sec is obtained by combined heating of 3.1 MW ICRF power and 9.6 MW NBI power for the plasma parameters of I_p = 2.5 MA, B_T = 4.05 T, q_{eff} = 4, $\bar{n}_e \sim$ 3 x 10^{19} m^{-3} and deuterium discharge. A remarkable tail of residual hydrogen ion is observed by an active charge exchange (CX) neutral particle analyzer (NPA).

BEAM ACCELERATION BY 5 ω_{CD} HEATING

Figure 8 shows various ion cyclotron resonance layers and NB trajectory at B_T = 2.86 T and 3.03 T, where 5 ω_{CD} and 6 ω_{CD} resonance layers are located astride the magnetic axis. NB trajectory is off-axis so that the NB deposition power is peaked at r/a ~ 0.3. In higher harmonic ICRF heating ICRF power couples more strongly to higher energetic ions. Thus, beam acceleration should be optimized by shifting the 5 ω_{CD} resonance layer so as to be closer to the peak position of NB deposition power. Figure 9 indicates CX neutral particle energy spectra measured by a tangential NPA for shots of B_T = 2.86 T and 3.03 T at Ip = 1.8 MA. The injection energy of D beam is 87 keV. Beam acceleration is observed in both shots. However, the ions with energy > 100 keV are generated much more at B_T = 2.86 T, where the 5 ω_{CD} resonance layer is closer to the peak position of NB deposition power, than at B_T = 3.03 T. This difference is supported by the simulation results using the experimental conditions by the code where ICRF wave equation and one-dimensional Fokker-Planck equation are solved in turn until the ion distribution function converges[15]. That is, the simulated results reveal that deuterium ion tail is formed more strongly at B_T = 2.86 T, especially in an energy range of > 100 keV.

COMBINED 2 ω_{CH} ICRF and NBI HEATING OF PELLET INJECTED PLASMA

Preliminary experiments of combined 2 ω_{CH} ICRF (2.5 MW) and NBI (D beam, ~ 6 MW) heating of pellet injected plasmas are carried out. Figure 10 shows a typical result of the combined heating comparing the result with NBI heating alone. Monster sawteeth and a very peaked electron temperature profile are produced by ICRF heating in the combined heating. The central electron temperature reaches 6.5 keV during the first Monster sawtooth (at 10 sec) while it stays at 4.2 keV in NBI heating alone. The diamagnetic stored energy also shows its maximum at around 10 sec in the combined heating while the peak appears much earlier (at 9.4 sec) in NBI heating alone. Thus pellet enhanced phase lasts before the crash of the first Monster sawtooth (up to 1.4 sec after pellet injection) in the combined heating. The Abel inverted SX signals, shown in Fig. 11, suggest that the density profile is obviously more peaked during the first stable period than during the second one since $T_e(0)$ is almost constant during these periods. The energy confinement time normalized by heating power in this phase is improved by 20 % by ICRF heating and neutron yield increased by 40 %.

SUMMARY

Basic studies of LHCD physics are performed in JT-60U with a low power injection. Observed CD efficiency is consistent with the previous results in JT-60. It is found that the slowing down of fast electrons accelerated by waves is dominant in JT-60U LHCD discharges. The effects of the wave accessibility condition on the LHCD plasmas are also clarified on this series of experiments. Based on these results, high power LHCD experiments will be performed from April. 1993.

Significantly large loading resistances are obtained for both phasing modes with the JT-60U ICRF antenna with a poloidal septum, especially ~ 5 Ω for (0,0) phasing modes even at a big separatrix-wall gap of 29 cm. This enables combined ICRF and NBI heating in the better conditions where both NB power deposition and ICRF resonance are near the magnetic axis and the ripple loss is decreased. Operation range for sawtooth stabilization by 2 w_{CH} heating alone is extended to relatively high density and low-q regime, $<n_e>/P_{tot}$ ~ 0.8 x 10^{19}m^{-3}/MW at q_{eff} = 4 and I_p = 2.4 MA. This is considered to be due to the fact that the second harmonic minority ion heating is favourable for producing energetic ions at high densities. Preliminary results of combined 2 ω_{CH} ICRF and NBI heating for pellet injected plasmas show encouraging improvement of plasma performance.

REFERENCES

1. T. Fujii et al., Proc. 16th Symp. on Fusion Tech. (London, 1990) Vol. 2, p.1171.
2. T. Imai et al., Proc. 19th Int. Conf. on Contr. Fusion and Plasma Phys. (1992) Vol. 16C, Part II, p.953.
3. K. Ushigusa et al., Nuclear Fusion **29**, 1052 (1989).
4. F. Leuterer et al., Nuclear Fusion **31**, 2315 (1991).
5. The JET Team, Proc. 14th Int. Conf. on Plasma Phys. and Contr. Nuclear Fusion Research (Würzburg, 1992) IAEA-CN-56/E-1-1.
6. T. Imai et al., Proc. 13th Int. Conf. on Plasma Phys. and Contr. Nuclear Fusion Research (Washington, 1991) Vol. 1, p. 613.
7. M. Seki et al., "Characteristics of a large multijunction launcher for high power LHCD experiments in JT-60U", in this conference.
8. D. Ress et al., Rev. Sci. Instrum. **61**, 2777 (1990).
9. K. Ushigusa et al., Nuclear Fusion **32**, 1977 (1992).
10. H. Kimura et al., Proc. 4th Int. Symp. on Heating in Toroidal Plasmas (Rome, 1984) Vol. 2, p.1128.
11. M. Saigusa et al., Nucl. Fusion **33**, 421 (1993).
12. H. Kimura et al., "Sawtooth stabilization experiments by ICRF heating alone and its combination with NBI or LHCD in JT-60U", in this conference.
13. V.P. Bhatnagar et al., Plasma Phys. Contr. Fusion **31**, 333 (1988).
14. D.J. Campbell et al., Proc. 15th Eur. Conf. on Contr. Fusion and Plasma Heating (Dubronik, EPS, 1988) Vol. 12B, Part I, p.377.
15. K. Hamamatsu et al., Nucl. Fusion **29**, 147 (1989).

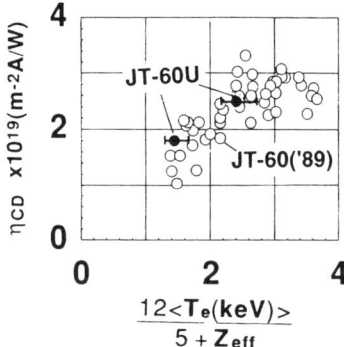

Fig. 1 Current drive efficiency against the JT-60 empirical scaling.

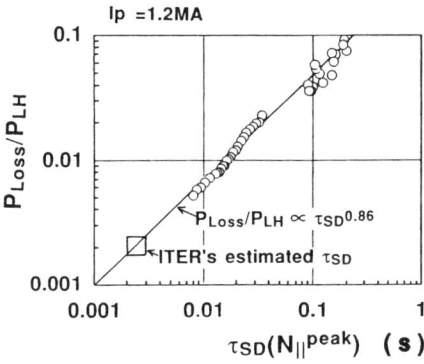

Fig. 2 The direct loss power rate against the slowing down time of fast electrons.

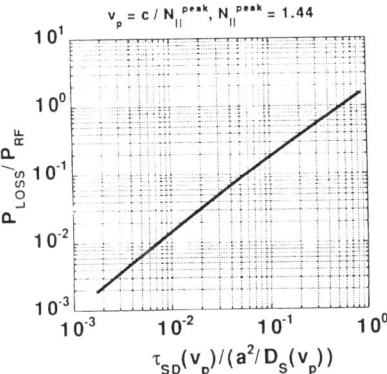

Fig. 3 Simulated direct loss rate as a function of the slowing down time of fast electrons normalized by the radial diffusion.

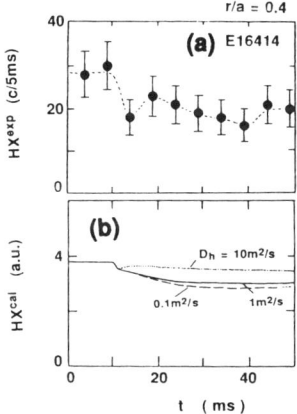

Fig. 4 Time evolution of the hard x-ray signal in pellet injected LHCD plasma. (a); experimental and (b); simulated results.

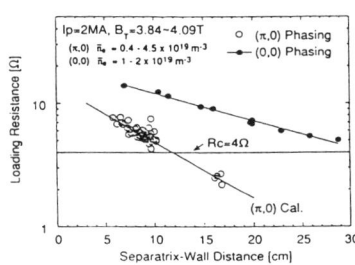

Fig. 5 Loading resistance as a function of separatrix-wall distance

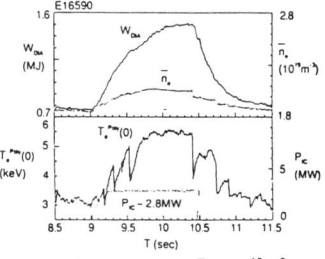

Fig. 6 Typical results of sawtooth stabilization by $2\omega_{CH}$ minority ion heating alone

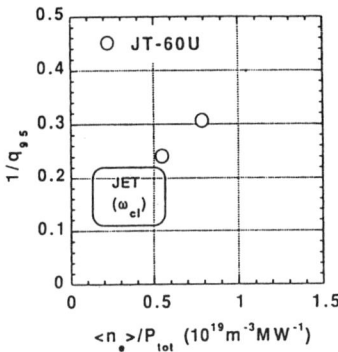

Fig. 7 Operation range for sawtooth stabilization by 2 ω_{CH} minority ion heating alone.

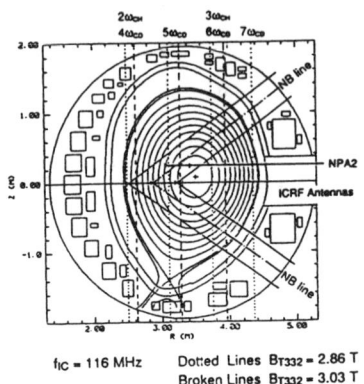

f_{IC} = 116 MHz   Dotted Lines B_{T332} = 2.86 T
Broken Lines B_{T332} = 3.03 T

Fig. 8 Various ion cyclotron resonance layers and NB trajectory at B_T = 2.86 T and 3.03 T.

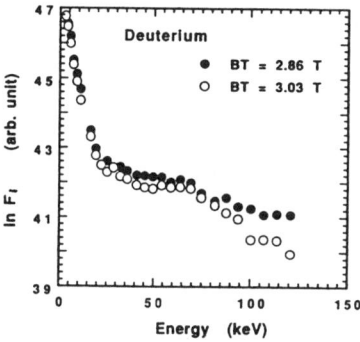

Fig. 9 CX neutral particle energy spectra measured by a tangential NPA for shots of B_T = 2.86 T and 3.03 T.

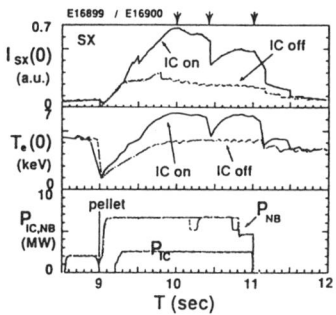

Fig. 10 Typical result of the combined heating comparing the result without ICRF heating.

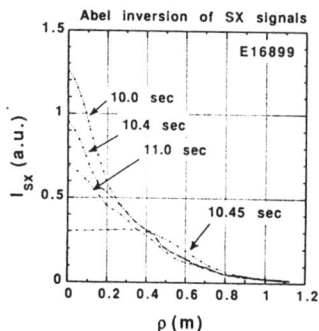

Fig. 11 Abel-inverted SX profiles for the combined heating shot shown in Fig. 10.

LOWER-HYBRID ENHANCED PERFORMANCE IN TORE SUPRA

G.T. Hoang, D. Moreau, E. Joffrin, X. Litaudon, Y. Peysson
and Tore Supra Team
Association Euratom-CEA-Département de Recherches sur la Fusion Contrôlée
CEN Cadarache, 13108 Saint Paul-lez-Durance, FRANCE
R.V. Budny, S. Kaye, S.A. Sabbagh[*)]
Plasma Physics Laboratory, Princeton, New Jersey 08543, USA.
V. Fuchs
Centre Canadien de Fusion Magnétique, Hydro-Québec, CANADA

ABSTRACT

The global energy confinement of the constant current discharges in Tore Supra is found to depend on the plasma density. Furthermore, the electron kinetic energy agrees with global Rebut-Lallia prediction[1]. Current ramp experiments showed an increase of the global energy confinement with increasing inductance (l_i) [1]. These results have been extended to the steady-state regime, in which the energy confinement time during the 12s LH pulse exceeds the usual L-mode scalings by 40%.

EXPERIMENTAL RESULTS

A series of experiments has been performed using LHCD in current ramp discharges at $I_p = 0.8MA$, $n_e(0)$ between 2.5×10^{19} m^{-3} and 3×10^{19} m^{-3}, and injected powers, P_{LH}, ranging from 2.5 to 2.9MW ($N_{//}$ being varied from 1.4 to 2.2). The LH pulse is applied just before the plasma current is ramped down from 1.7MA to a 0.8MA plateau at a rate of -1MA/s.

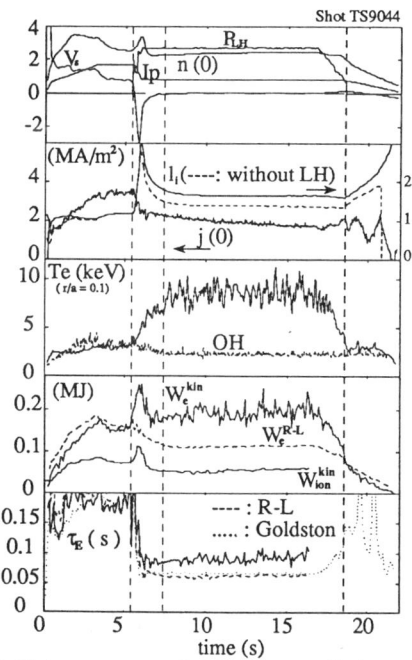

FIG. 1: Main parameters of a current ramp discharge
Plasma current, Ip(MA), loop voltage, Vs(V), central density, n(0)(10^{19} m^{-3}), LH power, PLH (MW), internal inductance, li, central current density from polarimetric measurements, electron temperature, Te, electron and ion kinetic energies (measured and Rebut-Lallia prediction), global energy confinement time (dashed curves:ohmic case)

The time evolution of a typical current ramp discharge is shown in Fig.1. With a 2.7MW/12s LH power application (at $N_{//} = 2.2$), the non-inductive current constitutes about 90% of the total plasma current ($\approx 0.7MA$), with a bootstrap current, calculated by the TRANSP code, of the order of 0.1MA. As soon as the LH power is applied, small sawteeth are suppressed, and no MHD activity is observed. The steady-state value of l_i reaches 1.7 instead of 1.4 obtained in the ohmic case. It can be seen in Fig.1 that the time scale over which l_i varies, and the relaxation time of the local current density at the center of the plasma are not the same. l_i becomes constant at t = 9s (i.e. 3.6s

© 1994 American Institute of Physics

after the LH application), while the central current density, j(0), continues to decrease slowly, and reaches a stationary value about 4s later (at t = 13s) due to a strong central electron heating. The global energy confinement time shows an enhancement of about 1.4 over both Rebut-Lallia and Goldston (with M = 1) L-mode predictions. The density and temperature profiles reach an equilibrium during the LH pulse. Bulk electrons are heated in the whole plasma volume, leading to an increase of the volume averaged temperature, $<T_e>$, by a factor of two (Fig.2). Only a weak increase of the ion temperature is observed (Fig.3); this can be explained by low collisionality due to low density and high electron temperature. In the transient phase (between 5.4s and 7s), the central electron temperature, $T_e(0)$, rises from 2keV to 6keV and reaches a value of about 10 keV at the beginning of the constant-l_i phase. During this constant-l_i phase, the electron temperature profile reaches its peak within the magnetic surface r/a < 0.3, with a peaking factor, $T_e(0)/<T_e>$, between 5 and 6 (Figs 2 and 3).

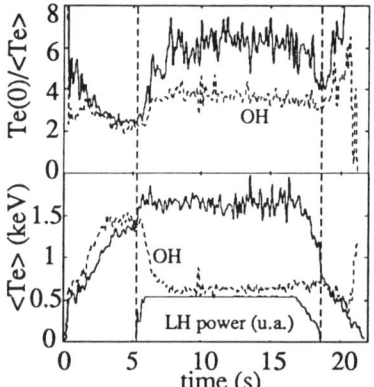

FIG.2: Volume-averaged electron temperature and peaking factor profile (shot TS9044)

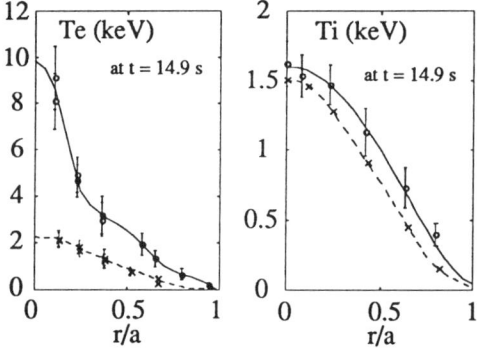

FIG. 3: Electron and ion temperature profiles

Figure 4 shows a constant current, improved confinement discharge. With the 3MW LH power (at $N_{//}$ = 2.2) applied directly on the flat-top , l_i becomes larger than in the ohmic phase but the current decreases on the axis. The central value of the safety factor, $q_\psi(0)$, computed by the equilibrium code IDENTD using the polarimetric, interferometric and magnetic measurements, increases from under 1 to about 2 (Fig. 5). The LH current profile is found to be flatter than the ohmic one in the central region.

During this improved confinement regime, the carbon (main impurity) reached a constant level with a concentration that is four times higher than that in the corresponding ohmic case. A comparison of the 'central' CVI radiance with the input carbon flux (from the emission of peripheral carbon lines) seems to indicate that the central impurity

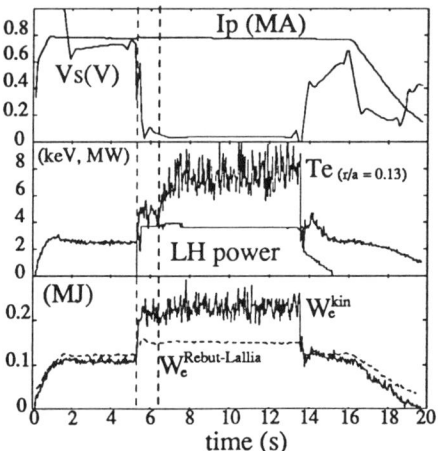

FIG. 4: Constant current improved confinement discharge (shot 9621)

confinement time does not change. It is interesting to note the low fraction of radiated power, Prad/Ptot ≈ 25%, which is down from 50% in the ohmic case.

CURRENT PROFILE AND TRANSPORT ANALYSIS

Current profile and local transport have been analysed for both high-l_i transient and steady-state phases (shot 9044). Figures 5(a) and 5(b) show that there is still no change in the magnetic shear in the central region during the transient phase (li ≈ 2), while an increase is observed in the confinement region (Fig. 5(b)). At the same time, the plasma pressure profile broadens (Fig. 5(c)). The magnetic shear then decreases within the surface $\rho < 0.3$, and probably becomes negative when the steady-state current profile is reached (t = 14.9s). This local behaviour of the q-profile is correlated with the peaked pressure profile, as in the PEP phase of JET. We refer to this as the "LHEP" (Lower Hybrid Enhanced Performance) regime. However, this further improvement of the plasma performance does not generally modify the volume integrated quantities such as the self-inductance and the total electron energy content. It should be noted that the transition to the "LHEP" phase is not necessary for routine high-l_i confinement improvement in steady-state discharges.

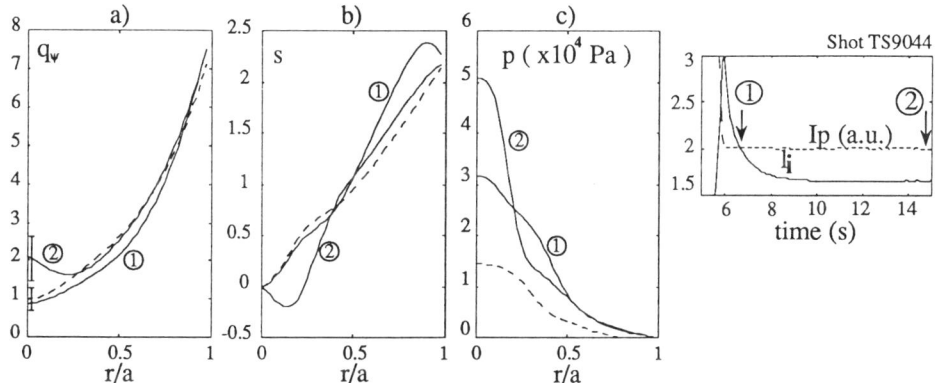

FIG. 5 Radial profiles of the safety factor ($q\psi$), magnetic shear (s), and measured plasma pressure (p) during the high-li transient (1) and stationary (2) phases. The dashed curves are for a corresponding ohmic discharge (9037).

The electron and ion thermal diffusivities are compared to those in an L-mode sawtooth suppressed LH discharge in Fig. 6. In the high-l_i transient phase, χ_e is found to be lower than the L-mode value for $\rho \leq 0.7$. The mean value of χ_e within the confinement zone ($0.4 \leq r/a \leq 0.7$) is 1.18 m²/s instead of 1.87 m²/s as in the L-mode sawtooth stabilized case, and there is no net change in local ion transport. The enhancement of the global confinement in the confinement zone is correlated with the increase of l_i, i.e. to the increase of the magnetic shear as shown on Fig.7. On this figure, the monster sawtooth discharges have been obtained with ICRH, where the q=1 surface increases leading to the increase of the magnetic shear in the gradient zone. As can be seen, the behaviour of the energy confinement for both high-l_i and monster sawtooth regimes are the same, i.e. decrease in χ_e in the confinement region due to the high value of magnetic shear.

For the steady-state phase, the decrease in χ_e in the central region (r/a < 0.3) can be linked to the possibility of attaining the second ideal ballooning stability zone, where the magnetic shear is lower than the threshold limit [2]. A rough estimate in the circular, large aspect ratio limit shows that the shear is less than this threshold value for $r/a \leq 0.35$. This value is consistent with the peaking of the pressure profile observed inside the central region. A similar conclusion has also been mentioned in DIII-D experiment [3]. It should be noted that the switch-off or interruption of the RF power generally leads to a collapse of the energy content. The central electron temperature drops to the initial level, while strong MHD activity appears. This phenomena, which is also observed during the LH assisted current ramp-up experiments, could be explained by the fact that the central magnetic shear rises, and therefore, the central zone of the discharge enters to the unstable ballooning region.

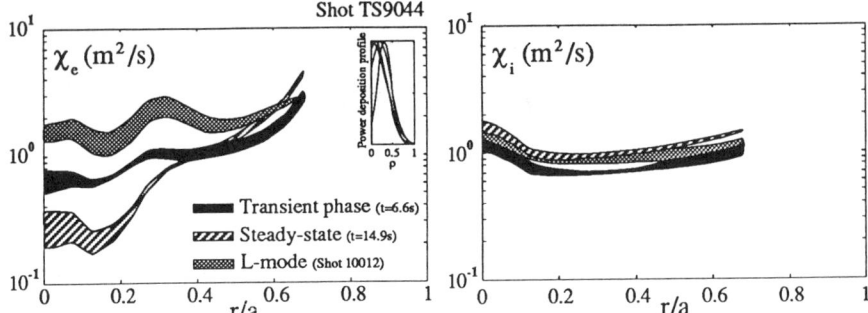

FIG. 6 Electron and ion thermal diffusivities for shot 9044 (at times indicated on Fig.5)

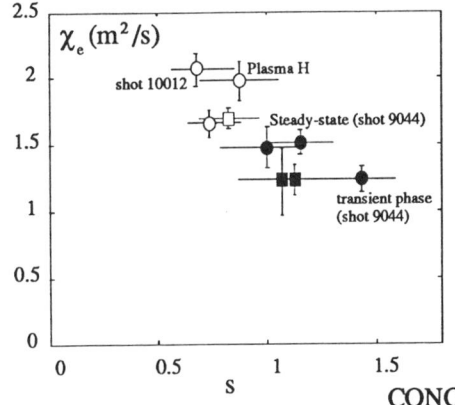

FIG. 7 Magnetic shear dependence of electron diffusivity, in the confinement zone (li is between 1.2 and 2, r/a = 0.5-0.7).

Circles: LHCD discharges
(open: L-mode, full: improved confinement regime)
squares: Ion cyclotron resonant heating (ICRH)
(open: L-mode, full: monster sawteeth)
RF power : 2.9 - 3.3MW

CONCLUSIONS

A stationary high-l_i, improved confinement regime has been routinely achieved with LHCD. The confinement time during the 12s LH pulse is about 40% higher than the usual L-mode scalings. In the high-l_i regime, the electron diffusivity is found to decrease with the magnetic shear in the confinement zone. In addition, very strong electron heating in the center ($\approx$10keV) leads to a peaked pressure profile which is correlated with the low (or negative) shear.

*) Columbia University, Department of Aplied Phisics, New York, Newyork 10027, USA

[1] G.T.Hoang et al., , in Proceedings of the Joint ICCP - 19th EPS, Innsbruck,1992, Vol. 16C, p I-27.
[2] Mercier, C., Plasma Physics 21 (1979) 589.
[3] The DIII-D Team (presented by Simonen T.C.) in Proceedings of the 14th International Conference, Würzburg, 1992.

MEASUREMENT OF THE PARALLEL DISTRIBUTION FUNCTION DURING LOWER-HYBRID CURRENT DRIVE IN TORE SUPRA

J. L. Ségui, G. Giruzzi, D. Vézard, W.D. Liu, X. Caron[a] and R.L Meyer[a]

Association Euratom-CEA sur la Fusion, Département de Recherche sur la Fusion Contrôlée. Centre d'Etudes de Cadarache. Saint Paul-lez-Durance (FRANCE)

a) *Laboratoire de Physique des Milieux Ionisés . Université de Nancy I - U.A. CNRS 835   Vandoeuvre-lès-Nancy (FRANCE)*

Abstract. A transmission diagnostic system working in the frequency range 77-109 GHz has been implemented in the Tore Supra tokamak. It makes use of an ordinary mode launched along a vertical diameter, i.e., propagating along a nearly constant magnetic field. The purpose of the diagnostic is to measure electron cyclotron absorption (ECA) below the electron cyclotron frequency f_c in the presence of the fast electron tails created during lower-hybrid current drive. Transmission spectra measured by this method are compared with spectra obtained by Fokker-Planck simulations and the relaxation properties of the tail are discussed.

Microwave transmission measurements below the electron gyrofrequency f_c are a powerful method for diagnosing the velocity distribution of suprathermal electrons in tokamaks[1,2]. The basic property used in such diagnostics is the proportionality between the local absorption coefficient and the parallel (with respect to the tokamak magnetic field) distribution function of the superthermal electron tail[3,4]. The ECA diagnostic set-up is described elsewhere[5,6]. Here, we present experimental results obtained in lower-hybrid (LH) driven discharges on Tore Supra.

The most challenging problem in the interpretation of a transmission experiment is the elimination of refraction effects. This problem actually sets a severe limit to the parameter range in which the diagnostic is applicable: the plasma density must be typically lower than 3/4 of the cut-off density of the mode used. Early experiments[1] were performed in low density plasmas, and the problem was avoided. Here, the refraction loss factor R_L (defined as the ratio between the transmitted power in the plasma and in vacuum, in the absence of absorption) is evaluated by means of accurate ray-tracing calculations, combined with measurements done at a reference frequency, for which there is no absorption. Such a frequency should be just above f_c: absorption at the first harmonic is forbidden by relativistic effects, whereas for the O-mode 2nd harmonic absorption is known to be negligible even in the presence of a dense electron tail[3]. The procedure used is the following: the frequency dependence of R_L is evaluated by a 3-D ray-tracing code, run with the actual plasma parameters, magnetic configuration and antenna pattern. The measurement at the reference frequency $f_{max} > f_c$ is used to compensate for the inaccuracies due to the error bars in the measured density profile and/or Shafranov shift. This method is applicable if the difference between the computed and the measured refraction loss at $f = f_{max}$ is small. In this regard, two regimes can be distinguished:

i) In the absence of sawteeth or intense MHD activity (as is usually the case in the LHCD phase), the agreement between the two determinations of $R_L(f_{max})$ is systematically found to be better than 5-10 %. This small error is efficiently corrected by using measurements at the reference frequency f_{max}.

ii) In the presence of sawteeth and/or large oscillations due to m = 1 modes[5] (which is the standard regime in Tore Supra Ohmic plasmas) the agreement is only rough, and the method is not so accurate. An extensive modelling of refraction effects during sawteeth and MHD activity is generally required.

Due to the presence of sawteeth, the elimination of refraction effects by simple comparison with an Ohmic phase with identical parameters is not generally possible. However, it was possible to exploit some differences in time scales in order to make such a comparison, at least in some cases. In fact, LH power modulation experiments[7] have shown that 150 keV electrons (roughly corresponding to the minimum frequency of the measurement range f_{min} = 77 GHz) in a similar discharge relax on a time scale shorter than 15 ms. Now, when the LH power is switched off after a long steady-state phase (with $V_{loop} \approx 0$), the sawtoothing behavior is recovered after 150 ms only, whereas the electron density and temperature vary on an intermediate time scale.

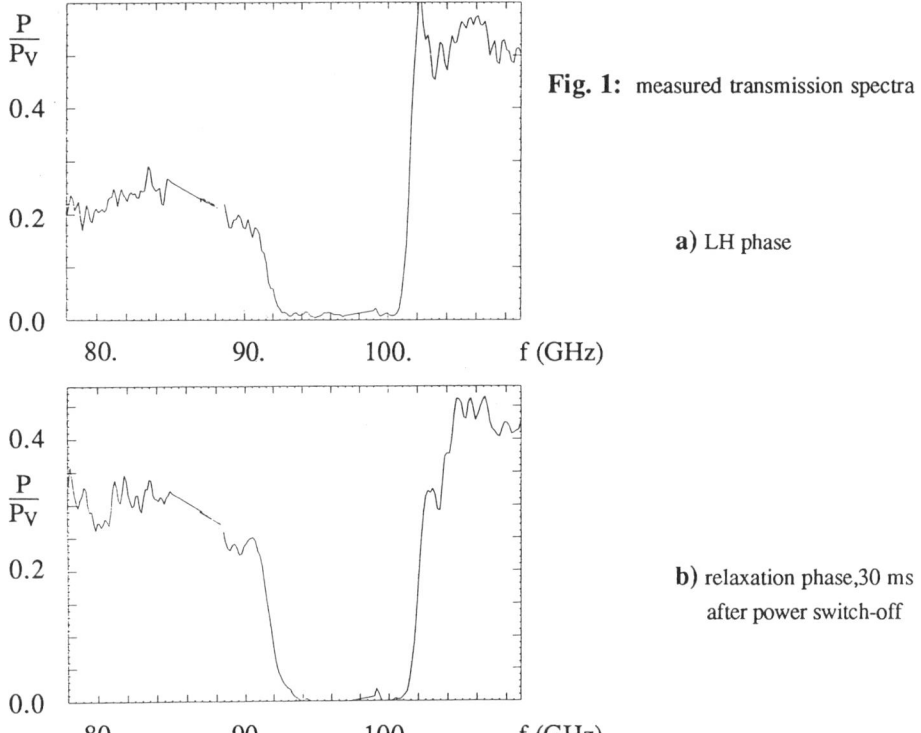

Fig. 1: measured transmission spectra

a) LH phase

b) relaxation phase, 30 ms after power switch-off

Thus, at a time $t = t_2$, 30 ms after the LH power has been switched off, the electron tail has relaxed, but the density profile has little changed with respect to the time $t = t_1$, representing the steady-state phase, just before the switching off. The relaxation of the electron tail is apparent in Figs. 1 (a) and (b), where the measured transmitted power, normalized to the power transmitted in vacuum P_V, is shown versus frequency, in the LH phase and after relaxation, respectively. The little difference between the two spectra in the high frequency part is due to little changes in the density profile, whereas the much larger difference in the low frequency side is the proof that the tail has relaxed. Therefore, the signal at $t = t_2$ can be used as an Ohmic reference in order to eliminate the effect of refraction. This procedure finally yields the experimental value of the optical depth, which is the quantity to be used in order to determine the parallel distribution of the superthermal tail.

The optical depth along the wave trajectory ℓ is defined as

$$\tau_o(f) = \int \alpha_o(f, \ell) \, d\ell \ . \tag{1}$$

For a slowly varying parallel tail, the absorption coefficient α_o is related to the parallel distribution by[3]

$$\alpha_o \approx \frac{\pi \omega_p^2 N_o}{2 \omega_c c} p_+(f) \, f_{\parallel}(p_+) \ , \tag{2}$$

where $p_+(f) = mc[(f_c/f)^2 - 1]^{1/2}$, N_o is the cold refractive index of the O-mode and ω_p is the plasma frequency. Since along a vertical chord the magnetic field is nearly constant, the measurement of the transmission coefficient yields the line-averaged value of the quantity $\omega_p^2 N_o f_{\parallel}$, thus, assuming the density profile is known, the line-averaged value of $f_{\parallel}$. If measurements on several chords are available, the radial structure of $f_{\parallel}(p_{\parallel}, r)$ can in principle be determined by Abel inversion[4]. Note that Eq. (2) is valid if the perpendicular energy is much smaller than the parallel energy. This is a good assumption for the high-energy part of the tail, but certainly not for the low-energy one. Thus, in the following, measurements of the relaxation time of the quantity $f_{\parallel}^* = (2\omega_c c \alpha_o)/(\pi \omega_p^2 N_o p_+)$ at several values of the resonant energy $E = mc^2(f_c/f - 1)$ will be presented, but identification of $f_{\parallel}^*$ and $f_{\parallel}$ is not possible in all the energy range. However, $f_{\parallel}^*$ is a good qualitative picture of $f_{\parallel}$, and its time evolution is very close to that of $f_{\parallel}$, as is found by comparison with Fokker-Planck simulations. We now consider the relaxation phase after the LH power has been switched off in discharge 10567. The measured transmission coefficient $\exp(-\tau_o)$ is shown in Fig. 2 during both the steady-state LH phase and the relaxation phase (6 ms after the power has been switched off); theoretical curves are also shown, obtained from the distribution function calculated by means of a 3-D Fokker-Planck code[8], including quasilinear diffusion and Coulomb collisions. It appears that both the shape and the level of the measured spectra are in substantially good agreement with the theoretical ones, although the relaxation is somewhat slower than predicted. This discrepancy is more evident in plots of the relaxation time of the parallel distribution, simply defined as $\tau^*(E) = [(d(ln f_{\parallel}^*)/dt)]^{-1}$, as shown in Fig. 3. The very good agreement in the shape of the curves shows that Coulomb collisions are indeed the dominant relaxation mechanism of the electron tail. It appears that at low energies τ^* decreases with E, in contrast with the simple picture of a relaxation time scaling as the single-particle slowing-down time, namely as v^3. In the simulations, the v^3 dependence is recovered at high energies, beyond those resonant with the LH spectrum, which are outside the measured range. The behavior of the relaxation time shows the importance of the detailed kinetic evolution with respect to single-particle effects. Similar results have been found in LHCD experiments on Versator and have been interpreted as an anomalous relaxation time[9]. The discrepancy in the magnitude of the relaxation time can be partly due to the effect of the dc electric field, which was not included in the simulations, since its radial structure during the relaxation phase is difficult to model. During the relaxation the dc field increases, accelerating the fast electrons, thus their relaxation time is expected to be enhanced. The second effect that could explain the discrepancy is the radial location of the superthermal electrons. In the simulations the LH power deposition profile was assumed to be peaked at $r/a \approx 1/3$ (see Ref. 8, Fig. 1), but this is not a directly measurable quantity and even the modeling of it is affected by large uncertainties. Moreover, the radial distribution of the superthermal electrons is determined by radial diffusion mechanisms, which are also unknown. For instance, if the fast electron profile is further shifted off-axis, the relevant slowing-down time (which is proportional to density) will be larger, because of the radial density profile. In this sense, comparison between the experimental and theoretical relaxation times could give indications about the radial profile of the fast electron distribution.

REFERENCES
1 - E.Mazzucato, et al., Nucl. Fusion **25**, 1681 (1985)
2 - R. Kirkwood, et al., Nucl. Fusion **30**, 431 (1990)
3 - G.Giruzzi, et al., Phys. Fluids **27**, 1704 (1984)
4 - I. Fidone, et al., Phys. Fluids **B3**, 2719 (1991)
5 - G.Giruzzi, J.L. Ségui, A.L. Pecquet, C.Gil, Nucl. Fusion **31**, 2158 (1991)
6 - J.L. Ségui, G. Giruzzi, in *Diagnostics for Contemporary Fusion Experiments* (SIF, Bologna, 1991), 795.
7 - D. Moreau, et al., Plasma Phys. Contr. Fusion **33**, 1621 (1991).
8 - G. Giruzzi, Plasma Phys. Contr. Fusion **35**, A123 (1993).
9 - R. Kirkwood, et al., Nuclear Fusion **31**, 1938 (1991).

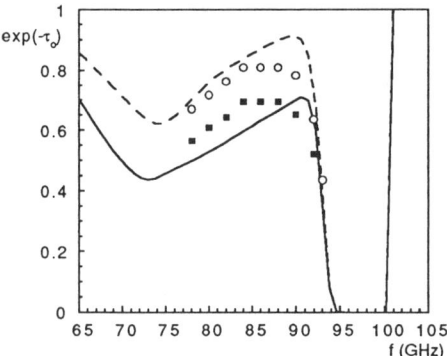

Fig. 2 : theoretical transmission spectra during the LH phase (solid line) and after 6 ms of relaxation (dashed line). Corresponding experimental points (squares and open circles, respectively)

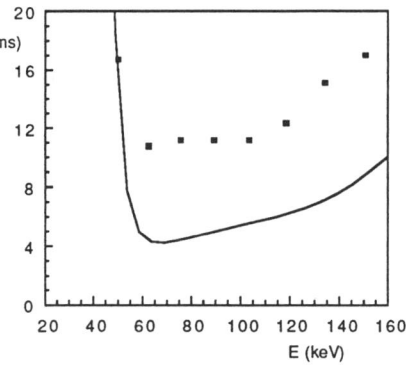

Fig. 3 : relaxation time vs energy. Theory (solid line) and experimental points

FIRST RESULTS OF THE 8 GHz LH EXPERIMENT IN FTU

F. Alladio, M.L. Apicella, G. Apruzzese, E. Barbato,
R. Bartiromo, F. Bombarda, G. Bracco, G. Buceti,
P. Buratti, A. Cardinali, R. Cesario, M. Ciotti,
V. Cocilovo, A. Coletti, I. Condrea, F. Crisanti,
R. De Angelis, F. De Marco, B. Esposito, T. Fortunato,
D. Frigione, L. Gabellieri, E. Giovannozzi, G. Granucci[+],
M. Grolli, S. Ide[*], A. Imparato, H. Kroegler, L. Lovisetto,
G. Maddaluno, M. Marinucci, G. Mazzitelli, D. McNeill,
P. Micozzi, F. Mirizzi, A. Moleti, F. Orsitto,
L. Panaccione, M. Panella, V. Pericoli, L. Pieroni,
S. Podda, G.B. Righetti, M. Roccon, D. Santi, F. Santini,
M. Sassi, E Sternini, G. Tonini, A.A. Tuccillo, O. Tudisco,
F. Valente, V. Vitale, V. Zanza, M. Zerbini

Associazione EURATOM-ENEA sulla Fusione, CRE Frascati,
C.P. 65, 00044 Frascati, Rome, Italy

ABSTRACT

The LH heating and current drive (CD) system of FTU started operation in the Autumn of 1992 with the first of its 9 gyrotrons power generators. The full performances of the tube (1 MW - 1 sec) have meanwhile been achieved. Operation at specific power higher than 10 kW/cm^2 is routinely carried out without any conditioning of the launcher. This has allowed preliminary investigation of CD and heating effects up to a density $n_e \sim 10^{20}$ m^{-3}.

THE LOWER HYBRID SYSTEM

The LH heating and CD system of FTU consists of nine identical transmitter modules. Each module makes use of a 8 GHz gyrotron, with power output of 1 MW and pulse length of 1 sec, powered by a 100 kV-30 A power supply through a modulator, with crow-bar protection, to regulate the gun-anode voltage. The gyrotron oscillation mode (TE$^0_{511}$) is converted at the tube output to the TE$^0_{01}$ which is transmitted via a standard circular (C$_{18}$) waveguide (WG), 30 m long, to a mode converting power splitter located in

---

[+] Istituto di Fisica del Plasma, CNR, Milano, Italy
[*] Naka Fusion Research Establishment, JAERI, 311-01 Japan

the torus hall. The output of this splitter consist of 12 rectangular WGs (WR_{137}) propagating in the TE_{10} mode. Each output is fed to a passive phase shifter and then divided by four, with two 3 db H-plane hybrids, to feed each column of the very compact grill [1]. This grill has 48 WGs in 4 rows with inner dimensions 4.2 x 28 mm^2 and 0.8 mm walls allowing the generation of directional and symmetric spectra with 1 <$N_{//}$ <3.8 and $\Delta N_{//}$ ~ 0.6 full width at half maximum; the value of the direct and reflected power is measured in all the WGs. A front windows block, 25 cm away from the plasma separates the pressurized part of the grill from the gold plated stainless-steel front end. Three grills are mounted together to form an antenna which is inserted in a FTU port. Each antenna has a radial stroke of 7 cm to match the right coupling density

FIRST RESULTS

After final commissioning of the gyrotron on the tokamak, full performances of the first transmitter have been reached and routine operations are carried out at its maximum output power with pulse duration from 0.5 to 1 second. This value corresponds to a power density, averaged on the 48 guides, higher than 10 kW/cm^2 obtained from the very first operation without any procedure of grill conditioning. In these conditions the total direct power amounts to 550-600 kW. The reduction with respect to the generator output is mainly due to losses in the 12 ways splitter and in the rectangular WGs.

The coupling of the power to the plasma was generally very good, as seen in Fig. 1 where a global reflection coefficient R < 10% is obtained when the antenna position is properly adjusted in the scrape-off plasma.

Preliminary CD experiments have been done in a limited range of parameters because of the low available power. In Fig.2 we report the relative loop voltage drop during the LH pulse when the plasma current is maintained constant by feedback control. We note the that the largest fraction of current is driven at n_e ~ $3.5*10^{19}$ m^{-3} with the smallest $N_{//}$ launched ($N_{//}$ ~ 1.6 at 75°). More data are needed at higher density to point-out the real $N_{//}$ dependence where this phasing approaches the accessibility condition. As density approach 10^{20} m^{-3}, the available power gives too small effects. Only the signature (Hard x-rays, EC emission) of the LH interaction with electrons is clearly detectable as already seen on FT.[2] It has to be noted that effects on loop voltage are hardly detected with opposite CD phasing that generally induce only a small increase. In all the experimental conditions we record only moderate increases of Z_{eff}, with the largest values occurring at lower density of Fig. 2, where it increases from the value of 3 in the OH target to the value of 4 with 0.5 MW of LH power coupled to the plasma. During CD substantial electron heating is also observed.[3,4] In Fig. 3 Thomson scattering profiles are shown for a typical case at low density. More than 1 keV increase in central temperature is recorded during LH giving peaking factors in excess of 3. This is generally accompanied by a slightly broadening of the density profile. From the rate of change of equilibrium energy, at the switch-on of the LH pulse, a preliminary value of the absorption coefficient ranging from 0.5 to 0.7 can be estimated. In the lower density discharges sawtooth is not present in the OH phase but can be triggered shortly after the LH onset. Sometimes only m=1 oscillations start suggesting vicinity of the power threshold for stabilization. In Fig. 4 the relative increase of the sawtooth period is reported for discharges that already exhibit the sawteeth in the OH phase. Larger effects are evident at lower density and a small decrease is seen with opposite CD phasing.

CONCLUSIONS

Very promising results have been produced during plasma experiments with the first gyrotron of the FTU LH system. The possibility to work at a power density higher than 10 kW/cm^2, suggested by the FT 8 GHz experiment [2] has been achieved routinely in a more complex launcher. Efficient power coupling to the plasma has been obtained in spite of the large power density at grill mouth. Clear results on heating and current drive have already been obtained although in a restricted range of parameters due to the limited power.

The authors gratefully thank S. Di Giovenale, M. Papalini and C. Poggi for their technical support.

FIGURE CAPTION

Fig. 1 - Global reflection coefficient versus line averaged

118 First Results of the 8 GHz LH Experiment in FTU

Fig. 2 - density for different phasing and plasma parameters (B_T = 6 T, P_{RF} = 450 - 500 kW).
Fig. 2 - Relative loop voltage drop versus line averaged density (B_T = 6 T, P_{RF} = 450 - 500 kW).
Fig. 3 - Thomson scattering profiles before and during LH pulse (IP = 300 KA, B_T = 6 T, n_e = 3.5x10^{19} m^{-3}, P_{RF} = 450 - 500 kW).
Fig. 4 - Relative increase of sawtooth period versus P_{RF}/n_e for different phasing.

REFERENCES

1. S. Di Giovenale et al., XVII SOFT, 1992, Roma.
2. F. Alladio et al., XII Int. Conference on Plasma Physics and Controlled Fusion, Nice, October 1988.
3. F. Alladio et al, LH Heating experiments in FT Tokamak, IX Int. Conf. on Plasma Physics, Baltimore, USA, 1982.
4. F.X. Soeldner et al., XIII Int. Conference on Plasma Physics and Controlled Fusion, Washington, October 1990.

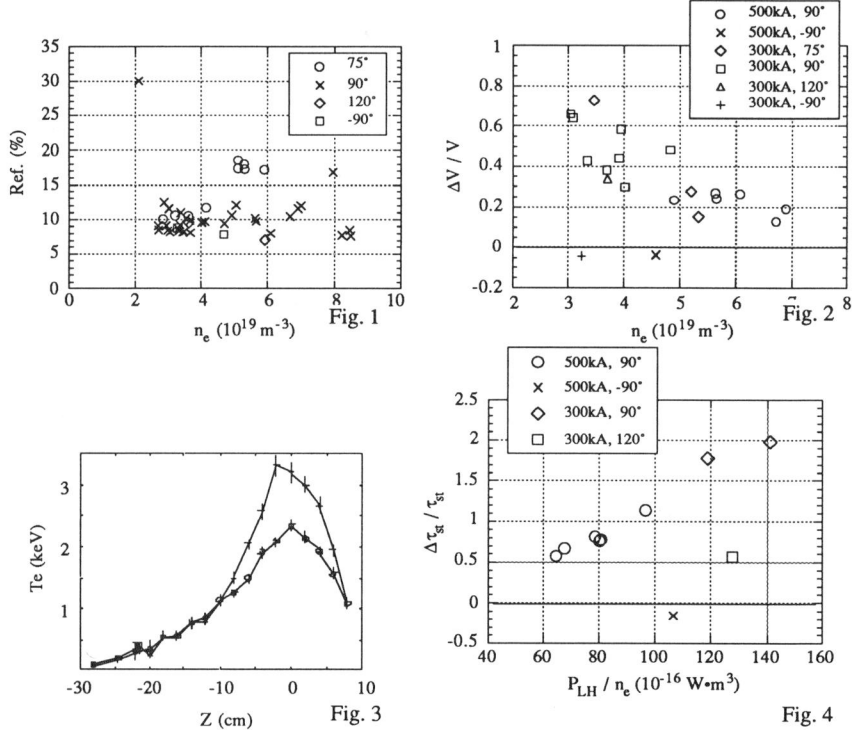

Fast Electron Current Density Profile and Diffusion Studies during LHCD in PBX-M

S.E. Jones, J. Kesner, S.C. Luckhardt, F. Paoletti,
Plasma Fusion Center, Massachusetts Institute of Technology, Cambridge, MA 02139

S. Bernabei, R. Kaita, S. von Goeler, and the PBX-M group,
Plasma Physics Laboratory, Princeton University, Princeton, NJ 08543

F. Rimini, *JET, Joint Undertaking, Abingdon, Oxfordshire OX14 3EA UK*

1. Introduction

Many advanced tokamak designs utilize current profile control for attaining such benefits as second stability, enhanced confinement, and sawtooth stabilization. PBX-M is an experiment dedicated to investigate this potential by using Lower Hybrid Current Drive (LHCD)[1]. Clearly any current profile control experiment using lower hybrid driven fast electrons requires knowledge of two key issues:

(1) the radial location of the fast electrons;
(2) the ability to maintain their spatial localization.

This paper reports on our efforts to address these two issues by using the novel 2-D hard x-ray imaging camera on PBX-M[2,3]. These images are produced from bremsstrahlung x-rays emitted by LHCD-driven fast electrons as they collide with the thermal plasma. Although the images are line-of-sight and energy integrated, an inversion procedure has been developed[3] to obtain a radial fast electron current density profile. Once these profiles are obtained, they can be fitted to theoretical predictions from a Fokker-Planck model for the fast electron distribution function, which includes a diffusion operator. From this procedure an effective fast electron diffusion constant can be computed.

The hard x-ray diagnostic on PBX-M was previously reported[2,3], but a brief description follows: The diagnostic is effectively an x-ray pinhole camera which views the plasma tangentially on the mid-plane. The detector itself is a two-dimensional medical x-ray image intensifier tube, and it is important to note the images are line-of-sight and energy integrated. The image is also time-integrated for as little as 4 msec, and 64 such images, or frames, from one discharge are stored. Thus, the time evolution of the image is obtained.

An example image from the camera is shown in Fig. 1 as a contour plot of detected x-ray intensity. This image was averaged over 32 frames at 4 msec/frame from the steady-state portion of discharge #298601. This 180 kA discharge at 14 kG was obtained using 310 kW of LHCD in conjunction with 500 kW of neutral beam injection. The loop voltage was 0.4 V, so the current drive was not 100%. Shortly after the auxiliary power was introduced, all sawteeth and MHD activity ceased.

Qualitatively, the "hollowness", or off-axis distribution of the energetic electrons, is reflected by the slight concavity and widening of contour lines just to the right of the central peak. This significant result demonstrates that steady-state off-axis current drive is possible. The rest of this paper focuses on this experimental image for extracting a radial fast electron current density profile and for computing an effective fast electron diffusion constant.

2. Fast Electron Current Density Profile for shot #298601

Since the x-ray images are line-of-sight and energy integrated, an inversion process is required to obtain the radial fast electron current density profile. This procedure has been previously reported[2], so the following description will be short. Rather than doing a difficult direct Abel-like inversion, we obtain the profile by modeling: using four variables to describe the fast electron velocity distribution function and

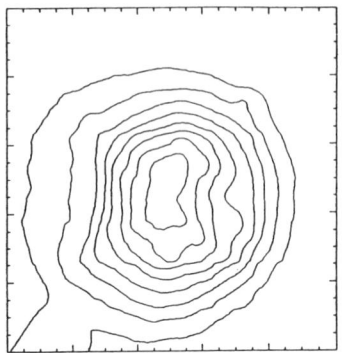 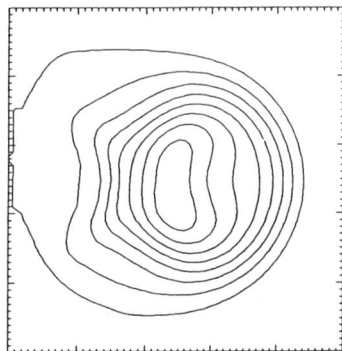

Figure 1. Contour plot of hard x-ray image during steady state portion of shot #298601. This image was averaged over 32 frames.

Figure 2. Simulated contour plot of the x-ray image for shot #298601, based on the radial fast electron current density profile in Fig. 3 and velocity space distribution in Fig. 4.

three variables to describe the superthermal electron (se) current density radial profile, the local hard x-ray emissivity is obtained. Specifically, we assume a separable fast electron distribution of the form

$$F_{se} = f_{se}(P_{||}, P_\perp) g(\Psi).$$

where

$$f_{se}(P_{||}, P_\perp) = \begin{cases} \exp(-P_{||}^2/2T_{||\,forward} - P_\perp^2/2T_\perp) & P_{||} > 0 \\ \exp(-P_{||}^2/2T_{||\,backward} - P_\perp^2/2T_\perp) & P_{||} < 0 \\ 0 & P_{||}^2/2 > E^* \end{cases}$$

and

$$g(\Psi) = \begin{array}{ll} \exp(-((\Psi-\Psi_{se})/\Delta\Psi_{seo})^2) & \Psi > \Psi_{se} \\ \exp(-((\Psi-\Psi_{se})/\Delta\Psi_{sei})^2) & \Psi < \Psi_{se}. \end{array}$$

Line-of-sight integrations are then performed, including the effects of various absorbers and the detector efficiency, yielding a simulated image which can be compared to the experimental image. The simulated image is recomputed using different parameters until a good comparison is found.

This procedure was applied to the discharge #298601 (Fig 1). A best-case simulated image is seen in Fig. 2, using the radial profile shown in Fig. 3 and velocity distribution function shown in Fig. 4. The location of the off-axis peak Ψ_{se} and the inboard (Ψ_{seo}) and outboard (Ψ_{sei}) scale lengths are sensitive to fitting to about a factor of 10%. This result clearly shows an off-axis location for the fast electrons, and since this shot was steady state this location was maintained.

3. Theoretical Model

A quantitative measure of the ability to maintain an off-axis fast electron current density profile is to calculate an effective diffusion constant and compare the diffusion time scale with the classical slowing-down time scale. Our approach is to assume a plausible physics model where the diffusion constant is a free parameter and the lower hybrid power absorption profile is extremely narrow. A diffusion constant is guessed initially and the model fast electron current density profile is calculated and compared to the

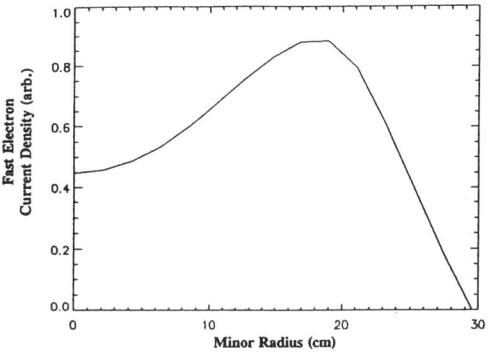

Figure 3. Model fast electron current density used for simulated Fig. 2.

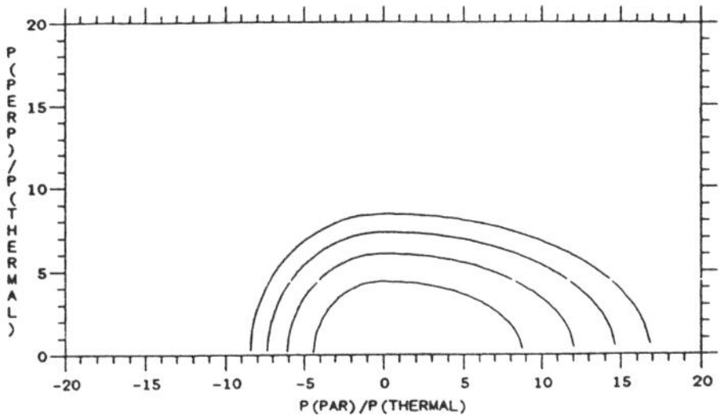

Figure 4. Model fast electron momentum space distribution used for simulating Fig. 2

experimental profile. The process is iterated using different diffusion constants until a good match is found. Since the lower hybrid power absorption profile might not be in reality as narrow as we assumed, this effective diffusion constant is an upper bound.

Our model is derived from the response function technique described by Rax and Moreau[4] which solves the following time dependent Fokker-Plank equation:

$$[\frac{\partial}{\partial t} - \frac{1}{p^2}\frac{\partial}{\partial p} + E\frac{\partial}{\partial p_{||}} - \frac{Z+1}{2p^3}\frac{\partial}{\partial \mu}(1-\mu^2)\frac{\partial}{\partial \mu} - \frac{D(p)}{r}\frac{\partial}{\partial r}r\frac{\partial}{\partial r}] K(p,p',\mu,\mu',r,r',t,t')$$

$$= d(p-p')d(r-r')d(\mu-\mu')d(t-t') / p^2 r \quad [1]$$

P is the normalized momentum variable, μ is the pitch angle variable, t is the normalized time variable, E is the normalized electric field, $p_{||}$ is the normalized parallel momentum, r is the normalized radial variable, Z is the effective charge, and D is the normalized effective diffusion constant. Since the Rax and Moreau solution use E=0, approximations are introduced to include a weak electric field, an important aspect of PBX-M. The full distribution function for the superthermal electrons (se) is written in terms of the response function as

$$F_{se}(p,\mu,r,t) = \int p'^2 r' dp' dr' d\mu' dt' W_a(p',\mu',r',t')[dp'd/dp + d\mu' d/d\mu'] K(p,p',\mu,\mu',r,r',t,t') \quad [2]$$

where W_a is a kernel describing the lower hybrid power absorption, and the fast electron current density can be written as

$$J(r) = q \int 2\pi^3 p^3 \mu F_{se}(p,\mu,r) \, dp d\mu \quad [3]$$

Our solution, including the weak electric field approximation, takes the form

$$K = \theta(t-t')\delta\left[\frac{p-p'}{\mu E} - \frac{1}{2(\mu E)^{3/2}}\ln\left[\frac{(1+p\sqrt{\mu E})(1-p'\sqrt{\mu E})}{(1+p'\sqrt{\mu E})(1-p\sqrt{\mu E})}\right] - t + t'\right]\sum (2l+1)$$

$$\times \exp(-k^2 H(p,p'))\left[\frac{\frac{p}{\sqrt{1-p^2\mu E}}}{\frac{p'}{\sqrt{1-p'^2\mu E}}}\right]^{l(l+1)(Z+1)/2} \frac{J_0(kr')J_0(kr)}{J_1(k)^2} P_l(\mu')P_l(\mu) \qquad [4]$$

where J_0 and J_1 are Bessel functions, P_l are Legendre polynomials, k are the zeros of J_0, the summation is over l and k, and

$$H(p,p') = \int_p^{p'} \frac{u^2 D(u)}{1-u^2} du. \qquad [5]$$

4. Results and Conclusion

This model was applied to shot #298601, and radial fast electron current density is shown in Fig. 3 assuming

$$W_a(p,\mu,r) = W_0 \frac{p'}{p}\delta(r-r')\delta(\mu-1) \quad p<p'$$
$$= 0 \quad p>p'$$

where $p'=0.5$ and $r'=0.57$. A best-fit profile from eq. 3 was obtained with $D_{eff} = 1.1$ m^2/sec, as shown in Fig. 4. Variations in the diffusion constant by a factor of two leads to clear departures from the experimental profile in Fig. 3. The sensitivity of this result was tested by varying the loop voltage, the lower hybrid absorption profile in velocity space, Z_{eff}, and adding up to 7 radially discrete sources to roughly model the effect of a broader radial absorption profile. In all cases, the best fit profile for the effective diffusion constant varied by less than about a factor of two. Note the diffusive time scale is 90 msec - much longer than the collisional slowing down time of about 7 msec.

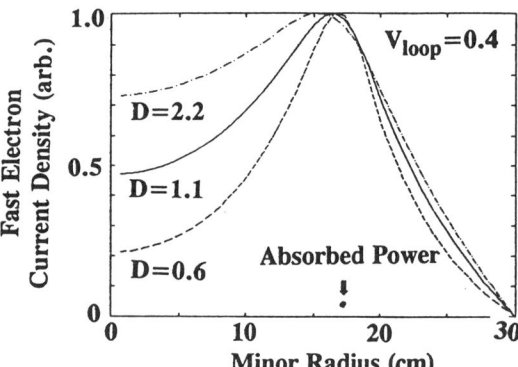

Figure 5. Theoretical results for the fast electron current density profile for 3 different values of an effective diffusion constant with a loop voltage of 0.4 v. Notice that $D^*=1.1$ gives the best-fit to the experimental profile seen in Fig. 3, and that factors of two larger or smaller lead to significant departures.

In conclusion, the hard x-ray camera has observed steady state off-axis current drive, and a radial fast electron current density profile has been obtained. A theoretical model simulating this shot produces an upper limit for an effective diffusion constant of about 1-2 m^2/sec. and sensitivity studies indicate that this result is good to within a factor of two.

We gratefully acknowledge useful discussions with J.M.Rax and N.J. Fisch. The MIT work is supported by USDOE Grant DE-FG02-91ER-54109 and the PPPL work by USDOE Grant DE-AC02-76-CH03073.
1. Chance, et al, PRL **51**, 1965 (1983).
2. R.Kiata,S.VonGoeler,S.Sesnic,S.Bernabei,E.Fredrickson, et. al., Rev. Sci. Instrum. **61** (10), 1990
3. S. Von Goeler, N.Asakura, R.Bell, S.Bernabei, M.Chance, et. al., Proceedings 19th EPS Conf. on Controlled Fusion and Plasma Physics (Innsbruck, 29 June - 3 July, 1992) v. 2, II-949.
4. J.M. Rax and D. Moreau, Nuclear Fusion, **29**, 10 (1989) 180

Determination of the Energy of Suprathermal Electrons during Lower-Hybrid Current Drive on PBX-M.

S. von Goeler, S. Bernabei, W. Davis, D. Ignat, S. Jones[†], R. Kaita,
G. Petravich[††], F. Rimini[†††], P. Roney, J. Stevens, A. Post-Zwicker[††††].

Plasma Physics Laboratory, Princeton University, Princeton NJ. 08543

1. Introduction. The Lower Hybrid Current Drive (LHCD) experiment on the PBX-M tokamak attempts to modify and optimize the radial current profile in order to find new tokamak operating regimes with a higher value of beta[1,2]. The lower hybrid waves interact with the plasma electrons and generate a suprathermal electron tail[3], presumably via Landau damping. In order to elucidate and test the physical mechanism of the wave-plasma interaction, we have installed on PBX-M a hard X-ray Camera[4,5] that produces images of the hard X-ray bremsstrahlung created in collisions of the suprathermal electrons with plasma ions. The first results from the Hard X-ray Camera were reported at the Innsbruck EPS conference[5]. In that paper, hard X-ray images from the camera are compared with computer-simulated images from the PBXRAY code. It was shown that the radial the location of the suprathermal electrons can be determined quite accurately. In particular it was found that LHCD with -90° or -105° phasing of the grill at high plasma densities generated a hollow ring of suprathermal electrons, a result that is considered crucial for the effort to modify the current profile. The present paper concentrates on the determination of the energy of the suprathermal electrons. We shall show that the suprathermal electrons in the high density regime have very low energies, (less than 100 keV). This result seems to support the notion that the electrons in the hollow discharges are in a regime where collisional slowing-down dominates acceleration by the electric field, and where low-$n_\parallel$ LH waves, that tend to accelerate electrons to high energies, cannot penetrate to the plasma center because of accessibility.

The paper is organized as follows: In Section 2 we discuss the absorber foil method for the determination of the electron energy, which is well known for soft X-rays, but - to our knowledge - has not been applied to hard X-ray measurements. Modeling with the PBXRAY code will illustrate its merits - and its subtleties. In Section 3 we present results for a PBX-M high density discharge.

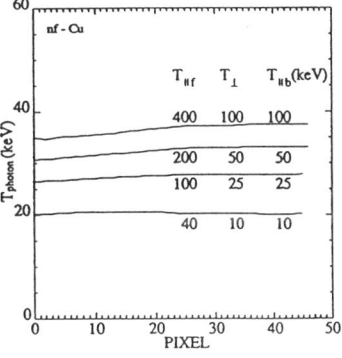

Fig. 1: Horizontal profiles of T_{photon} for different electron tail temperatures.

2. The Absorber Foil Method for Hard X-rays.
The Hard X-ray Camera has an imaging tube that integrates over photon energies. In order to determine the X-ray energy, we place various absorber foils in front of the pinhole of the camera in between shots, and compare the intensity I_{foil1} from a shot with absorber foil #1 to the intensity

---

[†] Plasma Fusion Center, MIT, Cambridge, MA 02139
[††] CRIP, Budapest, Hungary
[†††] JET, Joint Undertaking, Abingdon, Oxfordshire OX14 3EA U.K.
[††††] Oak Ridge National Laboratory, Oak Ridge, TN 37831
[1] R. Grimm, M.Chance, A.Todd, J. Manickam, M. Okabyashi, et al.: Nuclear Fusion 25, p.805 (1985)
[2] M. Chance, et al: Phys. Rev. Lett. 51, p.1965 (1983).
[3] S. von Goeler, J. Stevens, et al.: Proc 5th Topical APS Conf. Radio Freq. Plasma Heating, Madison, p.96 (1983)
[4] R. Kaita, S. von Goeler, S. Sesnic, S. Bernabei, E.Fredrickson, et al: Rev. Sci. Instrum. 61, p. 2756 (1990)
[5] S. vonGoeler, N. Asakura, R. Bell, S. Bernabei, T.K. Chu, et al.: Proc. 19th EPS Conf. Contr. Fusion and Plasma Physics, Innsbruck, Vol.II, p.949 (1992).

I_{foil2} from a shot with foil #2. In analogy with the well-known absorber foil method for soft X-rays, we plot an effective "temperature" of the photon spectrum T_{photon}

$$T_{photon} = \frac{E_{foil2} - E_{foil1}}{\ln(I_{foil1}) - \ln(I_{foil2})},$$

where E_{foil1} and E_{foil2} are the low energy cut-offs for foil #1 and foil #2. If the spectrum of the emitted bremsstrahlung falls off with energy like an exponential function, then T_{photon} is the negative reciprocal slope of the spectrum in a semilog plot. The absorber foils consist of copper (0.52 mm), molybdenum (0.95 mm), and silver (2.03 mm). Without any absorber foil, the aluminum vacuum window, the aluminum entrance window of the X-ray tube, and the connectic magnetic shielding foil in front of the imaging tube contribute to a low energy cut-off of 45 keV. The cutoff energies with an additional Cu, Mo, or Ag foil are 67 keV, 115 keV, and 176 keV, respectively.

Fig. 2: Contour plot of a hollow discharge.

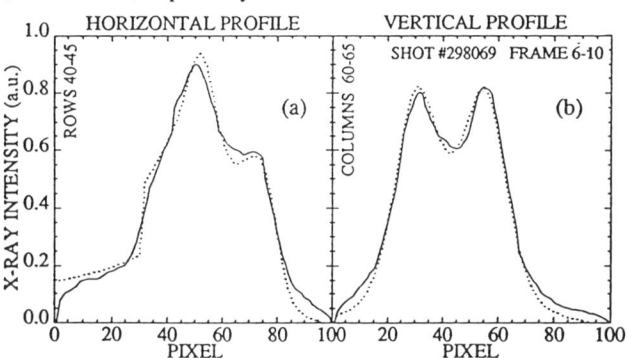

Fig. 3: Horizontal (a) and vertical (b) profile for a hollow discharge. Solid curves experiment, dotted curves simulations. Input for simulation is shown in Fig. 6. ["Horizontal profile" is our nomenclature for a horizontal slice through the center of an image]. In the computation that lead to Fig.1, the electron tail distribution function was assumed to be a Gaussian characterized by three parameters, the parallel forward temperature $T_{\|f}$, the perpendicular temperature $T_\perp$, and the parallel backward temperature $T_{\|b}$. Four each of the four curves, the tail temperatures were the same on all flux surfaces; the density of suprathermal electrons, though, changed. For the four differnt curves in Fig. 1, the temperatures varied by a factor of 10; however, the ratio of the three temperatures $T_{\|f}$, $T_\perp$, and $T_{\|b}$ was kept the same. It is remarkable how little the photon temperature changes across the image in Fig. 1, although the angle between the sight-line and the magnetic field varies significantly from the left side of the image to the right side. The photon temperature increases with increasing energy of the suprathermal electrons, however, the increase amounts only to a

The quantity T_{photon} is only indirectly related to the energy (or temperature) of the suprathermal electrons. We now want to use the PBXRAY code to establish such a relationship. Simulated horizontal profiles of photon temperature T_{photon} are shown in Fig. 1 for four different hot-electron velocity distributions.

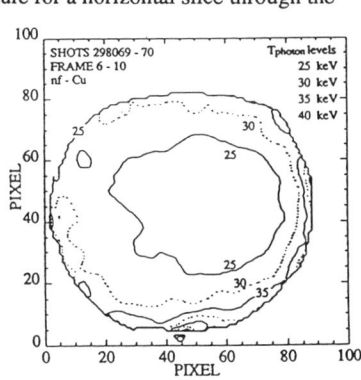

Fig. 4: Contour plot of the photon temperature in a hollow discharge.

factor 2, and the values for the photon temperature are considerably smaller than the values for the temperature of the suprathermal electrons. As a consequence, we believe that the photon temperature can be directly related in an approximate but simple fashion to the energy of the suprathermal electrons. However, a warning should be posted: The number of suprathermal electrons depends strongly on plasma density, and small changes in density may cause large variation in hard X-ray intensity. Therefore, the photon temperature can be measured only for plasma shots that are identical.

3. High Density Discharges.
In this section we want to discuss the analysis of three nearly identical discharges with a hollow hard X-ray radiation profile, PBX-M shots #298069 - 71. The plasma current for these shots was about I_p = 190 kA and the plasma density n_e = 1.6 × 10^{13} cm^{-3}. LHCD with -90° phasing and 185 kW of power took place from time 250 ms to time 650 ms. Simultaneously, neutral beams were injected with 2.2 MW power. As soon as the RF started, hard X-ray images were observed. The hard X-ray intensity rose quickly for 50 ms and then settled into an approximately stationary state. The hard X-ray vertical profile was very hollow initially and remained hollow, albeit somewhat flatter, for the rest of the LHCD period. A contour plot of an X-ray image from the early stage is shown Fig. 2; it is an average of 5 frames from time 270 to 295 ms of shot #298069. The figure exhibits the characteristic crescent-like shape that is typical for "hollow" discharges on PBX-M. Fig. 3a and 3b show the vertical and horizontal profiles from this image (solid curves). The experimental data are overlaid with computer simulations from the PBXRAY code (dotted curves).

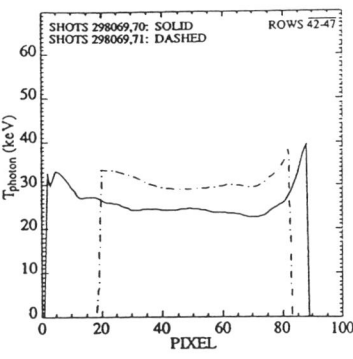

Fig. 5 Horizontal profile of T_{photon}.

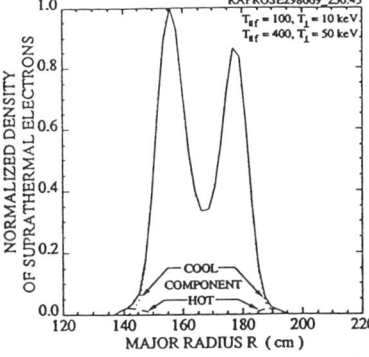

Fig. 6: Density of suprathermal electrons. The velocity distribution and radial profile of the hot electrons that went into the simulations, will be discussed below. A comparison of these results with the data shown at the Innsbruck Conference (Ref. 5), shows that there exists now a much better agreement between simulations and experiment. The improvement is mostly due to the fact that we have completed a calibration of the Hard X-ray Camera. In particular the right "shoulder" of the horizontal profile in Fig. 3a has now the correct height.

For shot #298070 and shot #298071, the copper (Cu) foil and the molybdenum (Mo) foil, respectively, were placed in front of the pinhole of the X-ray Camera. In Fig. 4, we show a contour plot of the photon temperature that was determined from the no-foil shot (nf) and the copper foil shot (Cu) during the time interval 270 -295 ms. The photon temperature is 25 keV in the central region of the discharge, and rises towards the outside to about 40 keV on the large major radius side. Near the edge, the X-ray intensity becomes very small, and the measurement was truncated when the signal became smaller than 3 bits. The photon temperature from the nf - Mo shots looks very much like Fig. 4. The hard X-ray intensity is much weaker for the Mo shot,

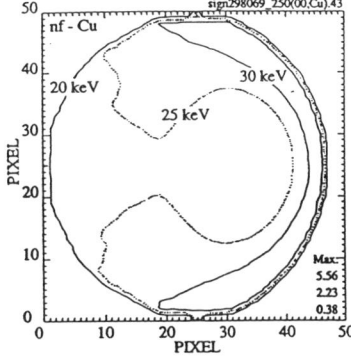

Fig. 7: Simulation of the photon temperature

however, and the measurement has to be truncated further inside. In Fig. 5, we show horizontal profiles of the photon temperature for the nf- Cu case and the nf-Mo case. The nf-Mo case gives slightly higher photon temperatures. The fact that the photon temperature rises towards the large major radius side means that we cannot simulate the discharge with only one distribution function, in other words, the distribution function changes as a function of radius.

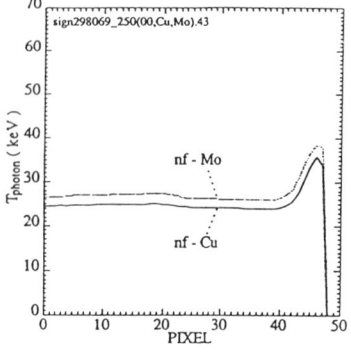

Fig. 8: Simulated horizontal profile of the photon temperature

We have attempted to model the discharge with two distribution functions. The radial profile factors of the two distributions are shown Fig. 6. In the plasma center a "cool" component ($T_{\|f}$ = 100 keV, $T_\perp$ = 10 keV, $T_{\|b}$ = 10 keV) dominates. At the plasma edge there seems to exist a small hot component ($T_{\|f}$ = 400 keV, $T_\perp$ = 50 keV, $T_{\|b}$ = 50 keV), which may be produced by electron runaway. The maximum density of the hot component is only 2 percent of the maximum density of the cool component. Therefore, the hot component makes only a small contribution to plasma current and energy. These distribution functions were also used to compute the two curves of the simulated hard X-ray intensity in Fig. 3a and 3b. In Fig. 7, we show a contour plot of the modeled photon temperature for the nf - Cu case, and, in Fig. 8, the horizontal profiles of the simulated photon temperature for the nf-Cu and the nf- Mo case. The simulations nicely reproduce the rise of the photon temperature on the right (large major radius) side of the image (Fig. 4 or Fig. 5). However up to now, we have no explanation for the small increase of the photon temperature on the left side of the image.

The analysis can now be done not only for one frame, but for a whole shot. In Fig. 9, we show the photon temperature in the central plasma region as a function of time for the nf-Cu case and the nf-Mo case. At time t=400 ms the photon temperature for the nf-Cu case seems to drop suddenly, whereas the nf-Mo temperature stays unchanged. At the time of the drop the plasma density for shot #298070 (the Cu shot) starts to deviate by 5% from the density for the other two shots. Therefore, we think that the drop is not a real change of the photon temperature, but reflects the sensitivity of the absorber foil method to density variations.

The fact that the photon temperature - and consequently the energy of the suprathermal electrons - is so low for the hollow discharge on PBX-M represents a significant new finding. As we mentioned in the introduction, these resulults depart from our former PLT measurements and from data of other machines. Of course the experimental techniques used on the earlier machines were also different. We plan to repeat the experiment using pulse-height-analysis techniques during the next PBX-M run.

Acknowledgments: .The support of Dr. M. Okabayashi and N. Sauthoff and many helpful discussions with members of the PBX-M team are gratefully acknowledged. We also want to thank J. Gorman, R. Such, S. Hosein, and D. Ciotti for the technical assistance in installing and maintaining the foil drive on PBX-M. This work has been supported under DOE contract No DE-AC02-76-CHO-3073

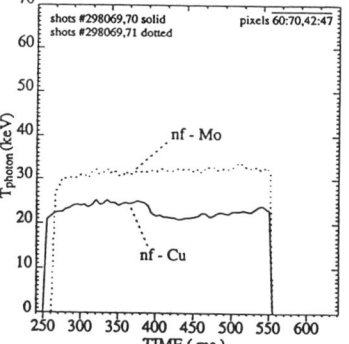

Fig. 9: Photon temperature vs. time.

COMPARISON OF X-RAY PINHOLE CAMERA IMAGES WITH CALCULATIONS BASED ON LOWER-HYBRID WAVE PHYSICS

D. W. Ignat, E. J. Valeo, and S. von Goeler
Princeton Plasma Physics Laboratory
Princeton, NJ   08543-0451

ABSTRACT

An x-ray pinhole camera helps the PBX-M experiment diagnose the location of lower hybrid current drive now being used to demonstrate deliberate modification of the current profile to improve plasma stability and beta. One computational model supporting the experiment is the Lower hybrid Simulation Code (LSC), which can predict the image in the camera after doing multiple ray tracing in general bean-shaped plasmas, and quasilinear damping of waves on electrons. We present experimental and computational images, and discuss the differences and similarities.

INTRODUCTION

Lower hybrid waves[1] have been applied to the PBX-M plasma with the long-range goal of adjusting the current profile to achieve high plasma pressures[2,3] above the Troyon-Gruber limit[4] and therefore in a second stable region. Experiments have shown that current is driven in the plasma. Images obtained from the x-ray pinhole camera show that the location of the fast electrons varies with conditions.[5] Calculations of the current drive in the actual geometry have been pursued with a Lower hybrid Simulation Code[6] (LSC), often in conjunction with the TRANSP[7] and TSC[8] codes which evolve the equilibrium and treat transport in the plasma.

The LSC does multiple ray-tracing in non-circular equilibria, and evolves a quasilinear diffusion coefficient and a consistent a distribution function in $v_\parallel$ on each flux surface. Currents are calculated, using the local dc electric field, with the response function formalism of Karney and Fisch.[9] The basics of this kind of approach are described by Karney.[10] Similar work has been undertaken by others.[11,12]

The distribution function found by LSC can be used to calculate the image on the x-ray camera[5] with the help of classic formulas for bremsstrahlung[13] and the inclusion of geometric details. This has been done, and we discuss here some of our findings. Although we take account of the electric field on the current driven,[9] we do not include effects having to do with electron inertia or diffusion of the high-velocity electrons. That is, electrons are not accelerated in the ambient dc electric field, and electrons remain on the flux surface where they interact with the rf fields.

© 1994 American Institute of Physics

DATA

The camera axis is tangent to a major radius about 10–15 cm inside the magnetic axis of the plasma. The image is formed by chords that sample various flux surfaces at various angles. The instrument produces a 2-dimensional image at many time points in a discharge. The image is usually dynamic and quite complex, so that a simplification in dealing with the large amount of data has been sought. The most common approach is to plot the intensity on a vertical strip of the image that corresponds to the major radius. Figure 1 presents that strip for a series of shots in which the phase separation on the lower hybrid grill launcher is scanned. A typical feature is the double humps with varied amount of fill-in in the center. Parameterized models of the fast electron distribution in space and velocity suggest that the humps are associated with a concentration of fast electrons at some intermediate radial location in the plasma.

CALCULATION

For the series of shots of Fig. 1 we construct a numerical equilibrium from magnetic and Thomson scattering data, and run the LSC, with appropriate aluminum absorber and screen sensitivity. A resultant 2-d image is shown in Fig. 2, and the vertical strip is given in Fig. 3, such that the solid line is the intended strip, and the dots show the result on a strip displaced to smaller major radius aiming point. A phase angle of 105° is assumed.

Note that there is qualitative agreement with the experimental result, in that there are two maxima, but that the center is less filled-in. Other phase angles give similar results, but the location of the peaks is not as constant as it is in the data. For this set of parameters, 150°, 75° and 105° give peaks located more or less as shown, but 120° is more central.

Figure 4 shows the absorbed power density as a function of square root of poloidal flux (roughly radius) in the plasma for the same calculation as in Fig. 3. Note that it shows remarkable structure compared to the vertical strip of the camera image, and also no power in the center at all. The comparison of Figs. 3 and 4 indicates the degree to which radial detail is lost in the camera integration.

The model has also been tried against the experimental density scan at constant phase of 90°. The experiment shows a strong central peak at lower densities and a filled-in double peak at higher densities. Our calculation does not exhibit the strong central peak at low density, but instead indicates double peaks at all densities for this set of experimental parameters.

DISCUSSION

In these discharges the low temperatures (near 1 keV) and low $n_\parallel$ of the launcher (2–3) mean that multiple bounces of the lower hybrid wave and subsequent upshift of $n_\parallel$ are required in our computational model for absorption. As a result, the location of the current driven does not have a simple dependence on the phase angle of the launcher, and, in fact, the location is to some degree independent of launcher phase angle. The x-ray camera calculation in LSC predicts double peaks on a vertical strip of the camera plane, with fairly constant location. This computed result is largely consistent with the persistent location of the double peaks in the measurement.

ACKNOWLEDGMENTS

This work was supported by the U. S. Department of Energy, not only under contract No. DE-AC02-76-CHO-3073, but also through a National Undergraduate Fellowship supporting D. P. Enright, who assisted in coding and installing the camera subroutines.

REFERENCES

[1] Nathaniel J. Fisch, Rev. Mod. Phys. **59** 175 (1987).

[2] M. S. Chance, S. C. Jardin, and T. H. Stix, Phys. Rev. Lett. **51** 1963 (1983).

[3] R. C. Grimm, M. S. Chance, A. M. M. Todd, et al., Nucl. Fusion **25** 805 (1985).

[4] F. Troyon and R. Gruber, Phys. Lett. **110A** 29 (1985).

[5] S. von Goeler, N. Asakura, R. Bell, et al., European Physical Society Meeting, 1992

[6] D. W. Ignat, E. J. Valeo, and S. C. Jardin, in preparation.

[7] R. Hawryluk, in *Physics of Plasmas Close to Thermonuclear Conditions*, (Varenna, 1979), **I** 61 (1979).

[8] S. C. Jardin, N. Pomphrey, and J. DeLucia, J. Comput. Phys. **66** 481 (1986).

[9] Charles F. F. Karney and Nathaniel J. Fisch, Phys. Fluids **29** 180 (1986).

[10] Charles F. F. Karney, Comput. Pys. Rep. **4** 183–244 (1986).

[11] P. T. Bonoli and R. C. Englade, Phys. Fluids **29** 2937 (1986).

[12] J. Kesner et al., to be published in Nucl. Fusion (1993).

[13] H. W. Koch and J. W. Motz, Rev. Mod. Phys. **31** 920 (1959).

[14] W. H. McMaster, et al., UCRL 50174, Sec. II rev. 1, May 1969.

FIGURES

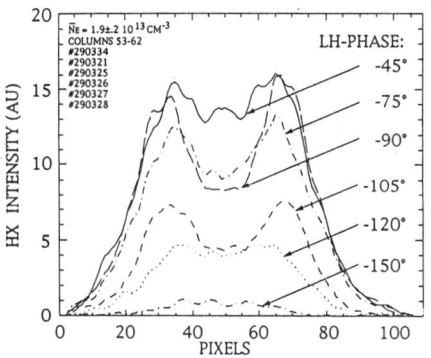

Fig. 1 Vertical strip on the hard x-ray camera in a phase scan showing persistent double-peaking and varied fill-in and amplitude

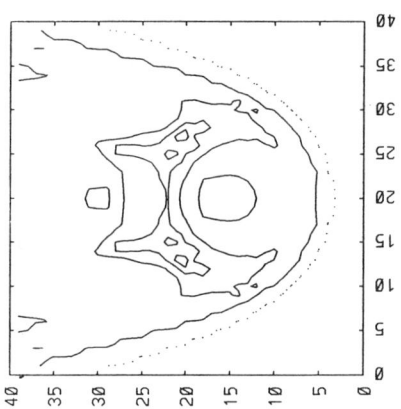

Fig. 2 Camera image computed by LSC as shown by contours of constant intensity. Phase angle 105°.

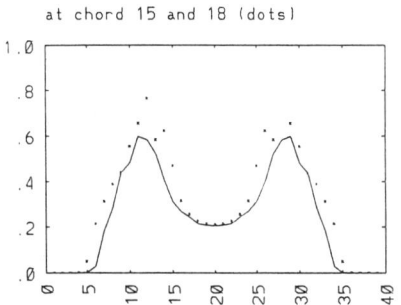

Fig. 3 Vertical strip in the computation. The line represents the strip of Fig. 1; dots a near-by strip. Phase angle 105°

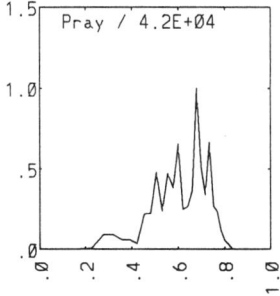

Fig. 4 Power density deposited in the plasma by lower hybrid waves, as a function of square root of poloidal flux.

MOTIONAL STARK EFFECT PLASMA EQUILIBRIA DURING LHCD EXPERIMENTS ON PBX-M

F. PAOLETTI[#], S. BATHA[+], S. BERNABEI, H. FISHMAN, R. HATCHER,
S. HIRSHMAN[*], D. IGNAT, S. JONES[#], R. KAITA, S. KAYE, J. KESNER[#],
C. KESSEL, B. LE BLANC, D. LEE[*], F. LEVINTON[+], S. LUCKHARDT[#],
M. OKABAYASHI, H. TAKAHASHI, S. SESNIC, Y. SUN, S. VON GOELER

Princeton Plasma Physics Laboratory, Princeton N.J. 08543
[#]Massachusetts Institute of Technology, Cambridge M.A. 02139
[+]Fusion Physics and Technology, Torrance C.A. 90503
[*]Oak Ridge National Laboratory, Oak Ridge T.N. 37831

ABSTRACT

The magnetic field pitch angle measurements from the multichannel Motional Stark Effect (MSE) polarimeter system, together with the external poloidal flux values, toroidal plasma current, coil currents and plasma pressure profile, as computed by the TRANSP code, are used to constrain the numerical solution of the Grad-Shafranov equation. The experimental results from the soft x-ray diagnostic and the hard x-ray bremsstrahlung radiation profiles show clear signatures of a change in the current density profile during Lower hybrid injection. Indications of a variation in the q-profile are supported by the behavior of the measured sawtooth inversion radius. Modifications in the current density profiles, induced by Lower Hybrid waves, are discussed in terms of their effect on q-profiles and volume-averaged quantities like l_i.

INTRODUCTION

Two numerical schemes are used on PBX-M for plasma equilibrium reconstruction, namely the FQ code[1, 2] and the VMEC code[3]. They both solve the MHD equilibrium Grad-Shafranov equation in a toroidally symmetric domain:

$$\Delta^* \psi = -\frac{4\pi}{c} R J_\Phi = R^2 \frac{dp}{d\psi} - g \frac{dg}{d\psi} \quad (1)$$

where ψ is the poloidal flux function, $p(\psi)$ is the pressure and $g(\psi) = RB_T$. The FQ code solves the equilibrium equation for ψ on a rectangular grid in R-Z space. In contrast, VMEC solves Eq.(1) in the inverse form and obtains the R and Z coordinates, directly as functions of ψ and the poloidal angle θ. In the FQ code, the ψ derivative of the pressure is expressed in terms of a polynomial expansion:

$$p'(\psi) = -[p_1 \alpha \psi^{\alpha-1} + p_2 \psi^{\delta_1}(1-\psi) + p_3 \psi^{\delta_2}(1-\psi)(\psi-1/2) + \\ + p_4 \psi^{\delta_3}(1-\psi)(\psi-2/3)(\psi-1/3)] \quad (2)$$

During the calculation, the exponents α, δ_1, δ_2, δ_3 are fixed. The code derives the new values for the coefficients p_1, p_2, p_3 and p_4 by fitting the measured electron pressure profile from the Thomson Scattering diagnostic[4]. This is an iterative process, since a solution of Eq.(1) is needed to map $R(\psi)$ and the match with the experimental pressure profile affects the equilibrium. Since both FQ and VMEC are free boundary codes, the plasma shape is determined self consistently from pressure balance in response to external coils and the plasma current. The neutral beam contribution to the total pressure, as computed by the TRANSP code[5], is also taken into account. The function gdg/dψ is expanded as a function of ψ in a similar way:

$$gg'(\psi) = -[g_1 \psi^\beta + g_2 \psi^{v_1}(1-\psi) + g_3 \psi^{v_2}(1-\psi)(\psi - 1/2) + \\ + g_4 \psi^{v_3}(1-\psi)(\psi - 2/3)(\psi - 1/3)] \quad (3)$$

Here the exponents β, v_1, v_2, v_3 are fixed during the calculation. The coefficients g_1, g_2, g_3 and g_4 are adjusted to provide a least squares fit to the measured values of:
- external poloidal flux differences between the measured values from pairs of solenoidal magnetic probes placed in different poloidal location outside the PBX-M plasma, (typically 4 of these differences are used)
- total plasma current and 10 external coil currents measurements
- vertical magnetic field profile on the mid plane, from MSE[6, 7] at 12 radial points.

In VMEC, the arbitrary free functions of ideal MHD are q(ψ) and p(ψ) and their profiles are changed incrementally through a conjugate-gradient iteration process, so that the final steady state is reached simultaneously for the equilibrium and the optimized profiles. The inverse of q, as well as p, are parametrized as a function of ψ using a Chebychev polynomial expansion:

$$\frac{1}{q(\psi)} = \sum_n i_n T_n(\psi) \quad ; \quad p(\psi) = \sum_n p_n T_n(\psi) \quad (4)$$

The i_n and p_n coefficients are self-consistently determined to give a best least squares fit to the experimental data. Typically n=5-7 coefficients are sufficient.

EXPERIMENTAL RESULTS

Lower Hybrid Current Drive (LHCD) through its change of the current density profile, can influence plasma properties in various ways depending on plasma shape and configuration. Different behavior has indeed been observed between a circular and a bean-shaped discharge. An example of a circular discharge is given in Fig.1, where we can highlight the following points: an LH power of 175 kW has been coupled to the plasma and a drop in the surface loop voltage of about 50% has been observed as evidence of current drive. The electron temperature profiles, with and without LH, are very similar showing that this drop is not due to a change in plasma resistivity. The electron density and temperature are:

$\langle n_e \rangle \cong 1.3 \times 10^{13} cm^{-3}$, $T_e(0) \cong 0.8 keV$.

The enhancement of 20% in the internal inductance l_i shows a peaking in the current density profile and suggests a

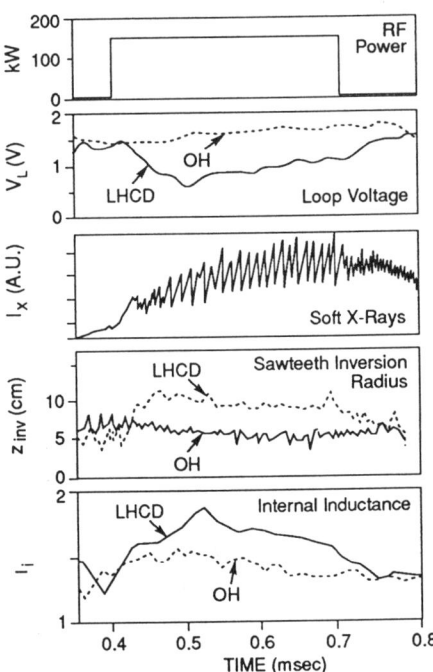

Fig.1 Circular plasmas and central LHCD

nearly central deposition. The hard x-ray emission profile is very peaked in the center, supporting the hypothesis that the fast electrons are produced in the central region and heavily accelerated by the remaining electric field (this is, in fact, a typical "run away" discharge). The sawtooth inversion radius is clearly moved outward during LH injection if we make a comparison with a similar ohmic discharge. There is no sawtooth stabilization or suppression and the period between sawtooth oscillations increases by a factor 4 with LHCD.

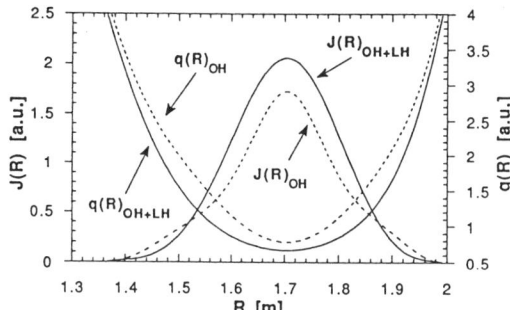

Fig.2 q(R) and J(R) for a circular discharge

The indication (from soft x-ray data) that the q=1 surface is enlarged in the presence of LHCD has been confirmed using the MSE data in the FQ and VMEC code to generate the q profiles in the mentioned cases (see Fig.2). The q profile is lower, in presence of LH by about 30% and it reaches the value q=1 at a larger minor radius than in the ohmic case. The relative change in the q=1 radius ($\approx$ 3.3 cm) quantitatively agrees with soft x-ray sawtooth inversion radius measurements.

A more peaked current density profile is observed, as an output of the reconstruction codes, in the LH case compared to the ohmic one. This supports the idea of a current driven in the central region.

As an example of the different behavior observed in a bean-shaped discharge, we refer to the case shown in Fig.3. Here, 183 kW of LH power was coupled to the plasma for a period of 400 ms, together with 500 kW of neutral beam injection over a time of 300 ms. The MSE measurements were taken when both the LH and NB were on. In this case, the observed drop in the surface loop voltage is about 40% and the current drive is reduced but not lost by the occurrence of an MHD mode (m/n=3/2) at t=325 ms. The enhancement in the internal inductance is very small, showing a more off-axis deposition. The sawtooth inversion radius is clearly moved inward during LH injection, in comparison with a similar OH + NBI shot. The soft x-ray data suggest that the area of the q=1 surface is reduced during LHCD. In this case, the electron density and temperature are:

$\langle n_e \rangle \cong 1.1 \times 10^{13} cm^{-3}$, $T_e(0) \cong 1.2 keV$

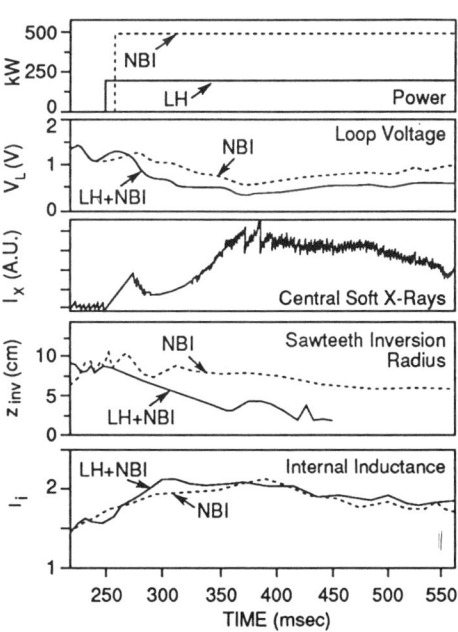

Fig.3 Bean-shaped plasmas and off-axis current drive

The behavior of the MSE-q (produced with FQ and VMEC) at t=450 ms clearly supports this hypothesis (see Fig.4).

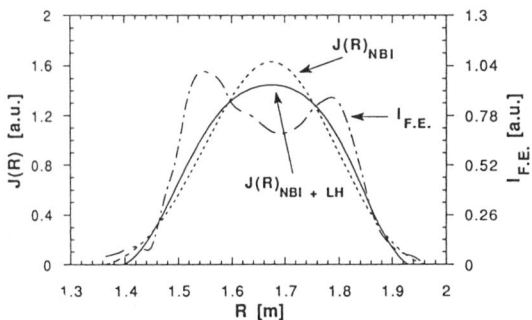

Fig.4 J(R) and $I_{F.E.}$ (fast electron current density) for a bean-shaped discharge

The q profile is higher, in presence of LH, by about 25% in the center, and the q=1 radius is smaller compared with a similar OH + NBI discharge. The current density profile is more broad in the LHCD case, and the radial position of the distortion in the j profile (due to the additional j_{LH}) is consistent with the inverted hard x-ray emission profile data (see Fig.4). The total hard x-ray intensity profile, on the other hand, is hollow initially and after the MHD mode appears, the hollowness is gradually reduced but still present at t=450 ms when the q profile is measured.

CONCLUSIONS

Two different numerical schemes, FQ and VMEC, have been successfully used to reconstruct magnetic equilibria on PBX-M. Cross-checking of the two computer codes has been performed on various discharges with differing plasma conditions. There has been good quantitative agreement between the two numerical methods. The results from both codes show a qualitatively different plasma response depending on plasma shape and other configurational parameters. In circular plasmas, a lowering of the q profile has been shown to occur in the presence of LHCD. In this case, centrally peaked current profiles, with substantial increase of l_i, is calculated. In bean-shaped plasmas, the q profile tends to rise only moderately with the application of the LH power and the q=1 surface moves inward, in agreement with the observations of the soft x-ray sawtooth inversion radius. The distortion in the current density profile induced by LH is consistent with the fast electron radial distribution from the inverted hard x-ray emission profile data. It has been demonstrated that the MSE diagnostic is indispensable for accurately reconstruction of the q profile and, in particular, for inferring the radius of the q=1 surface inside the plasma.

REFERENCES

1. C. E. Kessel et al., Bull. Am. Phys. Soc., **34**, 2050 (1989)
2. F. Paoletti et al., Bull. Am. Phys. Soc., **37**, 1573 (1992)
3. S. Hirshman and D. Lee, Comp. Phys. Comm., **39**, 161 (1986)
4. B. Le Blanc et al., Rev. Sci. Inst., **61**, 3566 (1990)
5. R. Kaita et al. Nucl. Fusion, **25**, 939 (1985)
6. F. Levinton et al., Phys. Rev. Lett., **63**, 2060 (1989)
7. D. W. Roberts et al., Rev. Sci. Inst., **61**, 2932 (1990)

Supported by USDOE Grants: MIT: DE-FG02-91E54109, PPPL: DE-AC02-76-CHO-3073, ORNL: DE-AC05-84OR21400

ELECTRON CYCLOTRON RANGE OF FREQUENCIES

HIGH-POWER 140 GHz ECRH EXPERIMENTS AT THE W7-AS STELLARATOR

V. Erckmann, R. Burhenn, T. Geist, H.J. Hartfuss, M. Kick,
H. Maassberg, W7-AS Team[1], NBI Team[2]
Max-Planck-Institut für Plasmaphysik, EURATOM Ass., D-8046 Garching, Germany

W. Kasparek, G.A. Müller, P.G. Schüller
Institut für Plasmaforschung, Universität Stuttgart, Germany

V.I.Il'in[*], V.I. Kurbatov, S. Malygin[**], V.I.Malygin[***]
[*] Kurchatov Inst., Moscow, Russia, [**] SALUT, Nizhny Novgorod, Russia,
[***] Inst. for Applied Physics, Nizhny Novgorod, Russia

ABSTRACT

A new 140 GHz ECRH system with 0.5 MW heating power and up to 1.1 s pulse duration was set into operation at the W7-AS Stellarator giving access to plasma experiments with ECRH at plasma densities up to 1.2×10^{20} m^{-3}, which is the X-mode cut-off density for this frequency. H-mode transitions, were observed for the first time in a stellarator during high density operation. The operational window for these transitions and the correlated phenomena are discussed in the light of the well known tokamak results. The onset of H-mode transitions at W7-AS was investigated with particular emphasis on the influence of the external gas feed. H-mode transitions in deuterium and hydrogen as a working gas are compared. The result of stimulated heat wave propagation by modulation of the ECRH power before and after the H-mode transition is discussed.

ECRH was combined with NBI at moderate and high densities with on- and off-axis heating. Density control was achieved for combined heating in long pulse operation despite the beam fuelling in contrast to discharges with pure NBI heating. Particle confinement degradation by profile changes inferred by ECRH is discussed as a possible mechanism. The impurity confinement is strongly affected and is discussed for on- and off axis combined heating conditions.

---

1) W7-AS Team: J. Baldzuhn, B. Bomba, R. Brakel, G. Cattanei, A. Dodhy, D. Dorst, A. Elsner, M. Endler, K. Engelhardt, V. Erckmann, U. Gasparino, S. Geißler, L. Giannone, P. Grigull, H. Hacker, O. Heinrich, G. Herre, D. Hildebrandt, J.V. Hofmann, R. Jaenicke, F. Karger, A. Kislyakov, H. Kroiss, G. Kuehner, A. Lazaros, C. Mahn, K. McCormick, H. Niedermeyer, W. Ohlendorf, P. Pech, H. Ringler, A. Rudyj, N. Ruhs, J. Saffert, F. Sardei, S. Sattler, F. Schneider, U. Schneider, G. Siller, U. Stroth, M. Tutter, E. Unger, F. Wagner, A. Weller, U. WenzeL, H. Wolff, E. Würsching, D. Zimmermann, M. Zippe, S. Zöpfel
2) NBI-Team: W. Ott, H.P. Penningsfeld, E. Speth

1. INTRODUCTION

The ECRH capability for plasma start-up from the neutral gas and highly effective electron heating assigns a key role to ECRH in the field of stellarator research. Major drawbacks of presently running high power ECRH systems, however, were the density restrictions given by the cut-off density for the present day microwave frequencies and the limited power per microwave source (typically 0.2 MW). The application of 140 GHz ECRH overcomes this density restrictions and opens a new parameter window for plasma physics investigations. The successful development of gyrotrons with built-in quasi optical mode converters allowed to increase the unit-power to approx. 1 MW and, at the same time simplifies the transmission system substantially, because a linearly polarized Gaussian microwave beam is generated by the gyrotron and can be transmitted by a simple mirror system. The new system, although not yet capable of CW operation, is a major qualitative step to qualify ECRH as an attractive bulk heating system for next step devices, because of its simplicity and wide variety of applications for plasma heating and control.

A brief description of the W7-AS experiment together with the 140 GHz gyrotron and transmission system is given in Section 2. First plasma experiments focused on the investigation of high density plasmas, which could be generated up to now with NBI only. The comparatively low densities accessible with 70 GHz ECRH provided clear experimental conditions for the investigation of electron heat transport, because the ions are only weakly coupled to the electrons and do not play a significant role in the overall power balance. At high densities, however, the collisional electron-ion coupling becomes strong and the ion confinement becomes important in the power balance analysis. Under high density conditions, H-mode transitions, which are well known from tokamak experiments [1-7] were observed for the first time in a stellarator. The operational window for these transitions and the correlated phenomena are discussed in Section 3. The basic features of the Stellarator H-mode were reported in previous papers [8,9] and are only briefly reviewed. Here we concentrate on a more detailed investigation of the influence of gas puffing on the H-transitions and a comparison of the H-mode transitions in hydrogen and deuterium.

In the second part of this paper, combined heating experiments with NBI are discussed, which is inherently related to high density operation. Typical high-power NBI-heated plasmas in W7-AS are non-stationary, because the combined effect of beam particle fuelling and recycling causes a steady density rise [10]. Such discharges are usually terminated by a thermal collapse induced by the increase of the impurity radiation with increasing density. Quasistationary behaviour could only be achieved in some corners of the W7-AS parameter range, i.e. operation in poor confinement regimes selected by an appropriate choice of the external rotational transform or directly after glow-discharge conditioning of the vessel walls. There was clear evidence from early experiments at W7-A [11,12] and W7-AS, that the density could be controlled by combining ECRH and NBI. A strong impact on the impurity confinement was found, which offers an experimental tool to control the impurities in properly chosen combined heating scenarios. These experiments suffered, however, from either short pulse lengths of ECRH and/or low density operation given by the 70 GHz cut-off constraints. Experiments on combined heating in long pulse operation and at high densities are presented in Sec. 4.

2. EXPERIMENTAL ARRANGEMENT

W7-AS is a modular stellarator with 5 magnetic field periods, low vacuum magnetic shear and high aspect ratio, the average minor and major radii are ≤ 0.18 and 2.0 m, respectively. The plasma cross section changes from nearly triangular to nearly elliptical within half a magnetic field period, indicating the dominant Fourier components (l = 2,3) of the magnetic configuration.

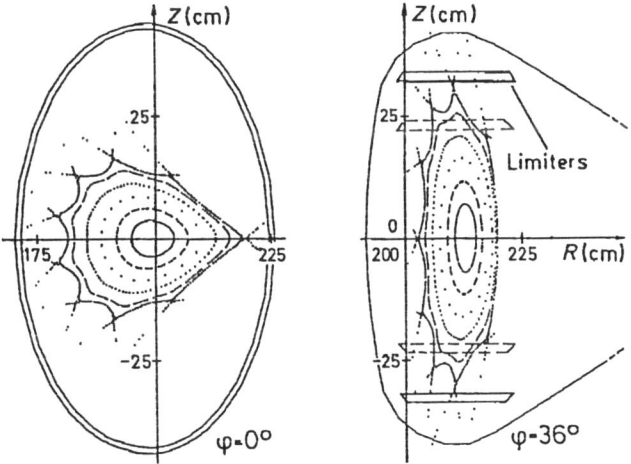

Fig. 1. Magnetic topology change within half a field period at ᴛ > 0.5 and extremal positions for the movable instrumented limiter.

Two movable instrumented limiters can be inserted to control the plasma edge or removed outside the last closed flux surface for sufficiently high rotational transform (ᴛ > 0.5) as shown in Fig. 1. The plasma is limited in this case by a natural separatrix, which is shaped by the relicts of 5/m natural magnetic islands originating from the non-axissymmetry of the configuration. For ᴛ < 0.5, the limiter always defines the last closed magnetic flux surface, because the poloidal cross section of the plasma increases.

Net current free plasma build-up and heating is achieved at W7-AS with up to 0.8 MW microwave power at 70 GHz in long pulse operation (< 3 s). The corresponding resonant magnetic induction is 2.5 T, with a cut-off density of $n_{e,crit} = 6.2 \times 10^{19}$ m^{-3} for 1st harmonic O-mode operation. Four microwave beams (0.2 MW each) are combined at one poloidal plane and each beam can be steered independently by movable launching mirrors inside the vacuum vessel.

A separate 140 GHz prototype gyrotron from the development line at the IAP-Nizhny Novgorod generates 0.5 MW power with up to 1.1 s pulse duration . The gyrotron operates alternatively at 0.9 MW for 0.3 s pulse duration and has a build in optical mode converter. The linearly polarized Gaussian output beam is transmitted to the plasma by an optical mirror system [13] with 10 mirrors. The overall transmission efficiency was measured calorimetrically at high power and is 80%. The main losses (15%) are coupling losses from the gyrotron output to the optical transmission system due to non-Gaussian emission (side lobes) from the gyrotron, because the internal mode converter is not yet fully optimized. Additional 5% are transmission line losses such as spill-over and resistive losses. The transmission system allows for an arbitrary choice of the wave polarization. The microwaves are launched similar to the 70 GHz system by two movable mirrors inside the vacuum vessel from the low magnetic field side in the equatorial plane of the torus and can be steered with an arbitrary toroidal and poloidal launch angle, respectively. The experiments were performed with perpendicular launch in the X-mode polarization at the 2nd harmonic. The cut-off density for this mode is 1.2×10^{20} m^{-3}. The microwave beams of both, the 70 and the 140 GHz systems are launched at equivalent poloidal planes with elliptical cross sections and an almost tokamak-like 1/R dependence of the magnetic induction. Thus the power deposition profile is very narrow with a typical spatial width of 3 cm.

The experiments reported here were performed at the EC-resonant magnetic induction of $B_0 = 2.5$ T, with $\bar{\iota}(a) \approx 0.5$. The plasma net current was feedback controlled by compensating the bootstrap current with the OH transformer ($I_p = 0$). All experiments were performed with boronized vacuum vessel walls and graphite armour of exposed in-vessel components.

3. H-MODE TRANSITIONS AND CORRELATED PHENOMENA

The plasma is generated and heated by a 70 GHz ECRH prepulse with sufficient pulse duration to ramp up the density and establish a quasi steady state phase just below the 70 GHz cut-off density at typically $4-5 \times 10^{19}$ m^{-3} by feed back controlled gas puffing. Then the 70 GHz heating pulse is turned off and followed by 140 GHz ECRH with approximately the same heating power of 0.4 MW. As seen from Fig. 2, the density is ramped up further in this phase until a preprogrammed central density of 0.9×10^{20} m^{-3} is reached. In the high density flat top phase a transition to a plasma state with better global confinement occurs as seen from the total stored plasma energy trace W in Fig. 2 for a hydrogen discharge. All phenomena reported later on are very similar for hydrogen and deuterium operation in contrast to tokamak results [7,14,15]. The density starts to increase further after the transition, although

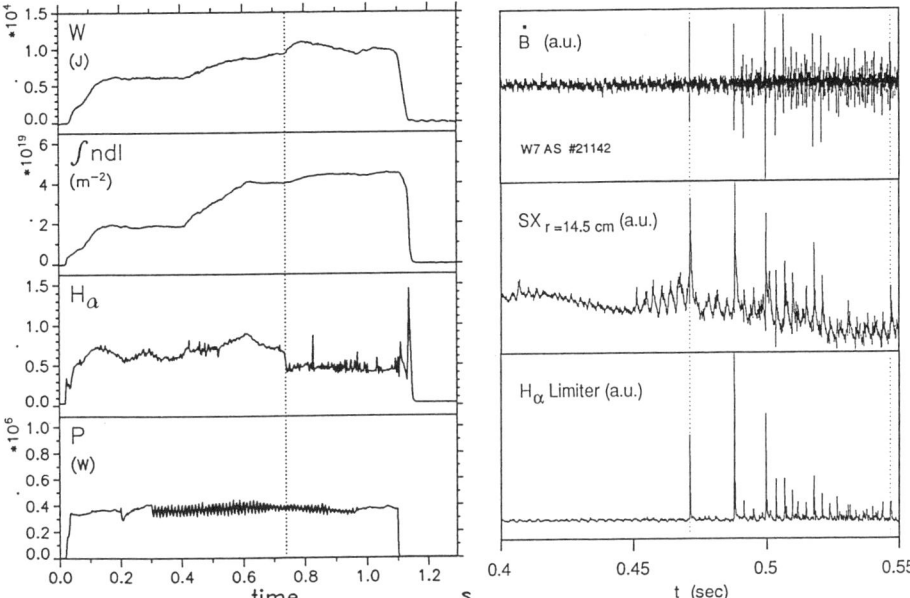

Fig. 2. Time evolution of the total stored plasma energy W, the line integrated density ∫ndl for the central chord, the H_α emission intensity at the top limiter and the ECRH input power P. The dotted line indicates the transition. Working gas is hydrogen.

Fig. 3. ELM activity from Mirnov coils (top), soft X-ray emission (middle) from a chord close to the plasma boundary and the correlated H_α signal (bottom) at the top limiter during the H-mode phase of a deuterium discharge.

the external gas feed is turned off by the feedback system, which indicates an improved particle confinement. The most pronounced signatures of the transition is a drop of the H_α line emission signals from both limiters (only one signal is displayed) indicating a reduced particle recycling. A reduction of the magnetic turbulence level is measured by Mirnov coils. Like in the H-mode of tokamaks, there is a back transition indicated by the sudden rise of the H_α emission after the ECRH-pulse is turned off. Some erratic bursts appear during the reduced recycling phase, which are correlated with magnetic fluctuations and SX-emission fluctuations as seen from Fig. 3. This type of instabilities is identified as the well known Edge Localized Modes (ELMs) appearing in tokamak H-modes.

The electron temperature profile as a whole increases after the transition with a pronounced steepening of the edge temperature gradient forming a temperature pedestal. The edge ion temperature measured by impurity line broadening increases also after the transition. These findings, together with the measured steepening of the edge density gradients give clear evidence for the development of a transport barrier close to the plasma edge.

An enhanced poloidal plasma rotation velocity at the plasma periphery of up to 3.5 km/s was measured in the H-mode like phase by B IV impurity line Doppler measurements and indicates an increase of the negative radial electric fields from typically -30 to -100 V/cm.

The total power emitted by impurity radiation rises after the transition and does not level off during the H-mode like phase of the discharge. The total radiated power measured by a 10 channel bolometer amounts up to about 50% of the input power. The impurity line emission related to the bulk plasma (e.g. O VIII, Fe XVI) increases, whereas the edge impurity line emission (e.g. O V, C III) remains essentially constant or even drops indicating no enhanced impurity influx. The impurity line emission from Al XII, which was inferred by laser blow-off techniques showed no drop within the available heating pulse duration. The spectroscopic measurements during the H-phase are compatible with impurity accumulation effects, a conclusive interpretation, however, is difficult, because of the nonstationary profiles of n_e and T_e.

The density has to be ramped up slow enough to avoid a serious deterioration of the discharge by strong edge cooling. The smoothness of the transition is dependent on the strength of the gas puffing in the density ramp-up phase and on the wall conditions (boronization). The development of the total stored plasma energy is shown in Fig. 4 as a function of $\int n dl$ for two series of discharges with a fast and a slow density ramp of 2 and 1×10^{20} m^{-3} s^{-1}, respectively. The transition occurs in both cases at approximately the same line integrated density, which corresponds to $n_{eo} = 5 \times 10^{19}$ m^{-3} and is the lower density limit for the transitions, also the energy gain is the same. With the slower density ramp, however, the transition is accompanied by a series of ELM's, which smooth out the transition until the H-mode is finally settled. The H_α signal is shown for both cases at the bottom of Fig. 4.

The quasistationary dynamic density scan shown in Fig. 4 confirms the global energy confinement-time scaling with density (all other parameters are constant)

$$\tau_e \propto n_e^{0.6} \cdot P^{-0.6} \cdot B^{0.6} \cdot r^{1.8} \cdot \tau^{0.3}$$

which was found from a multiple regression analysis of the W7-AS data base [16]. The increase of the confinement time in the H-mode is in the range of 10 - 30% and is still within the data scatter of the experimental data used for the regression analysis.

A comparison of the local heat diffusivity derived from stimulated heat wave propagation [17] before and after the transition was possible for discharges, which exhibit a sufficiently long steady state phase before the transition occurs, see for example Fig. 2. Such discharges could only be generated at the end of the campaign, where the effect of wall boronization became already week and the density threshold increased. For fresh boronized walls, the transition always occurs in the nonstationary density ramp-up phase (see Fig. 4) which makes an interpretation of the stimulated heat wave method very difficult if not impossible. The heat waves were exited by a square wave modulation (typically 15%) of the microwave power at 92 Hz and followed

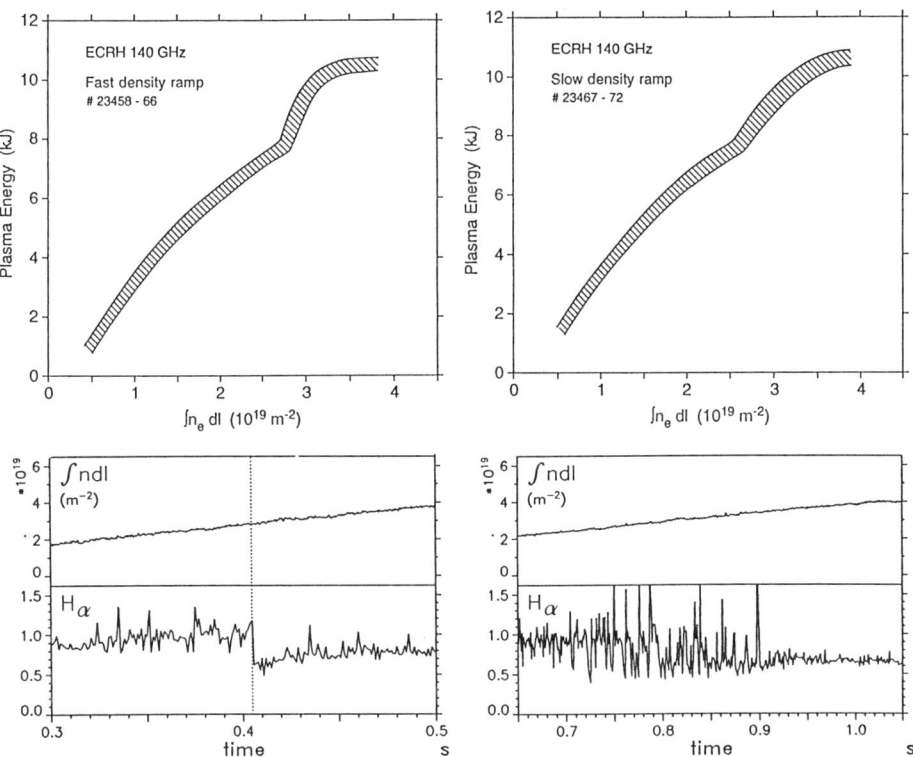

Fig. 4. Total stored plasma energy as a function of the line integrated density $\int n dl$ for fast (top left) and slow (top right) density ramp-up. The time development of $\int n dl$ and the H_α signal is given on bottom for both cases. Note, that the time intervals are different by a factor of two.

in phase and amplitude by a multichannel ECE radiometer. The phase and amplitude of the propagating heat wave is shown in Fig. 5 as a function of the effective radius averaged over a 200 ms time interval before and after the transition

As seen from Fig. 5, the heat wave is travelling slower in the H-mode phase of the discharge as compared to the pre-transition phase. The power deposition region is centred within ± 2.5 cm. The interpretation in terms of heat diffusivity is difficult, because at the high densities of 0.9×10^{20} m^{-3} the ions and the electrons are strongly collisionally coupled and the derived heat diffusivity characterizes both, the electron and ion heat diffusion. Furthermore, the slight density increase after the transition would already lead to a reduced heat diffusivity and has to be taken into account. A simple analysis of the heat wave propagation in cylindrical geometry and assuming constant χ in the analyzed radial interval gives a reduction from 1.0 m^2/s in the pre-transition phase to 0.7 m^2/s in the H-mode phase. A thorough analysis of the experimental results taking into account the density effects and the electron-

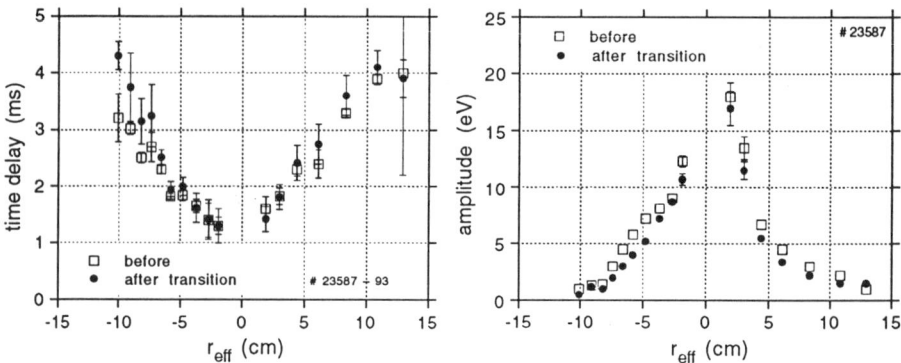

Fig. 5. Time delay (left) and amplitude decay (right) from heat wave propagation as a function of the effective radius in the pre-transition phase (squares) and after the transition (dots).

ion coupling is under way. Please note, that the heat diffusivity derived from heat wave experiments in the W7-AS stellarator in general does not deviate much from the steady state power balance analysis [17]. This is in contrast to the tokamak heat wave propagation experiments, where a discrepancy of typically a factor of 2 or more is found.

An examination of the experimental range for the achievement of the H-mode transitions indicates, that the edge value of the rotational transform, the density and the edge condition (limiter position) are leading parameters.

Below a line integrated density of 2×10^{19} m^{-2} ($n_{eo} \approx 4\text{-}5 \times 10^{19}$ m^{-3}) no transitions were observed, which explains a posteriori, that no clear H-mode phenomena were found in previous ECR-heated discharges, because of the density limitations given by the cut-off condition for 70 GHz heating.

The choice of the rotational transform is closely interlinked with the magnetic topology at the plasma boundary. Magnetic islands have significant influence on the edge structure due to the low shear in W7-AS. In all cases investigated so far (keeping in mind the limitation in the available heating power), the H-mode transitions occur only in a narrow parameter window around $\mathrm{t} = 0.52$.

The transition is restricted to cases where the limiter is not or marginally affecting the plasma edge (for topology, see Fig. 1). The power flow to both limiters is typically ≤ 30% of the total power in this cases. The limiter loading is concentrated to a small stripe of about 2 cm width along the limiter surface indicating that the power flow is mainly through the X-point of the separatrix. There is, however, some uncertainty about the exact position of the separatrix. The transition was suppressed completely, if the limiter was inserted about two centimetres from the last closed flux surface. The experimental boundary conditions for achievement of the transitions suggest, that an unperturbed and separatrix dominated plasma boundary is required to establish the H-mode like transitions with the available heating power.

No clear-cut results on the existence of a power threshold for H-mode transitions, which is found in tokamaks, could be obtained, because the heating power is already marginal to maintain the high density plasmas. A reduction of the heating power led to unstationary plasma behaviour and the interpretation becomes difficult.

4. COMBINATION OF ECRH AND NBI

Experiments with combined heating were performed in a wide parameter range at moderate densities around $n_{eo} \approx 5 \times 10^{19}$ m^{-3} and high densities around $n_{eo} \approx 1 \times 10^{20}$ m^{-3} with both, on- and off-axis power deposition of ECRH. In the following we concentrate on the effect of density clamping with combined heating. Experimental evidence for density clamping by ECRH was reported also from several other experiments [18,19]. An example is given in Fig. 6 with on axis ECRH at $n_{eo} \approx 5 \times 10^{19}$ m^{-3}. The discharge is started by a 70 GHz, 0.4 MW pre-pulse followed by a 140 GHz pulse, which is turned off at 0.95 s (the density is slightly ramped up in the early phase of this pulse). NBI is added with 0.35 MW power at 0.6 s and turned off at 1.15 s, i.e. the final phase of the discharge from 0.95 to 1.15 s is heated by NBI only.

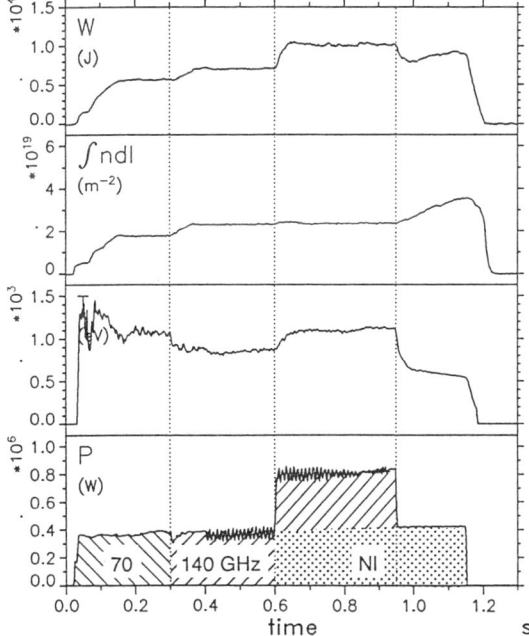

Fig. 6. Time evolution of the total stored plasma energy W, the line integrated density ∫ndl for the central chord, the central electron temperature from SX-diagnostics and the cumulative input power P for combined heating with ECRH and NBI. The ECRH power deposition is on axis.

During the ECR-heated prephase and the combined heating phase a pre-programmed constant density is maintained by feedback controlled gas puffing, whereas after switch off of ECRH the density increases, although the external gas feed is turned off by the feedback system. The density increase with pure NBI heating is a well known feature at W7-AS [10], where beam fuelling and recycling fluxes have a comparable contribution to the overall density rise. The beam fuelling is typically 1×10^{20} s^{-1} for 0.35 MW beam power. Despite the beam fuelling, the density can be controlled with combined heating. This is found also with off-axis ECRH power deposition at r/a = 0.5, at high plasma densities and with combination of 0.75 MW ECRH and 0.7 MW NBI heating. As a rule, the density control could be maintained only, if the ECRH power was approximately equal or higher than the NBI power. Two alternative explanations for the ECRH density clamping were discussed in the past, i.e. ECRH may affect the recycling by additional ionisation outside the plasma edge due to incomplete absorption in the plasma centre, or it may cause a degradation of the particle confinement in the bulk plasma [11]. The experiments reported here were performed with 2nd harmonic X-mode launch of ECRH. This mode is optically thick for the given plasma parameters and the absorption is complete in the plasma centre, which is confirmed by the heat wave experiments. It is unlikely therefore, that screening by a non-absorbed power fraction should have significant influence on the particle recycling. Additional evidence for degraded particle confinement in the bulk plasma comes from the profile development in the different heating cases. The density profile for combined heating is flat and the electron temperature profile is strongly peaked with on axis power deposition of ECRH. On the other hand, the density profile becomes slightly peaked and the T_e profile is flat in the central plasma zone within the deposition region for off-axis power deposition of ECRH. These results suggest, that the particle confinement is affected mainly by the profile changes inferred by the additional ECRH, i.e. nondiagonal terms such as thermodiffusion in the transport matrix may drive the particle diffusion. This would also explain the empirical finding, that a certain ECRH-power is needed to provide the density control. For the given high plasma densities sufficient ECRH power has to be added to shape the temperature profile properly. The impurity confinement was investigated by Al laser-blow off experiments, where the central impurity line emission from Al XII shows a decay time of typically 200 ms for the on axis combined heating case, whereas the decay time is much larger (and could not be quantified within the available discharge time interval) in the off-axis case as seen from Fig. 7. For both cases, the decay of the Al XII line emission is very similar in the ECRH prephase and in the combined heating phase.

The understanding of the measured behaviour is of crucial importance, because combined heating may provide tools for impurity control in steady state plasma operation. A thorough analysis of the measured profiles for the various heating scenarios is under way.

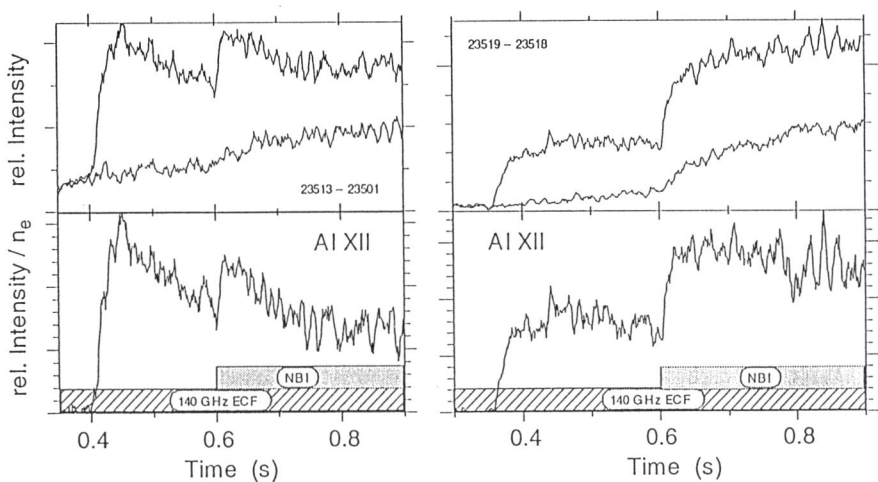

Fig. 7. Relative intensities of the Al XII line emission from laser blow off (top figure, upper trace) and the background emission (top figure, lower trace). The relative intensity (subtracted background) normalized to the density is given in the bottom figure for on-axis (left) and off-axis ECRH (right).

5. SUMMARY AND CONCLUSION

A new 140 GHz ECRH system with 0.5 MW microwave power and up to 1.1 s pulse duration was set into operation at W7-AS. The parameter range for ECR-heated discharges was extended to densities up to 1.1×10^{20} m^{-3} in long pulse operation.

We have observed spontaneous transitions to a moderately improved confinement of energy and particles (incl. impurities) under boronized wall conditions at W7-AS, which exhibit all major features of the H-mode transitions in tokamaks. A transport barrier at the plasma periphery is established and an increase of the radial electric fields was measured. ELM's were identified during the H-mode phase, which affect the energy and particle confinement. No significant difference was measured for hydrogen and deuterium operation. The H-mode transitions where observed only at densities above $4\text{-}5 \times 10^{19}$ m^{-3} and under optimum confinement conditions in the vicinity of $\tau = 0.5$, where the plasma is separatrix dominated. The H-mode transitions are quenched with the limiter inserted and with deteriorated edge magnetic configuration, which supports the explanation, that sufficiently long connection lengths between plasma edge and wall (limiter) are required to establish the H-mode under the given heating power restrictions. Heat wave experiments in the pretransition phase and the H-mode phase showed a reduction of the heat diffusivity in the central plasma region in the H-mode phase. The influence of the external gas puffing on the transition was investigated in detail. The transition occurs independent of the density ramp rate at about the same threshold density. With lower gas puffing, the transition is smoothed out by a series of ELMs.

A strong density clamping effect was measured in experiments with combined heating of ECRH and NBI. Whereas the density increases steadily in purely NBI heated discharges due to the beam fuelling and recycling, the density can be controlled in combined heating scenarios. There is experimental evidence, that the density clamping is mainly due to a particle confinement degradation driven by the profile changes inferred by ECRH. The global energy confinement time for discharges with combined heating fits well to the W7-AS scaling. The impurity confinement is strongly affected in the various combined heating scanarious. The heating power of ECRH has to be equal or larger than the NBI power to maintain the density control. Much more work is needed, however, to come to a conclusive explanation.

REFERENCES

[1] F. Wagner et al., Phys. Rev. Lett. 42 (1982) 1408
[2] K. Odajima et al., Plasma Physics and Contr. Nucl. Fusion Research 1986, Vol. 1, IAEA, Vienna (1987), p. 151
[3] S.M. Kaye, M.G. Bell, K. Bol, J. Nucl. Mat. 121 (1984) 115
[4] C.E. Bush et al., Phys. Rev. Lett. 65, 4, (1990) 424
[5] JET-team, Plasma Physics and Contr. Nucl. Fusion Research 1988, Vol. 1, IAEA, Vienna (1989), p. 159
[6] G.L. Jackson et al., Phys. Rev. Lett. 67 (1991) 3098
[7] ASDEX-team, Nucl. Fusion, Vol. 29, No. 11, (1989), p. 1959
[8] V. Erckmann, et al., Phys. Rev. Lett.70, 14 (1993), 2086
[9] V. Erckmann, et al., Proc. 14th Intern. Conf. on Plasma Physics and Contr. Nucl. Fusion Research, 1992, Germany, post deadline paper.
[10] H. Renner, et al., Plasma Physics and Contr. Nucl. Fusion Research 1990, Vol. 2, IAEA, Vienna (1991) p. 439
[11] H. Ringler, et al., Plasma Physics and Contr. Nucl. Fusion Research 1986, Vol. 2, IAEA, Vienna (1987) p. 603
[12] V. Erckmann, et al., Plasma Physics and Controlled Fusion, Vol. 28, No. 9A, p. 1277, (1986)
[13] W. Kasparek, Proc. of the 8th Joint Workshop on ECE and ECRH, Gut Ising, Germany, (1992), p. 423
[14] D.P. Schissel, et al., Nucl. Fusion, Vol. 29, (1989), p. 185
[15] O. Kardaun,, Proc. 14th Intern. Conf. on Plasma Physics and Contr. Nucl. Fusion Research, 1992, Germany, Paper IAEA-CN-56-F-1-3
[16] H. Ringler et al., Plasma Physics and Controlled Fusion, Vol. 32, No. 11, p. 933, (1990)
[17] L. Giannone, et al., Nucl. Fusion, Vol. 32, No. 11 (1992)
[18] V.V. Alikaev, et al., Plasma Physics and Contr. Nucl. Fusion Research 1986, Vol. 1, IAEA, Vienna (1987) p. 111
[19] K. Uo, et al., Plasma Physics and Contr. Nucl. Fusion Research 1986, Vol. 2, IAEA, Vienna (1987) p. 355

SUPPRESSION OF DISRUPTIONS WITH ECH ON JFT-2M

K. Hoshino, M. Mori, T. Yamamoto, H. Tamai, T. Shoji, Y. Miura,
H. Aikawa, S. Kasai, T. Kawakami, H. Kawashima, M. Maeno,
T. Matsuda, K. Nagashima, K. Oasa, K. Odajima, H. Ogawa, T. Ogawa,
T. Seike, T. Shiina, K. Uehara, T. Yamauchi, N. Suzuki, and H. Maeda
Dept. of Fusion Plasma Research, Japan Atomic Energy Research Institute, Tokai, Ibaraki 319-11, Japan

ABSTRACT

Typical MHD disruptions at surface safety factor $q_a=3$ have been avoided by off-central ECH in the JFT-2M Tokamak. For the suppression, the electron cyclotron resonance layer has to be placed radially in a very narrow region of width about 1 cm near the $q=2$ surface. Both low-field side heating and high-field side heating are effective for the suppression. Fourier spectrum of the magnetic fluctuation shows that mode amplitude decreases by ECH and disruption is avoided. Without ECH, the mode frequency decreases (mode locking) and disruption occurs. Pulse feedback modulation of the ECH power by a magnetic pickup signal shows that the effect of the ECH to reduce the fluctuation is observed only when O-point of the island is likely to be heated. The actual time scale of the suppression is much faster than the time scale of the change of the overall current profile. These observations show that ECH acts on the m=2 island. Further this heating is found to be effective to reduce the m=2 mode which appears at the density-limit-disruption.

INTRODUCTION

In the limiter discharges of JFT-2M tokamak (major radius R=1.31m, minor radius a=0.35m) with circular plasma cross section, the Mirnov oscillations which appear at the surface safety factor $q_a=3$ lead to the plasma disruption when the line-average plasma number density is in the region of $1.1 \times 10^{19} m^{-3} < \bar{n}_e < 2.5 \times 10^{19} m^{-3}$, with constant plasma current[1]. When the plasma density is below this density region, this MHD mode stays stationary with constant amplitude and constant frequency. This mode has been found to have poloidal mode number m=2, and toroidal mode number is supposed to be n=1. This m=2 mode grows as the density is raised and plasma disruption occurs when the density is increased to be larger than $\bar{n}_e=1.1 \times 10^{19} m^{-3}$. It is found that the disruption is avoided by local off-central (resonance layer location r/a=0.70) electron heating[1] by ECH (frequency 59.8 GHz, second harmonic extraordinary mode).

It has been pointed out that local heating of the m=2 island or local current drive can modify the local current gradient at the m=2 island and the mode can be suppressed[2,3,4]. We supposed in ref.1 that the effective resonance layer location should be on the m=2 island. To assure this, we modulated the ECH power[5] using a magnetic probe signal as reference.

In this paper, first we present an effect of ECH power modulation on the m=2 mode, and then we present such experimental topics as ; amplitude and frequency of the m=2 mode, effect of ECH to the density limit disruption. Also, we present the effect of NBH(neutral beam heating) to the m=2 mode at qa=3 for comparison with ECH.

Inside limiter bounded hydrogen circular plasma configurations are taken throughout these experiments.

ECH POWER MODULATION RESULTS

To ascertain that ECH of r/a=0.70 acts on the m=2 island, we did ECH power modulation using poloidal (B_θ) magnetic probe signal as reference.

First we fixed the ECR layer at r/a=-0.70(high-field-side *) and observed that continuous(CW) ECH decreases the amplitude of the m=2 mode. We changed the time delay of the modulated ECH pulses with respect to the probe signal. The m=2 mode frequency was 5kHz. Pulse width of a single ECH pulse is 0.1ms(10kHz, 100% modulation) as shown in **Fig.1**. Phase delay was changed from 0 deg. to 360 deg.s.

Suppression of the m=2 mode occurred only when the phase delay is in a particular region of angle width of 180 deg.s. In another region of phase delay, suppression was not observed. From the result we can draw a physical picture that heating of the O-point of the magnetic island is effective for the suppression, but heating of the X-point of the magnetic island is ineffective for the suppression.

Toroidal current flows in the island. Therefore, the zero cross points of the probe output voltage (proportional to dB/dt) should correspond to O-point (+ to -) and X-point (- to +). As the m=2 mode has two O-points and two X-points in cross section, the spatial angle difference between O-point and X-point of 90 deg.s corresponds to phase difference of 180 deg.s in probe signal. Then if we assume that heating around the O-point in total poloidal angle width of 90 deg.s brings suppression, the observed suppression in phase delay of width 180 deg.s is reasonable. For the more accurate study, we may have to reduce the ECH pulse width (now, the pulse width corresponds to 180 deg.s). In the experiment, no increase of the mode amplitude was observed by changing the phase.

It is noted that decrement of the m=2 mode amplitude was almost the same for CW ECH pulse (20-30%) and modulated pulses (15-26%). The peak power was the same P=92kW in both cases. The average power of modulated ECH power was P=45kW. CW ECH of P=50kW couldn't decrease the probe signal in this case. Therefore the suppression of the

*In ref.1, we adopted low-field-side heating (r/a=0.70).

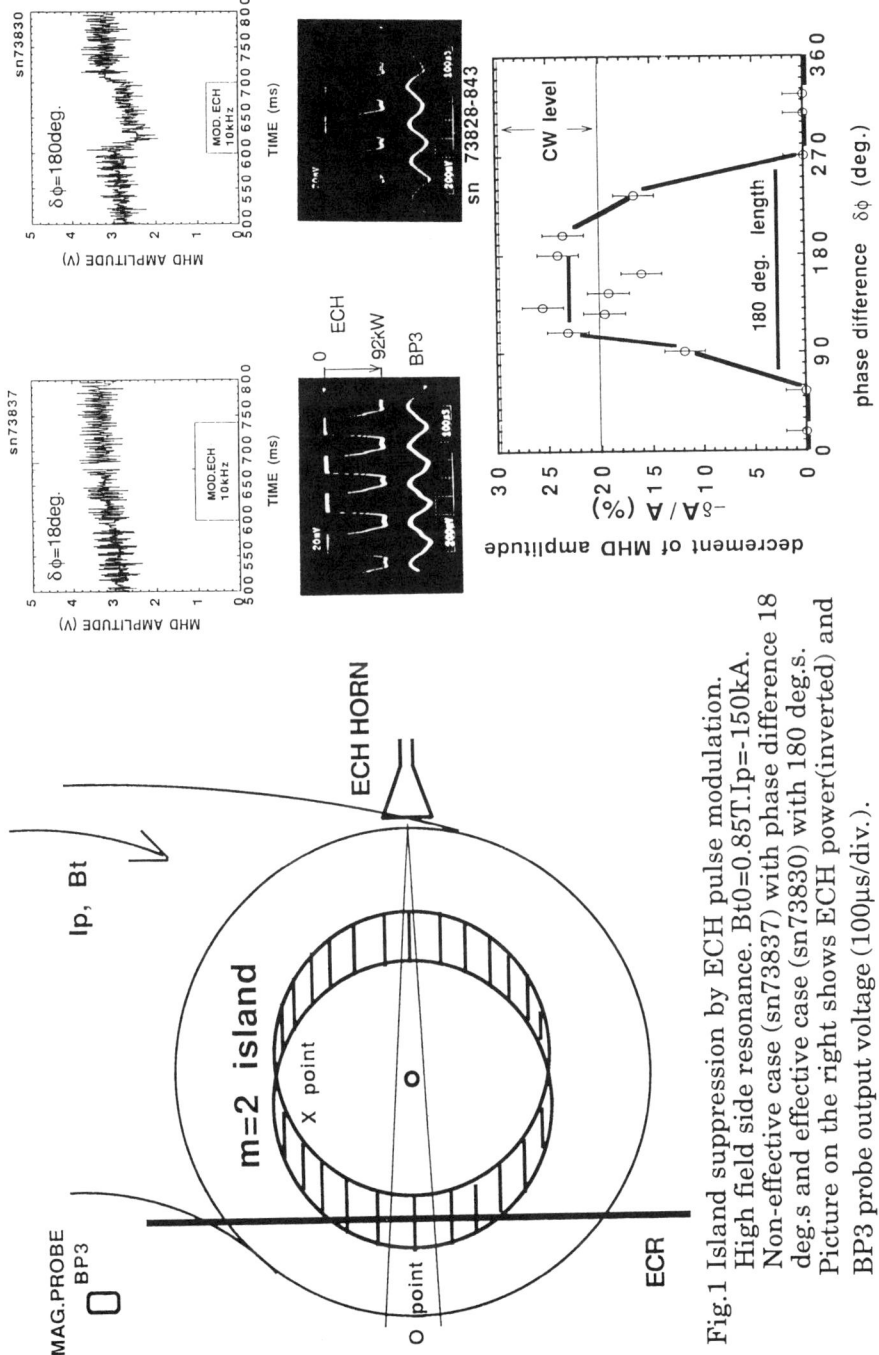

Fig.1 Island suppression by ECH pulse modulation. High field side resonance. Bt0=0.85T.Ip=-150kA. Non-effective case (sn73837) with phase difference 18 deg.s and effective case (sn73830) with 180 deg.s. Picture on the right shows ECH power(inverted) and BP3 probe output voltage (100μs/div.).

m=2 mode depends on the peaked power of the ECH short pulses rather than on the average power.

AVOIDANCE OF $q_a=3$ DISRUPTION BY ECH

The m=2 mode which appears at $q_a=3$ leads to plasma disruption when plasma density is raised above $1.1 \times 10^{19} m^{-3}$. **Figure 2** shows the time evolution of the frequency spectrum of the m=2 mode. Gas puff is applied from t=0.73 sec and plasma density rises, then mode frequency begins to decrease from 3.1kHz towards 0 kHz and mode amplitude increases. The probe output (dB/dt) decreases due to the decrease of mode frequency. Mode locking occurs and plasma disrupts at t=0.81sec. The magnetic field at disruption is about 20Gauss, which corresponds to the island current of about 2kA. Thus by raising the density, mode locking is induced and disruption occurs.

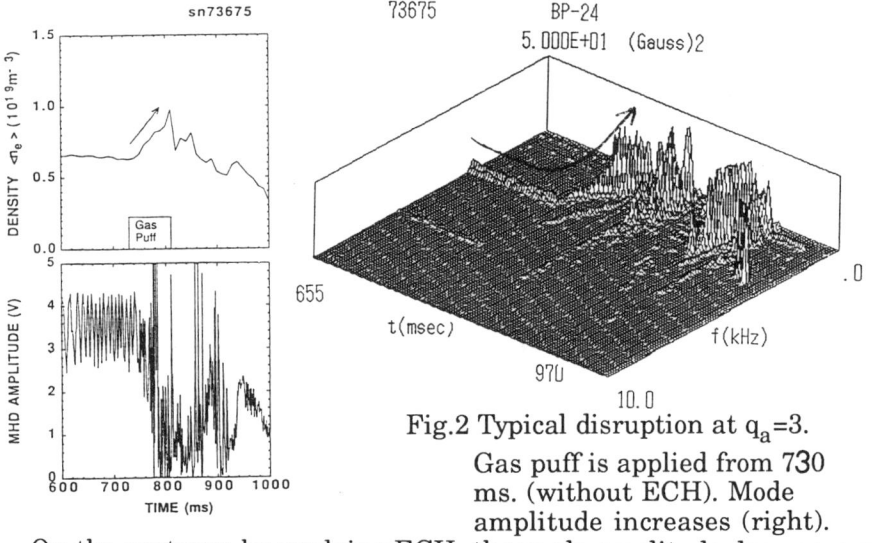

Fig.2 Typical disruption at $q_a=3$. Gas puff is applied from 730 ms. (without ECH). Mode amplitude increases (right).

On the contrary, by applying ECH, the mode amplitude decreases and mode frequency increases as shown in **Fig.3**. During the suppression of the m=2 mode by ECH, the output voltages of almost all of the poloidal magnetic probes decrease or stays constant (BP24, BP1) at t=650ms. But the probe output at the high-field side (BP24, BP1) increases after that. This increase cannot be explained by a simple inside displacement of the island.

The Fourier spectra of these probes tells that this m=2 mode has inside-outside (poloidal) asymmetry. Before ECH, the amplitude of the fluctuating magnetic field is 2 Gauss inside and 16 Gauss outside. And the wave form is distorted (as shown in the oscilloscope picture in Fig.1). It seems to have second harmonic 6kHz components (m=4) of 1.2 G (inside) and 1.8 G (outside). But, during the ECH (t=0.7s), the m=2 mode becomes symmetric (inside 2.5 G, outside 2.5 G). The shape of the probe

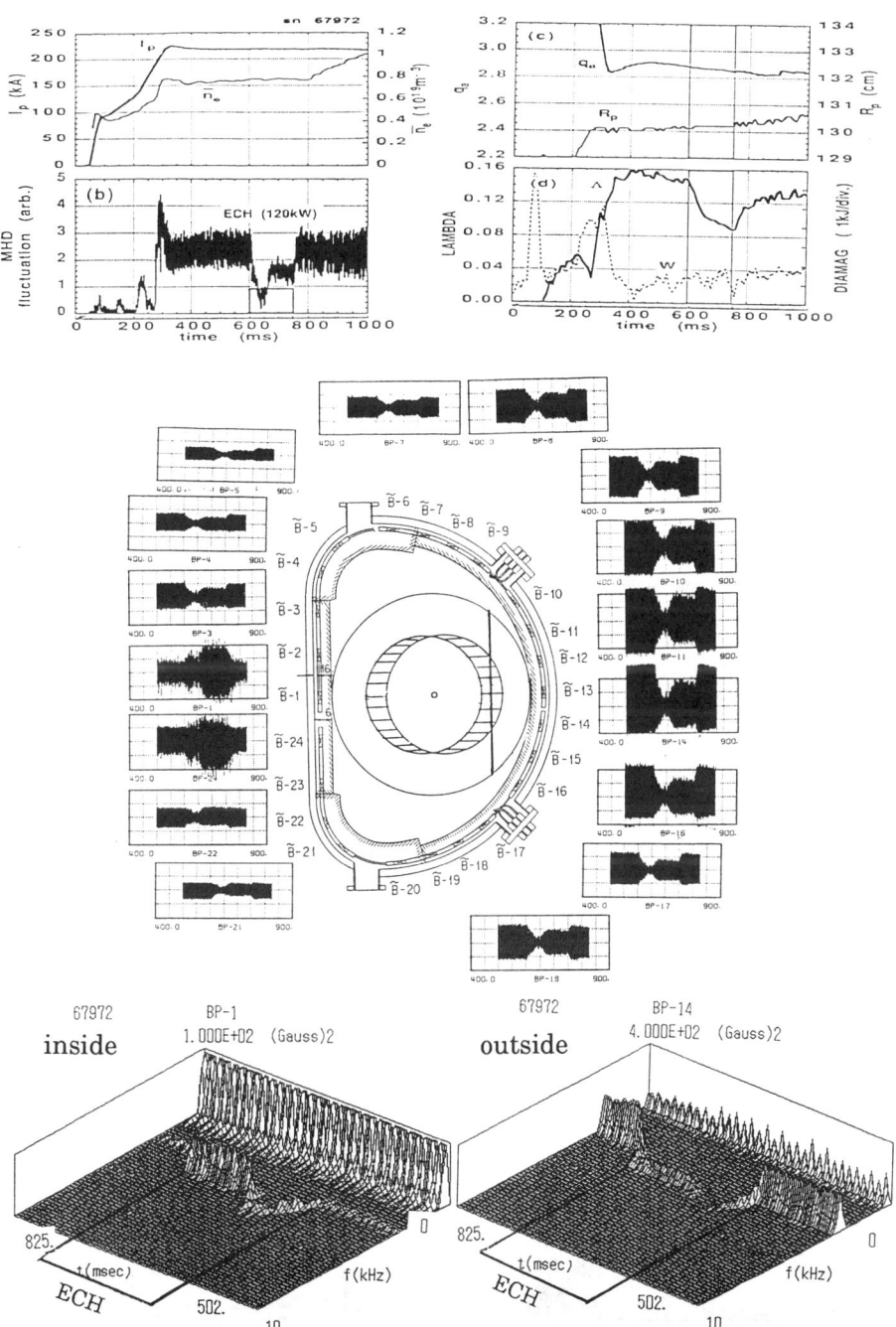

Fig.3 Suppression of m=2 mode at q_a=3 by ECH.
ECR layer is located at r/a=0.70 (low-field side).
B_θ probe signal and spectrums.

signal during ECH is similar inside and outside. It seems that poloidal asymmetry is symmetrized by ECH during the suppression.

We could raise the plasma density during the suppression of the m=2 mode as shown in **Fig.4**. Plasma disrupts (sn67974) when ECH is turned-off in region (B). Thus, disruption is avoided. Region (C) is found to be disruption free (sn67977). Without ECR layer at r0=0.7a, we could not avoid disruptions.

We found that the minimum ECH power required to avoid disruption at $q_a=3$ was 40+-5kW in the case of high-field side resonance. This power level is 40kW/221kW=18% of the joule heating power.

m=2 MODE WITH NEUTRAL BEAM HEATING

We investigated the effects of NBH to the m=2 mode at $q_a=3$(**Fig.5**). The joule plasma rotates to counter direction. Co-injection(direction of the plasma current) leads to mode locking and disruption even in region (A). Counter injection increases both of the mode amplitude and mode

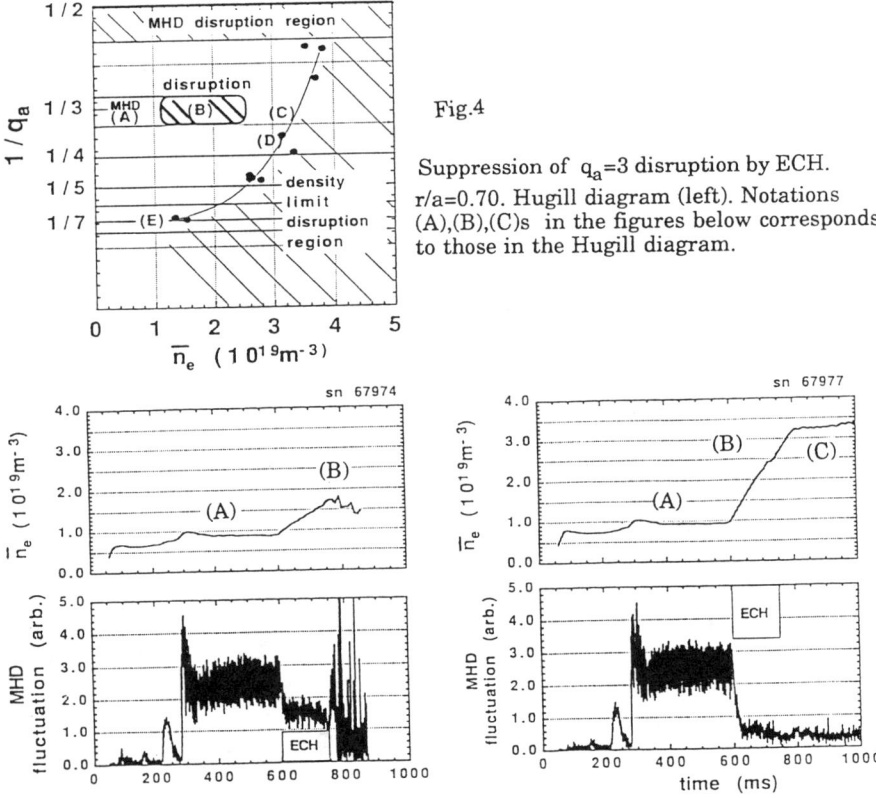

Fig.4

Suppression of $q_a=3$ disruption by ECH. r/a=0.70. Hugill diagram (left). Notations (A),(B),(C)s in the figures below corresponds to those in the Hugill diagram.

frequency. Mode locking appears after 0.1sec from the injection and then plasma disrupts in region (B). We increased the NBH power up to 788 kW, but disruption could not be avoided.

Thus it is known that localized electron heating on the island by ECH is important for the avoidance of disruption.

EFFECT OF ECH TO THE DENSITY LIMIT DISRUPTIONS

The ECH is found to reduce the m=2 mode which grows at the density limit(**Fig.6**, q_a=2.4). The ECR layer is set at 0.86a. The density limit increased from $2.5 \times 10^{19} m^{-3}$ to $2.8 \times 10^{19} m^{-3}$ (12% increment).

Probe spectrum shows no inside-outside asymmetry (inside 1.5 G, outside 1.7 G) before ECH. This is different from m=2 mode at q_a=3. During the suppression by ECH, mode frequency (5.3kHz) stays constant and mode amplitude is suppressed both inside and outside. But the suppression is only for first 40-50 ms during ECH. The low cutoff density ($2.1 \times 10^{19} m^{-3}$) of the wave or movement of the q=2 surface may limit the effectiveness of ECH.

CONCLUSIONS

(1) Heating on (or quite near) the m=2 island, which is confirmed by ECH power modulation experiment, is effective for the avoidance of disruption not only at q_a=3 but also at density limit.
(2) With ECH on the m=2 island, amplitude of the m=2 mode decreases. Then the mode locking and successive disruption are avoided.
(3) The m=2 mode which appears at q_a=3 is poloidally distorted. ECH recovers the symmetry during suppression.
(4) In order to use ECH as disruption suppression, a teqnique has to be developed to track the radial position of the m=2 mode and change the power deposition by ECH.

REFERENCES

1. K. Hoshino et al., Phys. Rev. Lett. 69, 2208 (1992).
2. V. S. Chan et al., Nucl. Fusion 22, 787 (1982).
3. K. Yoshioka et al., Nucl. Fusion 24, 565 (1984).
4. E. Westerhof et al., Plasma Phys. Controlled Fusion 30, 1691 (1988).
5. H. Hsuan, private communication. We are informed that such ECH power modulation study has been tried in the TFR tokamak.

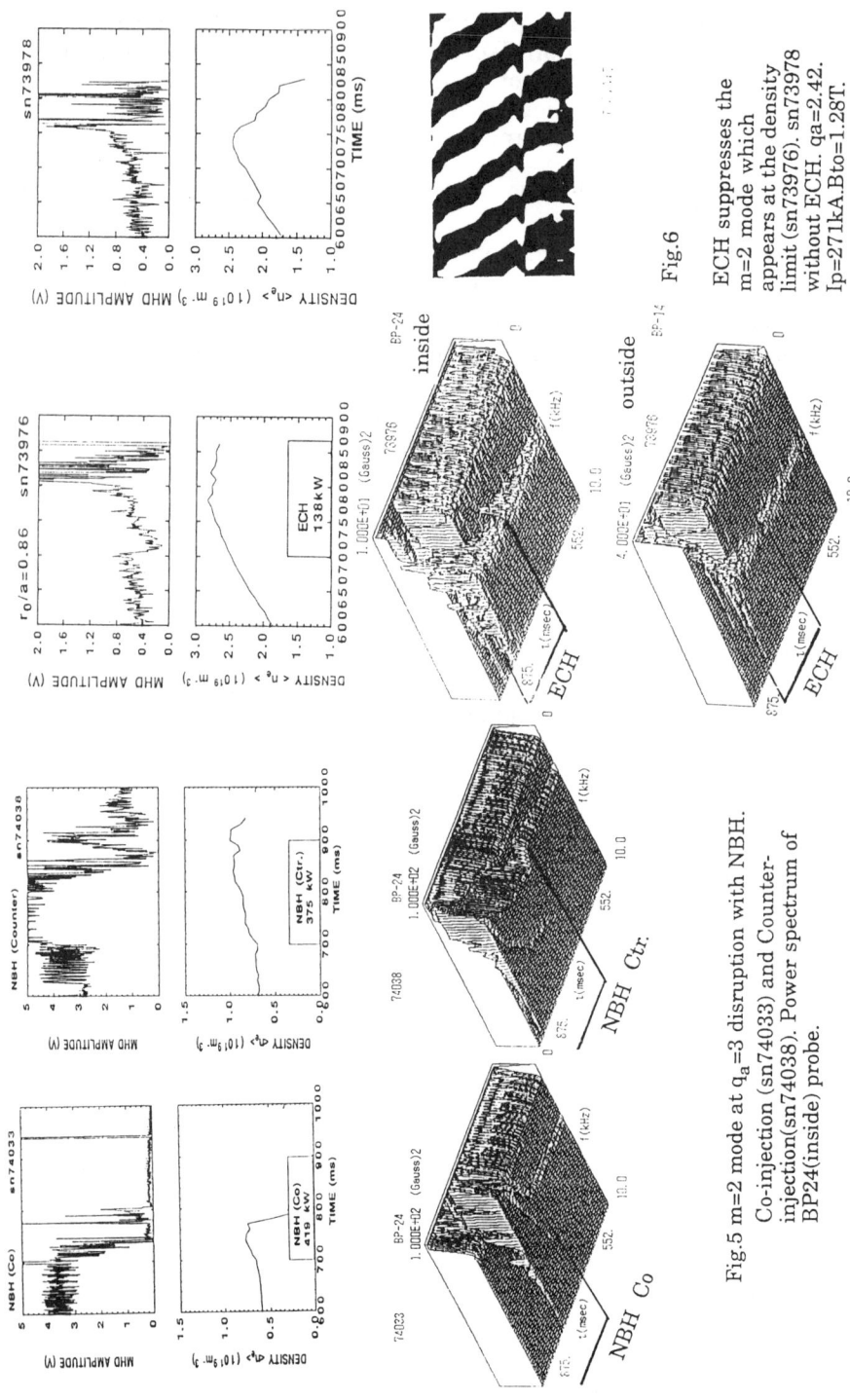

Fig.5 m=2 mode at q_a=3 disruption with NBH. Co-injection (sn74033) and Counter-injection(sn74038). Power spectrum of BP24(inside) probe.

Fig.6 ECH suppresses the m=2 mode which appears at the density limit (sn73976). sn73978 without ECH. qa=2.42. Ip=271kA.Bto=1.28T.

Experiments on Nonlinear Absorption of ECH Waves in MTX and a Comparison with Theory*

Barry W. Stallard, S.L. Allen, J.A. Byers, T.A. Casper, B.I. Cohen, R.H. Cohen, C.J. Lasnier,
M.E. Fenstermacher, J.H. Foote, E.B. Hooper, M.A. Makowski, W.H. Meyer, J.M. Moller,
B.W. Rice, T. D. Rognlien, G.R. Smith, K.I. Thomassen, R.D. Wood
Lawrence Livermore National Laboratory, Livermore, CA 94550
and
K. Hoshino, K. Oasa, K. Odajima, T. Ogawa (JAERI, Tokai, Japan),
T. Oda (Hiroshima University, Hiroshima, Japan),
and T. Ogo (Fukuoka University of Education, Fukuoka, Japan)

ABSTRACT

Intense pulse electron cyclotron heating (ECH) experiments have been carried out on the MTX tokamak. Rf pulses at 140 GHz with peak power of 1-2 GW and 25 ns pulse length were generated by the ETA-II / IMP FEL and transported quasi-optically to MTX for O-mode launch. Because of the intense rf electric fields (~250 kV/cm), reduction of plasma absorption by nonlinear effects was predicted and several rf beam geometries ($k_{\parallel}$ gradient) were investigated to study their effect on the absorption. Measurements of beam transmission showed increases, compared to low power (2 kW), which agreed with theory to within the data scatter. For these experiments x-ray, ECE, and Thomson diagnostics showed evidence for localized absorption at the cyclotron resonance and hot electron production. A comparison of these results with calculations from the orbit following code ORPAT will be presented.

INTRODUCTION

Experiments to study the non-linear absorption of intense pulse microwaves have been carried out in the MTX tokamak. The motivation of these experiments was to explore the efficacy of very high peak power and high average power plasma heating using new technology capabilities of the FEL. Results were in general agreement with nonlinear theory. They extrapolate to very high absorption at reactor parameters.

The LLNL FEL consists of the electron linear accelerator ETA-II and the IMP wiggler operating as an amplifier.[1] Typical beam pulses of 2 kA, 6 MeV energy, and 40 ns pulse length were injected into the wiggler (5 m length and 10 cm period), co-linear with approximately 7 kW of 140 GHz drive power provided by a gyrotron oscillator. This drive power coupled 2 to 3 kW of power to the TE_{11} mode in 3.25 cm waveguide. With optimized tapering of the wiggler field, we achieved a peak output power of 1 to 2 GW. The FEL output power was transported quasi-optically to the MTX tokamak over a 34 m distance using six mirrors mounted within a 50 cm diameter evacuated pipe. There was no vacuum window between the FEL output and the tokamak. The final mirror optic, located 242 cm from the input duct to MTX, focused the beam to a 2 cm diameter waist (1/e power diameter). The design transport efficiency of the system is 89%. Fig. 1 shows a typical rf pulse measured at the FEL output. For the experiments reported here the FEL was operated single pulse at a 0.5 Hz rate. We did brief testing of the FEL at a 2 kHz burst. Up to 50 pulses at 0.5 to 1 GW power were achieved.[2] These results showed the capability for high average power output with further refinements of the hardware.

In order to study the nonlinear heating physics, beams with two different geometries were injected into MTX as shown in Fig. 2. At the resonance the beam spot sizes were similar but the toroidal gradient in $k_{\parallel}$ differed by a large factor. For the first experiments a cir-

---

*This work was performed under the auspices of the U.S. Department of Energy by Lawrence Livermore National Laboratory under contract No. W-7405-Eng-48.

cular beam was injected into the smooth wall input duct of 2 cm width, 30 cm height, and 22 cm length. For this beam the diameter (1/e power) in vacuum at the plasma resonance was 3.6 cm and the variation of $k_{\parallel}$ across the beam was $\Delta k_{\parallel} / k \approx 0.1$. To test the predictions of theory for enhanced absorption with a large gradient in the toroidal $k_{\parallel}$, for a second series of experiments we injected the beam into a corrugated vertical side wall duct, tapered in the horizontal direction (the side wall normal to the horizontal electric field for the O-mode). The Gaussian-shaped input beam coupled to the HE_{11}-like mode and the length of the 3.4:1 down taper was set to reconstruct the HE_{11} mode at the output.[3] The beam size at the resonance in vacuum was elliptical and elongated in the horizontal direction with dimensions 4.8 x 3.6 cm. This beam had a smooth variation of $k_{\parallel}$ across the beam $\Delta k_{\parallel} / k \approx 0.35$.

NONLINEAR ABSORPTION OF INTENSE PULSE ECH

At 1 GW peak power the electric fields of the microwave beam at the resonance are $\tilde{E} \approx$ 250 kV/cm. Electrons are heated so rapidly that their relativistic mass increase causes the electron gyro motion and the wave fields to fall out of phase, and electrons become trapped in the wave (phase space "bucket"). As they stream through the beam the electrons perform rapid adiabatic excursions up and down in energy as their gyrophase changes. Upon exiting the beam the adiabaticity of the electrons is broken and approx-imately one half of these electrons retain high energy.[4] For an FEL beam with no gradients in $k_{\parallel}$, in a plasma with no gradients in magnetic field, the absorption is reduced from the predictions of linear theory. The situation is different if the gradients are non-zero. For a gradient in $k_{\parallel}$, electrons trapped within the wave about the doppler-shifted cyclotron resonance can be adibatically raised to higher energy (rising buckets) if gradients are not too large.[5] This effect increases heating. Under certain conditions the absorption can exceed the predictions of linear theory.

Other nonlinear effects can be important for intense electric fields. In MTX for 1 keV electron temperature and 1 GW power the electron quiver velocity is large, $\tilde{v}/v_e \approx 0.4$, where $v_e = (T_e/m)^{0.5}$ is the electron thermal velocity. Under these conditions, thresholds for parametric instabilities may be exceeded, and these may be important in determining plasma absorption and heating. Reflective (backscattering), forward scattering and absorptive instabilities have been investigated for MTX parameters.[6] These studies have shown that Brillouin backscattering near ion cyclotron frequencies is unstable and may be of most concern for these experiments. However, for the ~ 20 ns pulse length in our experiments the expected growth of the instabilities ($\Upsilon_i \tau_p \sim \Omega_{ci} \tau_p \sim 5$) is too small to be significant.

Nonlinear modeling

The code ORPAT[7] was used to model nonlinear absorption at the resonance. ORPAT follows the relativistic guiding center orbits of electrons as they stream through the FEL beam, keeping track of the gyro phase. The beam profile varied toroidally to model the experiment and a flat pulse of 20 ns duration approximated the beam pulse shape. For radial zones near the resonant surface, the heating of an ensemble of electrons from a Maxwellian target plasma was calculated as they transited the beam along the magnetic field. The beam amplitude was attenuated by the absorbed energy for calculations in successive zones, but the phase front curvature was assumed unchanged by the attenuation. Plasma refraction of the FEL beam in the vertical direction reduces the electric field intensity at the resonance. To take this into account, the ray tracing code TORCH was used to calculate a refraction factor R_1 giving the reduction in beam intensity at resonance compared to the vacuum value.[8] As described in the next section, for comparison between ORPAT and experimental beam transmission to the wall opposite the injection port, an additional factor R_2 was calculated, giving the refraction from the resonance to the wall. The resulting toroidally averaged beam transmission intensity near the beam vertical center is given by:

$$T_2 = R_1 R_2 \int T(z,y=0)\, dz,$$

where $T(z,y)$ is the absorption at the resonant layer and y (z) is the vertical (toroidal) direction. The integration in z models absorption on the calorimeter tiles described below.

EXPERIMENTAL RESULTS

In MTX experiments we determined the ECH absorption for central resonance (B_{tor}=5 T) from measurements of beam transmission. Because the plasma heating experiments were carried out for single FEL pulses with typical energy 20 J, bulk heating of the plasma could not be measured. However, for some of our experimental runs fast diagnostics were available and generation of hot electrons was observable on second harmonic ECE, soft x-rays (provided by JAERI), and Thomson scattering measurements of the hot electron tail. The beam transmission diagnostics consisted of a calorimeter on the inside wall opposite the input port, which was segmented poloidally to measure the transmitted beam vertical profile, and a single point measurement at the approximate beam center using a fundamental mode waveguide horn connected to an rf receiver. The input FEL beam pulse shape and spatial profile were measured at both the FEL output and at the tokamak input using similar waveguide horns (both fixed and movable probes) and calibrated rf receivers. A calorimeter (provided by JAERI) measured the total FEL beam pulse energy at the tokamak input. Using the measured pulse shape, the peak power was determined.

Injection with smooth wall

Fig. 3 shows the transmission, measured by the central absorbing tiles of the calorimeter, as density was scanned, for the FEL beam injected into the smooth wall duct. The electron temperature of the target plasma was nominally 1 keV but varied from about 0.8 to 1.6 keV. The peak power of the FEL pulse was about 1 ($\pm$0.2) GW. The inferred transmission is given by normalizing the measured transmission with plasma to the value without plasma. Each measured transmission value was also normalized to the total pulse energy for each shot. There is considerable scatter in the data but the transmission exceeds by a large factor the transmission at low power (~ 2 kW) obtained with the gyrotron driver only. The source of the scatter is not fully understood. There is scatter in the electron temperature of the target plasma but no apparent correlation of transmission with the electron temperature variations.

Also shown in the figure is the modeling of transmission using ORPAT for an average FEL power <P> = 0.75 GW. Although the peak power was ~1 GW, the lower value was used to take into account the time average of the FEL pulse. Within the scatter of the data the theory and measurements are in agreement. For the low power data the transmission curve for linear theory agrees closely with the measurements. For the high power experiments fast ECE and Thomson measurements were not available, but fast x-ray measurements at 180 deg. toroidal angle from the injection port showed no evidence for hot electrons.

Injection with tapered wall

Similar transmission measurements were made for the divergent beam. Shown in Fig. 4 are these measurements compared with the no taper data. For this beam geometry, greater absorption is inferred from the lower transmission. This data also has large scatter and also no apparent correlation with electron temperature. However, the data is clearly below the no taper data and also similar to the low power data, in reasonable agreement with ORPAT.

For the tapered wall experiments, fast diagnostic measurements are available which show clear evidence for hot electrons and localization of heating near the resonant surface. Thomson data measured on axis at the resonance is presented in Fig. 5 for target plasma parameters $n_{e0} = 0.65 \times 10^{20}$ m^{-3} and $T_{e0} = 1.5$ keV. The open diamonds show the measured data without the FEL, averaged over six shots. The dashed line curve is the

Maxwellian behavior (relativistically corrected[9]) that is expected for no FEL heating. The solid circles show the signal with the FEL. To increase the signal to noise ratio in the tail of the scattered spectrum, we again averaged six shots from similar discharges. The increased signal at high energy shows the presence of hot electrons.

These data are in agreement with the nonlinear theory. To model the Thomson scattering expected from the nonlinear heating, we used the ORPAT code to calculate the electron energy distribution at the Thomson observation port located 120 deg from the FEL port. The orbits of electron with a heated distribution $F(E)_0$ at the FEL port were followed as they drifted 120 deg around the torus to determine $F(E)_T$ at the Thomson port. The scattered spectrum (x's and solid line curve) was determined by applying a relativistic correction factor[9] to a 2-temperature Maxwellian fit to the ORPAT distribution. The fit is fairly insensitive to the assumed temperature of the hot component. For the solid curve in Fig. 5 the best fit parameters were T_{e0} = 1.55 keV and T_{eh} = 10 keV.

Inside and outside wall ECE measurements also confirm the presence of hot non-thermal electrons. A heterodyne receiver on the inside (JAERI) and a grating polychromter on the outside view the plasma. Typical data for the inside signal show an abrupt increase followed by a decay time of ~ 40 microsec; the outside signal shows a rapid increase and decrease with no slowly decaying signal. This behavior is expected for non-thermal electrons. Because of their relativistic mass increase, the non-thermal electrons radiate at a lower frequency than the local cyclotron frequency. This radiation is accessible to the inside view antenna but must pass through an absorbing resonance before reaching the outside antenna. From the ECE decay time and the electron density, the estimated mean hot electron energy is E_h ~10 keV, consistent with the modeling.

Fast x-ray measurements in Fig. 6 show that absorption is localized near the magnetic axis. For this data the x-ray diagnostic was now located at the FEL port. Plasma parameters are shown for a shot with FEL injection at 200 ms. Expanded fast x-ray traces for the central channel and several adjacent channels, spaced at 3.8 cm intervals, are shown. Only the central channel shows a signal which decays in about 25 microsec, similar to ECE decay rates.

For experiments with the tapered input duct we looked for evidence of backscattering from the plasma. The rf probe viewing the plasma from outside the MTX port was a small waveguide horn connected to a calibrated rf receiver and located near the FEL input axis. Only backscattered frequencies down shifted > 200 MHz from the FEL frequency could be distinguished by the available receiver (bandwidth ≈ 400 MHz). Backscattered signals near ion cyclotron frequencies f_{ci}≈38 MHz could be discriminated only by the two way transit time delay (~12 ns) between the probe and the plasma. We scanned the receiver for down-shifted signals ≤ 1.3 GHz (~f_{pi}). Our measurements showed no evidence for backscattering. From our measurements an upper limit for backscattered power is $P_{back}/P_{FEL} < 2 \times 10^{-5}$.

CONCLUSIONS

Our experiments clearly demonstrate the reduction in absorption for intense pulse ECH and the ability to increase the absorption by control of the beam $k_{\parallel}$ spectrum. The data are also in good agreement with nonlinear modeling, including the production of hot electrons. They extrapolate to very high absorption at reactor parameters. As expected from the theory for our parameters, we have seen no evidence for parametric instabilities. Initial testing of the FEL at high repetition rate indicates that high average power operation can be achieved and the prospects for operation at shorter wavelengths look promising. Bulk heating remains to be demonstrated.

REFERENCES

1. S.L. Allen and E.T. Scharlemann, to be published in the *Proc. of the 9th International Conf. on High Power Beams,* Washington, D.C. 1992.
2. C.J. Lasnier, S.L. Allen, B. Felker, et al, to be presented at the *Particle Accel. Conf.*

(IEEE), Washington, D.C. 1993.
3. J.L. Doane, International Jour. of IR and Millimeter Waves 5 ,737 (1984).
4. W.M. Nevins, T.D. Rognlien, and B.I. Cohen, Phys. Rev. Lett. 59 ,60 (1987).
5. R.H. Cohen and T.D. Rognlien, Phys. Fluids B3 ,3406 (1991).
6. M. Porkolab and B.I. Cohen, Nuclear Fusion 28 ,239 (1988).
7. T.D. Rognlien, Phys. Fluids 26 ,1545 (1983).
8. G.R. Smith, M.E. Fenstermacher, and E.B. Hooper, Nuclear Fusion 30 ,2505 (1990).
9. J. Sheffield, *Plasma Scattering of Electromagnetic Radiation*, Academic Press, 1975.

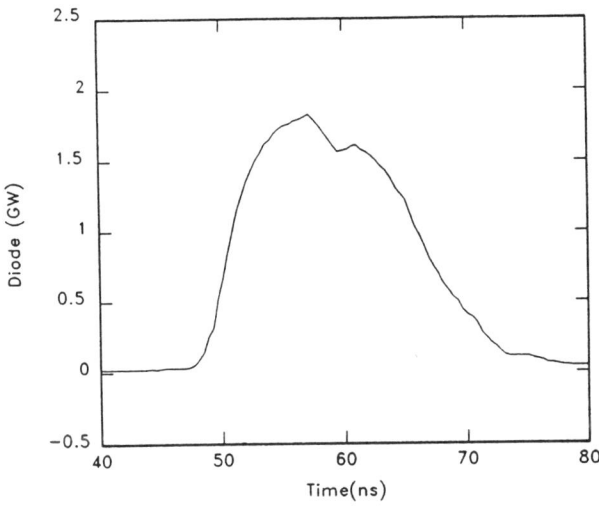

Fig. 1 Typical FEL output beam pulse for MTX ECH experiments.

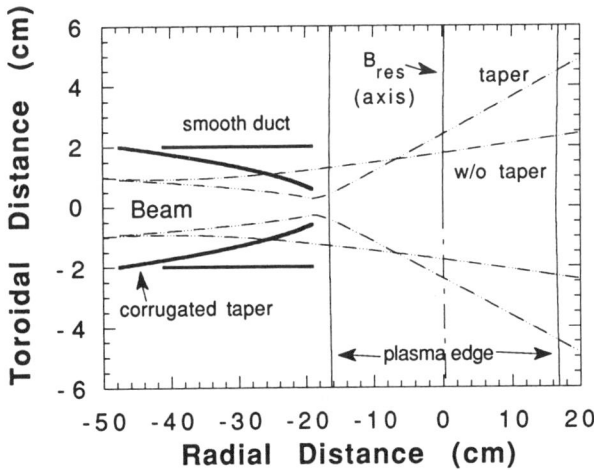

Fig. 2 FEL beam profiles in the toroidal direction with smooth duct input port and tapered duct input port.

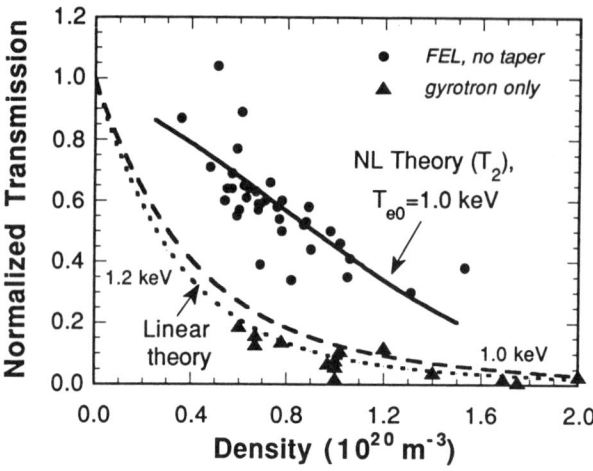

Fig. 3 Measured transmission with smooth duct input port of FEL beam (nominal 1 GW peak power) and FEL driver only ($\approx$ 2 kW power).

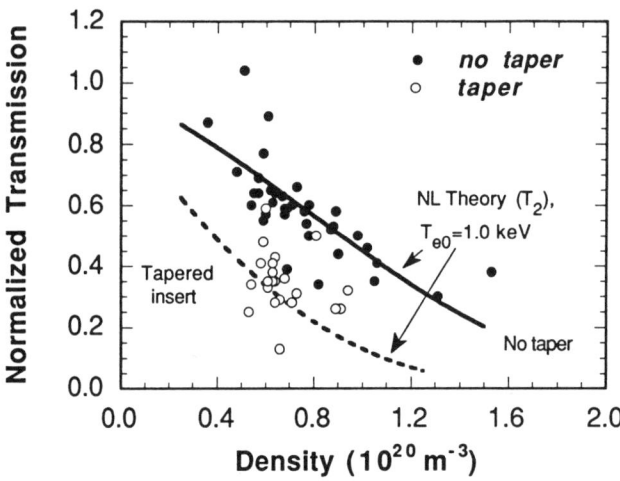

Fig. 4 Measured transmission of FEL beam (nominal 1 GW peak power) comparing tapered input port with smooth wall input port (no taper). The increased toroidal gradient in $k_{\parallel}$ with the taper increases the absorption.

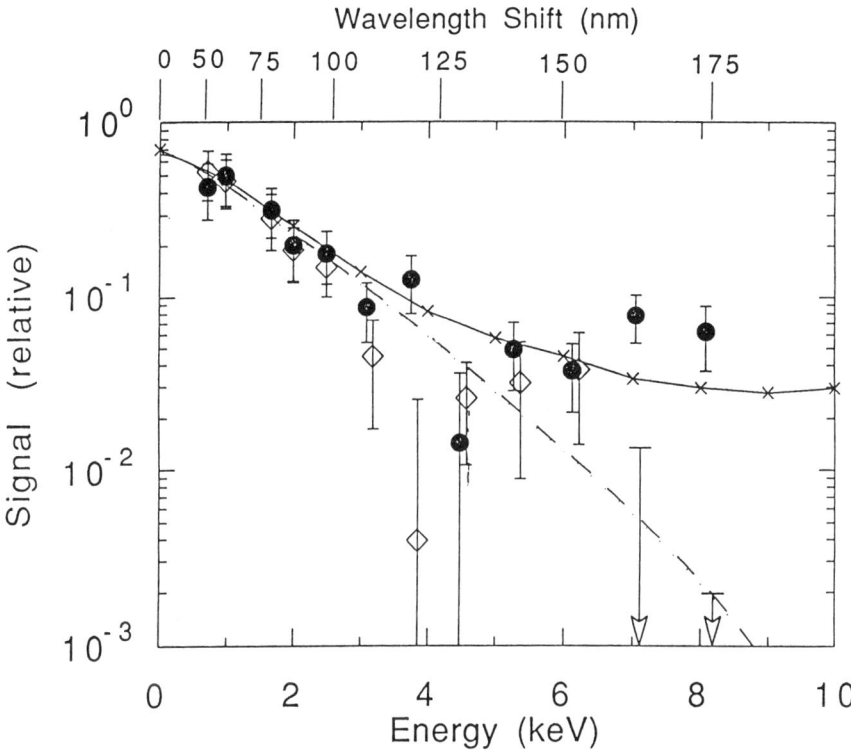

Fig. 5 Thomson scattering spectrum showing hot electron tail (solid dots) for FEL shot and target plasma without FEL (open diamonds). The data are 6 shot averages for plasma target parameters T_{eo} = 1.55 KeV and n_{e0} = 6.5 10^{19} m^{-3}. The solid line shows the ORPAT modeled spectrum for T_{eh} = 10 keV.

164 Experiments on Nonlinear Absorption of ECH Waves

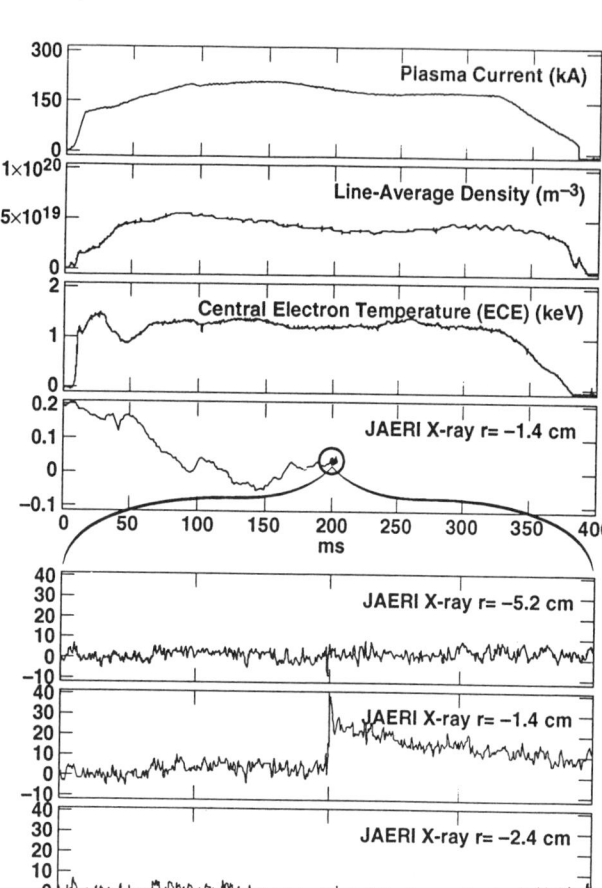

Fig. 6 Typical tokamak discharge parameters for FEL heating experiments. The expanded traces for the fast x-ray signal show the localization of heating.

DIMENSIONALLY SIMILAR DISCHARGES WITH CENTRAL RF HEATING ON THE DIII–D TOKAMAK

C.C Petty, T.C. Luce, and R.I. Pinsker
General Atomics, San Diego, CA 92186-9784

ABSTRACT

The scaling of L–mode heat transport with normalized gyroradius is investigated on the DIII–D tokamak using central rf heating. A toroidal field scan of dimensionally similar discharges with central ECH and/or fast wave heating show gyro-Bohm-like scaling both globally and locally. The main difference between these results and those using NBI heating on DIII–D is that with rf heating the deposition profile is not very sensitive to the plasma density. Therefore central heating can be utilized for both the low-B and high-B discharges, whereas for NBI the power deposition is decidedly off-axis for the high-B discharge (*i.e.*, high density).

INTRODUCTION

The diffusion of heat across magnetic field lines is of fundamental importance to tokamaks. Diffusion processes can be categorized as being either short wavelength or long wavelength. For "dimensionally similar" discharges, defined as having identical values for all dimensionless parameters except the relative gyroradius (*i.e.*, having ν^*, β, q, T_e/T_i, R/a, b/a, etc., the same but not ρ^*), the diffusivity has a very simple scaling: $\chi \propto B^{-1}a^{-1/2}$ for microturbulence or gyro-Bohm-like scaling, or $\chi \propto B^{-1/3}a^{1/3}$ for macroturbulence or Bohm-like scaling.[1] Nearly all theories for diffusion are gyro-Bohm-like; however, along a dimensionally similar path the empirical L–mode tokamak scalings (such as Goldston) have $\tau \propto B^0$ which is slightly worse than Bohm-like.

Comparing dimensionally similar discharges of the same size but different magnetic field should allow one to distinguish between a gyro-Bohm-like or a Bohm-like diffusion process. If the plasma transport is purely diffusive, then the global confinement should scale as $\tau \propto Ba^{5/2}$ for gyro-Bohm-like diffusion or $\tau \propto B^{1/3}a^{5/3}$ for Bohm-like diffusion. A technical difficulty with this experiment is keeping the auxiliary heating profile similar for the low-B and high-B discharges, especially with NBI heating. Central heating with rf power is ideally suited for this type of experiment since the deposition profile is relatively independent of the plasma density.

EXPERIMENTAL RESULTS

For DIII–D, two forms of central rf heating are available to directly heat electrons: 60 GHz ECH and 60 MHz fast wave (FW). The ECH is resonant centrally at 2 T for fundamental absorption and 1 T for second harmonic absorption. The FW in the direct electron heating regime is always absorbed near the plasma center, regardless of the magnetic field, since the damping increases with electron temperature.[2]

The fast wave current drive (FWCD) database was examined for pairs of shots which were dimensionally similar. Table I shows some dimensionless parameters for the best matched pair of rf discharges. These divertor plasmas are in L–mode and are

© 1994 American Institute of Physics

Table I
Dimensionless parameters for dimensionally similar discharges
with central rf heating

	$B_T = 1.07$ T	$B_T = 1.99$ T
R/a	2.7	2.7
b/a	1.7	1.8
q_{95}	6.9	6.7
ℓ_i	1.8	1.8
β_p	0.37	0.34
Z_{eff}	3.2	3.3

fueled by deuterium gas puffing. The high-B shot had co-ECCD and co-FWCD while the low-B shot had co-FWCD only. The ratio of rf-to-ohmic power was nearly the same (5:1) for these two plasmas. More than 70% of the total heating power was deposited inside $\rho = 1/3$ for both discharges.

Figure 1 shows the profiles of β, ν^*, T_e/T_i, and the absorbed power fraction for the two rf-heated dimensionally similar discharges. Ion temperature profiles were obtained by pulsing the neutral beams for approximately 10 msec to obtain CER data. The profiles are well matched throughout the plasma. Although not shown, the density and temperature profile scale lengths are also well matched for these two discharges.

Table II shows the heating power, stored energy, and global confinement for these two discharges. The confinement time scaled like $\tau \propto B^{1.2}$, which is slightly stronger than gyro-Bohm-like. Since the global confinement exhibits gyro-Bohm-like scaling, the local transport should also be gyro-Bohm-like. The local transport is determined from a power balance analysis code. The ECH and FW deposition profiles and current drive are calculated by ray tracing codes. For the $B_T = 1.99$ T discharge, a large fraction of the FW power is absorbed centrally by second harmonic heating of the $\approx 2\%$ hydrogen minority. This complicates a two-fluid analysis of the transport, since the partition of the FW power to ions and electrons is not accurately known. Therefore in order to eliminate this uncertainty, a one-fluid transport analysis is performed. The one-fluid heat diffusivity for each discharge was calculated using

$$\chi_{\text{eff}} = -\frac{q_e + q_i}{n_e \nabla T_e + n_i \nabla T_i} \ . \tag{1}$$

The ratio of the χ_{eff} in the 1.99 T discharge to that found in the 1.07 T discharge is shown in Fig. 2. Also shown are the ratios for Bohm-like and gyro-Bohm-like transport scaling. The experimental data for the rf centrally heated plasmas is again clearly much closer to gyro-Bohm-like than Bohm-like for all radii. The rf heating profiles were similar only outside $\rho = 1/3$, which may explain the deviation from gyro-Bohm-like scaling inside that radius, although sawtooth transport may also be an important effect.

DISCUSSION

Dimensionally similar discharges with central rf heating exhibit gyro-Bohm-like scaling both globally and locally. This is in contrast to previous results from DIII–D

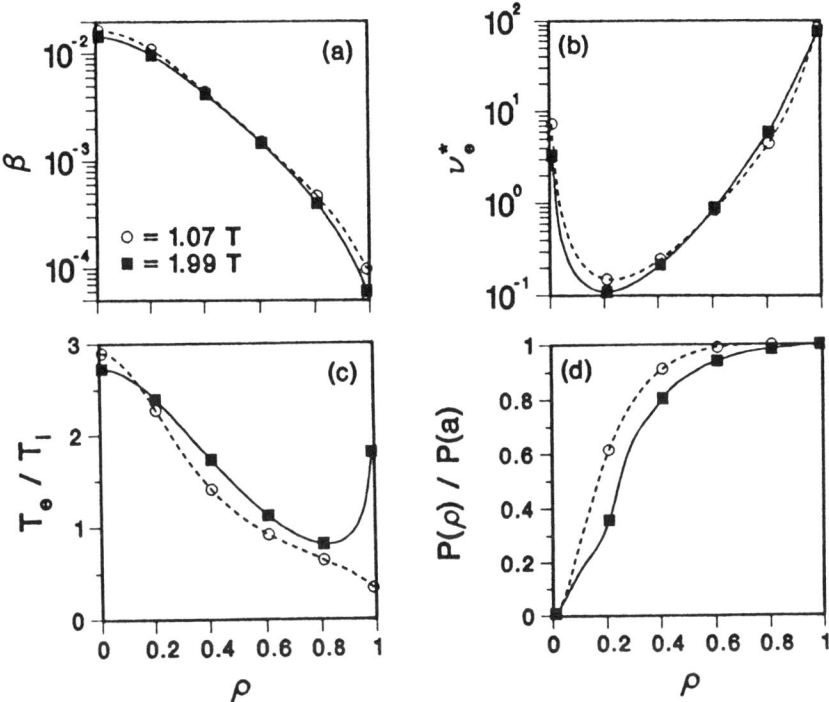

FIG. 1. Profiles of dimensionless parameters for two dimensionally similar discharges with central rf heating.

Table II
Global confinement parameters for dimensionally similar discharges with central rf heating

	$B_T = 1.07$ T	$B_T = 1.99$ T
P_{rf}	0.90 MW	1.45 MW
P_{oh}	0.16 MW	0.29 MW
W	39 kJ	131 kJ
τ	36 msec	75 msec

with NBI heating where the global confinement time followed Bohm-like scaling or worse for a similar L–mode B-scaling experiment.[1] One difference between these two experiments is that with rf heating the deposition profile is not very sensitive to the plasma density. Therefore central rf heating can be utilized for both the low-B and high-B discharges, whereas for NBI the power deposition is decidedly off-axis for the high-B discharge (i.e., high density). However, when dimensionally similar discharges

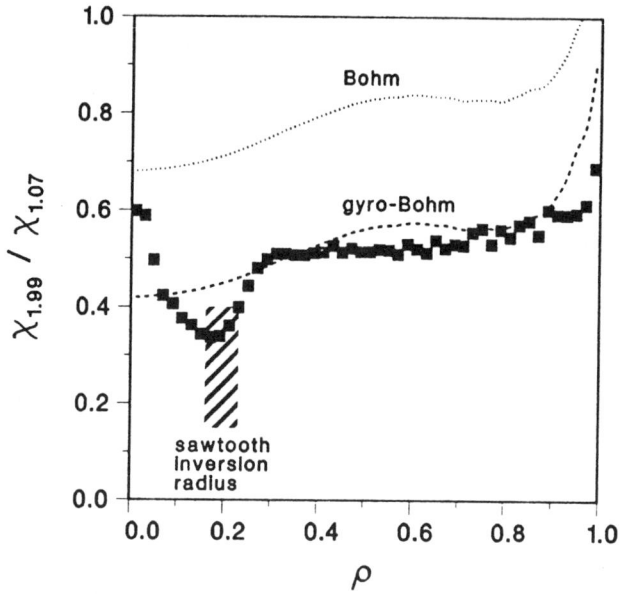

FIG. 2. Ratio of 1.99 T and 1.07 T one-fluid heat diffusivity versus normalized radius (solid squares).

with central ICRH were studied on JET, the transport still appeared to be Bohm-like.[3] The different experimental results between DIII-D and JET may be due to systematic differences in the operating regimes. For DIII-D the rf power directly heats the electrons resulting in T_e/T_i well above unity, whereas $T_e \approx T_i$ in the JET experiments. In addition, the rf experiments on DIII-D operated at higher safety factor and lower density, and thus higher Z_{eff} than the JET experiments. Future experiments on DIII-D at higher rf power will address these systematic differences.

ACKNOWLEDGMENT

This is a report of work sponsored by the U.S. Department of Energy under Contract No. DE-AC03-89ER51114.

REFERENCES

1. R.E. Waltz, J.C. DeBoo, and M.N. Rosenbluth, Phys. Rev. Lett. **65**, 2390 (1990).
2. C.C. Petty, R.I. Pinsker, M.J. Mayberry, et al., Phys. Rev. Lett. **69**, 289 (1992).
3. J.P. Christiansen, B. Balet, D. Boucher, et al., Plasma Phys. and Contr. Fusion **34**, 1881 (1992).

MODIFICATION OF ELECTRICAL CONDUCTIVITY IN T-10 BY ELECTRON CYCLOTRON HEATING

R.W. HARVEY, C.B. FOREST, O. SAUTER,* J. LOHR, and Y.R. LIN-LIU
General Atomics, San Diego, CA 92186-9784

ABSTRACT

The CQL3D Fokker-Planck code is used to investigate effects of quasilinear distortion of the electron tail in the T-10 second harmonic electron cyclotron current (ECCD) drive experiment. The experiment operates in a regime of substantial tail formation. Current drive efficiency may be doubled relative to low power cases. That portion of electric field-driven current which is synergetic with the rf-induced nonthermal tail may tend to cancel the ECCD current in relevant cases.

The T-10 second harmonic electron cyclotron current drive (ECCD) experiment[1] consists of injection of ~ 1 MW of power from the outside equatorial plane of the tokamak. Due to the small beam divergence, 3 degrees FWHM, of the injected power, very high absorbed power densities are obtained for current drive near the magnetic axis. These power densities $p_{EC} \sim 10$ W/cm^3 strongly exceed the criterion for quasilinear electron tail formation by ECH,[2] $p_{EC} > 0.5\, n_{13}^2$ W/cm^3 where n_{13} is the density in units of 10^{13} cm^{-3}. Consequently, substantial nonthermal electron tails have been anticipated, and observed, in the experiment.

The experiment is operated in a regime where the ECCD provides less than the total plasma current and thus a toroidal voltage continues to exist in the steady state plasma, Ohmically driving the remainder of the total plasma current. Typical total plasma current in the experiment is 80 to 140 kA. The nonthermal tails produced by the ECH can carry energy comparable to the bulk plasma, and thus it is also expected that there will be substantial additional current beyond what is calculated with with Spitzer conductivity due to the action of the toroidal electric field acting on the nonthermal tail. This additional current, beyond the simple Ohmic and ECCD currents, is referred to as "synergy current." As will be seen below for the T-10 cases, this synergy current can be as large as the ECCD current.

A further aspect of the experiment which must be considered is that the resistive time of the plasma may be longer than the EC pulse length. Consequently, the internal loop-voltages may differ significantly from the values observed at the plasma periphery. Thus, the resistive time can be estimated from

$$\tau_{\text{resistive}} = \frac{4\pi r^2}{\eta_\parallel c^2} = 1.0 \text{ sec} ,$$

assuming $r = 0.2\, a$, $T_e = 5$ keV, $Z_{\text{eff}} = 2$, whereas the EC pulse length is typically ~ 0.4 sec. In a consistent manner, ONETWO[3] transport code simulations of a typical discharge, using a simple model for sawtoothing instability which spreads the EC current drive so that the safety factor on axis, q_0, remains greater than 1.0, shows that the internal inductive effects remain for up to about 1.0 sec; that is, this period is the persistence time for the Lenz's law type shielding of the the applied rf current

---

*Permanent Address: Centre de Recherches en Physique des Plasmas/EPFL, Lausanne, Switzerland.

by induced toroidal electric field and Ohmic current. (A more elaborate modeling of sawtoothing T–10 discharges by Forest et al.,[4] accounting for Kadomtsev's conservation of helical flux[5] during the sawtooth crash, reveals a very interesting electric generator effect when the central rf driven current exceeds the the pre-rf Ohmic current. Quasi-steady negative voltages to ~ -0.5 V appear near the magnetic axis, for a co-current simulation.)

The CQL3D FP code[6] is well suited to exploring the effects of the interaction of rf and toroidal electric fields. The steady state of the electron distribution function f_e and the radial rf absorption profile are obtained by iteration between (1) the Guassian elimination solution of the Fokker-Planck equation for the steady state f_e at each flux surface

$$\frac{\partial f_e}{\partial t}(u_{\|0}, u_{\perp 0}, \rho, t) = \left\langle \frac{eE_{DC}}{m} \cdot \frac{\partial f}{\partial \underline{u}} \right\rangle + \langle C(f) \rangle + \langle Q(f) \rangle = 0 \quad , \qquad (1)$$

where $u_{\|0}$, $u_{\perp 0}$ are momentum-per-mass of electrons at the outer equatorial plane of each flux surface, $\langle \ \rangle$ indicates a bounce-average; and (2) the rf energy transport equation integrated along a ray

$$\nabla \cdot v_g \mathcal{E} = -\int d\underline{u} \, (\gamma - 1) \, mc^2 \, Q(f) \quad , \qquad (2)$$

where v_g is the ray group velocity, $\mathcal{E}$ is the energy density. In Eq. (1), C is the full collision operator linearized about a Maxwellian distribution shifted in the $u_\|$-direction to conserve momentum in the electron-electron collision process. The quasilinear operator Q is the full operator for the finite gyroradius cyclotron interaction generalized to include the effects of relativity. The launched spectrum of rf energy is discretized into a set of rays which are injected from the plasma periphery. The rays are further discretized into short length elements, each of which contributes to the operator Q, and the damping of the ray element is self-consistently obtained using Eq. (2). The damping rate agrees well with standard expressions.

Using the T–10 EC injection geometry — viz., injection from the outside equatorial plane at 21 deg from perpendicular at the plasma edge, 140 GHz, X–mode — a series of runs were made varying the plasma density, EC power, and loop voltage. The on-axis toroidal magnetic field was chosen to be 2.45 T so that the power deposition and current drive were strongly localized near the magnetic axis, and central temperature is 4 keV. The results from these runs are given in Figs. 1 and 2.

The current I_{rf}^* is the total plasma current driven by the combination of EC power and applied toroidal loop voltage (I_{total}) minus the Ohmic current (I_{ohmic}) obtained from the code with only the loop voltage turned on: $I_{rf}^* = I_{total} - I_{ohmic}$. Thus the current I_{rf}^* is the sum of the ECCD current and the toroidal electric field synergy current.

Figure 1 shows I_{rf}^* versus EC power, for several values of loop voltage V_L and for co- and counter-rf injection. Fig. 1(a) is for central density $n_{e0} = 1.05 \cdot 10^{13}$ cm^{-3} and loop voltages 0.0 and 0.4 V. Fig. 1(b) is for a less extreme case of density $n_{e0} = 1.3 \cdot 10^{13}$ cm^{-3} and loop voltage to 0.25 V. The dashed lines in Fig. 1 are linear extrapolations of the low power results given in the diagrams. Thus in Fig. 1(a) for the loop voltage $V_L = 0.0$ V-case, the driven current more than doubles at the highest powers over the linear results. This increase in current drive efficiency is due to formation of a nonthermal tail[2] within the central 15% of the minor radius. For

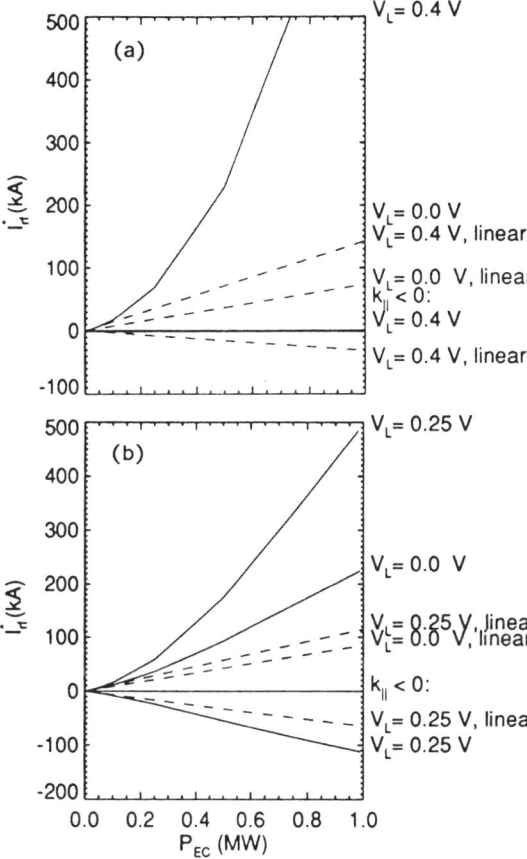

FIG. 1. The driven current I_{rf}^*, equal to the sum of the ECCD current and the toroidal electric field synergy current, versus EC power. The solid curves give code results and the dashed lines are extrapolation of the low power results for (a) central density $n_{e0} = 1.05 \times 10^{13}$ cm^{-3}, and (b) $n_{e0} = 1.3 \times 10^{13}$ cm^{-3}.

the co-current ECCD-case combined with $V_L = 0.4$, the tail electrons essentially run away at the high power and the resulting current depends on the maximum energy on the computational grid (250 keV in these cases). The linear extrapolation (dashed line) for this case underestimates the calculated current even at quite low power, and thus linear calculations of this "hot conductivity" effect will not be sufficient. For the counter-current drive case ($k_\parallel < 0$ and $V_L = 0.4$), the net current I_{rf}^* even changes sign above an rf power of 750 kW. This corresponds to the synergy current exceeding the ECCD in magnitude.

Figure 1(b) for the less extreme case $n_{e0} = 1.3 \cdot 10^{13}$ cm^{-3}, $V_L = 0.0, 0.25$ V still exhibits the doubling of the current drive efficiency beyond linear estimates, and the large synergy current as in the co-current case of Fig. 1(a). However, for the counter-current case, the synergy current does not lead to a reversal of the net driven current I_{rf}^*.

Figure 2 gives I_{rf}^*, the ECCD + synergy current, versus loop voltage. Thus, the extent to which I_{rf}^* is not equal to its value at $V_L = 0.0$ V gives the synergy current. In Fig. 2(a), the EC power P_{EC} is a parameter and density is $n_{e0} = 1.3 \cdot 10^{13}$ cm^{-3}. Negative loop voltage V_L corresponds to counter-current drive. We note that for $V_L \sim -0.4$ V the synergy and ECCD currents cancel quite independent of power from 200 to 800 kW.

In Fig. 2(b), density is the parameter with rf power held fixed at $P_{EC} = 400$ kW. The voltage for cancelling of the ECCD and synergy current, viz., the voltage at which $I_{rf}^* = 0.0$, is seen to be a quite strong function of density. Synergy current will be an important component in the present experiments.

It can be surmised from the results in Fig. 2 that the MHD-generated negative voltage ~ -0.5 V occurring in the simulation of T-10 reported by Forest et al.[3] will

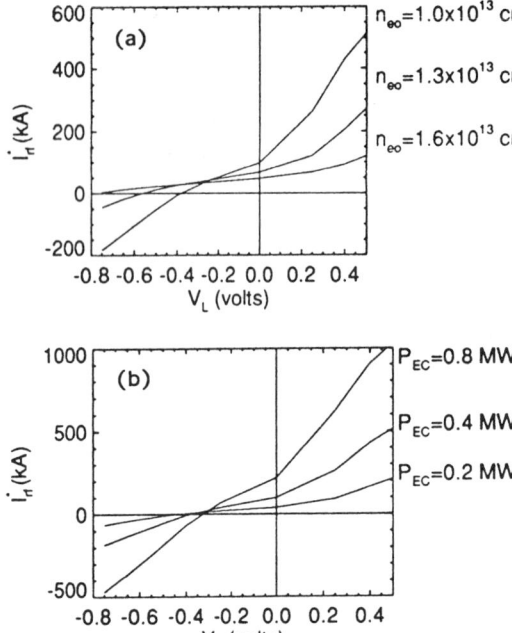

FIG. 2. I_{rf}^* versus toroidal loop voltage with (a) EC power as parameter, and (b) central density as parameter. The temperature is 4 keV.

be closer to an intermediate value ~ 0.25 V if the results presented here are accounted for.

In summary, an accurate interpretation of T–10 experimental voltage traces must take into account the inductive effect and the interaction of the toroidal electric field with the quasilinear distortion of the electron tail.

ACKNOWLEDGMENTS

This is a report of work sponsored by the U.S. Department of Energy under Contract No. DE-AC03-89ER51114. One of the authors (O.S.) was supported by the Swiss National Science Foundation.

REFERENCES

1. V.V. Alikaev et al., this conference.
2. R.W. Harvey, M.G. McCoy, and G.D. Kerbel, Phys. Rev. Lett. **62**, 426 (1989).
3. W.W. Pfeiffer, R.H. Davidson, R.L. Miller, and R.E. Waltz, General Atomics Report GA-A16178 (1980).
4. C.B. Forest et al., this conference.
5. W. Pfeiffer et al., Nucl. Fusion **25**, 655 (1985).
6. R.W. Harvey and M.G. McCoy, General Atomics Report GA-A20978 (1992), to be published in *Proc. of IAEA Technical Committee Meeting on Advances in Simulation and Modeling of Thermonuclear Plasmas, Montreal, 1992* (International Atomic Energy Agency, Vienna, 1993).

ECRH SCENARIOS AT HIGH MAGNETIC FIELD AND ELECTRON DENSITY ON THE FTU TOKAMAK

L.Argenti, A.Bruschi, S.Cirant, G.Granucci, S.Nowak, A.Simonetto, G.Solari
Istituto di Fisica del Plasma, Associazione EURATOM/ENEA/CNR
via Bassini 15, Milano, ITALY

ECRH AT HIGH MAGNETIC FIELD AND ELECTRON DENSITY

In most of the ECRH experiments performed so far, over a frequency range spanning from 28 to 140 GHz, absorption of Electron Cyclotron Waves is almost total provided that the electron temperature at resonance is above a few hundreds eV and the electron density along the ray path between the resonant layer and the launching antenna is not too close to cut-off. Even in the case of relatively low optical thickness [1], plasma single-pass absorption is in general much higher than absorptivity of the vessel walls, so that most of the power is damped inside the plasma after many reflections. However, approaching the electron density, temperature and magnetic field of fusion plasmas, the optical thickness becomes so high that the wave is strongly absorbed as it firstly comes close to the resonance. In addition, in the frequency range above 100 Ghz, the transverse size of the microwave beam can be much smaller than the plasma radius without suffering of severe diffraction effects. The great advantage of ECRH at high magnetic field is therefore the capability of full absorption in very localized plasma volumes.

The microwave system for ECRH at 140 GHz, 2 MW, on the FTU tokamak [2] has been designed to fully exploit the remarkable features of this heating technique. The absorbing volume can be set at plasma centre or at any point of the plasma radius either by moving the antennas or by increasing the magnetic field. The power deposition profile and the total absorption can be calculated [3] by using measured electron density profiles (ohmic target plasma) and a 3D code for ray tracing and wave damping. To facilitate the calculations, the profiles are described by analytical functions, best fitting the experimental values (Fig.1).

In both the regimes of peaked and flat density so far observed in FTU, at a typical central magnetic field of 5÷6 Tesla and electron temperature of 1.5 keV, central absorption is, as expected, strong and localized for a very wide density range (Fig.2u). Localized absorption at low density is maintained high by taking into account the reflection from the wall of the beam power left from the first pass.

Since more than 95% of the total launched power is absorbed within 10% of the minor radius from the resonance position, the analysis of thermal transport in high density plasmas both in steady state and in transient conditions can be performed with clear assumptions on the distribution of the source term. Stron localization allows the study of the stabilization of MHD instabilities either by local profile modification or by direct heating inside the magnetic islands.

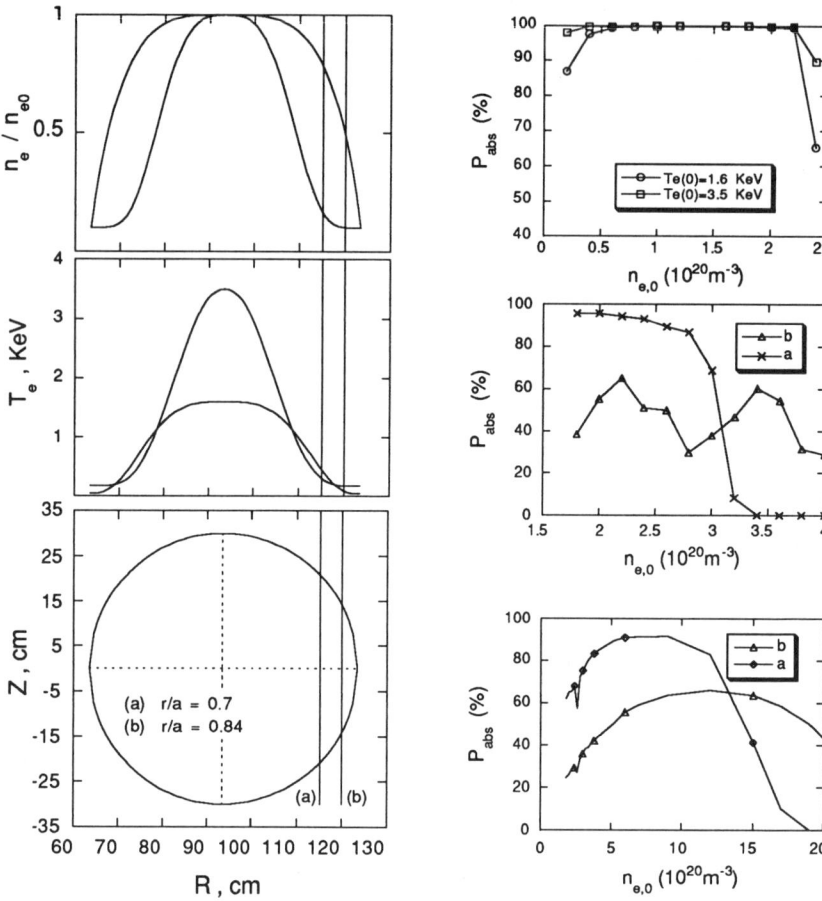

Fig.1 - Electron density (upper) and temperature (intermediate) profiles. a) and b) are the positions where peripheral heating has been calculated.

Fig.2 - Total absorption after two passes at resonance for central (upper) and peripheral heating with flat (intermediate) and peaked (lower) density profile

OFF-AXIS AND PERIPHERAL ECRH

Although high, the electron density during central ECRH at 140 GHz is lower than the maximum achievable in FTU ohmic plasmas at 5÷6 Tesla and at q-values of about 2÷3, which is around $4 \div 6 \cdot 10^{20}$ m^{-3}. However, if the resonance position is moved closer to the antennas by increasing the magnetic field, the central electron density can be accordingly increased, provided that cut-off is kept behind the resonance. In fact, the ratio between the central electron density $n_{e,0}$ and the periferal density (e.g. at r/a=0.8) can be as high as 1.8 and 8.7, respectively for the flat and peaked profiles shown in Fig.1.

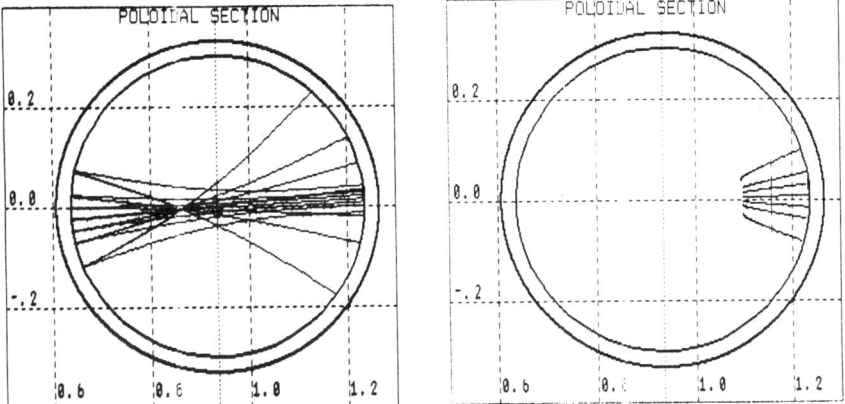

Fig.3 - Ray traces in central ($n_{e,0}=1.6 \cdot 10^{20}$ m^{-3}) and peripheral ($r_{res}=0.7$; $n_{e,0}=7 \cdot 10^{20}$ m^{-3}) heating.

To check for absorption in the case of far off-axis heating, the ray-tracing and absorption code has been run with the resonance set at r/a=0.7÷0.84. Figs.2i,2l shows that absorption remains in the order of 40% with central densities as high as $3.6 \cdot 10^{20}$ m^{-3} and $17 \cdot 10^{20}$ m^{-3} for flat and peaked density profiles respectively. Reflection at cut-off behind the resonance (Fig.3) helps in keeping high the absorption before occurrence of the first wall reflection. The calculations show that significant absorption of the EC waves at 140 GHz, localized in the edge region, is possible in FTU tokamak with central densities in principle higher than the ohmic limit. This allows the experimental study of the effect of plasma edge parameters on plasma stability and on the density limit in tokamaks.

THE ECRH SYTEM

The total microwave power of 2 MW will be delivered by 4 gyrotrons for 0.5 s, with an output 0.5 MW each in a gaussian beam. Each one of the four transmission lines consists of an oversized (88.9 mm i.d.) corrugated circular waveguide carrying the HE11 mode, and a few 90° mitre bends. A universal polarizer is inserted in the transmission line in order to inject into the plasma the proper polarization. To verify the effect of the alignment on the overall transmission efficiency, estimated at 80%, the first transmission line is mounted on adjustable stands to offset and tilt at each junction (Fig.4) within 0.1 mm and 0.5 mrad respectively.

The microwave beams are launched by a system of ellipsoidal mirrors (Fig.5) which refocus inside the plasma the e.m. power radiated from the truncated waveguides. The four beams can be moved independently for off axis heating and for oblique launch by tilting the final mirrors.

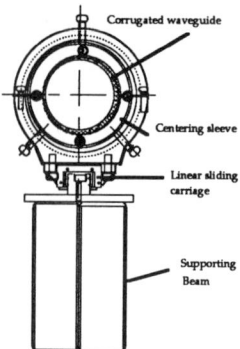

Fig.4 - Waveguide on adjustable stands and supporting beam.

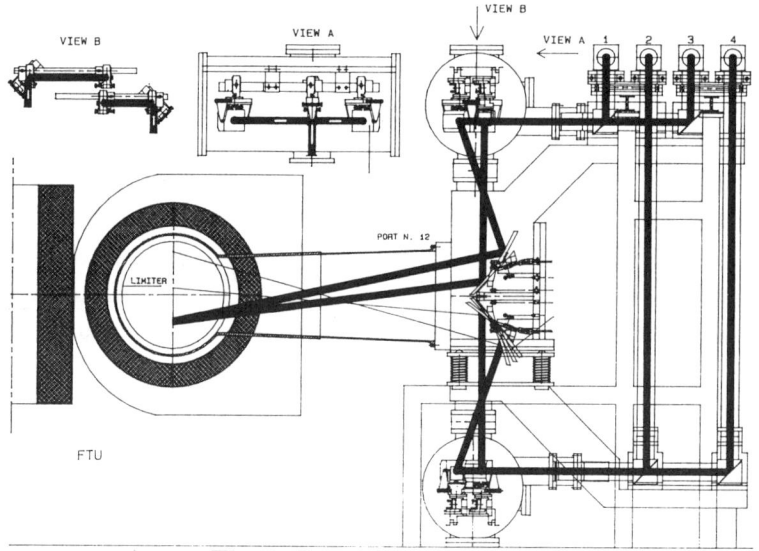

Fig.5 - Launching system

AKNOWLEDGMENTS

The authors greately acknowledge the FTU team for the encouragement given and for having made available all the necessary plasma parameters.

REFERENCES

1. A.Airoldi et al., Suprathermal effects in ECRH experiments on Thor tokamak, Plasma Phys. Contr. Fusion, **30**, 681 (1988)
2. L.Argenti et al., The 140 GHz, 1.6 MW, Electron Cyclotron Resonance Heating system for the Frascati Tokamak Upgrade, Proc. 17th Symposium on Fusion Technology, Rome, September 14-18, 1992
3. A.Airoldi, A.Orefice, G.Ramponi, Nuovo Cimento **6D**, 527 (1985)

CURRENT DRIVE
AND PROFILE CONTROL

REVIEW OF TOKAMAK EXPERIMENTS ON DIRECT ELECTRON HEATING AND CURRENT DRIVE WITH FAST WAVES

ROBERT I. PINSKER
General Atomics, San Diego, CA 92186-9784

ABSTRACT

Results from tokamak experiments on direct electron interaction with the compressional Alfvén wave ('fast wave') are reviewed. Experiments aimed at electron heating as well as those in which fast wave electron current drive was investigated are discussed. A distinction is drawn between experiments employing the lower hybrid range of frequencies, where both the lower hybrid wave ('slow wave') and the fast wave can propagate in much of the plasma, and those experiments using the fast wave in the range of moderate to high ion cyclotron harmonics, where only the fast wave can penetrate to the plasma core. Most of the early tokamak experiments were in the lower hybrid frequency regime, and the observed electron interaction appeared to be very similar to that obtained with the slow wave at the same frequency. In particular, electron interaction with the fast wave was observed only below a density limit nearly the same as the well known slow wave density limit. In the more recent lower frequency fast wave experiments, electron interaction (heating and current drive) is observed at the center of the discharge, where slow waves are not present.

INTRODUCTION

Over the past two decades, the heating of tokamak plasmas by injection of compressional Alfvén waves ('fast waves') has become a mature subject (see Ref. 1 for an interesting review of some aspects of this development). In almost all of these experiments, the mechanism for absorption of the wave energy is ion cyclotron damping, at either the fundamental resonance of a minority ion species ($\omega = \Omega_{min}$) or at a low order harmonic of the majority ($\omega = n\Omega_{maj}$, n =2 or 3). In recent years, an alternative mechanism for fast wave absorption in a collisionless plasma, the coherent combination of electron Landau damping and transit time damping, has been explored in a number of tokamak experiments. This absorption mechanism was discussed along with absorption at cyclotron resonances by Stix.[2] An important application of this method of interacting with electrons in a tokamak discharge is noninductive current drive; by transferring toroidal momentum from the fast wave to the electrons, a toroidal current can be produced, just as in the well established lower hybrid current drive (LHCD) technique. The theoretical advantages of fast wave current drive (FWCD) over LHCD in a tokamak reactor are well known: no density limit for electron interaction at a given FW frequency, excellent accessibility to the plasma core in high β, high T_e discharges, current drive efficiencies equal to or greater than that of LHCD, and existing efficient CW rf sources with high unit power ($\geqslant$ 2 MW/unit).

Non-inductive current drive is not the only use for direct electron interaction with fast waves. Another important application to the tokamak is as an electron heating method with several unique properties: the damping will always be strongly peaked at the point of maximum electron β, usually near the magnetic axis, and the power deposition profile shape will be nearly independent of the toroidal field and the density profile shape. Furthermore, since the power is not transferred to the electrons via an energetic ion tail, as is the case for low field side ICRH and for NBI, the power coupled to the electrons is not limited by fast ion energetics[3] or instabilities (TAE, fishbones, etc.).

The fast wave exists in a magnetized plasma over a very wide range of frequencies, from the MHD ($\omega \ll \Omega_i$) regime, where it is usually referred to as the compressional Alfvén wave*, through the ion cyclotron range of frequencies and the lower hybrid range of frequencies. While FWCD may be applied in a tokamak reactor at a frequency just below the lowest ion cyclotron resonance frequency to avoid damping on alpha particles, the lowest frequency regime of fast wave propagation has not yet been the subject of many experimental studies at high power levels in tokamaks, and therefore will not be discussed further in this review. Most of the physics of direct electron interaction with the fast wave just below Ω_i is the same as in the most extensively studied regime, the ICRF. While the properties of the fast wave itself are qualitatively similar in the ICRF and the LHRF, important differences between wave damping in these two regimes arise from the interaction between the fast wave and the other cold plasma mode in these frequency ranges, which is known as the slow wave, the electron plasma wave, or the lower hybrid wave. In the ICRF, the slow wave can propagate only at very low densities $\omega < \omega_{pi}$, so that interaction of the two waves is possible only very near the boundary of the plasma. The only role of the slow wave in ICRF fast wave experiments is thus as a possible parasitic loss process near the launching structure.[5] By contrast, in the LHRF, slow and fast waves at the same frequency and at the same wavelength along the toroidal field can both propagate throughout most of the plasma volume. Any mechanism that would permit coupling between the two modes, which are actually completely distinct only in an unbounded uniform quiescent plasma, would be expected to have a strong effect on the absorption of wave power on electrons, because the electron Landau damping of the slow mode is much stronger than that of the fast wave af the same $n_\| \equiv k_\| c/\omega$. Hence, appreciable unlike-mode coupling would be expected to result in otherwise anomalously strong electron damping of the launched fast wave.

The electron damping of the fast wave has been calculated by several authors[2,6-8] in various limits. For a Maxwellian electron distribution, the most general result[7] for the spatial damping rate due to electron damping may be written as[8]

$$2\,\text{Im}\,(k_\perp) = \text{Re}\,(k_\perp)\,\frac{\sqrt{\pi}}{2}\,\beta_e \xi e^{-\xi^2}\left[1+\frac{1}{\alpha^2}\right] , \qquad (1)$$

where

$$\alpha = \left(\frac{T_e}{m_i c^2}\right)\left(\frac{\omega^2 - \Omega_i^2}{\omega_{pi}^2}\right)(S - n_\|^2)|K_{zz}| ,$$

$$\xi = \frac{\omega}{k_\|}\sqrt{\frac{m_e}{2T_e}} \quad ; \quad \beta_e = \frac{8\pi n_e T_e}{B_0^2} \quad ; \quad S = 1 - \frac{\omega_{pi}^2}{\omega^2 - \Omega_i^2} + \frac{\omega_{pe}^2}{\Omega_e^2 - \omega^2} ,$$

$$K_{zz} = 1 + \frac{1}{k_\|^2 \lambda_{D^2}}\,[1+\xi Z(\xi)] .$$

If $n_\|^2 \ll |S| \simeq \omega_{pi}^2/(\omega^2 - \Omega_i^2)$, and taking the cold plasma limit of $|K_{zz}|$, α reduces to

---

*Perhaps the first proposal[4] for FWCD in a tokamak reactor was in this low frequency regime.

was mentioned in another paper[16] from JIPP T–IIU, which will be described later in this review.

Two different fast wave launching structures were compared in a set of experiments[17,18] on the Princeton Large Torus (PLT). These experiments were also performed at 800 MHz. In the first set of experiments, an array of six small loop antennas,[19] with no Faraday shield, was used to launch up to 150 kW for up to 0.3 sec. A second set of experiments used a 4 × 3 (toroidal × poloidal) array of dielectric-loaded waveguides to launch up to 340 kW of power. The coupling properties of the latter launching structure were studied in some detail theoretically[20,21] and experimentally.[21] The conclusion of the coupling study was that under the conditions used in the current drive experiments, only a small fraction (less than 5% for 180° phasing) of the launched power coupled directly to the slow wave, with the remainder being launched in the desired fast wave polarization. Yet for both the dielectric-loaded array coupler and the array of loop antennas the current drive efficiency was similar to the 800 MHz slow wave couplers already studied on PLT,[22] and the density limit for fast electron interaction was independent of the launched mode (Fig. 1). A detailed comparison of the 800 MHz slow and fast wave experiments concentrated on the phenomena associated with the density limit; no significant difference between the two launches was found.

Another detailed comparison of slow and fast wave launch at the same frequency on the same device was performed on the JFT–2M tokamak.[23] In this experiment, fast waves launched with a phased pair of loop antennas were compared with slow waves launched from a pair of waveguides, both at 750 MHz, $P_{rf} = 100$ kW. While the density limit for observable current drive was again identical for the two launches, the details of the phenomena just below the density limit were slightly different. Parametric decay activity was weaker for the fast wave launch in this density regime, and the maximum density for which a fast electron tail was observed was somewhat higher. It was concluded that at low density slow waves were responsible for the observed current drive, and that the small effect of FWCD was manifest only in the neighborhood of the density limit.

An explanation of these results was proposed by the present author,[18] based on ideas mentioned by Ando et al. (Ref. 16) and Andrews.[24,25] The fact that the fast wave electron damping is so much weaker than that of the slow wave at the same $n_\parallel$ implies that any power which is undergoing mode conversion back and forth between the two modes is much more likely to be absorbed as a slow wave than as a fast wave.[16] Furthermore, a simple statistical argument[18] shows that even without damping, given a *random* mode coupling mechanism, wave power launched as a fast wave will end up in the slow wave polarization, while the reverse is not true. The net result of these two phenomena is that wave power launched in the LHRF in either of the two polarizations is likely to be absorbed as a slow wave, and hence be subject to the slow wave density limit. All that is necessary for this argument to apply is the existence of a random mode coupling mechanism.

Two such mechanisms important in different regimes were first discussed in detail by Bonoli and Ott[26] in the context of slow wave launch. In a quiescent toroidal plasma with inverse aspect ratio $\gtrsim 0.2$, ray trajectories starting at $n_\parallel$ just above the critical $n_\parallel$ for accessibility in a slab geometry become ergodic. The resultant random shifting up and down of $n_\parallel$ along the ray trajectory along with the accompanying mode conversion constitutes a random mode coupling mechanism to which the statistical argument mentioned above applies. The other mode coupling mechanism applies even in a slab geometry: in a turbulent plasma in which the density fluctuation spectrum

improving the current drive efficiency by raising the frequency and operating at lower values of $n_\parallel$. By going to very high ion cyclotron harmonics, ion damping becomes negligible. For small tokamaks, the size of fast wave launching structures becomes relatively manageable at these frequencies. Finally, rf sources in this frequency range were already installed at several tokamak facilities by the early 1980s. For these reasons, the first tokamak experiments on FWCD were performed in the LHRF.

FWCD EXPERIMENTS IN THE LHRF

By the mid 1980s, slow wave current drive in the LHRF was well established experimentally in tokamaks.[9,10] The LHCD density limit scaling with frequency was well characterized experimentally[9,11,12] though not completely understood theoretically.[13,14] It was recognized that while both linear and nonlinear ion interaction can be avoided by operating with $\omega/\omega_{LH} \gtrsim 2$, the relatively high $n_\parallel$ necessary for accessibility to the center of a dense plasma would inevitably lead to strong Landau damping in the outer part of a large, hot $(T_e(0) \gtrsim 15$ keV) reactor plasma. For slow and fast wave propagation in a slab geometry at the same frequency, the accessibility limit (due to mode conversion between the two modes with concomitant reflection) is identical for the two modes. However, the fast wave is unaffected by the lower hybrid resonance, so that fast waves can be used at lower frequencies ($\omega/\omega_{LH} \lesssim 1$) for which the accessibility limit can be pushed down nearly to $n_\parallel = 1$. In the tokamaks of the mid 1980s, the fast wave damping was very weak even in the LHRF, particularly at the low values of $n_\parallel$ that were practical to launch and at which the current drive efficiency was expected to be observably large. Therefore, though the driven current per *dissipated* wave power is independent of the launched wave mode, the observable efficiency (driven current per *input* wave power) was expected to be much lower for the fast wave than for the slow wave at the same frequency and $n_\parallel$. Thus, the signature of fast wave current drive in the LHRF should be a weak direct electron interaction which should decrease only with power per particle as the density is raised through $\omega/\omega_{LH} = 2$ and then $\omega/\omega_{LH} = 1$.

In this context, the results from the experiments on FWCD in the LHRF were rather surprising. The first result reported[15] was from JIPP T-IIU, using an 800 MHz rf source. A four-element dipole antenna array with a double-layer Faraday shield was used to launch up to about 100 kW of rf power with a wide $n_\parallel$ spectrum centered around $n_\parallel \sim 2$ for 90° phasing. At the low density of $\bar{n} \simeq 3 \times 10^{12}$ cm^{-3}, the current drive efficiency was observed to be identical to that observed with slow wave launch at the same frequency on JIPP T-IIU. However, the same density limit of about $\bar{n} \simeq 8 \times 10^{12}$ cm^{-3} was observed for both slow and fast wave current drive. Furthermore, the phenomenology around the density limit — the observation of fast ion tails correlated the presence of parametric decay into ion-cyclotron quasimodes, where the cyclotron frequency of the decay waves corresponds to the field at the outboard midplane of the discharge — were similar for the two launched modes. Interestingly, the maximum electron interaction occurred at antenna phasing of 0°, and the decrease in electron interaction as the phase angle was increased was independent of the relative sign of the phase velocity and the direction of the plasma current. The JIPP T-IIU experimentalists speculated that the parametric decay resulted from mode conversion from the fast to the slow wave, and that the density limit therefore was due to channeling of wave energy into the ion tail via the parametric decay of the slow waves. They did not suggest, however, the possibility that the efficient current drive at lower densities was also due to damping of mode-converted slow waves. This possibility

$$\alpha \simeq \frac{T_e}{m_e c^2} \left(\frac{\omega_{pi}^2}{\omega^2} \right) ,$$

so that at low frequencies, high density, and/or high electron temperature, $1/\alpha^2$ is negligible compared to unity, and the damping length reduces to the well known formula

$$2 \, \text{Im}(k_\perp) \simeq \text{Re}(k_\perp) \, \frac{\sqrt{\pi}}{2} \, \beta_e \xi e^{-\xi^2} , \qquad (2)$$

where the dispersion for the fast wave is very nearly $\text{Re}(k_\perp) \simeq \omega/v_A$. The overall toroidal magnetic field dependence of the damping is quite strong: $\text{Im}(k_\perp) \propto B_0^{-3}$. The damping has a broad maximum around parallel phase velocities of $\xi = 1/\sqrt{2}$, or at a parallel index of refraction of

$$n_\parallel \simeq \frac{23}{\sqrt{T_e \, [\text{keV}]}} . \qquad (3)$$

At values of $n_\parallel$ which can be launched easily with ordinary antenna structures with a directive spectrum, Eq. (3) shows that the electron temperature required for strong single pass electron absorption is much higher than has been achieved in tokamaks to date. The long damping length predicted by Eq. (1) at fairly low $n_\parallel$ is the reason that the fast wave is well suited to driving current near the magnetic axis in a large tokamak reactor, while difficult to test in present day experimental devices. (Of course, the reverse applies to LHCD.) For the fast wave, the electron damping is generally too weak to produce a quasilinear flattening of the electron distribution with any reasonable fast wave power density, so that one cannot simultaneously access the efficient current drive regime of high phase velocities and maintain high single pass absorption. Therefore, the strategy employed in FWCD experiments generally has been to operate at the highest $n_\parallel$ which can be launched with a given antenna structure and simultaneously to maximize the electron temperature and hence the single pass absorption with other forms of electron heating. Nonetheless, in every experiment performed to date, the single pass electron damping has been quite weak, so that multiple pass absorption is expected. Consequently, the relationship between the launched and absorbed spectrum is not as straightforward as in heating and current drive schemes with strong first pass absorption.

In the ICRF, ion cyclotron harmonic damping tends to dominate direct electron damping, so that various methods have been used to minimize this competing damping mechanism. At high aspect ratio, it may be possible to operate without any cyclotron harmonic layers in the plasma, or at least near the magnetic axis. For proof-of-principle experiments, operation at moderate cyclotron harmonics, $\omega = (3 \text{ to } 5)\Omega_i$ is adequate to avoid strong ion damping, as long as no significant source of energetic ions (NBI) exists in the plasma. Whether such a scenario is applicable in a reactor is unclear, due to harmonic damping on alphas. A practical difficulty in this frequency range stems from the high $n_\parallel$ necessary for significant electron absorption: the antenna loading is relatively weak at high $n_\parallel$, due to the large evanescent region at the edge of the discharge for high $n_\parallel$. The low antenna loading can pose significant technical difficulties in launching high power levels.

By going towards higher frequencies, that is, towards the LHRF, some of these problems are alleviated. The $1/\alpha^2$ term in Eq. (1) dominates, so that the fast wave damping at the same $n_\parallel$ can be considerably stronger than in the low frequency limit [Eq. (2)]. Alternatively, one can maintain the same single pass absorption while

regime, and reached a central electron temperature of about 9 keV with strong NB heating ($P_{NB} = 24$ MW).

Two sets of experiments on the JIPP T–IIU tokamak on direct electron absorption of the FW in the ICRF have been reported. The first experiment[16] was performed at 40 MHz and at very low toroidal field so that $\omega \sim 13\Omega_H$. In this case, the plasma was produced by the rf power alone. A phased array of five loop antennas was used to launch fast waves with $n_\| \approx 7$ for $\Delta\phi = 90°$, at up to 400 kW. Though the direction of the current was determined by the polarity of the vertical field, as is often the case in startup experiments, the efficiency of the current drive did respond to the antenna phase angle roughly as expected. The electron temperature in this experiment did not exceed about 10 eV, so that the remarkable feature of this experiment was that any electron interaction was observed at all; under these conditions, $\xi \sim 23$, so direct electron absorption of the nominal launched $n_\|$ would be expected to be utterly negligible.

In the more recent experiments[34] on JIPP T–IIU, a 130 MHz rf system was used to power a four-element array of loop antennas mounted on the high field side of the torus. In this case, $\Delta\phi = 180°$, $n_\| = \pm 5$. The single pass absorption was predicted to be less than 0.1% for the experimental conditions. Conventional theories indicate that mode conversion to the IBW should be negligible. Nonetheless, significant electron heating was observed along with a tail in the deuterium neutral particle charge exchange spectrum. It was speculated that these results might be related to anomalously strong mode conversion near $\omega = 6\Omega_D$ and subsequent absorption of the IBW near the $\omega = 5\Omega_D$ resonance. The net absorption efficiency was estimated to be $\gtrsim 50\%$ despite the very low single pass absorption.

Another tokamak experiment in which unexpectedly strong absorption of the FW was observed has been carried out on the Phaedrus–T device.[35] Here a two-strap antenna was operated at 19 MHz, so that with $\Delta\phi = 90°$, the peak of the vacuum spectrum was $n_\| = 26$, and $\xi < 1$ for the peak of the vacuum spectrum. The antenna loading to these very short wavelengths however is very weak, so that the value of ξ characterizing the power that actually propagates in the plasma would be expected to be significantly larger. The expected weak damping was evident from the eigenmodes observed on both the antenna loading and on magnetic probes in the edge plasma as the density rose during high power rf injection. Nonetheless, the damping length measured in the edge plasma with magnetic probes appeared to be several times shorter than expected. Weak electron heating near the center of the discharge was observed for 90° antenna phasing, while no heating was found with 180° phasing, probably as a result of the extremely weak coupling to the high $n_\|$ excited by the latter phasing.

A recent experimental study[36,37] of direct electron damping on the DIII–D tokamak used a four-element phased antenna array and a 2 MW 60 MHz rf source. For the damping experiments 180° phasing was used to excite $n_\|$ in the neighborhood of 10. The upshift of $n_\|$ due to finite aspect ratio ($Rn_\| = $ constant) is important in the relatively low aspect ratio DIII–D device: $n_\|(\rho = 0) = n_\||_{ant}(R_0 + a)/R_0 = 1.4\,n_\||_{ant}$. In this experiment, though single pass absorption of only 5% to 25% was calculated using measured profiles, about 80% to 100% of the coupled rf power was absorbed (Fig. 3). Evidently efficient multiple pass absorption must occur. The central electron heating effectiveness was found to depend strongly on ξ evaluated at the center of the discharge: large increases in $T_e(0)$ were observed only if $\xi \lesssim 1.5$. The deposition profile was studied by modulation techniques similar to those employed in the previous experiments. A highly peaked deposition profile was found, in agreement the central electron heating reported there, though parametric decay of mode converted slow waves could still be an important energy loss mechanism. Central electron heating was observed only for $\xi \lesssim 4$. A more detailed study of electron absorption[30] performed by the same group employed a wider range of discharge conditions and used a FW amplitude modulation technique to deduce the radial deposition profile. The experimentally determined deposition profile was found to be strongly centrally peaked, in accordance with theoretical expectations (Fig. 2). Both the modulation technique and analysis of the rate of change of the plasma stored energy at FW turn-on and turn-off yielded an overall absorption efficiency of about 0.3 to 0.4 in the best case. It is claimed that this absorption coefficient is consistent with ray-tracing calculations of single pass absorption. These authors speculated that the power not absorbed on the first pass is collisionally dissipated in the edge plasma and/or the vacuum vessel walls; this view is supported by the observation of an anti-correlation between absorption efficiency and impurity influx.

A series of discharges in the JET tokamak with ICRF heating in a number of heating regimes were analyzed by Eriksson and Hellsten to determine the relative contributions of direct electron absorption and slowing down of an ion tail produced by ICRH to the observed electron heating.[31] In these cases, a further complication is that direct electron heating can result from either damping of the fast wave or of ion Bernstein waves (IBWs) produced by mode conversion. A painstaking analysis of the time dependence of the electron temperature enabled these authors to distinguish between the different components of electron heating. In cases with high minority concentrations, discharges where the electron heating is primarily due to direct electron absorption were identified. With 180° phasing of the two-element antenna arrays, the high $n_\parallel$ tends to minimize the contribution of mode converted IBWs and maximize direct electron absorption of the fast wave itself. With 0° antenna phasing, mode conversion becomes important, and the measured electron deposition profile becomes strongly peaked around the mode conversion layer.

A more recent set of experiments on JET intended to maximize direct electron absorption of the fast wave were reported by Start et al.[32] In contrast to the earlier JET experiment, here the lowest order ion cyclotron resonance layer was displaced inboard of the magnetic axis to create a substantial central volume of the plasma free of any cyclotron resonances. A square-wave modulation of the rf power was employed to measure the direct electron absorption in this central region from time dependent analysis of $T_e(t)$. In this situation, most of the power not absorbed on the first pass through the center is absorbed at the $\omega = 2\Omega_H$ layer on the high field side of the magnetic axis. Measurements showed approximately 22% of the input rf power was directly absorbed by electrons in the central region, in good agreement with full wave and ray tracing codes.

Some recent experiments in TFTR[33] at 47 MHz have explored similar regimes of FW direct damping as the JET work. Similar amplitude modulation techniques were used to study direct electron absorption in two regimes. In one regime ($B_T = 2.3$ T), the second harmonic of the majority ^{3}He passed through the center of the discharge; approximately 30% to 50% of the power appeared to be absorbed on electrons near the center of the discharge. In the other regime studied in detail, the fundamental resonance of ^{3}He passed through the center ($B_T = 4.8$ T), but no ^{3}He was introduced into the deuterium plasma. In this case, about 60% of the rf power was absorbed by electrons near the center of the of the plasma. This discharge was in the 'supershot'

has significant power in the range of $k_\perp$ equal to the difference in $k_\perp$ of the two wave modes near the accessibility limit of tokamaks for wave launching in the LHRF.

At high central densities such that $\omega \ll \omega_{LH}$ ($\rho = 0$), mode coupling is not possible near the center of the discharge, since the slow wave cannot propagate there. From this perspective, the essential difference between the FW experiments in the LHRF and those in the ICRF is the fraction of the plasma volume in which both modes can propagate. In the ICRF, this region of overlap can vanish for $n_\parallel$ high enough to obtain significant electron damping, because the FW cutoff density can be higher than the lower hybrid resonance density. Thus, at low frequency, wave power can propagate in the core of the discharge only as either a fast wave or an ion Bernstein wave. Hence, FWCD at low frequency will not be subject to a density limit related to wave accessibility.

For this reason, along with the fact that the high values of $n_\parallel$ necessary for reasonable single pass electron damping can be launched with large antenna structures more conveniently at low frequencies, FWCD experiments in the LHRF have been largely abandoned in favor of the ICRF in recent years. Work with fast wave launch in the LHRF has been carried on the Versator II tokamak at MIT, where FW launch with a slotted waveguide coupler at 2.45 GHz has been studied,[27] and a detailed experimental study of parametric decay activity in the LHRF in which 800 MHz slow wave, 2.45 GHz slow wave, and 800 MHz fast wave launch (the latter using a four element array of dielectric loaded waveguides) have been compared on the same machine.[28] From the point of view discussed above, one would expect for the fast wave launch, the mode converted power in the edge region would tend to raise the energy density in that region, and hence lower the threshold power for the parametric decay instability. In the experiment, parametric decay and associated edge electron heating with the remarkably low power threshold of about $P_{rf} = 50$ W are reported.

EXPERIMENTS ON DIRECT ELECTRON DAMPING OF THE ICRF FW

The first set of experiments on FW direct electron interaction in the ICRF were intended to establish a connection between the damping theory discussed in the introduction and experimental results, as a first step in the development of FWCD in this frequency regime. To make the electron damping significant, the highest $n_\parallel$ that can be launched must be used. Therefore the antenna phasing used for the multiple loop arrays in these experiments is generally 180°, which coincidentally is also the most favorable antenna phasing for minimizing impurities.[1] While operating at low frequencies raises $n_\parallel$ for a given antenna geometry, Eq. (1) shows that the damping increases with frequency at a fixed $n_\parallel$, through $\text{Re}(k_\perp) \propto f$ and $1/\alpha^2 \propto f^4$ in the cold plasma limit. Furthermore, operating at higher frequency helps to minimize ion cyclotron damping, which otherwise dominates. Hence these experiments tend to be done at the upper end of the available frequency range at each tokamak facility, and at low toroidal field [Eq. (1) shows the damping increases with decreasing toroidal field as B_0^{-3}].

The first detailed experimental study of FW direct electron absorption in this regime was carried out on the JFT-2M tokamak at 200 MHz. After a preliminary experiment[23] using the same two-loop antenna as had been used at 750 MHz, a new four-strap antenna and 4×200 kW 200 MHz rf system were constructed. In the first set of results reported from this system,[29] the lower hybrid resonance layer for 200 MHz was located at $r/a = 0.93$; clearly, slow waves have nothing to do with

with the predictions of a full-wave code. A steady-state ELMing H–mode was produced with FW direct electron heating as the sole auxiliary heating source. The fact that the power threshold for the H–mode transition was approximately the same as for H–modes produced by ECH or NBI implies that nearly 100% of the fast wave power must be absorbed in the discharge.

Assuming that the fundamental theory is correct, the good absorption generally observed in the experiments even when the predicted single pass absorption is very poor implies that multiple pass absorption must occur. Though for heating purposes, multiple pass absorption should not present a problem as long as rf/wall interactions are minimized, such a large number of passes may imply loss of control over the absorbed spectrum. For efficient current drive, then, good single pass absorption might still be necessary. This fact, along with the formidable technical problems associated with launch of a high $n_\parallel$ spectrum with good directivity, makes FWCD experiments in the ICRF the most challenging of the experiments discussed in this review.

FWCD EXPERIMENTS IN THE ICRF

Before successful experiments on FWCD in the ICRF could be performed, solutions had to be found for the aforementioned technical challenges. A complete discussion of the work that has been carried out towards the solution of these problems is beyond the scope of this review; this effort has been described in some of the references.[38–40] Continuing work in this area is discussed in several papers in these proceedings.

A consequence of these technical problems has been that until recently, launching of significant ICRF power with a *directive* spectrum peaked at a value of $|n_\parallel|$ high enough to yield single pass damping in even the 5% to 10% range has been extremely difficult. Therefore, no conclusive evidence of substantial current driven by direct electron absorption of ICRF fast waves was obtained prior to the DIII–D results described in Ref. 41. The same four-element antenna array that had been used for the DIII–D FW direct electron heating studies was used for the FWCD experiments; in the latter case, antenna phasing of $(0, \pi/2, \pi, 3\pi/2)$ was used to launch a directive spectrum peaked between $n_\parallel = 5$ to 7 at 60 MHz. Up to 1.1 MW was coupled with this phasing into relatively low density discharges ($\bar{n}_e \sim 1 \times 10^{13}$ cm^{-3}). Preheating with 60 GHz ECH ($2\Omega_e$, X–mode) was used to raise the single pass absorption of the FW to about 10%. Studies of the global confinement properties of the discharges in which ECH and FW with directive antenna phasing were compared again indicated that nearly 100% of the FW power was absorbed, evidently by multiple passes of the FW through the hot core of the plasma.

In the best case with EC preheating, up to 160 kA of the 400 kA total plasma current were sustained noninductively, while without preheating, 135 kA of noninductive current were observed. In both cases, 40 kA of the noninductive current could be attributed to the neoclassical bootstrap current. The current drive efficiency was found to increase roughly linearly with the central electron temperature (Fig. 4). The observed current drive efficiency in the customary units could be expressed as $\gamma \equiv \bar{n}_e I_{rf} R_0 / P_{FW} \simeq 5.5 \times 10^{17} T_e(0)$ [keV] A/W/m^2, with the best result obtained to date being $\gamma \simeq 0.020 \times 10^{20}$ A/W/m^2, obtained at a central $T_e(0) \simeq 3$ keV. The increase in the neoclassical bootstrap current caused by the rf has been subtracted from the noninductive current to obtain this number. Higher rf power levels are expected to produce higher electron temperatures, and hence higher current drive efficiencies in the near future. FWCD modified the magnetic structure (sawtooth behavior) near the magnetic axis, with quite different effects observed for counter-FWCD than for co-FWCD. A strong effect of this kind is expected as a result of the highly centrally-peaked FW deposition profile. Generally, the DIII–D FWCD results are consistent with multiple pass ray tracing calculations, as well as with full-wave code calculations.[42]

CONCLUSIONS AND OUTLOOK

Tokamak experiments on electron heating and current drive by direct electron absorption of fast waves have made significant advances in the last eight years. The results of early experiments with fast waves in the lower hybrid range of frequencies can be understood in terms of mode conversion of the launched fast waves to lower hybrid slow waves and subsequent efficient absorption of the slow waves. Direct electron absorption experiments in the ICRF have demonstrated that this technique of heating electrons is a useful addition to the already well established ICRF plasma heating methods of minority and second harmonic heating. Noninductive currents in excess of 0.1 MA have been driven by the absorption of fast waves with a directive spectrum and the predicted favorable temperature scaling of the FWCD efficiency has been observed.

In the next few years, this favorable temperature scaling along with improved solutions to the technical problems of directive FW launch should permit extension of the FWCD regime to the ~ 1 MA level, with complete sustainment of the plasma current by fast wave power alone. An upgrade of the DIII–D FWCD system to 6 MW in the 30 to 120 MHz range is in progress; with additional electron heating, driven currents in the 0.5 to 1.0 MA range are expected. On JFT–2M, increased ECH power and improved 200 MHz FW antennas should permit unambiguous observation of FWCD in the next two years. New four-element antenna arrays are being installed in JET at this writing; these antennas along with an improved phase control system will enable FWCD experiments at truly reactor relevant power levels (up to 32 MW of generated power) and central electron temperatures. Other tokamaks for which FWCD experiments have been discussed include Tore Supra, JT–60U, TFTR, and Alcator C–Mod. In the next few years, FWCD should be further validated by these near-term experiments and will therefore become the central current drive method of choice for future devices as tokamak plasmas become more reactor-like. Indeed, this is the environment in which fast wave current drive has always been expected to operate most effectively.

ACKNOWLEDGMENT

This is a report of work sponsored by the U.S. Department of Energy under Contract No. DE-AC03-89ER51114.

REFERENCES

1. J.-M. Noterdaeme, *Radio-Frequency Power in Plasmas: Ninth Topical Conference, Charleston, SC 1991* (American Institute of Physics, New York, 1992), p. 71.
2. T.H. Stix, Nucl. Fusion **15**, 737 (1975).
3. J.G. Cordey et al., in Proceedings of the 18th European Conference on Controlled Fusion and Plasma Physics, (Berlin, 1991) **15C**, Part III, (EPS, 1991), p. 385.
4. D.J.H. Wort, Plasma Phys. **13**, 258 (1971).
5. E.A. Berro and G.J. Morales, IEEE Trans. Plas. Sci. **18**, 142 (1990).

6. D. Moreau, M.R. O'Brien, M. Cox, and D.F.H. Start, in Proceedings of the 14th European Conference on Controlled Fusion and Plasma Physics, (Madrid, 1987) **11D**, Part III, (EPS, 1987) p. 1007.
7. S.C. Chiu, V.S. Chan, R.W. Harvey, and M. Porkolab, Nucl. Fusion **29**, 2175 (1989).
8. M. Porkolab, *Radio-Frequency Power in Plasmas: Ninth Topical Conference, Charleston, SC 1991* (American Institute of Physics, New York, 1992), p. 197.
9. W. Hooke, Plasma Phys. and Contr. Fusion **26**, Vol. 1A, 133 (1984).
10. N.J. Fisch, Rev. Mod. Phys. **59**, 175 (1987).
11. F. Alladio, E. Barbato, G. Bardotti, et al., Nucl. Fusion **24**, 725 (1984).
12. M.J. Mayberry, M. Porkolab, K.-I. Chen, et al., Phys. Rev. Lett. **55**, 829 (1985).
13. J.-G. Wegrowe and F. Engelmann, Comments Plasma Phys. and Contr. Fusion **8**, 211 (1984).
14. V.S. Chan and C.S. Liu, Fusion Technol. **7**, 288 (1985).
15. K. Ohkubo, Y. Hamada, Y. Ogawa, et al., Phys. Rev. Lett. **56**, 2040 (1986).
16. R. Ando, E. Kako, Y. Ogawa and T. Watari, Nucl. Fusion **26**, 1619 (1986).
17. R.I. Pinsker, P.L. Colestock, S. Bernabei, et al., *Applications of Radio-Frequency Power to Plasmas: Seventh Topical Conference, Kissimmee, FL 1987* (American Institute of Physics, New York, 1987), p. 175.
18. R.I. Pinsker, *Fast Wave Current Drive Experiments on the Princeton Large Torus*, Ph. D dissertation, Princeton University (1988).
19. P.L. Colestock, J.E. Stevens, J.C. Hosea, et al., *RF Plasma Heating: Sixth Topical Conference, Callaway Gardens, GA 1985* (American Institute of Physics, New York, 1985), p. 48.
20. R.I. Pinsker, R.E. Duvall, C.M. Fortgang and P.L. Colestock, Nucl. Fusion **26**, 941 (1986).
21. R.I. Pinsker and P.L. Colestock, Nucl. Fusion **32**, 1789 (1992).
22. J.E. Stevens, R.E. Bell, S. Bernabei, et al., Nucl. Fusion **28**, 217 (1988).
23. Y. Uesugi, T. Yamamoto, K. Hoshino, et al., in Proceedings of the 14th European Conference on Controlled Fusion and Plasma Physics (Madrid, 1987), **11D**, Part III (EPS, 1987) p. 942.
24. P.L. Andrews, Phys. Rev. Lett. **54**, 2022 (1985).
25. P.L. Andrews, Bull. Am. Phys. Soc. **32**, 1793 (1987).
26. P.T. Bonoli and E. Ott, Phys. Fluids **25**, 359 (1982).
27. J.A. Colborn, R.R. Parker, S.C. Luckhardt, et al., Nucl. Fusion **31**, 960 (1991).
28. J. Villaseñor, M. Porkolab, et al., these proceedings.
29. T. Yamamoto, Y. Uesugi, H. Kawashima, et al., Phys. Rev. Lett. **63**, 1148 (1989).
30. Y. Uesugi, T. Yamamoto, H. Kawashima, et al., Nucl. Fusion **30**, 831 (1990).
31. L.-G. Eriksson and T. Hellsten, Nucl. Fusion **29**, 875 (1989).
32. D.F.H. Start, D.V. Bartlett, V.P. Bhatnagar, et al., Nucl. Fusion **30**, 2170 (1990).
33. J.R. Wilson, M.G. Bell, H. Biglari, et al., "ICRF Heating on TFTR — Effect on Stability and Performance," presented at 14th Int. Conf. on Plasma Phys. and Controlled Fusion Research, Würzburg, Germany, 1992, paper IAEA-CN-56/E-2-2.
34. Y. Takase, T. Watari, R. Kumazawa, et al., Nucl. Fusion **30**, 1585 (1990).
35. N. Hershkowitz, R. Majeski, P. Probert, et al., *Radio-Frequency Power in Plasmas: Ninth Topical Conference, Charleston, SC 1991* (American Institute of Physics, New York, 1992), p. 267.

36. C.C. Petty, R.I. Pinsker, M.J. Mayberry, et al., *Radio-Frequency Power in Plasmas: Ninth Topical Conference, Charleston, SC 1991* (American Institute of Physics, New York, 1992), p. 96.
37. C.C. Petty, R.I. Pinsker, M.J. Mayberry, et al., Phys. Rev. Lett. **69**, 289 (1992).
38. F.W. Baity et al., in *Fusion Technology 1990* (Proc. 16th Symp., London, 1990) (North-Holland, Amsterdam, 1991), Vol. 2, p. 1035.
39. R.I. Pinsker et al., Proc. of 14th IEEE/npss Symp. on Fusion Engineering (San Diego, CA, 1991) (IEEE, Piscataway, NJ, 1992), Vol. 1, p. 115.
40. R.H. Goulding et al., *Radio-Frequency Power in Plasmas: Ninth Topical Conference, Charleston, SC 1991* (American Institute of Physics, New York, 1992), p. 287.
41. R.I. Pinsker, C.C. Petty, M. Porkolab, F.W. Baity, et al., "Direct Electron Heating and Current Drive with Fast Waves in DIII-D," presented at 14th Int. Conf. on Plasma Physics and Controlled Fusion Research, Würzburg, Germany, 1992, paper IAEA-CN-56/E-2-4.
42. P.T. Bonoli, "Status and Comparison of Codes Used for Fast Wave Current Drive," these proceedings.

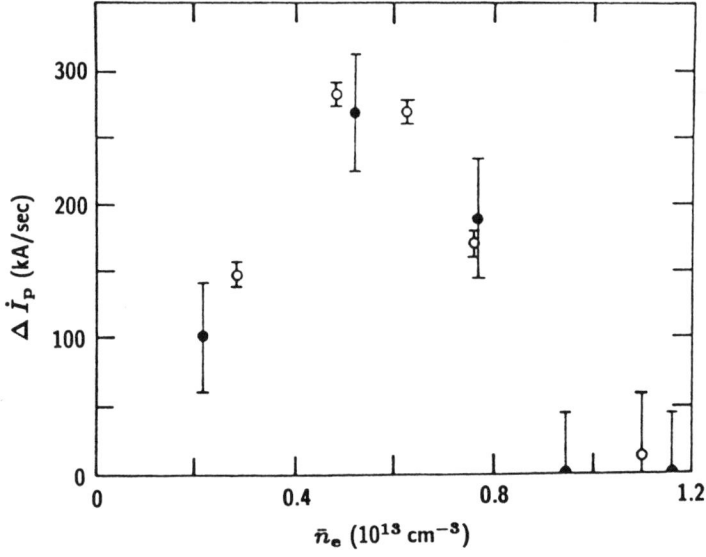

FIG. 1. Comparison of the change in the decay rate of the plasma current as a function of line-averaged density for two 800 MHz couplers on PLT: o = six-waveguide slow wave array, • = 4 × 3 dielectric-loaded waveguide fast wave launcher. This quantity is a rough measure of the noninductively driven current (Ref. 17).

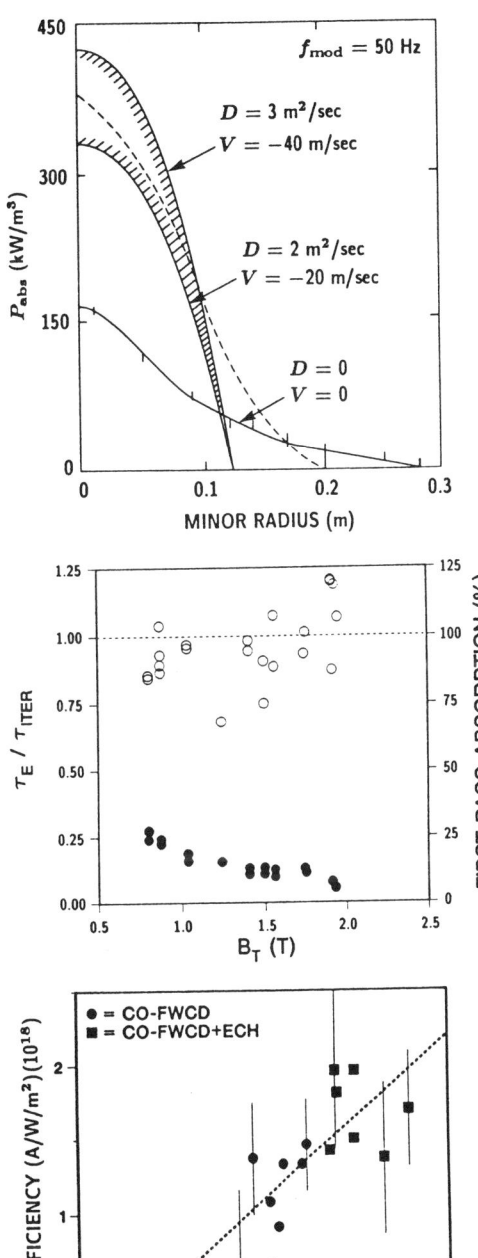

FIG. 2. 200 MHz FW power deposition profile measured in JFT–2M with a modulation technique (shaded region) and calculated from *single pass* ray tracing (dashed curve) (from Ref. 30).

FIG. 3. The measured global absorption efficiency in DIII–D (o) with 60 MHz FW direct electron heating as a function of toroidal field. Also shown is the calculated first-pass absorption (•) (Ref. 37).

FIG. 4. 60 MHz FWCD efficiency measured in DIII–D with preheating (squares) and without (circles), as a function of the central electron temperature. Also indicated is the best-fitting line passing through the origin, with a slope of 5.5×10^{17} A/W/m^2/keV (Ref. 42).

STATUS AND COMPARISON OF CODES USED FOR FAST WAVE CURRENT DRIVE *

Paul T. Bonoli
Massachusetts Institute of Technology, Cambridge, MA 02139

ABSTRACT

The status of computer models for fast wave current drive in the ion cyclotron range of frequencies is reviewed in this paper. The treatments of wave propagation, wave absorption, and current drive efficiency in the various models are discussed and the important physics issues in each of these areas are emphasized. The predictions for electron heating and current drive among these models is reviewed, especially as related to the recent DIII-D fast wave experiments and to the proposed Tokamak Physics Experiment (TPX). Finally, areas requiring further research in these models will be identified.

I. INTRODUCTION

In recent years a number of rather sophisticated numerical models have been developed for the study of fast wave electron heating and current drive in the ion cyclotron range of frequencies. The status of these models will be reviewed in this paper and the different approaches used to calculate wave propagation, wave absorption, and current drive efficiency will be discussed. The treatment of wave propagation using full-wave and toroidal ray tracing techniques will be reviewed in Sec. II. The treatment of wave absorption in the various models will also be discussed in Sec. II. Methods used for the computation of the driven rf current density in these models will be reviewed in Sec. III. Predictions for electron heating and current drive from several models will be compared, especially as related to the recent DIII-D fast wave ICRF experiments and the proposed Tokamak Physics Experiment. Finally, areas requiring further research in these models will be identified.

II. WAVE PROPAGATION AND ABSORPTION

A. Wave Propagation

Many fast wave current drive models use full-wave codes to compute ICRF wave propagation [FISIC[1], PICES[2], FASTWA[3], ALCYON[4], Fukuyama[5]]. An

---

* Work supported by the US Department of Energy Contract No. DE-AC02-78ET-51013

attractive feature of this approach is that diffraction effects (important near the antenna) are modelled properly. Also, the ICRF wave fields are reconstructed properly after multiple reflections of the ICRF wavefront. The starting point for the full-wave analysis is Maxwell's equations written in the form :[6]

$$\vec{\nabla} \times \vec{\nabla} \times \vec{E} = \frac{\omega^2}{c^2} \underline{\underline{\epsilon}} \cdot \vec{E} + \frac{4\pi i}{\omega} \vec{J}_{ANT}, \qquad (1)$$

where $\vec{E}$ is the ICRF electric field, $\underline{\underline{\epsilon}}(\vec{k},\omega)$ is the dielectric tensor (expanded to second order in the ion Larmor radius-ρ_i), and $\vec{J}_{ANT}$ is the current density flowing in the antenna. A spectral representation for $\vec{E}$ is often used to solve the wave equation system above and has the form :[1,6]

$$\vec{E}(\vec{x}) = \sum_{m,n_\phi} \vec{\tilde{e}}(\rho)^{(m,n_\phi)} \exp(im\theta + in_\phi\phi), \qquad (2)$$

where m (n_ϕ) are the poloidal (toroidal) mode numbers and ρ is the radial or flux surface variable. The toroidal mode number is a conserved quantity, thus the antenna loading and power balance can be treated separately for each n_ϕ-mode and the results added. Enough poloidal modes must be retained in the sum for $\vec{E}$ in order to resolve the shortest perpendicular wavelength that can propagate. In the case of a full-wave code[1] which solves the fourth order wave equation (fast wave plus ion Bernstein wave) a large number of poloidal modes must be retained to resolve $\lambda_\perp^{IBW} \lesssim \rho_i$. The spectral representation (Eq.(2)) allows unperturbed orbit integrals to be done analytically and with appropriate assumptions about phase decorrelation, the effects of the poloidal magnetic field (B_θ) can be retained.[7] Thus these codes include the calculation of toroidal effects on the parallel wavenumber $k_\| = \underline{k} \cdot \underline{B}/B \equiv k_\|^{(m,n_\phi)}$. The two main disadvantages to the full-wave approach are that the analysis breaks down if $k_\perp \rho_i > 1$ too near the ion-ion hybrid resonance layer and solutions can be computationally expensive in the presence of a strongly propagating ion Bernstein wave. The number of poloidal modes required in the representation for $\vec{E}$ can be reduced greatly by implementing an order reduction technique[2,3,8] to reduce the wave equation system from fourth to second order. This is accomplished by writing $\underline{\underline{\epsilon}}(\vec{k},\omega) \approx \underline{\underline{\epsilon}}^{(0)}(\omega) + i\tilde{k}_\perp \underline{\underline{\epsilon}}^{(1)}(\rho_i) - \tilde{k}_\perp^2 \underline{\underline{\epsilon}}^{(2)}(\rho_i^2)$, where $\underline{\underline{\epsilon}}^{(0)}$, $\underline{\underline{\epsilon}}^{(1)}$, $\underline{\underline{\epsilon}}^{(2)}$ are respectively zero, first, and second order in ρ_i and $\tilde{k}_\perp$ is the value of perpendicular wavenumber obtained from the exact Vlasov dispersion relation, evaluated for arbitrary $\tilde{k}_\perp \rho_i$. The disadvantage to the order reduction scheme is that information about mode conversion is eliminated.

Toroidal ray tracing techniques[9-14] are also useful for studying ICRF wave propagation in larger, higher density tokamaks where $\lambda_\perp < a$. In this approach the ray equations of geometrical optics are solved numerically using standard predictor-corrector algorithms. Ray tracing techniques offer a straightforward evaluation of toroidal effects on the parallel wavenumber ($k_\parallel$) and the ion Bernstein wave can be followed since $\lambda_\perp^{IBW} \ll a$. Two important disadvantages of ray tracing methods are they ignore diffraction effects (which can be important owing to the finite poloidal extent of the ICRF antenna) and proper reconstruction of the ICRF wavefront after multiple reflections is difficult. The latter difficulty is quite important in the case where single pass electron (or ion) damping is weak and the ICRF wave undergoes multiple transits before complete absorption occurs.

B. Wave Absorption

The mechanism for ICRF wave absorption by electrons is a combination of electron Landau damping (ELD) and transit time magnetic pumping (TTMP).[15-18] In the limit of arbitrary wave phase speeds ($\omega \simeq k_\parallel v_{te}$) it has been shown for a Maxwellian distribution function that the absorbed electron power can be written as[16] $k_{\perp Im} = k_{\perp Re}(\pi^{1/2}/2)\beta_e \zeta_e \exp(-\zeta_e^2)$, where $\zeta_e = \omega/k_\parallel v_{te}$. A second case has also been distinguished in which the electron distribution function is characterized by a hot anisotropic tail ($T_\parallel \gg T_\perp$) possibly due to the presence of energetic electrons generated by lower hybrid rf current drive (i.e. lower hybrid-fast wave synergy experiments). This case has been treated numerically[5,12] using combined 3-D Fokker Planck and ICRF wave propagation models.

The mechanism for wave absorption by ions can be ion cyclotron damping at $\omega = \omega_{ci}$ and $\omega = 2\omega_{ci}$. Higher harmonic damping at $\omega \geq 3\omega_{ci}$ is included in some models.[2,3,10,17] However, these higher harmonic damping calculations all neglect mode conversion.

III. CURRENT DRIVE COMPUTATION

A. Fokker Planck Analysis

It is found that for fast wave current drive (FWCD) with $\omega \simeq k_\parallel v_{te}$ and $v_\perp \sim v_{te}$, the distortion in $f_e(r, v_\parallel)$ due to the quasilinear interaction is weak.[4] It is therefore appropriate to solve the linearized Fokker Planck equation with an adjoint technique.[19] The resulting expression for the current drive efficiency is given by

$$\left(\frac{J_{rf}}{S_{rf}}\right)_{IC} = \frac{\int d^3p \vec{\Gamma} \cdot (\partial \chi / \partial \vec{p})}{\int d^3p \vec{\Gamma} \cdot \vec{v}}, \qquad (3)$$

where $\chi(\vec{p})$ is the Spitzer-Härm function and is the solution to the adjoint problem. The wave-induced rf flux $\vec{\Gamma} = -\underline{\underline{D}}_{RF} \cdot (\partial f_e / \partial \vec{p})$ is calculated assuming f_e

is Maxwellian and the quasilinear diffusion tensor is calculated following Refs. (13) and (17) as $\underline{\underline{D}}_{RF} = \underline{e}_{\|}\underline{e}_{\|}D_w$, where D_W can be written in terms of the wave polarizations and $|E_y|^2$. The ICRF electric field $E_y(\vec{x})$ is calculated from either the full-wave or ray tracing solutions. The rf current density is then formulated in terms of the current drive efficiency and electron power dissipation (S_{ELD}) as:[2,20]

$$J_{IC}^{(n_\phi)}(r) = \int_0^{2\pi} d\theta \sum_m G^{(n_\phi,m)} \sum_{m'} S_{ELD}^{(n_\phi,m,m')}(r,\theta) \qquad (4a)$$

$$G^{(n_\phi,m)} \equiv (J_{rf}/S_{rf})_{IC}^{(n_\phi,m)}, \qquad (4b)$$

where S_{ELD} is known from full-wave or ray tracing methods. The double m-sum in Eq.(4a) arises from the expression for absorbed power which is $\frac{1}{2}Re(\vec{E}^* \cdot \vec{J})$. Placement of the current drive efficiency in the outer m-sum is an attempt to approximately model the effect of toroidicity [$k_\|^{(m,n_\phi)}$ variations] on the driven rf current density. Many current drive models[2-4,10,11,20] use a parameterization[21] for the current drive efficiency (G) in order to reduce computational expense, where $G^{(n_\phi,m)} = G(Z_{eff}, \epsilon, \theta, v_\|^{(m,n_\phi)}/v_{te})$. Here $\epsilon = r/R$, θ is the poloidal angle, and $v_\| = \omega/k_\|^{(m,n_\phi)}$. Thus, the effects of particle trapping are included in this parameterization for G.

B. MHD Equilibrium

Fast wave current drive models calculate the ICRF current density from a target equilibrium described by a moments solution of the Grad Shafranov equation or a direct numerical solution $\psi(R,Z)$, where ψ is the poloidal flux function. In addition to FWCD, these models frequently calculate other current density source terms due to injected lower hybrid waves (J_{LH}), neutral beam injection (J_{BD}), and neoclassical effects such as the bootstrap current (J_{BS}). Some models iterate between the MHD solver and current drive packages to obtain self-consistent MHD equilibria.[11,22,28]

IV. MODEL RESULTS AND COMPARISONS

A. Fast Wave Electron Heating in DIII-D

Fast wave electron heating experiments have been carried out in the DIII-D tokamak[23,24] at $B_o = (0.8 - 2.0)$T, $f_o = 60$ MHz, and $(0 - \pi - 0 - \pi)$ phasing of a four strap antenna, corresponding to a peak $n_\| \simeq 10$ in the coupled power spectrum. Here $n_\| = k_\| c/w$ is the parallel refractive index of the fast wave. A comparison of the predicted rf power density profile (due to ELD) from the full-wave code FISIC[1] and modulation experiments in DIII-D is shown in Fig. 1. The experimental and predicted profiles agree remarkably well. The parameters used in this comparison were $B_o = 2$T, $f_o = 60$ MHz, $I_p = 1$ MA, $\kappa = 1.99$, $n_e(0) = 2.7 \times 10^{19} \mathrm{m}^{-3}$, $T_e(0) = 2.43$ keV, deuterium majority component - (97%), $T_D(0) = 1.33$ keV, hydrogen minority component - (3%), $T_H(0) = 1.33$ keV, $a = 0.63$m, and $R_o = 1.71$m. The antenna spectrum was modelled in the full-wave code using 7 toroidal modes $n_\phi = (17, 22, 26, 30, 35, 40, 47)$ and 65 poloidal modes were employed in the spectral representation for $\vec{E}(-32 \leq m \leq 32)$. Although the single pass absorption predicted in this case is low ($\approx 5\%$), the full-wave code computes 100% absorption of the ICRF power after multiple reflections of the wavefront. Approximately 80% of the power is absorbed via ELD and the remaining 20% is absorbed by cyclotron damping at the second harmonic hydrogen cyclotron layer which lies near the plasma center.

B. Fast Wave Current Drive in DIII-D

Fast wave current drive experiments have also been carried out recently in the DIII-D tokamak[25] at $B_o = (1-2)$T, $f_o = 60$ MHz, and $(0 - \pi/2 - \pi - 3\pi/2)$ phasing of the four strap antenna array, corresponding to a peak $n_\parallel \simeq 5.5$ in the coupled power spectrum. A comparison of the fast wave current drive efficiency (γ_{FW}) of three FWCD models with the experimental data is shown in Fig. 2, where $\gamma_{FW} =< n_e(10^{20}\mathrm{m}^{-3}) > I_{FW}(A)R_o(m)/P_{FW}(W)$. The parameters used for these comparisons were $B_o = 1$T, $Z_{eff} = 3.5$, $I_p = 0.40$ MA, $n_e(0) = 2.7 \times 10^{19} \mathrm{m}^{-3}$, $\kappa = 1.99$, deuterium majority component - (98%), hydrogen minority component - (2%), $a = 0.63$m, and $R_o = 1.71$m. The experimental data points in Fig. 2 demonstrate the increase in γ_{FW} as $T_e(0)$ increases due to the increased single pass absorption of the fast wave at higher electron beta (β_e). The open circles correspond to experimental data with CO-injection of fast wave power and the open squares correspond to CO-injection of fast waves plus electron cyclotron heating (ECH). Results from the PICES code[26] (solid squares) were obtained using 100 toroidal modes to model the antenna spectrum. Two simulation points from the FISIC code[20] and a single point from the ALCYON model[4] are also shown for a single dominant toroidal mode ($n_\phi = 17$). There is generally good agreement between the model predictions and experiment, especially for the detailed comparisons done with PICES.

C. Fast Wave Current Drive Predictions for TPX

The conceptual design activity for the Tokamak Physics Experiment (TPX) has resulted in extensive cross-checking and benchmarking among the various FWCD models in the US. A scan of the FWCD efficiency versus frequency predicted for TPX by three different models[27] is shown in Fig. 3. Parameters used in this comparison were $B_o = 4$T, $I_p = 2$ MA, $a = 0.5$m, $R_o = 2.25$m,

$\kappa = 2.0$, deuterium majority plasma - (97%), hydrogen minority plasma - (3%), $n_e(0) = 1 \times 10^{20} \text{m}^{-3}$, $T_e(0) = T_D(0) = T_H(0) = 10$ keV, $n_e(\psi) = n_e(0)(1-\psi)^{\alpha_n}$, $T(\psi) = T(0)(1-\psi)^{\alpha_T}$, $\alpha_N = 0.25$, $\alpha_T = 1.0$, and $40 \leq f_o$ (MHz) ≤ 100. Here ψ is the normalized poloidal flux function. A distribution function for fast deuterons associated with 15 MW of neutral beam injection power is also included in the cyclotron harmonic absorption calculation. The value of $\zeta_e = \omega/k_\parallel v_{te}$ was held constant in this scan by varying the toroidal mode number as the wave frequency was changed. These calculations indicate favorable frequency regimes for FWCD in TPX at 45 MHz, 80 MHz, and 100 MHz. All parasitic ion resonances are absent from the plasma at 45 MHz, resulting in only ELD in this regime. Some parasitic absorption occurs at 80 MHz via ion cyclotron damping at $\omega = 3\omega_{CD}$. About 6% of the power is absorbed by thermal deuterons, 27% of the power is absorbed by fast neutral beam deuterons, and 64% is absorbed via ELD. The strong reduction in γ_{FW} near 60 MHz is due to FW absorption via minority hydrogen cyclotron damping at the plasma center.

Fast wave current drive results form the ACCOME/FISIC model[28] are shown in Fig. 4. These results correspond to the high bootstrap current operating scenario in the first stability regime planned for TPX (i.e., "ARIES-I" mode). The parameters used are the same as in the previous TPX example with $T_e(0) = T_D(0) = 9.4$ keV and $f_o = 45$ MHz. In this case 0.19 MA of core current is supplied by 6 MW of neutral beam injection power and 0.18 MA of FW seed current is generated by 2.8 MW of ICRF power. The bootstrap current is 0.99 MA with $I_p = 1.36$ MA and $f_{BS} = 0.73$. The values of safety factor at the axis and 95% flux surfaces are respectively $q(0) = 1.41$ and $q(95) = 5.65$. The final profiles of current density and safety factor from the ACCOME model are the result of five iterations between the current drive packages and equilibrium solver.

V. CONCLUSIONS AND AREAS FOR FUTURE RESEARCH

In conclusion, the modelling of fast wave current drive experiments has benefited from the variety of numerical models in existence. Ray tracing and full-wave approaches used in FWCD codes tend to complete each other, resulting in a more complete picture of wave propagation and absorption. The use of exact Fokker Planck solutions, application of the adjoint method to solve for the current drive efficiency, and utilization of parameterization techniques to represent $(J_{rf}/S_{rf})_{IC}$ have all resulted in accurate and computationally attractive options for evaluating the driven ICRF current. Preliminary comparisons with electron heating and FWCD data from DIII-D are encouraging. Electron power deposition profiles from code models are consistent with the experiment, despite the weak single pass absorption. Also the predicted ICRF currents are consistent with the required currents inferred from time-dependent transport analyses of DIII-D discharges.

Several areas in FWCD modelling require further research, however. An attractive scenario for avoiding parasitic ion resonances during FWCD is one in which $f_o < f_{ci}$. In this case, it is possible to excite the shear Alfven wave at the plasma edge on the high field side of the tokamak, where $n_\parallel^2 \simeq S$ is satisfied. One primary goal in this area would be to incorporate the predicted [29-31] electron absorption at this location into the existing FWCD models. A second topic for future research is the treatment of wave propagation and absorption at $k_\perp \rho_i > 1$. Recall that the dielectric tensor used in the wave equation (Eq. 1) was derived to second order in $k_\perp \rho_i$ and in the limit of $k_\perp \rho_i < 1$. This limit may not be satisfied for the mode converted wave at the ion-ion hybrid resonance layer. Furthermore, the calculation of wave absorption at $\omega \geq 3\omega_{ci}$ usually neglects mode conversion because of difficulties related to $k_\perp \rho_i > 1$ in the wave analysis.[2,3,10,17] A promising technique for resolving this problem has been developed by Sauter.[32,33], however the analysis was done in slab geometry. One final area for future work is the evaluation of the quasilinear diffusion tensor directly from wavefields computed by a wave propagation code. This would allow wave polarization effects to be included in the computation of the driven ICRF current. A parameterization for $(J_{rf}/S_{rf})_{IC}$ would not be used in this case. The current drive efficiency would be evaluated directly from Eq.(3) using the exact solution (or a parameterization) for $\chi(\vec{p})$. Explicit evaluation of $\underline{\underline{D}}_{RF}$ has already been carried out in early work by Karney[13], in the ALCYON model,[4] and in LH-FW synergy models.[12]

ACKNOWLEDGEMENTS

The author wishes to thank Professor Miklos Porkolab for his collaborative efforts in the work that was presented. It is also a pleasure to thank Dr. D.B. Batchelor, Dr. E.F. Jaeger, Dr. P.E. Moroz, Dr. T.K. Mau, Dr. A.K. Ram, Dr. C.C. Petty, and Dr. R.I. Pinsker for making certain results of their work available for reproduction in this paper.

FIGURE CAPTIONS

Fig. 1: Fast wave electron heating in DIII-D [$B_o = 2$T, $f_o = 60$ MHz, $I_p = 1$ MA, $T_e(0) = 2.43$ keV, $n_e(0) = 2.7 \times 10^{19} \text{m}^{-3}$]. Comparison of experimental (open circles and squares) and theoretical (solid line) power deposition profiles.

Fig. 2: Fast wave current drive in DIII-D [$B_o = 1$T, $f_o = 60$ MHz, $I_p = 0.4$ MA, $T_e(0) = 1.5$ keV, $n_e(0) = 2.7 \times 10^{19} \text{m}^{-3}$]. Comparison of experimental and theoretical current drive efficiencies (γ_{FW}) versus $T_e(0)$. Open circles correspond to CO-injection of FW power and open squares correspond to CO-injection of FW power plus ECH. The solid symbols were obtained from the PICES (squares), FISIC (triangles), and ALCYON (diamond) codes.

Fig. 3: Fast wave current drive efficiency (γ_{FW}) versus frequency in TPX [$B_o = 4$T, $I_p = 2$ MA, $T_e(0) = T_D(0) = 10$ keV, $n_e(0) = 1 \times 10^{20}$m^{-3}]. Comparison for three FWCD models.

Fig. 4: ACCOME/FISIC predictions for FWCD in TPX high bootstrap current scenario [$B_o = 4$T, $I_p = 1.36$ MA, $T_e(0) = T_D(0) = 9.4$ keV, $n_e(0) = 1 \times 10^{20}$m^{-3}, $f_o = 45$ MHz, $P_{FW} = 2.8$ MW, $P_{NB} = 6$ MW]. (a) Current densities (A/m^2) versus square root toroidal flux (ρ). Curves labelled FW, BD, BS, and TOT correspond respectively to the fast wave, beam driven, bootstrap, and total current density profiles.

REFERENCES

[1] M. Brambilla, T. Krücken, Nucl. Fusion **28**, 1813 (1988); P.T. Bonoli, M. Porkolab, Radio Frequency Power in Plasmas, AIP Conf. Proc. 244, ed. D.B. Batchelor (AIP, NY, 1992) p. 155.

[2] E.F. Jaeger, D.B. Batchelor in Radio Frequency Power in Plasmas, AIP Conf. Proc. 244, ed. D.B. Batchelor (AIP, NY, 1992) p. 159.

[3] P.E. Moroz, P.L. Colestock, Plasma Physics and Contr. Fusion **33**, 417 (1991).

[4] A. Becoulet, D. Moreau, General Atomics Report GA-A20846 UC-420 (April, 1992).

[5] A. Fukuyama, US Japan Workshop on Non-Inductive Current Drive and Profile Control, Princeton Plasma Physics Laboratory (1992) p. 214.

[6] M. Brambilla, Plasma Physics and Contr. Fusion **35**, A141 (1993).

[7] D. Smithe et al., Phys. Rev. Lett. **60**, 801 (1988).

[8] D.N. Smithe et al., Nucl. Fusion **27**, 1319 (1987).

[9] M. Brambilla, Computer Physics Reports 4, 71 (1986).

[10] T.K. Mau et al., EPS Topical Conf. on Radio-Frequency Heating and Current Drive of Fusion Devices (EPS, 1992) Vol. 16E, p. 181.

[11] M.-Y. Hsiao, D.A. Ehst, K. Evans, Nucl. Fusion **29**, 49 (1989).

[12] A.K. Ram et al., EPS Topical Conf. on Radio-frequency Heating and Current Drive of Fusion Devices (EPS, 1992) Vol. 16E, p. 201.

[13] C.F.F. Karney, Computer Physics Reports 4, 183 (1986).

[14] V. Bhatnagar, J. Jacquinot, D.F.H. Start, Proc. of IAEA Tech. Comm. Meeting on Fast Wave Current Drive in Reactor Scale Tokamaks, Arles, France, 1991 (EURATOM-CEA, France) p. 110.

[15] T.H. Stix, Nucl. Fusion **15**, 737 (1975).

[16] M. Porkolab, Radio Frequency Power in Plasmas, AIP Conf. Proc. 244, ed. D.B. Batchelor (AIP, NY, 1992) p. 201.

[17] S.C. Chiu et al., Nucl. Fusion **29**, 2175 (1989).

[18] P.E. Moroz et al., Phys. Fluids B **4**, 2915 (1992).

[19] C.F.F. Karney, N.J. Fisch, Phys. Fluids **28**, 116 (1985).

[20] P.T. Bonoli, R.C. Englade, M. Porkolab, M.E. Fenstermacher, EPS Topical Conf. on Radio-Frequency Heating and Current Drive of Fusion Devices (EPS, 1992) Vol. 16E, p 169.

[21] D.A. Ehst, C.F.F. Karney, Nucl. Fusion **31**, 1933 (1991).

[22] R.S. Devoto et al., Nucl. Fusion **32**, 773 (1992).

[23] C.C. Petty et al., Phys. Rev. Lett. **69**, 289 (1992).

[24] C.C. Petty et al., Radio Frequency Power in Plasmas, AIP Conf. Prof. 244, ed. D.B. Batchelor (AIP, NY, 1992) p. 96.

[25] R.I. Pinsker et al., 14th Int. Conf. on Plasma Physics and Contr. Nucl. Fusion Research (Wurzburg, Germany, 1992), paper IAEA-CN-56/E-2-4.

[26] E.F. Jaeger, D.B. Batchelor, M. Murakami, Tenth RF Topical Conference on Radio-Frequency Power in Plasmas (Boston, MA) April 1-3, 1993, paper A15.

[27] P.E. Moroz, D.B. Batchelor, E.F. Jaeger, T.K. Mau, D. Mikkelsen, M. Porkolab, Tenth Topical Conference on Radio-Frequency Power in Plasmas (Boston, MA) April 1-3, 1993, paper C21.

[28] P.T. Bonoli et al., Bull. Am. Phys. Soc. **37**, 1601 (1992).

[29] C.F.F. Karney, F.W. Perkins, Y.-C. Sun, Phys. Rev. Lett. **42**, 1621 (1978).

[30] M. Porkolab, "Mechanisms for Edge Absorption," ICRF Workshop (Boulder, CO, 1990).

[31] J.A. Heikkinen, T. Hellsten, M.J. Alava, Nucl. Fusion **31**, 417 (1991).

[32] O. Sauter, "Nonlocal Analysis of Electrostatic and Electromagnetic Waves in Hot, Magnetized, Nonuniform, Bounded Plasmas," Phd Dissertation, (CRPP, Lausanne) Report LRP 457/92, May, 1992.

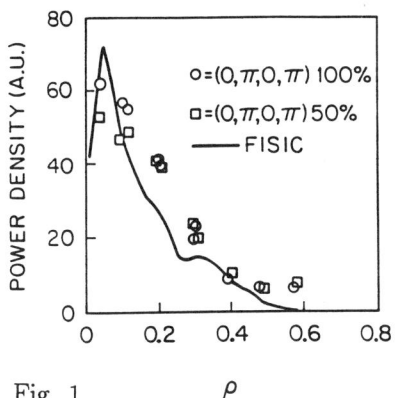

Fig. 1

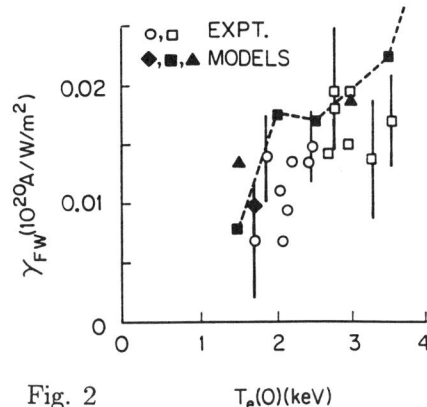

Fig. 2

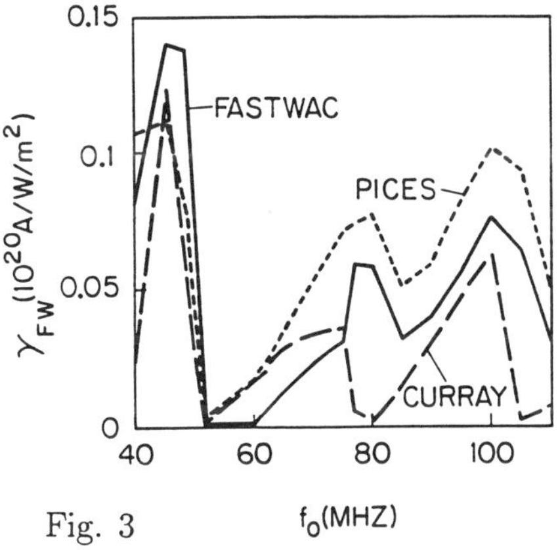

Fig. 3

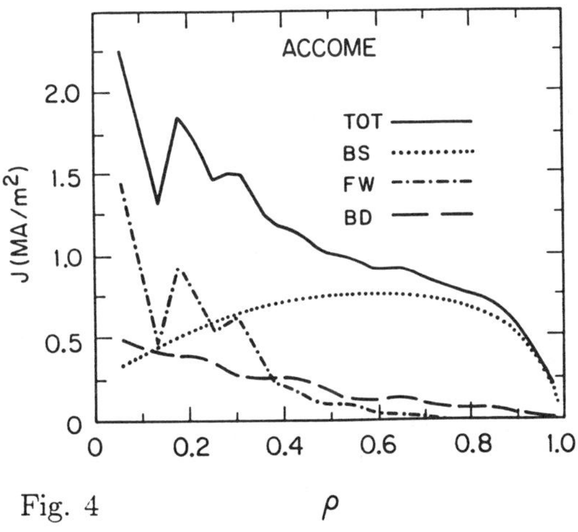

Fig. 4

CONTROL OF THE CURRENT DENSITY PROFILE WITH LOWER-HYBRID CURRENT DRIVE ON PBX-M

R. E. Bell, S. Bernabei, L. Blush,[e] T. K. Chu, R. Doerner,[e] J. Dunlap,[d]
A. England,[d] G. Gettelfinger, N. Greenough, J. Harris,[d] R. Hatcher, S. Hirshman,[d]
D. Ignat, R. Isler,[d] S. Jardin, R. Kaita, S. Kaye, T. Kozub, H. Kugel, B. LeBlanc,
S. Jones,[c] J. Kesner,[c] D. Lee,[d] F. Levinton,[a] S. Luckhardt,[c] M. Okabayashi,
F. Paoletti,[c] S. Paul, F. Rimini, N. Sauthoff, S. Sesnic, L. Schmitz,[e] Y. Sun,
H. Takahashi, W. Tighe, G. Tynan,[e] E. Valeo, S. von Goeler

Princeton Plasma Physics Laboratory, Princeton University, P.O. Box 451, Princeton, NJ 08543
[a]*Fusion Physics and Technology, 3547 Voyager St., Suite 104, Torrance, CA 90503-1673*
[b]*JET Joint Undertaking, Abingdon Oxfordshire, OX143EA, United Kingdom*
[c]*Massachusetts Institute of Technology, 77 Massachusetts Ave., Cambridge, MA 02139*
[d]*Oak Ridge National Laboratory, Fusion Energy Division, P.O. Box 2009,
Oak Ridge, TN 37831-8072*
[e]*Institute of Plasma and Fusion Research, University of California-Los Angeles,
Los Angeles, CA 90024-1597*

ABSTRACT

Lower hybrid current drive (LHCD) is being explored as a means to control the current density profile on PBX-M with the goal of raising the central safety factor q(0) to values of 1.5-2 to facilitate access to a full-volume second stable regime. Initial experiments have been conducted with up to 400 kW of 4.6 GHz LH power in circular and indented plasmas with modest parameters. A tangential-viewing two-dimensional hard x-ray imaging diagnostic has been used to observe the bremsstrahlung emission from the suprathermal electrons generated during LHCD. Hollow hard x-ray images have indicated off-axis localization of the driven current. A serious obstacle to the control of the current density profile with LHCD is the concomitant generation of MHD activity, which can seriously degrade the confinement of suprathermal electrons. By combining neutral beam injection with LHCD, an MHD-free condition has been obtained where q(0) is raised above 1.

INTRODUCTION

The Princeton Beta Experiment-Modified (PBX-M) is evaluating the use of lower hybrid current drive (LHCD) as a tool to alter the current density profile to improve confinement and stability. The operation of a tokamak reactor at a high plasma pressure (high β) is economically attractive. The PBX-M experiment is directed toward achieving high β operation by accessing the second stable regime.[1,2] Access to the second stable region was achieved in the outer portion of the PBX-M plasma with the use of high indentation to improve the magnetic well.[3] The requirements of high indentation are relaxed if the current density profile can be altered such that the safety factor, q, is raised in the plasma center. Obtaining q(0) = 1.5-2 with suitably flat current density profiles near the edge are desired to place the full plasma volume in the second stable regime.[4]

MHD activity can be altered by the introduction of LH driven current. Depending on the lower hybrid power, the waveguide phasing and the plasma density, MHD can be either stabilized or destabilized. LH induced MHD activity can seriously degrade the confinement of suprathermal electrons. This paper will examine the MHD activity during LHCD in circular and indented plasmas, with neutral beam injection (NBI) and without.

PBX-M is a medium sized tokamak with a major radius of 1.65 m and a minor radius of 0.3 m and is capable of plasma indentations up to 28%.[3] The plasma is surrounded by a series of passive plates which make up a conducting shell to stabilize surface kink modes. The lower hybrid system on PBX-M operates at a frequency of 4.6 GHz. The coupler contains 32 waveguides each 6 cm high × 0.5 cm wide. The phase of each waveguide can be individually adjusted during the LH pulse (at a rate of 20°/ms). By adjusting the relative phases ($\Delta\phi$) of the 32 waveguides, the parallel refractive index, $n_\parallel = ck_\parallel/\omega$, can be varied up to $n_\parallel = 4.1$ ($\Delta\phi = 180°$). The coupler is located 8.7 cm below the midplane and is curved in two dimensions to approximate the edge contours of the plasma. A second coupler capable of a slightly higher $n_\parallel$ spectrum will soon be added above the midplane.

Initial LH experiments[5,6] have been carried out at low LH power, $P_{LH} \leq 400$ kW. Consequently, plasmas with relatively low average electron density, $n_e = 0.5 - 3.0 \times 10^{19}$ m^{-3}, electron temperature, $T_e(0) = 0.8 - 1.2$ keV, and plasma current, $I_p = 100 - 250$ kA, have been used to study the effects of LHCD. An $n_\parallel = 2.1$ ($\Delta\phi = -90°$) was used for the plasmas described in this paper. A tangential-viewing two dimensional hard x-ray (HXR) camera was constructed and installed on PBX-M to measure the bremsstrahlung emissions of the suprathermal electrons produced during LHCD.[7] Hard x-ray images with a 3 ms time resolution and a 3 cm spatial resolution allow the study of suprathermal production and confinement. Hard x-ray images have been correlated with MHD activity measured by a toroidal array of Mirnov coils and a soft x-ray (SXR) array. A Motional Stark Effect (MSE) diagnostic[8] was used to measure changes in q(r) during LHCD.

LHCD INTO CIRCULAR PLASMAS

Many of the initial experiments with LHCD at low power were conducted with circular plasmas with relatively low electron density and plasma current. The low density improved the efficiency of current drive and the low current made it easier to study changes in the current density profile at low P_{LH}. At the lower densities, however, there was a tendency for the production of runaway electrons with LH when there was still a significant loop voltage present. Line average densities of $1.5-2 \times 10^{19}$ m^{-3} were sufficient to eliminate the runaway conditions.

Figure 1 shows some plasma parameters for a circular plasmas with and without LH power. With LHCD, the loop voltage (V_{LOOP}) was reduced and large sawteeth were produced. The sawtooth inversion radius increased with the addition of LHCD, indicating that the radius of the q=1 surface was increasing. This was consistent with the damping of the LH waves inside the q=1 surface. The vertical HXR profile also showed a central peaking on axis.

Figure 2 shows the peaked HXR signal measured along a vertical chord of the image with a tangency radius near the plasma center plotted versus the time of each image. The measured q profile from the MSE diagnostic for these plasmas is shown in Fig. 3. With LHCD, q(0) was reduced as the q=1 radius increased. By comparing profiles from HXR and SXR measurements on the same vertical coordinates, some insight can be gained as to the location of the suprathermal population with respect to the MHD activity in the plasma. Plotted in Fig. 4 are the HXR profile and the amplitude of the m=1 sawteeth precursor oscillations as seen by the SXR array. The HXR intensity was peaked well inside the q=1 surface which occurred near the outer edge of the envelope of the m=1 precursor oscillations. For this circular plasma case (with $\kappa = 1$), this vertical position is in agreement with the radial q measurements from MSE (≈ 15

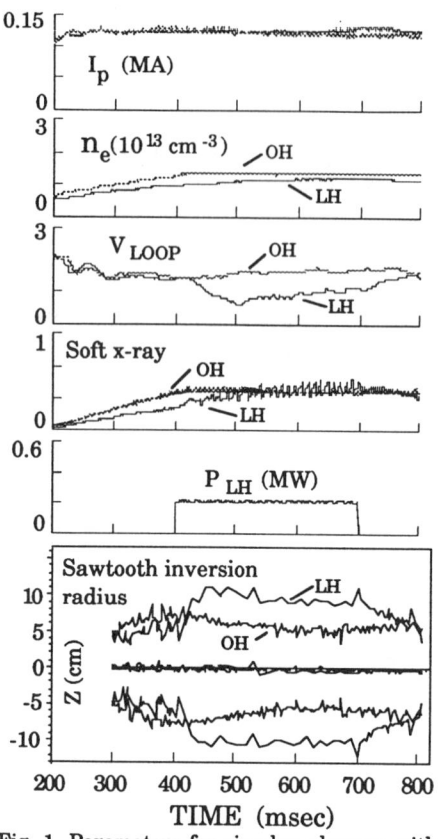

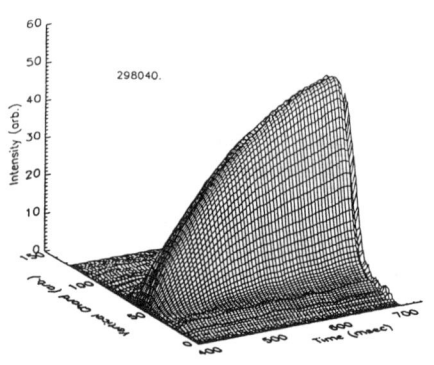

Fig. 2. Time evolution of the hard x-ray intensity along a vertical chord through the magnetic axis for shot 298040.

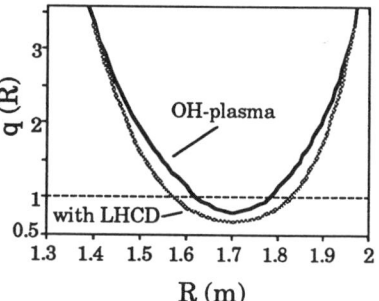

Fig. 1. Parameters for circular plasmas with (#298040) and without (#298036) LHCD.

Fig. 3. Measured q profile showing reduction of central q with LHCD into a circular plasma.

cm from center). When q measurements are not available, such MHD activity can be used as a guide to the location of rational q surfaces. In this way, with even elongated plasmas, MHD from the vertical SXR array can be compared along analogous vertical coordinates to vertical HXR profiles which are affected least by the line integration.

The large sawtooth amplitude and increasing sawtooth inversion radius are typical of LHCD into circular ohmic plasmas. For lower densities, a runaway condition would often develop with a greatly increasing hard x-ray intensity. Also, for higher LH power and/or lower density a large n=1 mode would frequently occur, which would eventually lock, and cause a disruption. The maximum electron density for these low current circular plasmas was limited, but this was overcome with the addition of neutral beam injection (NBI).

A small amount of NB power (0.5 MW) was added with a net LH power of 0.3 MW to a circular discharge similar to the one shown above (with the timing of LHCD as above and coincident NBI) to increase the electron density of the plasma. As the density increased, the intensity of the HXR signal decreased as the efficiency of the LHCD was reduced (see Fig. 5). With the increase in density, the HXR profile was no longer peaked. Two distinct lobes appeared, and the distance between them increased as the profile became more hollow. Figure 6 shows contours of the HXR intensity along a vertical chord versus time and the peak positions of the lobes.

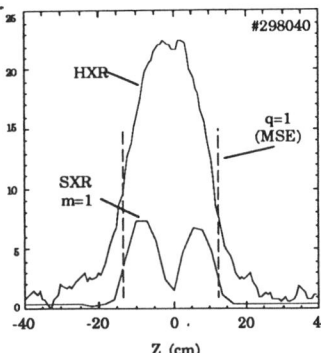

Fig. 4. Hard x-ray profile and m=1 oscillations from the SXR array versus vertical position with approximate q=1 radius from MSE.

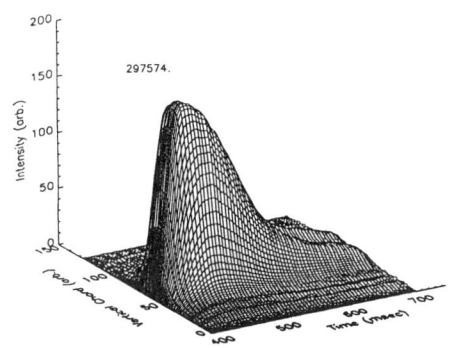

Fig. 5. Time evolution of hard x-ray intensity for circular plasma with LHCD and NBI.

It is useful to quantify this "hollowness" with a parameter that takes into account the shape and depth of the central region with respect to the width of the profile. The parameter for hollowness used here has a value of -1 for a centrally peaked parabolic profile, 0 for a flat central region, and positive values for increasing hollowness. Figure 7 shows this hollowness parameter increasing with electron density. Normalized HXR profiles are shown in Fig. 8 at an early and late time to illustrate the degree of hollowness that is represented by the quantity plotted in Fig. 7. This hollowness parameter can reflect changes in the HXR profile which are often brought about by MHD activity. A decreasing hollowness is often an indication of an increasing diffusion of the suprathermal electrons. It could also indicate a change in LH absorption or the creation of runaway electrons. The large sawteeth which can be seen on the line averaged electron density in Fig. 7 have an effect on the suprathermal distribution. At each sawtooth event, a corresponding drop in the hollowness parameter can be seen. The sawteeth, while momentarily decreasing the hollowness, do not impede a rising trend in hollowness.

The effect on the HXR profile during a sawtooth cycle is shown in Fig. 9. Just after a sawtooth crash, the HXR profile was generally flat in the central region. As the cycle proceeded with the corresponding increases in

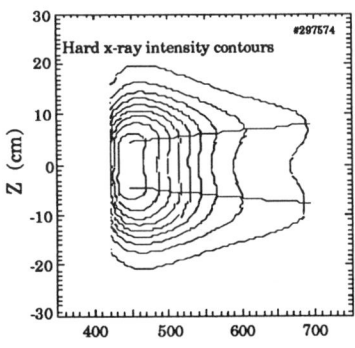

Fig. 6. Contours of hard x-ray intensity. Lines indicate position of maximum values at each time slice.

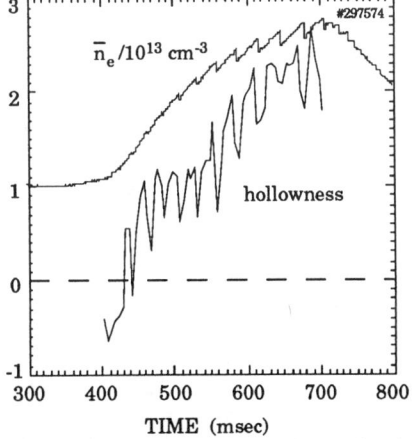

Fig. 7. Hollowness parameter increasing in time with electron density. A decrease in hollowness occurs at each sawtooth crash.

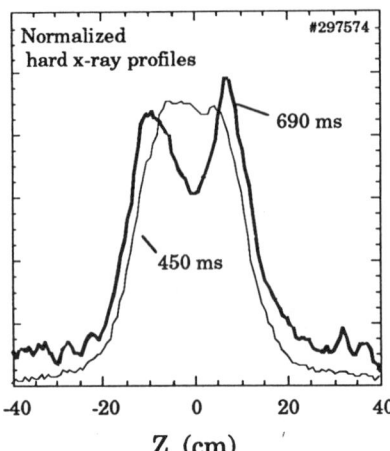

Fig. 8. Normalized HXR profiles early and late during the LHCD pulse showing increasing hollowness.

electron density and temperature, the hollowness increased. This was seen as a reduction in the HXR intensity in the central portion of the profile. This hollowing out of the central region may indicate that the diffusion of suprathermals into this region was reduced during the period between sawtooth crashes or a changing LH absorption. At the sawtooth crash the lobes of the profile were eliminated; the reconnection of the magnetic field lines were adversely affecting the confinement of the suprathermal electrons.

If the fast electron diffusion was small, the position of the lobes on the HXR profile (or just outside the lobes due to the chord averaging), should be approximately the location where the suprathermal production was occurring, i.e. where current was being driven by LH. In the first example of LHCD in a circular discharge, no hollowness was seen, and a central peaking was consistent with the large sawteeth amplitude, the increased sawtooth inversion radius, and LH current deposition within the $q=1$ surface. In the second example with NBI and higher density, two distinct lobes appeared which moved outward as the density increased. In the latter case, the sawtooth amplitude also increased suggesting that LH current was still being deposited inside the $q=1$ surface. Soft x-ray data were not available for these LHCD plus NBI plasmas, but the sawtooth inversion radius, and therefore the $q=1$ surface, could be estimated from Thomson scattering profiles. The position of the HXR lobes were within this radius. The expanding lobe position is probably reflecting the accessibility of the LH wave into the central region of the plasma. At higher central densities, the LH wave does not reach the center and damps off axis but still within the $q=1$ surface.

LHCD INTO INDENTED PLASMAS

By shaping the plasma, higher values of plasma current and electron density can routinely be reached. For LHCD experiments plasmas with indentations from 10-14% and plasma currents in the range 180-250 kA were typical. Hollow HXR profiles were common, but a variety of MHD modes could be destabilized by LHCD often resulting in a reduction in hollowness or a loss of suprathermal electrons measured as a loss in HXR intensity.

The dominant MHD modes were $m=1/n=1$, $m=3/n=2$ together with $m=2/n=2$, and $m=2/n=1$. The toroidal mode number, n, was determined from a toroidal array of Mirnov loops. The poloidal mode

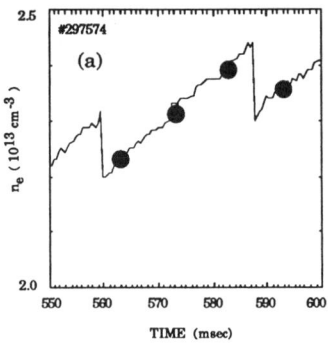

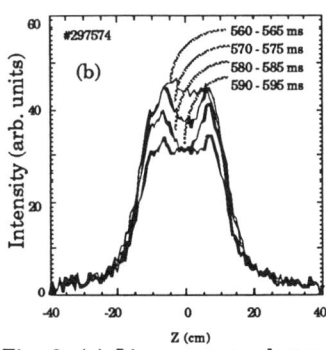

Fig. 9. (a) Line average electron density with sawtooth signature.
(b) Evolution of HXR profile during a sawtooth cycle in circular plasma with LHCD and NBI.

number, m, was determined from the SXR array. For all MHD modes observed, the hollowness of the HXR profile could be adversely affected when the modes had sufficient amplitude. An m=2/n=1 mode coupled to a m=1/n=1 mode could destroy suprathermal confinement as indicated by a sudden drop in HXR intensity. These coupled modes would often lock causing a disruption. With LH current deposition inside of the q=1 surface, MHD activity would often be seen to increase. With the LH current deposition outside of the q=1 surface and sufficient LH current, the q=1 radius was observed to decrease.

Two cases with LH current deposition outside of the q=1 surface will be considered here with I_p = 180 kA, q_{edge} ≈ 8, κ = 1.7, and an indentation of 12%. The first case had LHCD alone and the second case had LHCD plus NBI. The time evolution of the HXR intensities for these two cases are shown in Fig. 10 (LHCD) and Fig. 11 (LHCD + NBI). In both cases, the q=1 radius decreased with the application of LHCD.

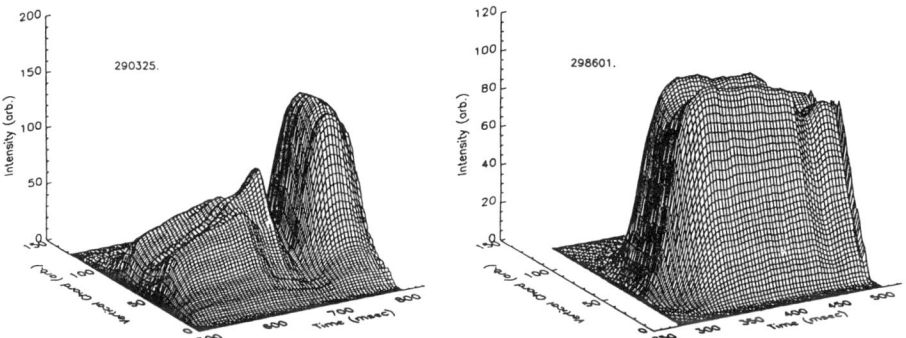

Fig. 10. Time evolution of hard x-ray intensity for indented plasma with LHCD.

Fig. 11 Time evolution of hard x-ray intensity for indented plasma with LHCD and NBI

For the first case of Fig. 10, a net LH power of 360 kW was applied to a plasma with a line averaged density of 2.2×10^{19} m^{-3}. Some parameters for this discharge are shown in Fig. 12. Shown are the line-averaged electron density, a central SXR signal, loop voltage, the hollowness parameter from the HXR image, and a Mirnov signal.

Contours of the amplitude of the oscillations measured with the SXR array between 4 and 8 kHz can be seen at the bottom of Fig. 12. Both m=1 sawteeth precursors and bursts of m=1 oscillations are present. The radius of the q=1 surface as indicated by these oscillations decreased in time after the application of LHCD. The last sawtooth event occurred at 625 ms. With the disappearance of all m=1 activity at about 635 ms, an m=2 mode began to grow. At about this time the q=1 surface is believed to have disappeared. The m=2/n=1 mode increased in amplitude leading to a disruption at 700 ms. This m=2/n=1 mode may have been destabilized due to LH current deposition near the q=2 surface or perhaps more indirectly through change in the q profile in which q' was flattened near q=2. Figure 13 shows the HXR profiles and SXR MHD amplitudes for the m=1 mode at 600ms and the m=2 mode at 650 ms. The lobe position was outside of the m=1 position consistent with deposition of LH current outside of the radius of q=1.

As the m=2/n=1 mode grew, there was a peaking and a narrowing of the HXR profile which can be seen in the contours of HXR intensity and in the hollowness parameter of Fig. 12. The drop in HXR intensity at the location of the m=2 island indicates a loss of suprathermal confinement (see Fig. 12 center and Fig. 13). The subsequent disruption may have resulted from the change in the q profile associated with the loss of the current-carrying suprathermal population due to the m=2 mode. The large HXR signal after the disruption in Fig. 10 was due to a runaway condition because of the low electron density.

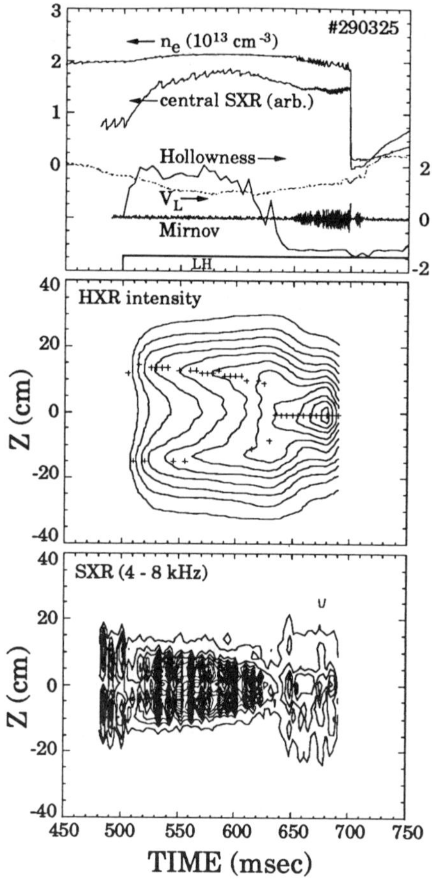

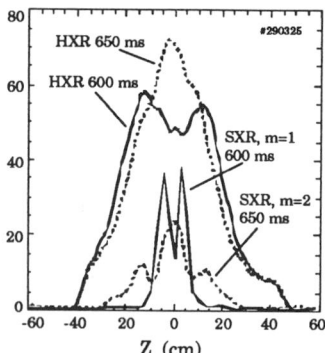

Fig. 12. LHCD into indented plasma.

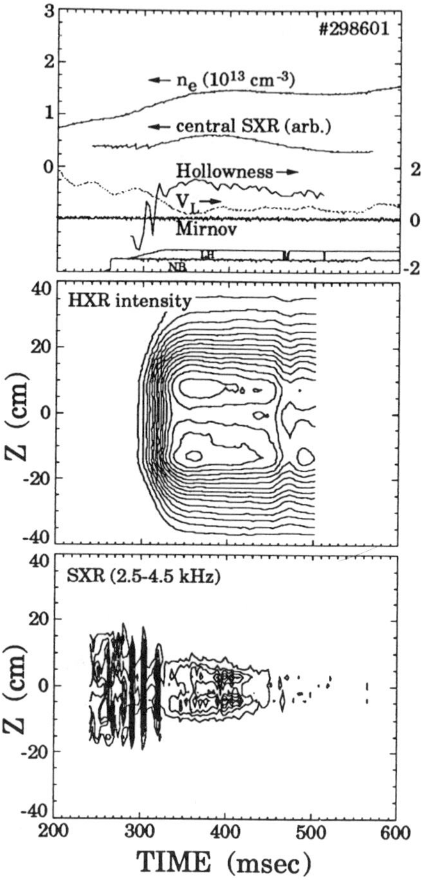

Fig. 13. Comparison of HXR profiles and SXR MHD amplitudes during LHCD.

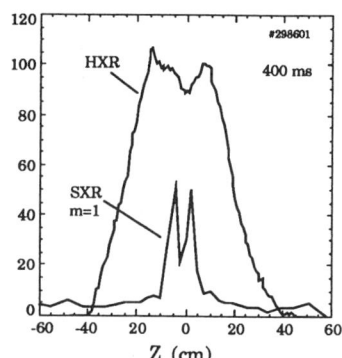

Fig. 14. LHCD plus NBI into indented plasma.

Fig. 15. Comparison of HXR profile and m=1 during LHCD plus NBI.

Even before the onset of the m=2/n=1 mode, there is a gradual reduction in the hollowness of the HXR profile. The central region of the HXR profile was observed to fill in and become peaked. The reason for this is unclear. There was ongoing sawteeth and m=1 activity during this time, as before, and the q=1 surface was shrinking. This seems to be analogous to the behavior of the HXR profile during sawteeth for the circular plasma described above. The filling in of the HXR profile, just in the central region, seems to indicate that the suprathermal confinement (or the LH damping) was affected by the reconnection process occuring at the sawtooth crash.

For the case shown in Fig. 11, the addition of small amount of NBI power to LHCD resulted in an MHD-free discharge. Shown in Fig. 14 are parameters from a discharge with a net LH power of 334 kW, 0.56 MW of NBI and a line averaged density of 1.5 × 10^{19} m^{-3}. NBI was begun early in the discharge, and sawteeth were quickly suppressed as the LH power was ramped up. An m=1 oscillation remained for another 120 ms until it too disappeared, after which no MHD were present. Figure 15 shows the HXR profile at 400 ms compared to the location of the m=1 mode which was still present at that time. The MHD-free condition was maintained until the end of the LHCD and NBI period at 750 ms. The decreasing radius of the m=1 mode (see bottom of Fig. 14) with constant I_p indicates a broadening of the current profile. A measurement of the q profile with MSE for this discharge indicated that q(0) rose above 1 by t=530 ms. It is not yet understood what role NBI plays in suppressing MHD modes that are present with LHCD alone. An increase in electron temperature affecting wave damping or a change in the shear due to NB driven current are possible explanations for the observed improvement.

The HXR intensity remained relatively constant and a hollow profile was maintained. The position of the lobes of the HXR profile remained unchanged as the radius of the q=1 surface was decreasing, suggesting that the location of LH current deposition also remained constant.

In summary, the HXR images from suprathermal electrons produced during LHCD have been correlated with MHD activity in PBX-M. In particular, the location of the LH current deposition correlates with observed changes in the q profile. The effect of the sawtooth crash on the HXR profile indicate that the reconnection process affects suprathermal confinement. MHD modes can be induced by LHCD, and can adversely affect suprathermal confinement. The addition of NBI to LHCD can suppress low n modes that otherwise would destroy localization of LH current.

This work was supported by U.S. Dept. of Energy, Contract Nos. DE-AC02-76-CH0-3073 (PPPL), DE-AC05-84OR21400 (ORNL), DE-FG02-91ER54109 (MIT), and DE-FG02-89ER51121(UCLA).

REFERENCES

[1] M. S. Chance, S. C. Jardin, and T. H. Stix, Phys. Rev. Lett., 51, 1963 (1983).

[2] R.C. Grimm, M. S. Chance, A. M. M. Todd, et al., Nucl. Fusion, 25, 805(1985).

[3] R. E. Bell, N. Asakura, S. Bernabei, et al., Phys. Fluids B 2, 1271 (1990).

[4] M. S. Chance, Y.-C. Sun, S. C. Jardin, C. E. Kessel, and M. Okabayashi, Proceedings of the Second Symposium on Plasma Dynamics:Theory and Applications, University of Trieste, Italy (July 8-10, 1992).

[5] R. Kaita, N. Asakura, S. Batha, et al., Proceedings of the Fourteenth International Conference on Plasma Physics and Controlled Nuclear Fusion Research, (Wurzburg, Germany, September 1992)(International Atomic Energy Agency, Vienna, Austria, Paper IAEA-CN-56/H-1-2.

[6] S. Bernabei, R. Bell, M. Chance, et al., Phys. Fluids B, in press.

[7] S. von Goeler, N. Asakura, R. Bell, et al., in Plasma Physics and Controlled Nuclear Fusion Research, Proceedings of 18th European Conference, Innsbruck, Austria, 1992, (European Physical Society, Geneva,1991), Vol. II, 949.

[8] F. Levinton, R. J. Fonck, G. M. Gammel, R. Kaita, H. W. Kugel, E. T. Powell, and D. W. Roberts, Phys. Rev. Lett., 63, 2060 (1989).

CURRENT PROFILE CONTROL AND STABILITY STUDIES IN THE TOKAMAK PHYSICS EXPERIMENT (TPX)

M. Porkolab, P.T. Bonoli, J.J. Ramos
Massachusetts Institute of Technology, Cambridge, MA 02139

M.E. Fenstermacher, Lawrence Livermore Nat. Lab., Livermore, CA 94550

ABSTRACT

Using the ACCOME and PEST II codes, noninductive current drive scenarios and their stability are studied in the proposed TPX experiment. The current drive techniques include neutral beams as well as lower hybrid and fast magnetosonic waves. The non-standard cases studied include inverted $q(r)$ profiles (reversed-shear scenario), reversed edge current profiles, and second-stable scenarios ($q(0) > 2$). The MHD stability of some of these scenarios has been assessed. The power requirements for TPX are established.

I. INTRODUCTION

Maintaining steady state plasma currents and controlling their profiles by non-ohmic means is one of the key objectives of the Tokamak Physics Experiment (TPX).[1] The TPX parameters are as follows: major radius $R = 2.25$ m, minor radius $a = 0.5$ m, aspect ratio $R/a = 4.5$, elongation $\kappa = 1.80$, triangularity $\delta = 0.5$, plasma current $I_p \lesssim 2.0$ MA, density $n_e(0) \lesssim 1 \times 10^{20} \mathrm{m}^{-3}$, temperatures $T_e(0) \simeq T_i(0) \lesssim 10$ keV, and magnetic field $B_T \lesssim 4.0$ T. The parameters have been varied about the nominal values by factor of two to scope out the achievable performance of the device. The available powers in the baseline, and potential upgrades are as follows: $P_{NBI} = 8(24)$ MW at 120 kV, $P_{ICRF} = 8(18)$ MW at 40-80 MHz, $P_{LH} = 1.5(3.0)$ MW at 3.7 GHz. The role of the various powers are as follows: NBI is used for core current drive and ion heating (beta enhancement), ICRF in the form of fast wave current drive (FWCD) is used for central (core) current drive and for heating electrons, and lower hybrid waves (LHCD) are used for off-axis current drive (profile control) as well as for heating electrons. The launcher designs are discussed elsewhere in these Proceedings.[2,3]

II. RESULTS WITH THE ACCOME CODE

The ACCOME code[4] is a simulation model for computing the self-consistent MHD equilibria in the presence of NBCD, LHCD, FWCD, ohmic currents and bootstrap currents. The lower-hybrid package includes a ray tracing model and a Fokker-Planck model, including diffusion of fast electrons. The ICRF fast wave package is based on a full wave model and a Karney-Ehst type of current drive calculation.[5] The MHD solver (SELENE) computes a free boundary solution of the Grad-Shafranov equation. The density and temperature profiles are assumed to be given. Typical current drive efficiencies obtained are as follows: $\gamma_{FW} \simeq 0.10$ at $T_e(0) = 10$ keV, $\gamma_{NBI} \simeq 0.06$ at $T_e(0) = 10$ keV, $\gamma_{LHCD} \simeq 0.2 \pm 0.1$ at $T_e \simeq 5 - 10$ keV. Here we defined the current drive "figure of merit," $\gamma = \langle n_{e20} \rangle I_{CD}(MA)R(m)/P(MW)$, where $\langle n_{e20} \rangle$ is the volume averaged density measured in units of 10^{20}m^{-3}. In addition, we assume density and temperature profiles of the form $(1 - \psi)^\alpha$, where α_n designates the density profile and α_T designates the temperature profile. Typical energies of resonant electrons are $\varepsilon \simeq 50 - 100$ keV, for $T_e(0) \simeq 10$ keV, corresponding to a launched lower-hybrid spectra of $N_\parallel \simeq 2.2 - 2.8$. Therefore the slowing-down time of these electrons is a few ms, considerably shorter than their assumed confinement times ($\tau_E \simeq 20 - 100$ ms). Therefore, profile control is assured under the present conditions. Furthermore, ray tracing calculations indicate strong single-pass absorption of the lower hybrid waves, with well behaved ray trajectories.

An example of a high-bootstrap current scenario, still in the first stability regime, is the so-called "ARIES-I" scenario (see Figs. 1,2). The plasma parameters are $n_e(0) = 1.0 \times 10^{20} \text{m}^{-3}$, $T_e(0) = T_D(0) = 9.4$ keV, $\alpha_n = 0.25$, $\alpha_T = 1.0$, $f = 80$ MHz, $B = 4.0$ T. In this case $P_{NBI} = 6$ MW, $P_{LH} = 0$ MW, and $P_{FW} = 2.8$ MW to drive the currents, and $I_{NBI} = 192$ kA, $I_{BS} = 987$ kA ($f_{BS} = 0.73$) for a total current of $I_p = 1.36$ MA. In this case $\beta_T = 1.75\%$, $\beta_p = 2$, $\beta_N = 2.6$, $q_o(0) = 1.41$, $q_{95} = 5.6$, and stability calculations indicate a mild unstable situation to low n external kinks which are stabilized by a conducting wall at $R/a = 1.3$. Additional heating power may be necessary to achieve these parameters.

An example of an inverted q-profile (reversed shear) scenario is shown in Figs. 3,4. This is maintained by the combined current drive powers $P_{LH} = 1.5$ MW and $P_{NBI} = 4.0$ MW. The plasma parameters were $n_e(0) = 1.0 \times 10^{20} \text{m}^{-3}$, $T_e(0) = T_D(0) = 10$ keV, $B = 4$ T, and the unique pressure profile was $p(\psi) = p(0) = (1-\psi)^2$, with particular density and temperatue profiles $T(\psi)/T(0) = 0.667(1-\psi)^{3.5} + 0.333(1-\psi^8)^{1.5}$ (giving a temperature pedestal near the edge, characteristic of H-mode profiles). The driven currents are $I_{NBI} = 207$ kA, $I_{LH} = 308$ kA, $I_{BS} = 990$ kA ($f_{BS} = 0.66$), for a total current of $I_p = 1.50$ MA. The beta values are $\beta_T = 1.48\%$, $\beta_p = 0.93$, $\beta_N = 1.97$. We note that $q(0) = 3.05$, $q_{min}(0.6) \simeq 2.33$ and $q(95) = 4.54$. We have also examined an extrapolation of this scenario to $T_e(0) = T_D(0) = 20$ keV. We find that the same kind of q profile is maintained with $P_{NBI} = P_{FW} = 0$, $P_{LH} = 0.60$ MW, $I_{LH} = 184$ kA, $I_{BS} = 1.995$ MA ($f_{BS} = 0.92$), $q(0) = 2.39$, $q(95) = 3.09$, $q_{min} = 1.79$, $\beta_T = 4.90\%$, $\beta_N = 4.50\%$, $\beta_p = 2.73$. Heating power around the 30 MW level may be necessary to achieve these parameters. The stability of the inverted q-profile scenarios have not been examined yet in any detail. We expect that a conducting wall at $R/a = 1.3$ may be sufficient for stability.

An example of reverse LH current drive in the plasma periphery is shown in Fig. 5. The parameters are $B_o = 3.35$ T, $n_e(0) = 6.6 \times 10^{19} \text{m}^{-3}$, $T_e(0) = T_D(0) = 12.6$ keV, $a_n = 0.75$, $a_{T_e} = a_{T_D} = 1.0$, $P_{NB} = 2.4$ MW, and $P_{LH} = 0.36$ MW. The power spectrum for the reverse LHCD power is characterized by

$N_\parallel$ components in the range $-8 \lesssim N_\parallel \lesssim -4$ and the rf frequency is 3.7 GHz. The total reverse LH current in Fig. 5 is $I_{LH} = -0.21$ MA, the neutral beam seed current is $I_{NB} = 0.17$ MA and $I_p = 0.96$ MA (note that $I_{BS}/I_p > 1$). The LHRF power deposition profile in Fig. 5 demonstrates wave penetration and absorption to $\psi \gtrsim 0.8$. In summary, for a parabolic pressure profile the available power (1.5 MW) is sufficient to drive $\sim 10\%$ of the total current in the counter-direction in the edge region (the power greatly varies with the edge density profile and temperature pedestal). The consequences of such a current on MHD stability are substantial. For example, Fig. 6 shows the results of 10% negative edge current. Here we considered $p(\psi) = p_o(1-\psi)^2$, $\beta_T = 0.7\%$, $\beta_p = 2.8$, $\beta_N = 2.0$, $I/aB = 0.36$ MA/mT and $q_o = 2.4$ (such that $n = \infty$ ballooning modes are stable, with free access to the second stability region) The first equilibrium, without a negative edge current, is unstable to the $n = 1$ external mode (its stability limit is $\beta_n = 1.5$). The second equilibrium, with 10% negative edge current, is stable to the $n = 1$ mode without a conducting wall up to $\beta_N = 2.0$ (in these equilibria p/p_o and q_oI/aB were held constant). These results demonstrate the importance of profile control for improved MHD stability.

III. CONCLUSIONS

We have found that FWCD by the magnetosonic fast wave is strongly peaking on axis, and that profile control by off-axis current drive is important for maximizing the bootstrap current and optimizing beta. At present the reversed shear scenario is favored by the MHD community for its high bootstrap fraction and relatively low edge q (i.e., high current). Off-axis current drive by lower-hybrid waves is essential to achieve this equilibrium. This technique is also important to raise q_o above 2.0 (i.e., access to second stability). Finally, reversed edge current drive by LHCD is helpful to reduce the edge bootstrap currents in H and VH mode profiles and thereby stabilize external kinks.

ACKNOWLEDGEMENT

Work supported by the US Department of Energy Contract No. DE-AC02-78ET-51013 at MIT and under Contract No. W-7405-Eng-48 at LLNL.

REFERENCES

[1] M. Porkolab et al., Radiofrequency Heating and Current Drive of Fusion Devices, ECA, **16E**, p.209, EPS (eds. C. Gormezano et al.).

[2] D. Swain et al., presented in these proceedings

[3] A. Hubbard et al., presented in these proceedings

[4] R.S. Devoto, P.T. Bonoli, et al., Nucl. Fusion **32**, 773 (1992).

[5] P.T. Bonoli, presented in these proceedings.

[6] M. Fenstermacher et al., presented in these proceedings.

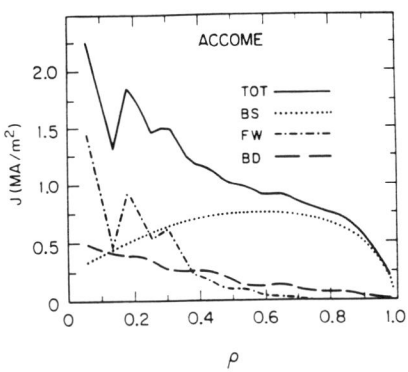

Fig. 1. Aries-I scenario current profiles

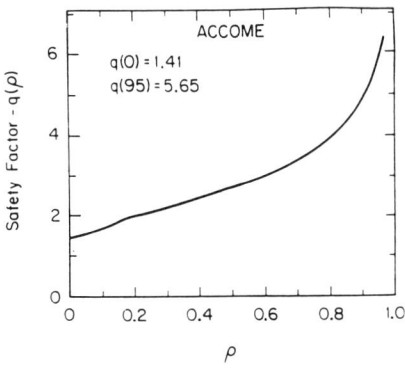

Fig. 2. Aries-I scenario q profile

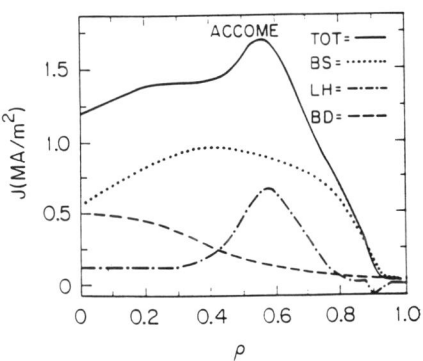

Fig. 3. Inverted q current profiles

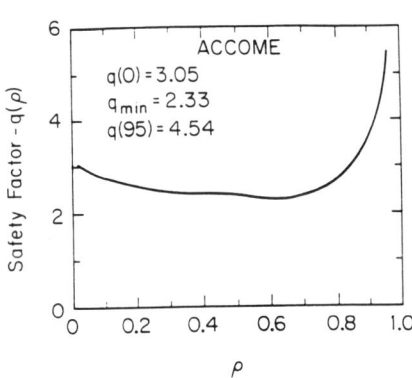

Fig. 4. Inverted q-profiles

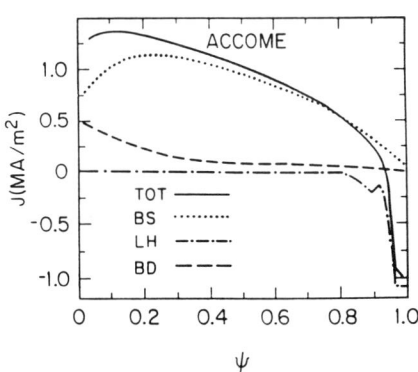

Fig. 5. Reversed edge current profiles

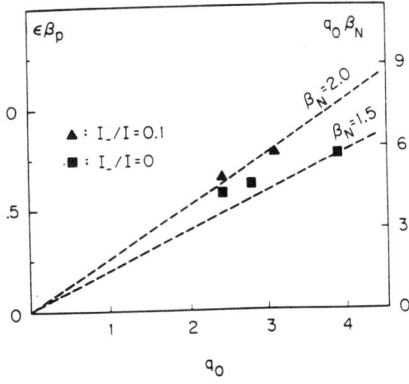

Fig. 6. MHD stability boundary versus β_N

ADVANCED TOKAMAK OPERATIONS WITH ICRF AND LOWER-HYBRID POWER[1]

T.K. Mau, B.J. Lee
University of California, Los Angeles, CA 90024

D.A. Ehst
Argonne National Laboratory, Argonne, IL 60439

Abstract

Advanced tokamak operating modes based on high bootstrap current, first- and second-stability regime plasmas are examined in the context of the TPX experiment and the ARIES reactors, using a combination of ICRF fast wave and lower hybrid power. The main method of analysis entails the alignment of driven current density profiles with those required for stability. In most of the cases studied, the required power levels and launched spectra are found to be reasonable.

Introduction

For a steady-state tokamak reactor to be economically attractive [1], it must have, among other features, low recirculating power fraction, and high mass power density that enables compact power units to be built. The first feature may be achieved by choosing plasma equilibria with high $\epsilon\beta_p$ to raise the bootstrap current fraction, f_{BS}, so as to reduce the seed current drive powwr. Higher power densities in the form of higher plasma β may be accessed through detailed local control of the current and pressure profiles, which may also lead to higher confinement times and higher f_{BS}. Throughout the ARIES reactor study [1], two such advanced tokamak operating modes, ARIES-I and -IV, were identified. In both cases, a preferred combination of ICRF fast wave and lower hybrid power is used to drive the required seed current profiles needed for ballooning stability. In the proposed TPX experiment [2], advanced tokamak modes modeled after ARIES-I and -IV will be examined. This paper describes how these two modes may be accessed via RF power alone, in both the TPX device and the ARIES reactors. Estimates of the power requirements are obtained.

Ray Tracing Analysis

Two ray tracing codes have been used to model the wave propagation and absorption in the toroidal geometry for the fast wave (FW) and lower hybrid (LH) wave. The bulk of the analysis is performed with the CURRAY code [3] that is coupled to a calculated free boundary equilibrium. In solving the ray equations, the cold plasma dispersion relation is used for the FW, while thermal effects are included for the LH wave. Damping of the wave power, $P(s)$, along the ray is given by

$$dP(s)/ds = -(\gamma/\vec{N}\cdot\vec{\Gamma})P(s), \qquad (1)$$

---

[1] Work supported by USDOE via PPPL Subcontract S-03265-G and Grant DE-FG03-86ER-52126.

where s is the normalized arclength, $\vec{N}$ is the refractive index, and γ, $\vec{\Gamma}$ are normalized damping decrement and power flow, respectively. Defining $\hat{e} = \vec{E}/|\vec{E}| = e_\perp \hat{x} + e_\eta \hat{y} + e_\| \hat{z}$, and $\bar{K}$ the hot dielectric tensor, γ is given by $\gamma = \gamma_e + \sum_i \gamma_i$, with

$$\gamma_e = Im(K^e_{yy})|e_\eta|^2 + 2Re(K^e_{yz})Im(e^*_\eta e_\|) + Im(k^e_{zz})|e_\||^2 \tag{2}$$

$$\gamma_i = Im(K^i_{xx})|e_\perp|^2 + 2Re(K^i_{xy})Im(e^*_\perp e_\eta) + Im(K^i_{yy})|e_\eta|^2, \tag{3}$$

and $\vec{N} \cdot \vec{\Gamma} = N_\perp^2(1 - |e_\perp|^2) + N_\|^2(1 - |e_\||^2) - 2N_\| N_\perp Re(e^*_\perp e_\|)$. The polarization factors, $(e_\perp, e_\eta, e_\|)$, are obtained by including plasma thermal effects [4]. Electron Landau and TTMP effects are accounted for in Eq. 2, while harmonic ion cyclotron resonances without mode conversion effects are assumed in Eq. 3. In the case of LH wave damping, quasilinear velocity diffusion effect is modeled approximately by multiplying γ_e in Eq. 2 by the factor $[1 + 1.4(c/N_\| v_e)^3]^{-1}$ [5], while spatial diffusion of resonant electrons is found to be negligible. Once the power deposition density is known, the local driven current density, j_{CD}, is computed using an established empirical formula for current drive efficiency [6]. In the CURRAY analysis, the seed current profile, j_{sd}, is first calculated given a prescribed equilibrium, with $j_{sd} = j_{eq} - j_{BS}$, and incident FW and LH power spectra are determined for matching J_{CD} with j_{sd}.

ARIES-I Advanced Mode

The ARIES-I reactor concept is based on high magnetic field and high aspect ratio in a first-stability plasma to allow for a low plasma current that gives rise to a high bootstrap current fraction. This fraction is further enhanced by raising the on-axis safety factor, q_0, to 1.3 with current profile tailoring using FW near the magnetic axis and LH power in the periphery. A list of reference parameters are given in Table 1. Current drive analysis is done by the RIP ray tracing code [7] that couples to an equilibrium consistent with the driven (RF+BS) current. In this case, 92 MW of FW power at 148 MHz, and 5 MW of LH power at 7 GHz are required to generate the 3.3 MA of seed current, resulting in a normalized efficiency of $\gamma_{CD} \equiv \langle n_{20} \rangle I_{CD}(A) R(m)/P_{CD}(W) = 0.33$. The ICRF power is launched from above the midplane in two poloidally stacked antenna arrays which are phased to give an incident spectrum centered at $N_\| \approx 1.5$-2.5 and $N_\theta \sim 4$-8, in order to achieve satisfactory current profile alignment, as shown in Fig. 1(a).

The CURRAY code is used to analyze the current drive scenario for the TPX/A-I mode, with parameters given in Table 1. Using a prescribed equilibrium, a bootstrap overdriven current of 0.05 MA is found near the plasma edge, that needs to be cancelled using LH waves. Here, about 4.2 MW of ICRF power at 45 MHz centered at $N_\| = 3.25$ for central drive, and 1.4 MW of LH power at 3.7 GHz and $N_\| = 3.0$-4.5 for reverse drive, are required. The resultant matching between j_{CD} and j_{sd} is quite reasonable, except for a modest overdrive on axis, due to the peakedness of the FW-driven current profile.

ARIES-IV Advanced Mode

The ARIES-IV reactor employs a second-regime stability plasma that is sustained with current profile tailoring to obtain $q_0=2$, high $\epsilon\beta_p$, and high f_{BS}, as listed in Table 1, with an achieved β-value of 3.4%. The seed current consists of three components: central, overdrive and edge. From CURRAY calculations, the power requirements are

Table 1:
Key ARIES and TPX Parameters

	ARIES-I	TPX/A-I	ARIES-IV	TPX/A-IV
Major radius, R (m)	6.75	2.25	6.12	2.25
Aspect ratio, A	4.5	4.5	4.0	4.5
On-axis magnetic field, B_0 (T)	11.3	4.0	7.7	4.0
Plasma current, I_p (MA)	10.2	1.15	6.75	0.95
On-axis safety factor, q_0	1.30	1.30	2.0	2.0
Edge safety factor, q_*	3.91	2.37	4.60	5.00
Peak density, n_{e0} (10^{20} m^{-3})	1.88	1.0	2.54	1.0
Ave. density, $\langle n_e \rangle$ (10^{20} m^{-3})	1.45	0.81	2.27	0.74
Peak temperature, T_{e0} (keV)	36.6	13.3	27.3	16.3
Ave. temperature, $\langle T_e \rangle$ (keV)	17.4	4.82	10.3	4.60
Toroidal beta, β (%)	1.9	2.0	3.4	2.0
Poloidal beta, β_p	2.81	3.55	5.40	5.36
Effective charge, Z_{eff}	1.65	1.82	1.65	1.82
Bootstrap fraction, f_{BS}	0.68	0.73	0.96	1.16
Current drive power, P_{CD} (MW)	97.0	5.6	70.2	>9.1

found to be: 20.5 MW of ICRF power at 124 MHz and $N_\parallel$=2.75 for central drive, 34.1(15.6) MW of LH power at 8 GHz and $N_\parallel$=2.1–4.1(-7) for reverse (edge forward) drive. The result is displayed in Fig. 2(a), where the profile match is excellent.

Applying the ARIES-IV concept to TPX results in a reference scenario as given in Table 1 under TPX/A-IV. Shown in Fig. 2(b), using 0.1 MW of FW power at 45 MHz and $N_\parallel$=3.25, an amount of current equal to the central seed (0.1 MA) is driven, but the profile alignment is poor. With a launched power of 2.61 MW at 8 GHz and $N_\parallel$=2.5–5.5, 78% of the overdrive current can be cancelled. For forward drive near the edge, 6.4 MW of LH power at 8 GHz and $N_\parallel$=10-15 is required, at an extremely low efficiency of 0.009 A/W.

Discussions

For the ARIES-I and ARIES-IV reactors and their counterparts in TPX, reasonable RF power and spectrum requirements are found for sustaining the advanced tokamak modes. The case of TPX/A-IV may need further study, including better definition of the edge plasma profiles. The high $N_\parallel$ required and the low efficiency for edge current drive are concerns that need to be addressed. Based on self-consistent MHD and current drive studies [8], results from the CURRAY analyses may be taken to represent upper bounds for the power requirements.

References

[1] The ARIES Team, Report UCLA-PPG-1323, Vol. II (1991)
[2] W.M. Nevins, et al., Proc. 14th IAEA Conf. on Plasma Phys. and Controlled Nucl. Fusion Research, Würzburg, Germany, Paper F-1-5 (1992).
[3] T.K. Mau, et al., Abst. EPS Topical Conf. on Radiofrequency Heating and Current Drive of Fusion Devices, Brussels, Belgium (1992) 181.

[4] S.C. Chiu, et al., Nucl. Fusion **29** (1989) 2175.
[5] P.T. Bonoli, IEEE Trans. Plasma Sci. **12** (1984) 95.
[6] D.A. Ehst, C.F.F. Karney, Nucl. Fusion **31** (1991) 1933.
[7] D.A. Ehst, K. Evans, Jr., Nucl. Fusion **27** (1987) 1267.
[8] M. Porkolab, et al., Abst. EPS Topical Conf. on Radiofrequency Heating and Current Drive of Fusion Devices, Brussels, Belgium (1922) 209.

Figures

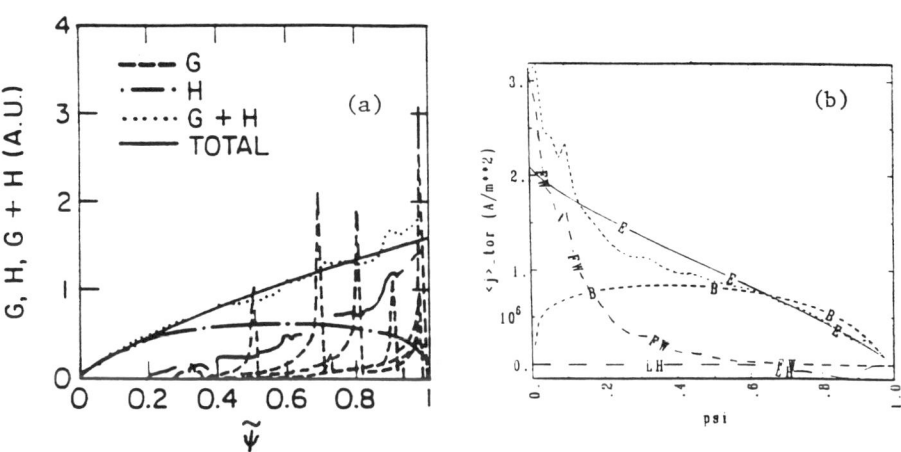

Fig. 1(a): Converged, stable ARIES-I equilibrium with RF current drive [G] and bootstrap [H] contributions. Fig. 1(b): Current density profiles for TPX/A-I: equilibrium [E], bootstrap [B], fast wave [FW], lower hybrid [LH], and total driven [dotted].

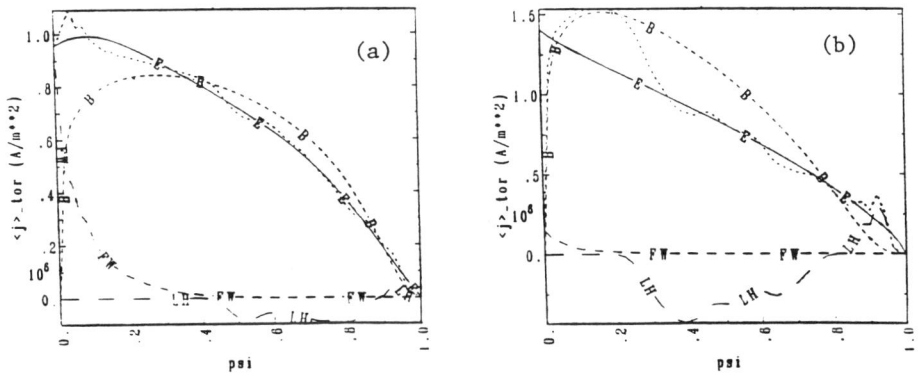

Fig. 2: Current density profiles for (a) ARIES-IV and (b) TPX/A-IV: equilibrium [E], bootstrap [B], fast wave [FW], lower hybrid [LH], and total driven [dotted].

Parasitic effects of ion absorption on fast wave current drive in TPX[*].

P.E. Moroz, D.B. Batchelor[*], E.F. Jaeger[*], T.K. Mau[♦], D.R. Mikkelsen[#], M. Porkolab[°]

University of Wisconsin, Madison, WI 53706
[*]Oak Ridge National Laboratory, Oak Ridge, TN 37831
[♦]University of California, Los Angeles, CA 90024
[#]Princeton Plasma Physics Laboratory, P.O. Box 451, Princeton, NJ 08543
[°]Massachusetts Institute of Technology, Cambridge, MA 02139

Abstract. Parasitic effects of ion absorption on fast wave current drive (FWCD) in TPX have been studied analytically and numerically. Main emphasis has been given to FWCD at frequencies, f = 40 - 110 MHz, in deuterium plasma. The general ion cyclotron harmonic resonances of all plasma species (including neutral injected fast ions) were considered. Fast wave power deposition, power partition between various plasma components, and the resulting current drive efficiency were calculated. The results presented show that the current drive efficiency can be adversely affected by parasitic ion absorption. Favorable current drive scenarios were identified.

Introduction. Parasitic effects of ion absorption on FWCD is an important factor since wave energy absorbed by ions is lost for the current drive. It is well known that even small concentration of minority ion species in some cases can give a strong wave absorption. This work reports on theoretical and numerical studies of various absorption mechanisms that can effect current drive efficiency for a particular case of the TPX tokamak. The initial complement of auxiliary power includes 8 MW of NBI, 8 MW of ICRF/FWCD, and 1.5 MW of LHCD power for heating and current drive. Additional power may be added at a later date as upgrades. Significant wave power can be absorbed by the beam injected fast ions due to ion cyclotron harmonic resonances, even if their concentration is small (only a few percent). To model this absorption the numerical full wave codes, FASTWAC and PICES, the 1-D code, FAST1D, and the ray tracing code, CURRAY, have been amended to include general ion cyclotron harmonics and many species plasmas. The density and effective temperature distributions of fast ions have been computed via the SURVEY code. The following typical parameters of TPX plasma have been considered: D-plasma with 3% H, $n_e(0) = 10^{20}$ m^{-3}, $T_e(0)=T_i(0)= 10$ keV, $B_0 = 4$ T, $\alpha_T = 1$, $\alpha_N = 0.25$, where the parameters α define the temperature and density profiles that are taken to be proportional to $(1-\rho^2/a^2)^\alpha$.

Fast ion distribution. The deposition of neutral beams was calculated with a multi 'pencil' code, SURVEY, in a geometry which includes shifted magnetic flux surfaces with realistic elongation and triangularity. The finite size of the ion sources and the beam divergence is represented by a bundle of parallel beamlets. The calculations of the neutral beam deposition used the Freeman-Jones fits to electron ionization and hydrogenic charge exchange, Gryzinski's formula for proton ionization, and the Olson and Salop impurity electron loss formula. The fast ion distribution was calculated by solving the Fokker-Planck equation with energy diffusion included. However, the rf power absorbed by the beam was not included in the Fokker-Planck calculations reported here. As an example, the results of computations, with the SURVEY code, of the distribution of fast ion density, N_f, and effective temperature, $T_{eff} = 2/3 <E>$, for the TPX parameters listed above are presented in Fig. 1.

---
[*]*This work is supported by the U. S. Department of Energy.*

Wave absorption mechanisms. Fast wave energy can be absorbed by electrons directly or indirectly through the mode conversion process or via the energy transfer from ions due to collisions. In the numerical models presented below we assumed that only the direct absorption of fast waves by electrons contributed to the RF driven current. All other absorption mechanisms have been considered as parasitic for the current drive, although, in principle, they can supply some additional current drive mechanisms[1]. Also, they can contribute to increasing the electron temperature which can increase the fast wave absorption and current drive efficiency[2,3].

The frequency region under consideration, f = 40 - 110 MHz, includes various regimes of wave plasma interaction. One can easily see this fact from Fig. 2 where the position of ion cyclotron resonances for H and D ions at various frequencies is shown. For proper consideration of ion absorption the high cyclotron harmonics (up to $\omega = 4\Omega_D$) have been included into the analysis. Direct absorption of fast waves by electrons can be estimated from the expressions given in Ref. 4-6 and has been calculated, in the codes presented, as a combination of Landau damping, transit-time magnetic pumping, and the cross-term effect. Ion cyclotron harmonic absorption was included into calculations by incorporating corrections to K_{11}, K_{12}, K_{21}, and K_{22} dielectric tensor components:

$$\Delta\varepsilon_n = \frac{\omega_{pi}^2}{\omega k_\| v_i} \cdot \frac{n^2}{\lambda_i} \cdot I_n(\lambda_i) \, e^{-\lambda_i} \cdot Z(\zeta_{ni}) , \qquad (1)$$

where we have used definitions: $\zeta_{ni} = (\omega - n\Omega_i)/k_\| v_i$, $\lambda_i = \frac{1}{2}k_{\perp F}^2 \rho_i^2$, $k_{\perp F}$ is the fast wave root of the dispersion relation, and Z is the plasma dispersion function. Power absorbed at the ion harmonic resonance is proportional to the imaginary part of $\Delta\varepsilon_n$, or in terms of density and temperature, proportional to $\beta_i T_i^{n-2}$, where β_i is the resonant ion contribution to the standard ion beta. Due to this fact the high harmonic absorption is most significant for the hot plasma component, or in the case considered, for the beam injected fast ions. The fast wave energy can be absorbed near the cyclotron harmonic resonance or undergo mode conversion to the ion Bernstein wave. However, in the case considered for harmonic resonance absorption by fast ions, our estimates show that mode conversion is suppressed. This can be seen, for example, from the fact that the width of cyclotron absorption region, $\Delta x_i = L/\zeta_{oi}$, is much larger than the width of the mode conversion region,

$$\Delta x_{nc} \approx L \frac{(n-1)^2}{2n!} \left(\frac{n}{2}\right)^{2n-2} (2.5n+1)\beta_i^{n-1} , \qquad (2)$$

where $L = |B/\nabla B|$ is the magnetic field scale length ($L \approx R_0$ in a tokamak).

Brief description of RF codes. The full wave codes, PICES[3,7] and FASTWAC[5,9], use poloidal and toroidal mode expansions to solve the reduced order wave equations in an arbitrary tokamak plasma equilibrium. All toroidal effects are included. The hot plasma dielectric tensor, used in calculations, includes all important kinetic effects for electrons and ions, and collisions. The codes calculate spatial distribution of RF field components and power absorbed by various plasma species. Current drive computations are based on the approximate expression for the local current drive efficiency[8] relating the current driven with the local power absorbed directly by electrons. The 1-D code, FAST1D[9], presents the same physics as the FASTWAC code but for 1-D case only. It gives valuable information for choosing suitable parameters for full wave calculations. Another reason for developing the FAST1D code was the intention to compare results with the more developed theory of 1-D wave penetration and mode conversion, and with other 1-D codes. The CURRAY ray tracing code[10] which is essentially an extension of the LH code RAYLH[11], can cover frequencies from ICRF to LH. The ray equations are solved in flux surface coordinates on a non-circular plasma geometry generated by the free-boundary equilibrium code. The cold plasma dispersion relation is used for the fast wave. The damping of the wave power along each ray incorporates kinetic effects of electrons and ions and is given by $dP(s)/ds = -(\gamma_k/N \cdot \Gamma) P(s)$, where s is the normalized arc length, N is the refractive index, and γ_k, Γ are the normalized damping decrement and power flow, respectively. Typical ray traces are shown in Fig. 3. The local driven current was computed by using the results of Refs. 8, 12 for the current drive efficiency.

Results of calculations. To identify the favorable current drive scenarios, the frequency scan has been obtained with the FASTWAC, PICES and CURRAY codes. As an example, the results of computations with the FASTWAC code with various absorption mechanisms are presented in Fig.4 (similar calculations with the PICES code are presented in Ref. 7). For the scan the value of $\zeta_{oe}(0)=1$ has been kept constant, where $\zeta_{oe} = \omega/k_\| v_e$. Curves E, H, B, and D denote respectively absorption by electrons, H-ions, beam ions, and plasma D-ions. The current drive efficiency factor, $\gamma = <n_e> R_o I_{rf} / P_{in}$, obtained via all three codes is shown in Fig. 5. The results stress the three regions of greatest interest for the fast wave current drive in TPX: $f \approx 45$ MHz, 80 MHz, and 100 MHz. Parasitic ion absorption was one of the main factors limiting the current drive efficiency. The difference in results obtained with different codes is due to slightly different density and temperature profiles especially near the plasma edge and in the scrape off layer, and the different plasma equilibrium. Calculations for different regimes showed the importance of the factor, $\zeta_{oe}(0)$, which has to be close to 1 for the effective current drive. An example of computations with the FASTWAC code of the current in TPX, incorporating high harmonic ion absorption, is presented in Fig. 6. The total current is driven by the fast wave absorption (P_{rf} = 8 MW, f = 80 MHz, $\zeta_{oe}(0)$ = 1), neutral beam injection (calculated by the SURVEY code for P_{nb} = 8 MW), and bootstrap current generation.

References.
1. P. E. Moroz, Submitted to Plasma Phys. and Contr. Fusion.
2. T. C. Luce et al., Proc. 9th Top. Conf. RF Power in Pl., Charleston, p.271 (1991).
3. E. F. Jaeger, D. B. Batchelor, Proc. IAEA Tech. Comm. Meet., Arles, p. 86 (1991).
4. M. Porkolab, Proc. 9th Top. Conf. RF Power in Pl., Charleston, AIP, p. 197 (1991).
5. P. E. Moroz, P. L. Colestock, Plasma Phys. and Contr. Fusion **33**, 417 (1991).
6. P. E. Moroz et al., Phys. Fluids B **4**, 2915 (1992).
7. E. F. Jaeger, D. B. Batchelor, M. Murakami, this conference.
8. D. A. Ehst, C. F. F. Karney, Nucl. Fusion **31**, 1933 (1991).
9. P. E. Moroz, Univ. Wisconsin Report PLR-91-17 (1991).
10. T. K. Mau, et al., Proc. EPS Conf. on RF Heat. and CD, Brussels, 181 (1992).
11. M. Brambilla, A. Cardinali, Plasma Phys. **24**, 1187 (1982).
12. S. C. Chiu et al., Proc. EPS Conf. on RF Heat. and CD, Brussels, 173 (1992).

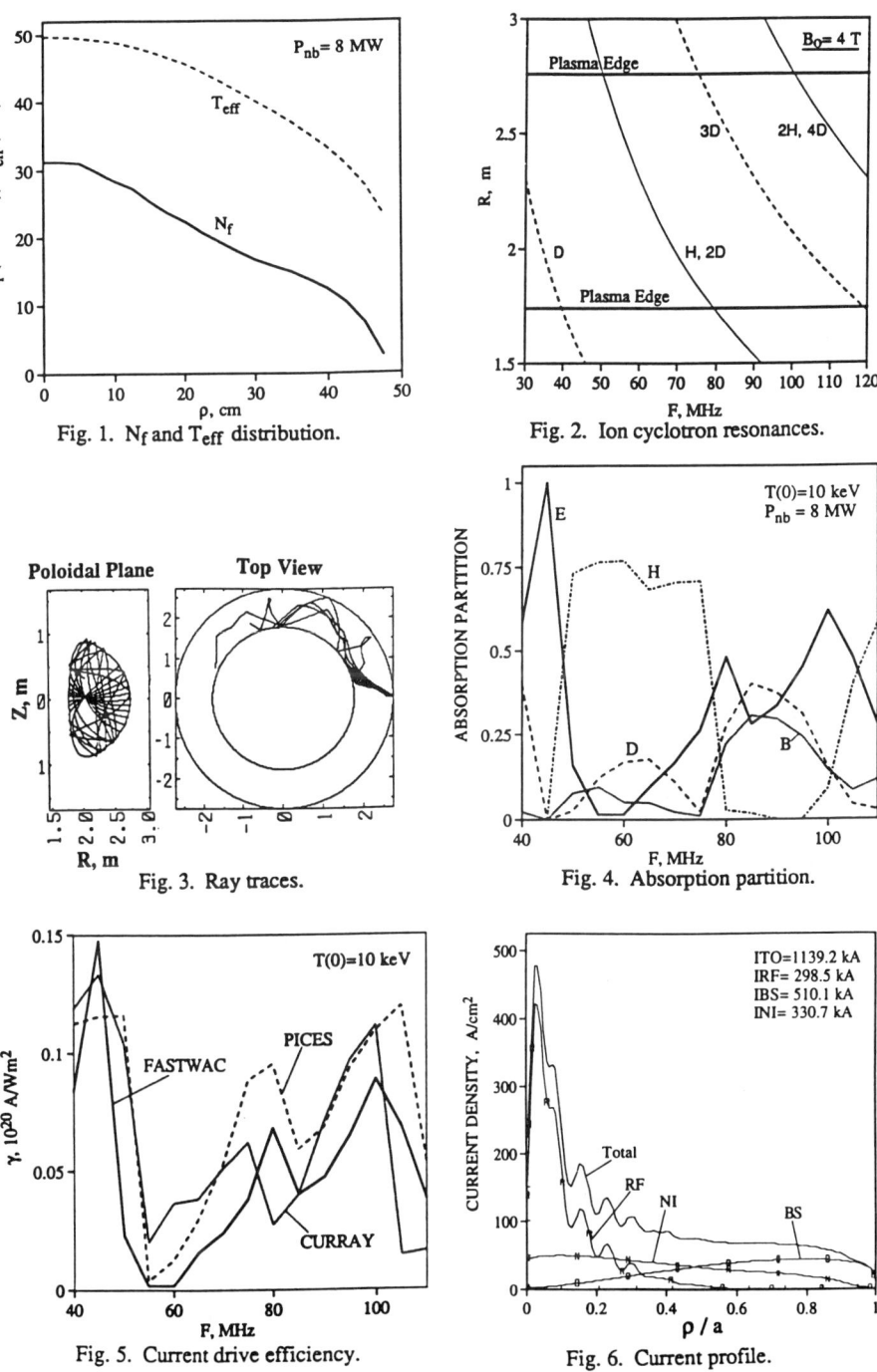

Fig. 1. N_f and T_{eff} distribution.

Fig. 2. Ion cyclotron resonances.

Fig. 3. Ray traces.

Fig. 4. Absorption partition.

Fig. 5. Current drive efficiency.

Fig. 6. Current profile.

Lower-Hybrid Counter Current Drive for Edge Current Density Modification in DIII-D*

M.E. Fenstermacher and W.M. Nevins
Lawrence Livermore National Laboratory, Livermore, CA 94550

M. Porkolab and P.T. Bonoli
Plasma Fusion Center, MIT, Cambridge, MA 02139

R.W. Harvey
General Atomics, San Diego, CA 92186

ABSTRACT

Each of the Advanced Tokamak operating modes in DIII-D is thought to have a distinctive current density profile. So far these modes have only been achieved transiently through experiments which ramp the plasma current and shape. Extension of these modes to steady state requires non-inductive current profile control, e.g. with lower hybrid current drive (LHCD). Calculations of LHCD have been done for DIII-D using the ACCOME [1] and CQL3D [2] codes, showing that counter driven current at the plasma edge can cancel some of the undesireable edge bootstrap current and potentially extend the VH-mode. Results will be presented for scenarios using 2.45 GHz LH waves launched from both the midplane and off-axis ports. The sensitivity of the results to injected power, n_e and T_e, and launched wave spectrum will also be shown.

INTRODUCTION

Recent experiments in DIII-D have focussed on four modes of plasma operation which have produced either enhanced confinement and/or high beta stablity under transient conditions. Known as the VH-mode, the high l_i H-mode, the second stable core mode, and the high $\epsilon\beta_p$ mode, each is thought to have a distinct current density profile. In particular, large bootstrap current density at the plasma edge builds up during the VH-mode and may need to be cancelled by a non-inductive counter current drive system to extend this mode to steady state.

The existing current drive program on DIII-D, including Neutral Beam (NB) injection, Electron Cyclotron (EC) harmonic waves, and Ion Cyclotron (ICRF) Fast Waves (FW), allows the current density profile in the plasma core (near the plasma axis and out to about 2/3 of the plasma radius) to be modified. However, none of these systems has shown a high enough current drive efficiency to affect the current density profile near the plasma edge where the temperature is comparatively low.

LHCD has comparatively high efficiency in this region, the waves can be strongly damped there with proper choice of launched spectrum, and there exist simulation codes that have been benchmarked to previous experimental results. Also, the power sources and other components of the necessary LHCD system for this DIII-D application are readily available and have been proven on previous experiments at ASDEX and PLT. This paper presents ray-tracing, power absorption and counter current drive calculations for LHCD in DIII-D VH-mode-like plasmas.

---

* Work performed for the U. S. DoE by LLNL under contract W-7405-ENG-48 and by GA under contract DE-AC03-89ER53277.

VH-MODE MODEL PLASMA

The plasma conditions assumed for the LH ray-tracing and CD calculations are those of DIII-D shot 75121 at 2540 ms. This was a double null (vertically symmetric) plasma with $B_T = 2.13$ T, $I_p = 1.63$ MA, $R_o = 1.67$ m, $a = 0.62$ m, $\kappa_x = 2.1$, $\delta_x = 0.89$, and $Z_{eff} = 2.0$. The shot progressed from L-mode to the H-mode and finally into the VH-mode at approximately 2400 ms. Analysis of the VH-mode phase has shown that the edge bootstrap current density continually increased throughout the period until it terminated at approximately 2600 ms. Plasma profiles for the fully developed VH-mode at 2540 ms are shown in Fig. 1. Also given is the profile of the bootstrap current density, showing a maximum of 0.42 A/cm^2 at $\rho = 0.9$. The integrated bootstrap current in the region $0.75 < \rho < 1.0$ was estimated to be 130 kA. It is not clear how much of this edge bootstrap current must be cancelled to extend the VH-mode. The objective of the LHCD calculations was to determine whether counter current could be driven in this region and to calculate the CD efficiency there.

CQL3D RESULTS

In addition to the efficiency and radial location of current drive for a given input power spectrum, an important question is whether the LH driven electron tail leads to appreciable Ohmic conductivity enhancement ("hot conductivity") which for the counter-LHCD case reduces the overall efficiency of the edge current control. The CQL3D Fokker-Planck code[2] is a comprehensive 2D-in-velocity-space, 1D-in-radius computational model which addresses these problems. It incorporates a LH ray tracing code, includes self-consistent damping of the energy flowing along the rays, the effects of trapping, and the toroidal electric field. Here it is used with the noncircular equilibria of the shot described above.

Fig. 2 shows current profiles obtained with CQL3D, using the experimental 0.1 V loop-voltage and 2 MW of LH power incident from the plasma periphery on the outer equatorial plane, with $N_\parallel$ from 4.0 to 6.0. The net current density was substantially flattened out near the plasma edge by the LHCD. Damping in these cases is essentially single pass.

Varying $N_\parallel$ from 3.5 to 7.0 (central value) gave driven rf currents from 160. to 70. kA, respectively, peaked in the range of radial ρ values from 0.75 to 0.975. The optimal situation for cancelling the bootstrap current near the plasma edge was with the $N_\parallel$ spectrum from 4.0 to 6.0, which gave 100 kA rf current peaked at $\rho = 0.85$. For the values of loop voltage obtained in the experiment ($V_\phi = 0.1 - 0.2$ volts) there was no appreciable synergy between the LHCD and the toroidal electric field. That is, the Ohmic and rf currents simply added.

Again using CQL3D, a series of code runs were performed to evaluate possiblities for increasing the LHCD efficiency using combinations of two injected spectra. The higher $N_\parallel$ spectrum can set the radial position of peak damping near the plasma edge, and the lower spectrum will increase the tail at that location.[3] Combining one-third power in $N_\parallel$=3-4 with two-thirds power in $N_\parallel$=4-6 for a total of 2MW input power, gave 120 kA peaked at $\rho = 0.85$, to be compared with 100 kA for the pure $N_\parallel$=4-6 spectra. Thus efficiency is increased by 20 percent. Using a half-and-half split of the power in the above two spectra, the total current was 130 kA, still peaked at $\rho = 0.85$, but the radial width of the current generation became greater than the extent of the bootstrap hump. If the spectra do not touch each other in $N_\parallel$, the damping locations decouple. Thus a compound spectra (two independent antenna) capability adds flexibility to the experiment, although the useful efficiency increase is only of order 20-30 percent.

ACCOME RESULTS

The ACCOME/SELENE code package[1] calculates a self-consistent 2D free boundary plasma equilibrium and non-inductively driven currents from NB, ICRF and LH injection and bootstrap currents. The LH ray-tracing and current drive package[4] uses a 1-D Fokker-Planck equation with an effective perpendicular temperature. Quasi-linear damping is calculated and the driven current is obtained from the adjoint formulation of Karney and Fisch.[5]

The plasma profiles are input to ACCOME in terms of normalized poloidal flux, ψ as $X(\psi) = (X_o - X_a)(1 - \psi^{\beta_x})^{\alpha_x} + X_a$ where $X = (n_e, T_e,$ or $T_i)$, X_o and X_a are the central and separatrix values. The VH-mode data from shot 75121 were fit with these functions using $X_o, X_a, \beta_x,$ and α_x as fitting variables. The weighting function emphasized the outer data more strongly than the core so that the edge bootstrap current density calculated by ACCOME would accurately reflect the experiment.

Rays were launched from the plasma separatrix adjacent to the off axis DIII-D port above the midplane. This was done in an attempt to achieve the maximum toroidal upshift of the launched $N_\parallel$ spectrum as the LH wave cone propagated into the plasma. This has the desireable effect of reducing the required $N_\parallel$ of the spectrum launched at the antenna which simplifies the antenna design.

For the baseline calculations, the absorbed power in the plasma was $P_{lh} = 3.0$ MW, two Gaussian peaks were used in the spectrum with central refractive indices, $N_{\parallel o} = 4.5$ and 6.0, and $1/e$ width, $\Delta N_\parallel = 0.33$. The counter LH current, $I_{LH} = -86$ kA, was driven in the outer region, $0.75 < \psi < 0.95$. The efficiency was, $\eta_{lh} = -0.028$ A/W.

Given the $N_\parallel$ upshift due to the off-midplane injection, the comparable case with CQL3D has $N_\parallel = 5-7$ and an efficiency, $\eta_{lh} = 0.038$, about 35% higher than the result above. This enhancement is likely due to both the broad $N_\parallel$ spectrum used (reaching higher T_e) and to 2-D velocity space effects not adequately modeled in ACCOME with the assumption of $T_{e\perp} = 2.5 T_{e,th}$. Benchmarking of the codes continues.

Results of an ACCOME scan of absorbed power using the baseline spectrum values are given in Fig. 3, showing a constant current drive efficiency. Also, in anticipation of density control with the DIII-D divertor cryopump, cases for higher plasma temperature and lower density (constant pressure) were done for baseline profile shape parameters, $P_{lh} = 3$ MW, and $N_\parallel$ adjusted to keep the counter LH current at approximately the baseline radial location. Results are also shown in Fig. 3. The current density profiles, for the case with $T_e = 9.2$ keV, are shown in Fig. 4.

SUMMARY

It has been shown that LH waves, with a spectrum launched in the counter current direction can modify the edge current density profile in DIII-D VH-mode plasmas. LH counter current drive efficiency in the region of undesireable bootstrap current, $0.8 < \rho < 1.0$, was in the range $0.028 < \eta < 0.065$ for the calculations done with the present VH-mode plasma. The higher values were obtained with the full Fokker-Planck treatment and a compound wave spectrum launched from the midplane port. Scans at constant plasma pressure showed a factor of 2.2 efficiency increase when the temperature (density) increased (decreased) by 50%.

REFERENCES

[1] K. Tani, et. al., J. Comput. Phys. **98**, 332, (1992).
[2] R.W. Harvey and M.G. McCoy, to be published in Proc. of IAEA TCM on Advances in Simulation & Modeling of Thermonuclear Plasmas, Montreal, 1992.
[3] F. X. Söldner et. al., in Proc. of 17th European Conf. on Contr. Fusion and Plasma Physics, Amsterdam (1990).
[4] P.T. Bonoli and R.C. Englade, **PF**, **29**, 2937, 1986.
[5] C.F.F. Karney and N.J. Fisch, **PF**, **28**, 116, 1985.

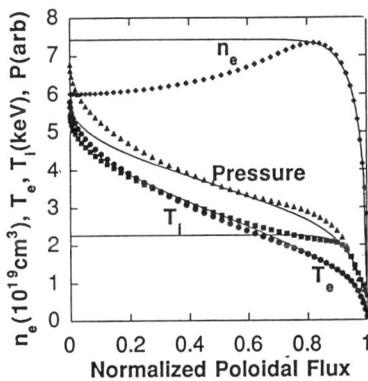

Fig. 1 Profiles of n_e, T_e, T_i, and pressure (points) and fits used in ACCOME (lines).

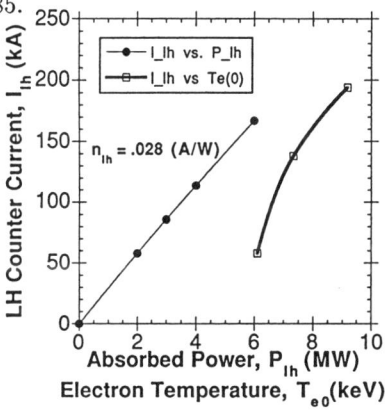

Fig. 3 Lower hybrid counter current vs. P_{lh} and $T_e(1/n_e)$ from ACCOME sensitivity scans.

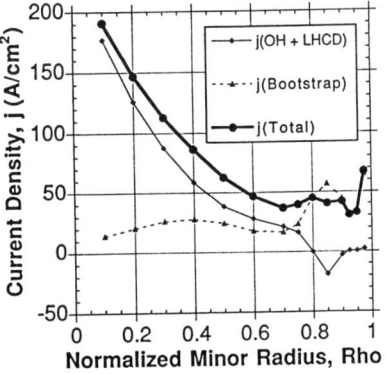

Fig. 2 Current density profiles from CQL3D for $N_\parallel = 4 - 6$, $T_{eo} = 6.1$ keV, and $P_{lh} = 2$ MW.

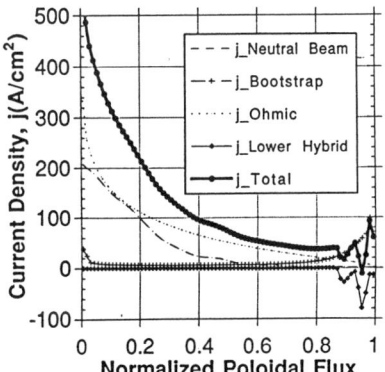

Fig. 4 Current density profiles from ACCOME for $N_\parallel = 3.7, 5.4$, $T_{eo} = 9.2$ keV, and $P_{lh} = 3$ MW.

EXPERIMENTAL STUDIES OF HIGH-ENERGY X-RAY EMISSION AND BOOTSTRAP CURRENT GENERATION IN HIGH $\epsilon\beta_p$ LOWER-HYBRID DRIVEN PLASMAS*

J.P. Squire[†], M. Porkolab, J.A. Colborn and J. Villaseñor
Plasma Fusion Center and Research Laboratory of Electronics, Massachusetts Institute of Technology, Cambridge, MA 02139 USA

ABSTRACT

High poloidal beta ($\epsilon\beta_p$~1) plasma equilibria have been produced by injection of lower-hybrid waves with both asymmetric (current-drive) and symmetric (heating) spectra in the Versator II tokamak. Asymmetric and symmetric injection both generate nearly the same plasma current (5kA), with the loop voltage measured as zero. We studied the rf-created high energy electron distribution with a radial and tangential array of X-ray spectrometers (1"x3" NaI), along with a single soft X-ray spectrometer (SiLi). Analysis of the X-ray data is carried out using a bremsstrahlung emission code. We find that nearly all of the plasma current in the current-drive case can be accounted for by the asymmetric electron tail. In the heating case, the soft X-ray data indicates the presence of an enhanced population of intermediate energy electrons (~10keV) consistent with that of the launched LH wave spectrum. Our estimates show that bootstrap current generated by these electrons can account for a majority of the plasma current.

INTRODUCTION

A novel technique for producing high $\epsilon\beta_p$~1 equilibria was pioneered at Versator by using the pressure supplied by a lower-hybrid driven energetic electron tail.[1] It has been demonstrated that substantial poloidal beta can be achieved if the rf-driven current (I_{rf}) is reduced to below the Alfvén current (I_A~17kA), with β_p scaling as I_A/I_{rf}. With high values of β_p a large fraction of bootstrap current is expected, although the fraction may depend on details of the electron distribution function because of the trapped electron population. The rf-driven energetic tail electrons emit X-rays via bremsstrahlung, and this can be measured to gain information about the electron tail distribution.[2] In this paper we discuss X-ray measurements taken during low and high $\epsilon\beta_p$ discharges and compare the results between current drive ($\pi/2$) and symmetric (π) injection of lower-hybrid waves. The rf-driven electron distribution function is expected to be substantially different for the two launched wave power spectra, therefore the amount of bootstrap current generated could also be substantially different.

© 1994 American Institute of Physics

EXPERIMENTAL RESULTS

Versator II is a modest sized tokamak with a circular cross section, major radius $R_o=0.405$m, and minor radius $a_L= 0.13$m. For plasmas discussed in this paper the toroidal field is $B_o \simeq 1.0$T, and the density is typically $\bar{n}_e \simeq 6 \times 10^{18}$m^{-3}. The lower-hybrid system has a source frequency of 2.45GHz with up to 100kW of power, and the coupler is a four-waveguide grill with relative phasing set at either $\pi/2$ (asymmetric or current-drive) or π (symmetric). The plasma is initiated with an ohmic heating (OH) transformer and the OH circuit is then opened. For the high $\epsilon\beta_p \sim 1$ experiments the plasma current is allowed to decay and subsequently the rf power is applied. The plasma current becomes constant at $I_p \simeq 5$kA, and the loop voltage is zero (see Fig. 1). Higher current ($I_p \sim 14$kA) plasma equilibria are used as a lower $\epsilon\beta_p(\sim 1/2)$ comparison.

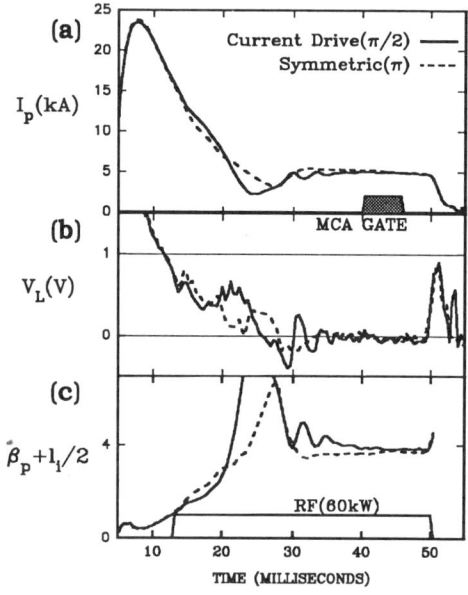

Fig. 1. A comparison of plasma parameters at high $\epsilon\beta_p$ for current-drive and symmetric RF injection. a) plasma current with the multi-channel analyzer gate shown. b) loop voltage. c) $\beta_p+l_i/2$ calculated from the plasma current and the vertical field.

The rf-created high energy electron distribution is studied using a radial and a tangential array of hard X-ray (20-500keV) spectrometers (1x3inch NaI).[3] The radial array consists of seven vertically viewing detectors that can be moved between plasma shots. The tangential array uses the same detectors, which are subsequently reconfigured to view the plasma from the side of the tokamak at a full range of angles to the toroidal magnetic field. In addition, a single movable soft X-ray (2-30keV) spectrometer (SiLi) is used.[4] Spectra are collected during the steady-state phase when the loop voltage is zero (see Fig. 1a, MCA gate).

First current-drive steady-state equilibria are obtained, and then the waveguide relative phasing is changed from $\pi/2$ to π for subsequent shots. During the low $\epsilon\beta_p$ equilibria we observe the usual strong asymmetric tangential X-ray emission indicating a substantial unidirectional electron tail, which is formed by traveling lower-hybrid waves. When attempting current-drive with π phasing, a small applied loop voltage (~0.2volts) is required to maintain the plasma current.

A large unidirectional tail is again observed. In contrast, when trying the same procedure at high $\epsilon\beta_p$, we find that nearly the same current is generated by π phasing as with $\pi/2$ (see Fig. 1), with no significant change in the loop voltage ($\Delta V_{loop}<0.02V$). The X-ray and density profiles show outward radial displacements consistent with the expected Shafranov shift. The tangential X-ray data at high $\epsilon\beta_p$ indicate a high energy electron tail for both phasings, but the hard X-ray flux for the π case is less than that for $\pi/2$ phasing by a factor of four (see Fig. 2). In contrast, the soft X-ray flux changes by less than a factor of two between the two phasings.

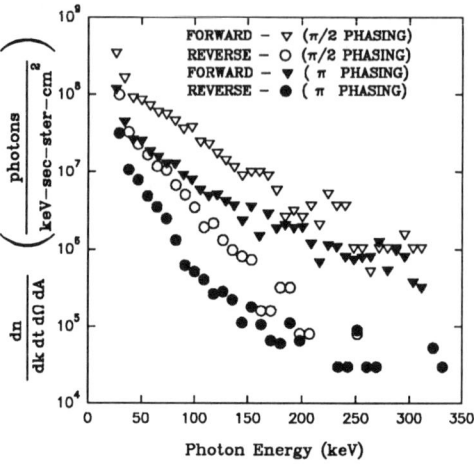

Fig. 2. A comparison of the tangential X-ray emission for $\pi/2$ (open symbols) and π (filled symbols) phasing at high $\epsilon\beta_p$

DISCUSSION

We carry out a quantitative analysis of the tangential X-ray emission in order to obtain information about the energetic electron velocity distribution. The analysis is done using a computer code based on a code developed by Texter.[2] A three-temperature Maxwellian model is used in an electron-ion bremsstrahlung emission calculation. The model includes perpendicular ($T_\perp$), parallel backward ($T_{\|b}$) and parallel forward ($T_{\|f}$) temperatures relative to the toroidal magnetic field, with a high energy cutoff (T_{max}) in the forward direction. For simplicity, we assume that the velocity distribution is independent of the plasma radius, and we use the radial X-ray emission profile to determine the spatial effect on the tangential emission. We obtain a least-squares fit to the emission by interation.

The results for the high $\epsilon\beta_p$ current-drive case are: $T_\perp=60\pm15$keV, $T_{\|b}=55\pm14$keV, $T_{\|f}>400$keV and $T_{max}=150\pm35$keV. Assuming the electron tail carries all the plasma current, we find the central tail density $n_{tail}(0)/n_{ebulk}(0)\sim1.3\%$ and $Z_{eff}\sim4.6$. We had no independent means of determining Z_{eff}, but this value is consistent with past estimates.[4] This model also accounts for a majority of the soft X-ray emission. For the lower $\epsilon\beta_p$ current-drive case we find similar results except the emission profile is broader and the tail is more energetic.

In the case of symmetric injection at lower $\epsilon\beta_p$ with a small applied loop voltage, the modeled electron tail resembles a runaway type ($T_{max} > 1$MeV). With the same Z_{eff} as the comparable $\pi/2$ case, this tail accounts for a majority of the plasma current. For the high $\epsilon\beta_p$ case, the high energy tail modeling results are: $T_\perp$=75±18keV, $T_{\|b}$=40±10keV, $T_{\|f}$=195±48keV and T_{max}>500keV. Unlike the lower $\epsilon\beta_p$ case, we find that the tail only carries 1/3 of the total current, I_{tail}=1.7±0.4kA. In contrast to the current-drive case, the calculated emission from this high energy model accounts for less than one half the observed soft X-ray emission (< 30keV).

To reconcile the difference in the soft X-ray emission, an isotropic Maxwellian electron distribution with an intermediate temperature is added to the high energy model. A best fit to the soft X-ray emission gives: T_{int}=15±4keV and $n_{int}(0)/n_{ebulk}(0)$~1%, when using a profile with a Gaussian half width of W_{int}≈0.07m determined from the soft X-ray profile. This intermediate electron energy is in the range expected to be resonant with the lower phase velocity waves launched by the π phasing. When this intermediate energy tail is used in a calculation for the total bootstrap current[5,6], we find $I_{bootstrap}$=4±1kA. We hypothesize that the high energy electron tail contributes little in the form of bootstrap current because of its highly anisotropic nature. The bootstrap current combined with the high energy tail current accounts for the total plasma current.

CONCLUSION

We find that symmetric (π phasing) lower-hybrid wave injection at high values of $\epsilon\beta_p$ generates nearly the same plasma current as asymmetric wave injection (current-drive, $\pi/2$ phasing). By means of X-ray analysis we determine that in the π phasing case only one third of the plasma current can be accounted for by the energetic electrons. The soft X-ray emission indicates an enhanced intermediate energy tail population of electrons. Calculations show that this tail can be carrying a majority of the plasma current in the form of bootstrap current.

* Supported by U.S. DOE Contract No. DE-AC02-78ET51013
† Presently at General Atomics and supported by an appointment to the U.S. DOE Fusion Energy Postdoctoral Research Program administered by ORAU.

REFERENCES

1. S.C. Luckhardt, K.-I. Chen, S. Coda, et al., Phys. Rev. Lett. 62 1508 (1989).
2. S. Texter, S. Knowlton, M. Porkolab, et al., Nucl. Fusion 26, 1279 (1986).
3. J.P. Squire, M. Porkolab, P.T. Bonoli, et al., 1992 Int. Conf. on Plasma Physics (Innsbruck, June 1992) Part II, pg. 965.
4. M.J. Mayberry, PhD thesis, Dept. of Physics, Mass. Inst. of Tech. (1986).
5. S.P. Hirshman, Phys. Fluids 31, 3150 (1988).
6. W.M. Nevins, Private communication (1992).

LOWER-HYBRID CURRENT DRIVE, HARD X-RAY EMISSION AND FLUCTUATIONS

Linda Vahala
Dept. ECE, Old Dominion University, Norfolk, VA. 23529

George Vahala
Dept. Physics, College of William & Mary, Williamsburg, VA. 23185

Paul. T. Bonoli [*]
Plasma Fusion Center, M. I. T, Cambridge. MA. 02139

ABSTRACT

In lower hybrid (LH) current drive, suprathermal electrons are excited, and, on collisions with the background ions, emit hard X-rays. These hard X-rays are readily detected. From the Monte Carlo solution of the wave kinetic equation for LH wave propagation in toroidal geometry with plasma fluctuations, it is shown that the hard X-ray data from JT-60 are indicative of the presence of fluctuations. These results are then extrapolated to what could be expected from LH current drive on Alcator C-Mod plasmas.

THE WAVE KINETIC EQUATION

The wave kinetic equation of Bonoli & Ott[1] for scattering from density fluctuations has been extended to incorporate the effects of wave scattering from both density and magnetic fluctuations[2]. The kinetic equation for the wave energy density F is given by[2]

$$\left(\frac{dF}{dt}\right)_{ray} = \sum_{\sigma=s,f} \int_0^{2\pi} d\beta \, [F(\phi+\beta) - F(\phi)] [S^n(\kappa_\sigma) K^n + S^B_{\alpha\gamma}(\kappa_\sigma) K^B_{\alpha\gamma}], \quad (1)$$

where the LHS of Eq. (1) gives the time evolution of F along the ray trajectory in a toroidal tokamak equilibrium, accounting for wave damping due to electron and ion Landau damping as well as collisional damping. The RHS of Eq. (1) gives the effects of fluctuations on the wave propagation. S^n and S^B are the spectral density and magnetic spectra, which have a slowly varying spatial dependence due to the inhomogeneous plasma equilibrium. κ_σ is the fluctuation wavenumber for either slow/fast wave scattering. K^n and K^B are complicated scattering kernels[2]. When fluctuations are present, the integral equation (1) is solved by Monte Carlo methods, using 100 random number iterations for each $n_{||} = c\, k_{||}/\omega$. Typically, 25 different $n_{||}$ are chosen from a Gaussian distribution and various launch angles are considered (from the low field side, $\theta = 0°$, down to a bottom launch angle of $\theta = -90°$).

---

[*]work supported by U.S. DOE Contract No. DE-AC02-78ET51013.

COMPARISON WITH JT-60 HARD X-RAY DATA

The hard X-rays that arise from the collisions of energetic electrons (induced by LH current drive) with the plasma ions has been detected[3] on JT-60 for various $n_{\parallel}$. The radial dependence of the hard X-ray spectra on the initial $n_{\parallel}$ of the LH wave is shown in Fig. 1. For each $n_{\parallel}$, Abel transforms were used to ascertain 3 radial chord positions.

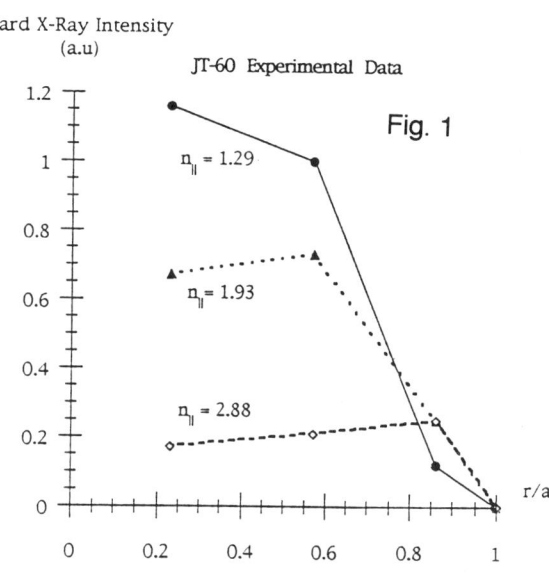

Fig. 1

The radial power density absorbed by the electrons in the LH current drive can be determined from Eq. (1). If one ignores fluctuation effects [i.e., sets the r.h.s of Eq. (1) to zero] and just performs a ray tracing calculation, then the electron power density is as shown in Fig. 2. In Fig. 2, we have included the effects of a Gaussian spread in the initial $n_{\parallel}$: e.g., G(1.93, 0.1) represents a Gaussian with mean 1.93 and a deviation of 0.1.

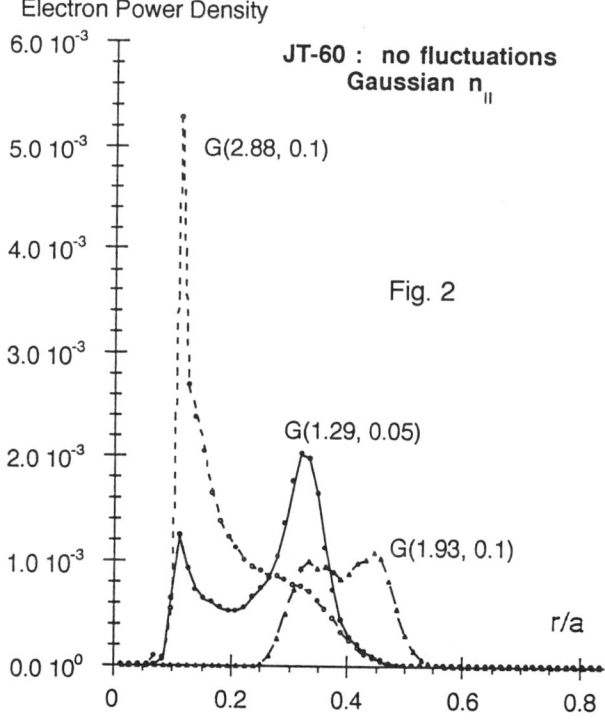

Fig. 2

Even allowing for some uncertainty in the experimental radial positions, there is little correlation between these results for no fluctuations and Fig. 1. Note in particular, the very strong internal peak for $n_{\parallel} = 2.88$, which is absent in the experimental data.

However, If one includes a 10% density fluctuation level (peaked at the plasma edge) in the wave scattering, then the absorbed electron power density is in very good agreement with Fig. 1. This result holds not only for δ-function initial $n_{\parallel}$, as reported in Ref. 2, but also for Gaussian $n_{\parallel}$, Fig. 3. Morevoer, these results are not sensitive to the poloidal launch angle. In Figs 3 & 4, we show the power deposition for two different launch angles : θ = 0° (Fig. 3) and θ = - 90° (Fig. 4)

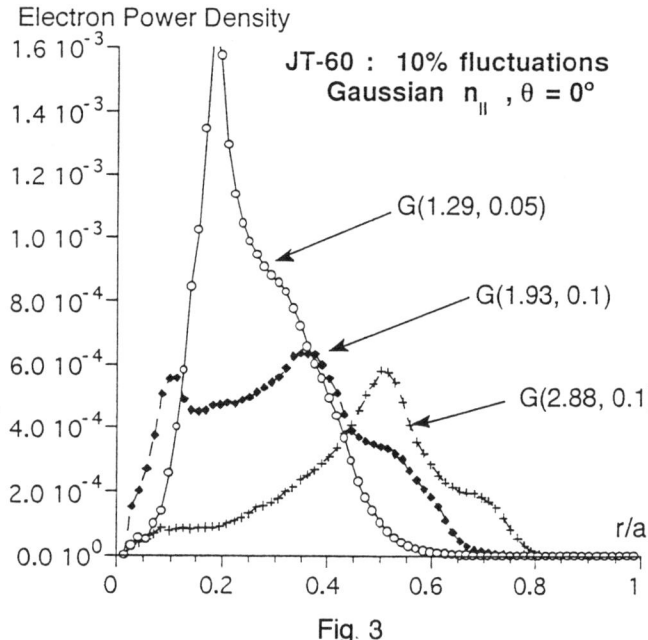

Fig. 3

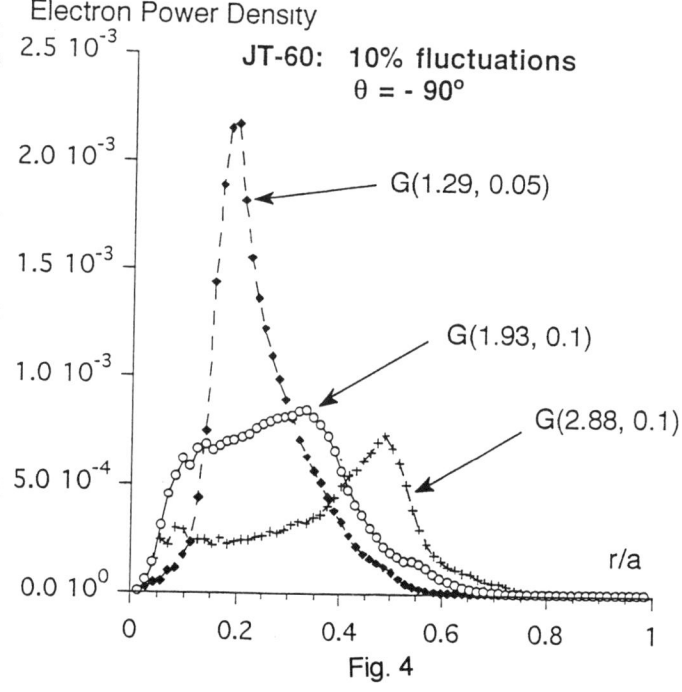

Fig. 4

ALCATOR C-MOD

It is of some interest to predict the analogous signals for Alcator C-Mod parameters, with LH wave frequency of 4.5 GHz. If one allows for 1% density fluctuations peaked at the edge [e.g., running C-Mod in a divertor mode], then the power density profiles are as shown in Fig. 5.

However, if the density fluctuation level is 10%, then the corresponding profiles are shown in Fig. 6.

From our results on JT-60, we thus expect that if hard X-ray profiles are measured on Alcator C-Mod, one could discern information on the edge density fluctuation levels.

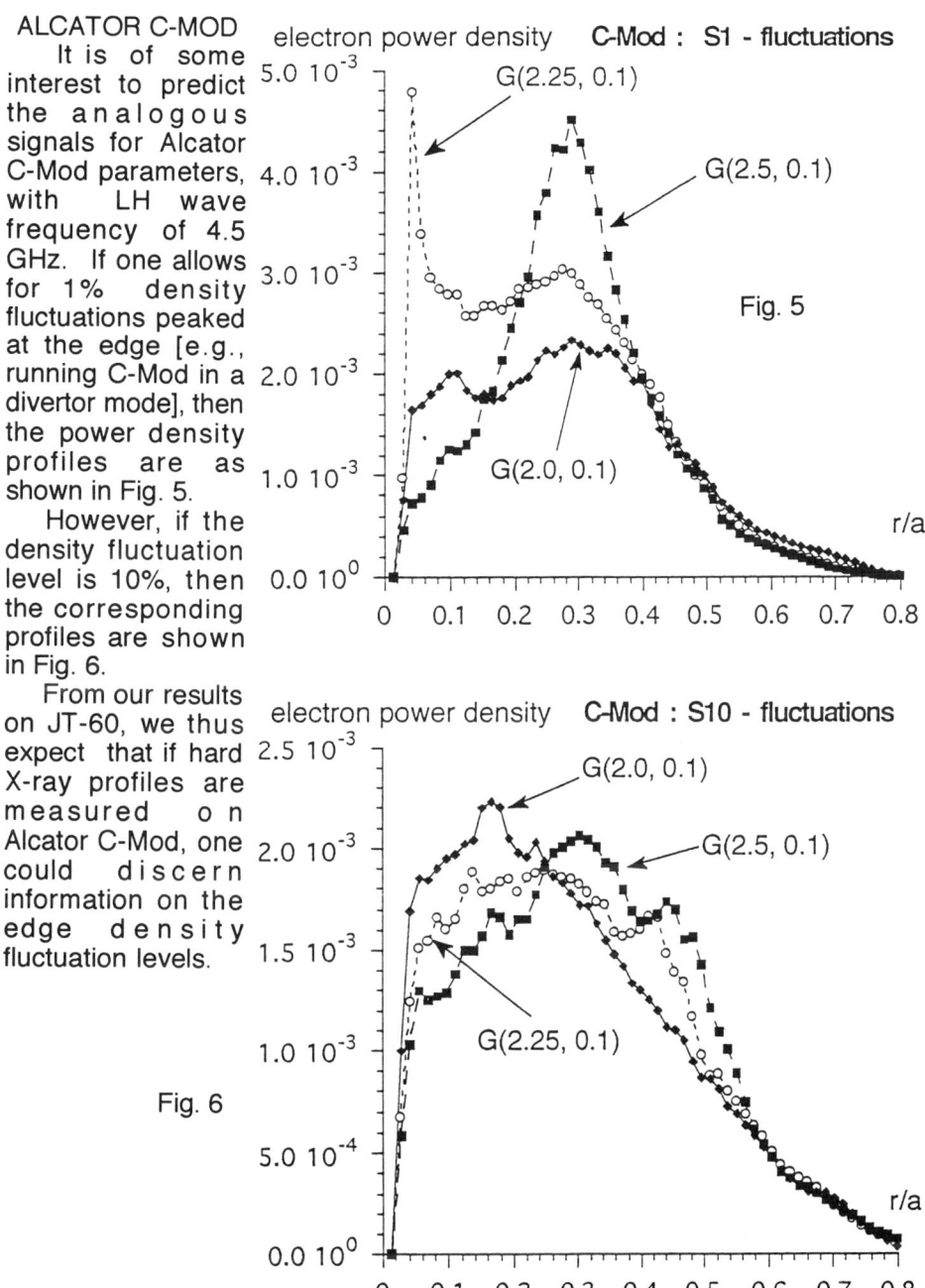

Fig. 5

Fig. 6

[1] P. T. Bonoli and E. Ott, Phys. Fluids **25**, 359 (1982)
[2] G. Vahala, L. Vahala and P. T. Bonoli, Phys. Fluids **B4**, 4033 (1992)

CURRENT DRIVE WITH THE SECOND ECR HARMONIC ON T–10

V. V. Alikaev, A. A. Bagdasarov, A. A. Borshegovskij, M. M. Dremin,
Yu. V. Esipchuk, Yu. A. Gorelov, N. V. Ivanov, A. Y. Kislov,
L. K. Kuznetsova, G. E. Notkin, Yu. D. Pavlov, K. A. Razumova,
I. N. Roy, N. L. Vasin, V. A. Vershkov, and the T–10 Team
Russian Research Center (Kurchatov Institute), Moscow, Russia

C. B. Forest, J. Lohr, T. C. Luce, and R. W. Harvey
General Atomics, San Diego, California, U.S.A.

ABSTRACT

The experiments on ECCD on the second harmonic were done. Current about 35 kA was generated. The efficiency of ECCD and its dependencies on plasma parameters were measured. Not all observed phenomena may be explained by the predictions of linear theory.

ECR current drive at the fundamental harmonic was observed in tokamaks[1,2] with efficiency η near that predicted by linear theory:

$$\eta_{\rm CD} \equiv \frac{I_{\rm CD}}{P_{\rm ab}} = \frac{T_e}{n_e(Z_{\rm eff}+5)} \cdot \langle \xi \rangle \qquad (1)$$

Here $I_{\rm CD}$ is generated current, $P_{\rm ab}$ absorbed power, T_e, n_e electron temperature and density, $Z_{\rm eff}$ effective plasma charge, and $\langle \xi \rangle$ a coefficient depending on the HFCD generation method. For experiments with high plasma density, the EC second harmonic CD is preferable. The mechanisms of the plasma HF power absorption for the fundamental and the second harmonic are different, as wave damping is proportional to $V_\parallel V_\perp$ and $V_\perp^2$, respectively. Analytic estimation gives a two times higher value for $\eta_{\rm CD}$ for the fundamental harmonic than for the second.

In T–10 experiments it was necessary
(1) to demonstrate the fact of ECCD at the second harmonic,
(2) to measure the value of $\eta_{\rm CD}$,
(3) to find its dependence on n_e, T_e, $P_{\rm ab}$, and resonance position inside the plasma.

Four gyrotrons with frequency $f = 140$ GHz and $P_{\rm HF} \leq 1.2$ MW were used. A metallic mirror turned the waves with a 24 deg angle to the major radius R at the outermost flux surface. $I_{\rm CD}$ was determined by comparison of loop voltages registered in two opposite directions of the plasma current for a constant direction of $I_{\rm CD}$ (co- and counter- experiments). The analysis assumed that the OH regimes are identical, which was verified experimentally, and that $I_{\rm CD}$ does not depend on co- and counter-directions (this may not be valid when $j_{\rm CD}$ is high and changes the MHD plasma structure). The model calculations were made using experimental profiles of plasma parameters and the $j_{\rm CD}(r)$ was chosen in a form which reproduces the measured time behavior of experimental loop voltage $U_\ell(t)$ and internal inductance $\ell_i(t)$ in both directions. Figure 1 demonstrates the difference in (a) U_ℓ and (b) ℓ_i behavior in co- and counter-experiments. The ℓ_i increase is due to $j(0)$ increasing during co-ECCD in the central part of the plasma. Sawtooth fluctuations were generated in this case and were absent in the counter case. P_{ab} was measured from the derivative of plasma diamagnetism at the start of the EC pulse and from the derivative of $\beta_p + \ell_i/2$ (inferred from the control fields for plasma equilibrium) at the same time. Simultaneously the injected HF power (P_{HF}) was measured calorimetrically at the ceramic window. Taking into account that P_{HF} increases 15% to 20% during the HF pulse, we can say that P_{ab} was about $0.8\, P_{HF}$.

Figure 2 demonstrates the dependence of I_{CD} and normalized $\hat{\eta} = \eta_{CD} \cdot (n_e/T_e)$ on $n_e(0)$ when the "cold" resonance is near the plasma center. $\hat{\eta}$ does not depend on $n_e(0)$ when $n_e(0) \geq 1.5 \times 10^{13}\,\mathrm{cm}^{-3}$ and decreases when $n_e(0)$ decreases. The same was found with the fundamental harmonic experiments. We suggest that this is connected with the comparable values of the collision time for resonance electrons τ_{col} and their confinement time τ_e. In this case the coefficient $1/[1+(\tau_{col}/\tau_e)]$ must be used in relation (1).

Figure 3 represents the I_{CD} and $\hat{\eta}$ dependence on P_{ab}. Note that in accordance with linear theory $\hat{\eta}(P) = $ const, however an increase of $\hat{\eta}$ with power is clearly indicated.

In Fig. 4 one can see the dependence of P_{ab}, $T_e(0)$, I_{CD}, and $\hat{\eta}$ on the B_z value, i.e., on the "cold" resonance position. At $B_z < 2.30$ T, the position of "relativistic cutoff" (≈ 14 cm from the cold resonance position) is located inside the plasma axis

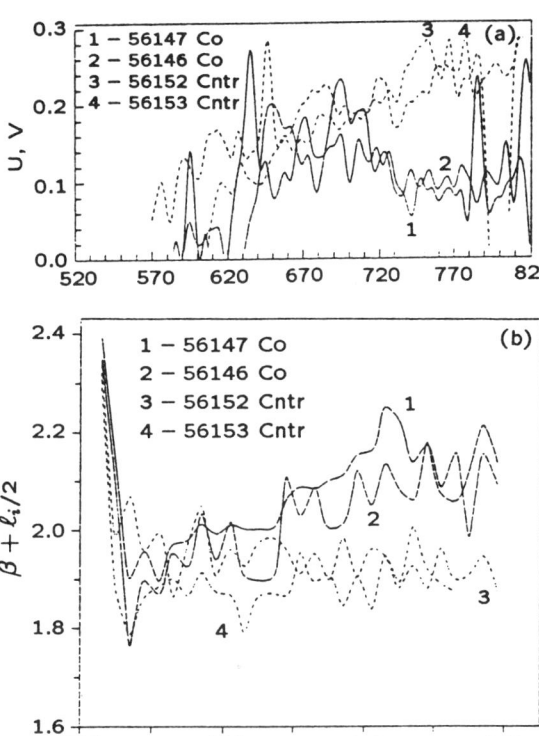

FIG. 1. (a) U_p behavior during ECCD for co- and counter- I_p directions. (b) The same for internal inductance ℓ_i. $I = 75$ kA, $n_e(0) = 1.7 \times 10^{13}\,\mathrm{cm}^{-3}$, $B_z = 2.47$ T, $P_{HF} = 0.6$ MW.

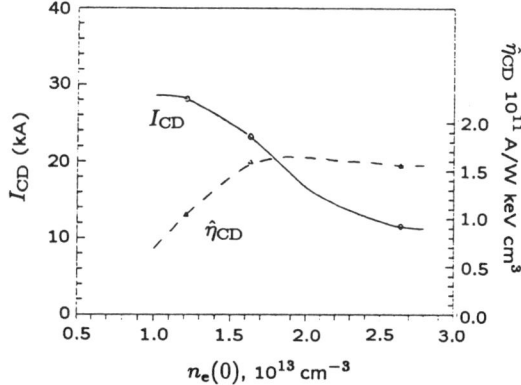

FIG. 2. Dependence of I_{CD} and $\hat{\eta} = (I_{CD}/P_{ab}) \times [n_e(0)/T_e(0)]$ on $n_e(0)$. $I_p = 140$ kA, $B_z = 2.48$ T.

therefore, the hot central part of the plasma cannot take part in the wave absorption and P_{ab} decreases. However, the P_{ab} decrease at the cold resonance position in the outer part of the plasma ($\Delta R = +6$ cm) cannot be so simply explained. Figure 4 shows that $\hat{\eta}_{max}$ occurs at $\Delta R = -12$ cm. Here, $T_e(0)$ and $n_e(0)$ were used for η normalization (dashed curve). The solid curve corresponds to the "cold" resonance T_e, n_e normalization. This may be explained by the increasing role in absorption of the tail electrons with high $V_\parallel$ values (due to wave-plasma interaction in its center). Suprathermal SXR emission from the plasma center also increases with inside resonance position displacement of up to $\Delta R = -12$ cm. The features of the P_{ab} distribution can be seen from Fig. 5. The derivative dT_e/dt at the end of the ECH pulse is proportional to the local HF-power deposition [$n_e(r)$ changes slowly in the central region]. Figure 5(a) demonstrates dT_e/dt determined from the SXR pinhole chamber for two pulses: (1) $T_e(0) = 3$ keV, (2) $T_e(0) = 2$ keV. The $dT(r)/dt$ profile shows that for $T_e(0) = 2$ keV an appreciable amount of HF power is absorbed behind the cold resonance position. Figure 5(b) represents the results of ray tracing calculations for three different $T_e(0)$ values: (1) $T_e(0) = 3$ keV, (2) $T_e(0) = 2$ keV, and (3) $T_e(0) = 1$ keV. For all the cases, the absorption is nearly full before the resonance. So the question is why the experimental dependence of absorption on resonance position

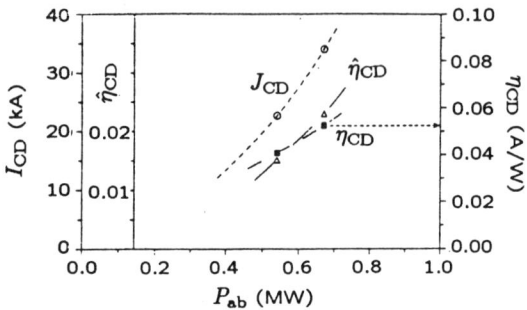

FIG. 3. I_{CD} dependence on P_{ab}. $I_p = 75$ kA, $n_e(0) = 1.7 \times 10^{13}$ cm^{-3}, $B_z = 2.48$ T.

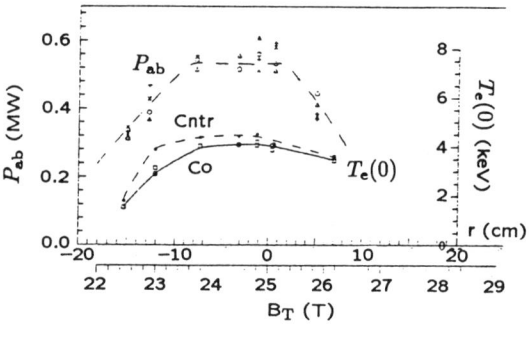

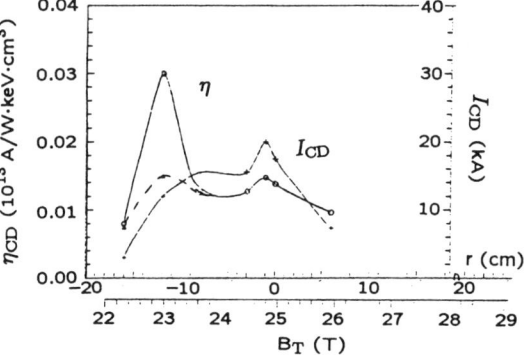

FIG. 4. Dependence of P_{ab}, $T_e(0)$, I_{CD}, and $\hat{\eta}$ on B_z. $I_p = 140$ kA, $n_e(0) = 1.7 \times 10^{13}$ cm^{-3}. Solid curve corresponds to η normalization on T_e and n_e for cold resonance position; dashed curve represents $T_e(0)$, $n_e(0)$ normalization.

$\Delta R = +6$ cm is lower than the calculation predicts (Figs. 4 and 5).

As was expected, the absolute value of measured η_{CD} with central ECCD is less for the second harmonic experiment than for the fundamental case by approximately a factor of 1.6.

SUMMARY

1. The possibility of second harmonic ECCD has been demonstrated. $I_{CD} \approx 35$ kA was achieved.
2. Second harmonic efficiency is about one-half of the fundamental efficiency for I_{CD} generation near the discharge center.
3. Second harmonic η_{CD} dependence on absorbed power is not in accordance with linear theory predictions.
4. The results permit planning of experiments with high n_e and therefore high β using $j(r)$ profiling with the second harmonic ECCD.

ACKNOWLEDGMENT

This is a report of work sponsored in part by the U.S. Department of Energy under Contract DE-AC03-89ER51114.

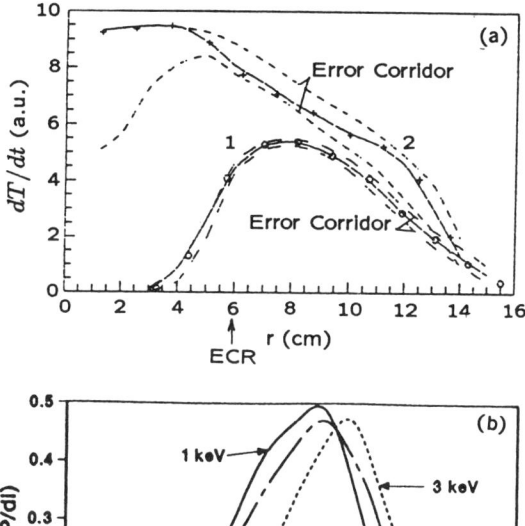

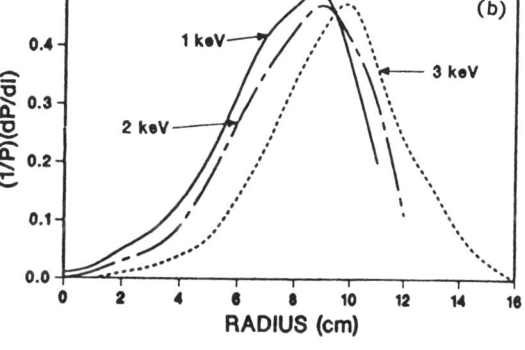

FIG. 5. (a) The derivative dT/dt at end of ECH pulse for two experiments: (1) $T_e(0) = 3$ keV, (2) $T_e(0) = 2$ keV. Arbitrary units differ for each curve. $I_p = 140$ kA, $n_e(0) = 1.7 \times 10^{13}$ cm^{-3}, $B_z = 2.6$ T ($\Delta R = +6$ cm). (b) $P_{ab}(r)$ profile from ray tracing calculations. (1) $T_e(0) = 3$ keV, (2) $T_e(0) = 2$ keV, (3) $T_e(0) = 1$ keV. $T_e(r)$ and $n_e(r)$ are the same as for (a).

REFERENCES

1. V. V. Alikaev et al., Nucl. Fusion **32**, 1881 (1992).
2. T. C. Luce et al., in *Plasma Physics and Controlled Nuclear Fusion Research*, IAEA, Vienna (1991), Vol. 1, p. 631.

INDUCTIVE EFFECTS DURING SECOND HARMONIC CURRENT DRIVE EXPERIMENTS ON T–10

C.B. FOREST, R.W. HARVEY, JOHN LOHR, T.C. LUCE, and Y.R. LIN-LIU
General Atomics, San Diego, CA 92186-9784

YU. ESIPCHUK, G. NOTKIN, K. RAZUMOVA, and THE T-10 TEAM
Russian Scientific Center, Kurchatov Institute, Moscow, Russia

ABSTRACT

Current drive during second harmonic, electron cyclotron heating experiments performed on the T–10 tokamak have been simulated with the ONETWO transport code to determine the effects of induction on the time evolution of the loop voltage and current density profile. Ray tracing shows the well focused rf power can generate centrally peaked current densities which exceed the Ohmic current densities by a factor of five, causing very peaked plasma current profiles which will be unstable to sawteeth. A Kadomtsev model of the sawtooth shows that a limit cycle is quickly reached which maintains a broad current profile and requires generation of a negative dc component of the loop voltage localized near the magnetic axis. This negative electric field effectively reduces the measured current drive efficiency. A broader profile of driven current, as in the fundamental current drive experiment on T–10, would not suffer this effect.

INTRODUCTION

Recent experiments performed on the T–10 tokamak have shown the existence of an rf driven current during second harmonic electron cyclotron heating [see the paper by V.V. Alikaev et al., this conference]. The efficiency of current drive, as determined by differences in loop voltage measurements during co- and counter-rf current drive discharges, was lower than that of similar experiments performed at the first harmonic [V.V. Alikaev et al., Nucl. Fusion **32**, 1811 (1991)] where results agreed with theoretical predictions. If it is assumed the absorption and current generation mechanisms are similar at first and second harmonic, the primary differences between these experiments are related to the difference in rf frequency (81 GHz at first harmonic and 140 GHz for the second harmonic). The launching structures were similar for each experiment; however, the angular divergence (approximately ±1.5 deg) of the injected rf power for the second harmonic experiment was approximately half that of the first harmonic case. For centrally peaked damping, the width of the power deposition profile as estimated from the Doppler resonance condition is $\Delta R \approx v_t R_0 \Delta k_\parallel / \omega_{rf}$; thus, second harmonic power density should be a factor of four greater than in the first harmonic case.

The tokamak transport code ONETWO coupled with the ray tracing package TORAY has been used to analyze the evolution of the current profile for the second harmonic T–10 experiment. Time-dependent profiles of electron density and temperature are estimated from experimental data and are used for calculations. In particular, the tokamak equilibrium is evolved self-consistently with the pressure profile (determined by experimental data) and current profile (determined by TORAY and the Ampère-Faraday equations with neoclassical resistivity with fixed boundary). A Kadomtsev sawtooth model is also included which redistributes current and poloidal flux when various trigger criteria are met {e.g., q_0 decreasing below a predetermined value [W. Pfeiffer et al., Nucl. Fusion **25**, 655 (1985)]}.

MODELING OF EXPERIMENT

Due to the strong localization of rf power, the rf driven current density at the center of the plasma is predicted to be a factor of 5 larger than the Ohmic current density. Figure 1 shows electron density and temperature profiles before and after the application of 650 kW of rf power to an 80 kA T–10 discharge. Also shown in Fig. 1 is the rf power deposition profile as determined by a TORAY calculation for a bundle of rays launched obliquely with an angle of 60 degrees to the magnetic axis with a Gaussian power profile of half-power width ±1.5 degrees. During the rf phase, the magnetic axis of the plasma shifts from 151 to 155 cm, which places the cold electron cyclotron resonance at a major radius 7 cm smaller than the major radius of the magnetic axis, resulting in centrally peaked power deposition. Figure 2 shows the Ohmic current profile, and the calculated ECCD current profile as determined by a model including trapping [R.H. Cohen, Phys. Fluids **30**, 2442 (1987) and Phys. Fluids **31**, 421 (1987)].

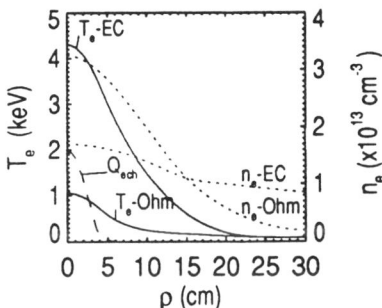

FIG. 1. Electron temperature and density profiles before and during ECH.

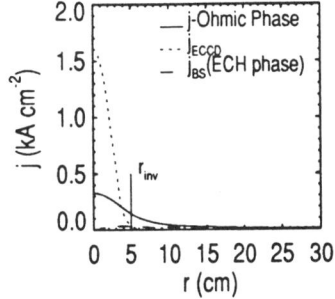

FIG. 2. Profiles of Ohmic current before ECH, driven electron cyclotron current, and bootstrap current during ECH. Also shown is sawtooth inversion radius used in simulation.

The value Z_{eff} is determined by matching the loop voltage during the Ohmic phase and is assumed constant throughout the ECH phase (for this discharge $Z_{eff} \simeq 2$). The simulation begins in the Ohmic phase, and the initial equilibrium has a current profile consistent with the electron temperature profile. The temperature and density profiles change during the switching-on period of the rf power. The rapid heating of the plasma and the changing conductivity profile can strongly affect the loop voltage measured on the boundary of the plasma and the internal parallel electric field.

Agreement with theory is determined by comparing the measured loop voltage and internal inductance with the predictions of ONETWO. The simulation has been performed for co-current drive with and without sawteeth, and without current drive (to show effects of changing conductivity profile). Distinctly different results were obtained with and without sawteeth. In particular, without sawteeth the quantity ℓ_i was approximately twice the measured value and the $q(0) \approx 0.2$ due to the extremely peaked current current density profile. Figure 3 shows experimental data and simulation values for the loop voltage and Fig. 4 for $\beta_p + \ell_i/2$. For both parameters, the agreement is within 10% if the sawtooth model is included. The initial response of the experimental loop voltage is an artifact of the T–10 vertical position feedback circuit responding to the initial outward shift of the plasma and should be ignored. For

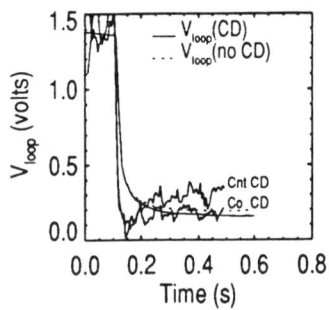

FIG. 3. The experimental and simulated loop voltages (with sawteeth) both with and without ECCD.

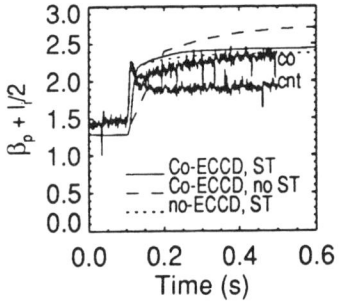

FIG. 4. The experimental values of $\beta_p + \ell_i/2$ for co- and counter-current drive, and simulated values with sawteeth (with and without current drive) and without sawteeth.

of the plasma and should be ignored. For this case, the ONETWO simulation shows that the total current consists of 30 kA of ECCD, 20 kA of bootstrap current and 20 kA of Ohmic current. In spite of a relatively large rf driven current, the difference in loop voltage with and without current drive is small. This is a result of the sawteeth redistributing the current.

SAWTOOTH LIMIT CYCLE

The conditions in this experiment were unique in that local rf-generated current density on axis could be many times the current density in the Ohmic phase. The simulation shows that, as the current profile begins to peak, a large negative electric field is generated in the central region of the plasma to slow the rise of central current density — a result of Lenz's law. In the absence of sawteeth the simulation shows that the loop voltage on the magnetic axis asymptotically approaches the value on the boundary of the plasma on a magnetic diffusion time scale. With sawteeth, a different effect is observed.

In the Kadomtsev model of the sawtooth cycle, the current profile immediately following the sawtooth crash is such that the value of $q_0 \sim 1$. As the current peaks (q_0 decays) the profile becomes unstable to sawteeth. This is numerically determined by the $q = 1$ surface reaching a minor radius corresponding to the sawtooth inversion radius at which time the sawtooth occurs, redistributing the observed plasma current and raising q_0 to 1. With rf current drive, this paradigm remains valid with the additional effects of a peaked rf current source and Lenz's law generating a strong back emf. In this case we note that the cycle must include a negative dc electric field in the region near the magnetic axis to oppose the large rf current density. Immediately following the sawtooth crash an electric field will appear as the rf driven current is generated, to maintain the current density profile. As this large negative electric field begins to decay, the current density begins to peak until the profile becomes unstable to the sawtooth instability. Figure 5 shows the calculated evolution of the electric field profile interior to the plasma. In contrast with the case without sawteeth, the time scale on which the plasma evolves to a quasi-steady state is much shorter because of the redistribution of current by the sawtooth.

Figure 6 shows the time evolution of the simulated loop voltage on the boundary of the plasma and ECE from a central resonance showing the onset of sawteeth. Note that the loop voltage is decaying with time before the sawteeth and quickly reaches a steady-state when the sawtoothing starts.

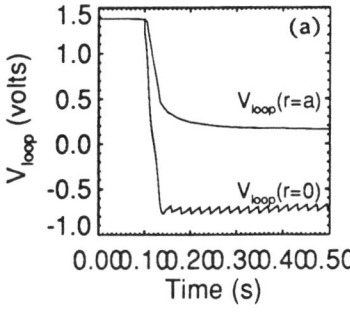
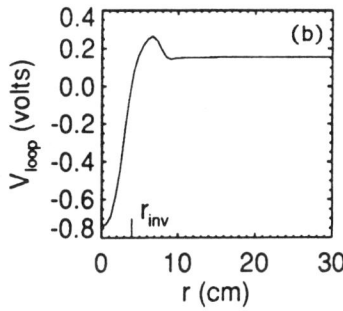

FIG. 5. (a) Calculated loop voltage evolution at the plasma boundary and at the magnetic axis with sawteeth. (b) $V_{\text{loop}}(r)$ immediately following sawtooth crash.

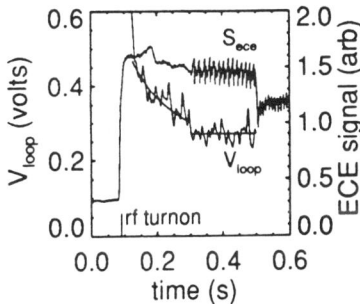

FIG. 6. Evolution of loop voltage and ECE showing onset of sawteeth and quasi-steady state loop voltage.

Quasilinear Fokker-Planck calculations have shown that negative electric fields strongly degrade the net current drive efficiency [see Harvey et al., this Proceedings]. With sawteeth, the rf driven current would thus be reduced due to the negative axial loop voltage; the self-consistent negative loop voltage would also be reduced. This emphasizes the importance of correctly estimating the effects of rf enhanced (or degraded) conductivity when evaluating current drive efficiencies.

CONCLUSIONS

The T-10 discharge is in a unique regime in which the rf-driven current density on axis can greatly exceed the current density of the Ohmic phase. For these discharges the ECCD as estimated by measurement of the loop voltage on the boundary of the plasma can underestimate the correct level of ECCD since much of the current can be cancelled by a sawtooth generated, negative dc loop voltage near the magnetic axis. Furthermore, the effects of rf enhanced (or degraded) conductivity require a self-consistent calculation to accurately determine the real value of ECCD. The close numerical agreement between experimental data and simulations with sawteeth suggest that the differences between first and second harmonic ECCD can be explained by the peaked deposition profile of the second harmonic experiments. Future experiments are planned to examine the effect of spreading of the EC beam. The results emphasize the need to match non-inductive current drive profiles to profiles which are stable to MHD instabilities if the full efficiency of the current drive method is to be realized.

ACKNOWLEDGMENT

This is a report of work sponsored by the U.S. Department of Energy under Contract No. DE-AC03-89ER51114.

On rf "Helicity Injection" and Alfvén Wave Current Drive

C. Litwin and N. Hershkowitz

Department of Nuclear Engineering and Engineering Physics
University of Wisconsin-Madison

We discuss the dynamo effect due to low-frequency waves in connection with recently proposed current-drive schemes. We focus on nonresonant thermal effects and discuss the role of wave helicity in two-fluid theory.

INTRODUCTION

Recently, Ohkawa [1] proposed to drive current by augmenting the plasma helicity via injection of helical Alfvén waves. Following this suggestion, Mett and Tataronis [2] and Taylor [3] showed that current can be driven in a nonresonant, ohmic-like manner by the dynamo "electric" field $\Xi \equiv \langle \tilde{u} \times \tilde{B}\rangle/c$ where $\tilde{u}$ and $\tilde{B}$ are fluctuating mass flow velocity and magnetic field, and $\langle \rangle$ denotes the time average. If the wave gives rise to the α-effect, i.e., if $\Xi_\parallel \neq 0$ (subscript refers to the direction of the mean magnetic field B_0), it can generate a steady current [3] which can be determined from the mean parallel Ohm's law

$$E_\parallel + \Xi_\parallel = \eta J_\parallel \qquad (1)$$

where η denotes the resistivity.

This nonresonant current drive mechanism is not limited by trapped particle effects [5, 6] since it acts on all electrons equally. However, the dynamo field of the wave is small, as it is proportional to resistivity. For plane Alfvén waves the dynamo field is given by

$$\Xi_\parallel = \frac{k^2 D_R}{2 k_\parallel c} \frac{i \tilde{B}^* \times \tilde{B} \cdot b_0}{B_0} \qquad (2)$$

where $D_R = \eta c^2/4\pi$ and $b_0 = B_0/B_0$. The resulting loop voltage modification is much smaller than typical Ohmic values for realistic wave amplitudes.

Thus, a crucial question is whether a larger dynamo effect can be found outside the MHD framework. It has been the subject of several past investigations [7, 8, 9, 10]. Of particular interest is the question of nonresonant thermal effects that have been reported [8, 9] to produce a collisionless α-effect. In this paper, we first review the collisional two-fluid picture of the dynamo effect emphasizing the role of wave helicity. Next, we discuss the discrepancy between results of the double-adiabatic approach [10] and the results reported in refs. [8] and [9].

COLLISIONAL TWO-FLUID DYNAMO

In two-fluid theory, plasma dynamics averaged over the oscillation time scale is governed by the momentum balance equations modified by ponderomotive forces [11, 12]. From these two-fluid equations, one-fluid theory can be derived in the standard manner. The parallel mean-field Ohm's law then has form (1) with $\Xi = f/en_0$ where n_0 and f denotes, respectively, mean density and electron ponderomotive force density. We omit the small contribution of the ion ponderomotive force.

In collisionless cold-fluid theory the ponderomotive force density has the form [11]

$$\mathbf{f} = -n_0 \nabla \Phi + \mathbf{B}_0 \times \nabla \times \mathbf{M} \quad (3)$$

where Φ is the ponderomotive potential and $\mathbf{M}$ is the rf induced magnetization. The parallel dynamo field due to the ponderomotive force, $\Xi_\parallel = -\nabla_\parallel \Phi/e$, vanishes upon averaging over the flux surface. Thus the ponderomotive force, in this limit, does not induce any change in the loop voltage [13].

Ponderomotive force is modified by pressure and collisional effects [11, 12]. Let us first consider the case of isotropic pressure. We shall restrict our attention to relations between the density n and pressure p of the form $n = n(p)$ which includes, in particular, isothermal (considered by Lee and Parks [12]), adiabatic and polytropic equations of state.

For such equations of state, the dynamo field has the form

$$\Xi = \frac{1}{c}\tilde{\mathbf{v}} \times \tilde{\mathbf{B}} + \frac{m}{e}\tilde{\mathbf{v}} \cdot \nabla \tilde{\mathbf{v}}$$
$$- \nabla \left(\frac{\tilde{p}^2}{2e} \frac{dn_0^{-1}}{dp_0} \right) \quad (4)$$

where $\tilde{\mathbf{v}}$ is the electron flow velocity. Note that pressure enters the above equation as the gradient of a scalar that modifies the ponderomotive potential. Also, the assumed equation of state does not modify the relation between $\tilde{\mathbf{B}}$ and $\tilde{\mathbf{v}}$ that follows from the linearized momentum balance equation and Faraday's law. Therefore, Ξ, expressed in terms of fluctuating velocities (rather than in terms of oscillatory electric field as in ref. [12]), has the same form as in the cold plasma theory [11] plus a term due to collisions. As before, the parallel ponderomotive force contribution vanishes upon averaging over the flux surface so that the flux-surface averaged parallel dynamo field becomes

$$\bar{\Xi}_\parallel = -\frac{1}{2e\omega}\text{Re}\left(i\tilde{\mathbf{v}}^* \times \nabla \times \tilde{\mathbf{C}} \cdot \mathbf{b}_0\right) \quad (5)$$

where $\mathbf{C}$ is the collisional momentum loss rate and the star denotes complex conjugation. Thus the dynamo field vanishes in the absence of collisions.

In the case of collisions with the neutral gas the collision operator can be taken in form $\mathbf{C} = -\nu m \mathbf{v}$. Then, for plane waves such that $\mathbf{k} \cdot \mathbf{v} = 0$ (and, in particular, for Alfvén waves), the parallel component of the dynamo field becomes

$$\bar{\Xi}_\parallel = \frac{\nu}{2e\omega} m |\tilde{v}|^2 k_\parallel \quad (6)$$

Observe that Eq. (6), unlike Eq. (2), predicts that even linearly polarized, zero-helicity waves give rise to the α-effect. This discrepancy is caused by different momentum loss mechanisms. In the fully ionized plasma the momentum loss due to Coulomb collisions with ions is more correctly described by $\mathbf{C} = e\eta \mathbf{J}$. Then, with the aid of Ampère's law and neglecting the displacement current, Eq. (5) becomes

$$\bar{\Xi}_\parallel = \frac{\eta c}{2\pi\omega}\text{Re}\left(i\tilde{\mathbf{v}}^* \times \nabla^2 \tilde{\mathbf{B}}\right) \cdot \mathbf{b}_0 \quad (7)$$

For low-frequency oscillations, with $\tilde{E}_\parallel = 0$, electrons drift with $E \times B$ velocity which for plane waves reduces Eq. (7) to the MHD result, Eq. (2).

Thus in the case of electron-ion collisions and $\tilde{E}_\parallel = 0$, only waves with nonvanishing helicity give rise to the α-effect. The induced current magnitude is the same but its direction opposite to the current obtained from the naïve helicity conservation. The coincidence of current magnitudes disappears if viscosity is taken into account, as first pointed out by Taylor [3], although the nonvanishing wave helicity is still required to produce the α-effect.

If $\tilde{E}_\parallel$ and finite frequency effects are retained, even the requirement of nonvanishing helicity is relaxed. In fact, in two-fluid theory, not only magnetic but also purely electrostatic fluctuations can produce the α-effect. Retaining the ion polarization drift we find that the dynamo field of a plane electrostatic wave has the form

$$\bar{\Xi}_\parallel = -\frac{\omega_{ci}/\omega}{1-(\omega/\omega_{ci})^2}\frac{k_\parallel c D_R}{2v_A^2}\frac{\tilde{E}_\perp^2}{B_0} \quad (8)$$

Thus electrostatic waves propagating obliquely with respect to the magnetic field produce a nonvanishing parallel dynamo field and can drive current.

DYNAMO IN DOUBLE-ADIABATIC THEORY

The isotropic pressure assumption is valid when the electron isotropization time is shorter that the oscillation period which in general is not the case in experiments. In fact, usually the contrary is true. Therefore the opposite, collisionless limit is more applicable and it is appropriate to describe electrons in the double-adiabatic approximation [14]. This approach should yield the same result as the nonresonant limit of kinetic theory considered in refs. [8, 9].

To lowest order in the electron Larmor radius, the electron pressure tensor takes the form $\mathbf{P} = p_\perp \mathbf{I} + (p_\parallel - p_\perp)\mathbf{bb}$ where $\mathbf{b} = \mathbf{B}/B$ is the unit vector along the total magnetic field (including the field of fluctuations). Elements $p_\perp$ and $p_\parallel$ of the pressure tensor satisfy Chew-Goldberger-Low equations of state [14]:

$$\frac{d}{dt}\left(\frac{p_\perp^2 p_\parallel}{n^5}\right) = 0 \quad (9)$$

$$\frac{d}{dt}\left(\frac{p_\perp}{nB}\right) = 0 \quad (10)$$

The mean force density on electrons, $\mathbf{f}$ is again obtained by averaging the momentum balance equation over time. By neglecting a small term due to the electron inertia and for frequencies high compared to the electron diamagnetic frequency it can be expressed as the sum $\mathbf{f} = \mathbf{f}_{em} + \mathbf{f}_{nl}$ where

$$\mathbf{f}_{em} = -e\left\langle \tilde{n}\tilde{\mathbf{E}} + \frac{1}{c}\tilde{n}\tilde{\mathbf{v}}\times\mathbf{B}_0 \right.$$
$$\left. + \frac{1}{c}n_0\tilde{\mathbf{v}}\times\tilde{\mathbf{B}} \right\rangle \quad (11)$$

is the mean electromagnetic force and

$$\mathbf{f}_{nl} = -\nabla \cdot \mathbf{P}_{nl} \quad (12)$$

is the pressure force is due to the nonlinear modification of the mean distribution function by electromagnetic oscillations [15]. Fukuyama et al. [8] showed that the nonresonant, flux-surface averaged, contribution to $\mathbf{f}_{em}$ is given by

$$\bar{\mathbf{f}}_{em} = -\mathbf{k}\nabla\cdot\left[\frac{\partial \chi_{ij}^H}{\partial \mathbf{k}}\frac{\tilde{E}_i^* \tilde{E}_j}{16\pi}\right] \quad (13)$$

where χ^H is the hermitian part of the susceptibility tensor. Thus, if $k_\parallel \neq 0$ and the wave field is inhomogeneous, the spatial dispersion due to thermal effects would lead to the α-effect [8], were the contribution of $\mathbf{f}_{nl}$ omitted as in refs. [8] and [9]. However, this omission is unjustified: retaining this term leads to drastically different conclusions [10].

To simplify the problem we shall assume that the time-averaged pressure is both uniform and isotropic and that both density and magnetic field profiles are constant: $p_{\perp 0} = p_{\parallel 0} = p_0 = const.$,

$n_0 = const.$ and $\mathbf{B}_0 = const.$. The oscillating wave field is, however, assumed inhomogeneous, e.g., as in a waveguide, and the fluctuating pressure remains anisotropic, consistent with Eqs. (9) and (10).

With the above simplifications, the contribution of nonlinear pressure modification to the parallel ponderomotive force becomes

$$f_{nl\parallel} = \nabla \cdot \left\langle \tilde{\mathbf{B}}_\perp (\tilde{p}_\parallel - \tilde{p}_\perp) \right\rangle / B_0 \quad (14)$$

For small plasma β this contribution might appear small compared to, e.g., the third term on the rhs of Eq. (11), However, large cancellation takes place between the terms on the rhs of Eq. (11) as we discussed in the preceding section.

With the aid of Maxwell's equations and the linearized equation of continuity and motion one can eliminate the oscillatory electric and magnetic fields and density from Eqs. (11) and (14) and express f_{em} and f_{nl} in terms of fluctuating velocities only. The flux surface averages of these two contributions do not vanish separately. However their sum can be shown [10] to have form

$$\bar{\Xi}_\parallel = -\frac{1}{e}\nabla_\parallel \Phi \quad (15)$$

where Φ is a positive definite binomial in $\tilde{v}$. It vanishes upon integration over closed flux surfaces and therefore does not give rise to a net current.

Thus contrary to the results in Refs. [8] and [9] in which the nonlinear pressure contribution was neglected, we find that thermal effects do not give rise to the nonresonant α-effect, if the mean-state inhomogeneity length scale is much longer than the rf amplitude gradient length scale and the mean pressure is isotropic (similar to the assumptions made in ref. [8]). The omitted term cancels exactly the contribution of the time-averaged electromagnetic force that was computed in the above-mentioned references.

This research has been supported by U.S. Department of Energy grant no. DE-FG02-88ER53264.

REFERENCES

[1] T. Ohkawa, Comm. Plasma Phys. Contr. Fusion 12, 165 (1989)
[2] R.R. Mett and J.A. Tataronis, Phys. Rev. Lett. 63, 1380 (1989)
[3] J.B. Taylor, Phys. Rev. Lett. 63, 1384 (1989)
[4] H.K. Moffatt, *Magnetic Field Generation in Electrically Conducting Fluids*, (Cambridge University Press, Cambridge 1978)
[5] N.J. Fisch and C.F.F. Karney, Phys. Fluid 24, 27 (1981)
[6] D.A. Ehst and C.F.F. Karney, Phys. Fluids 31, 1933 (1991)
[7] V.S. Chan, R.L. Miller and T. Ohkawa, Phys. Fluids B 2, 944 (1990)
[8] A. Fukuyama, S.I. Itoh, K. Itoh et al., Proc. IAEA Conf. Contr. Thermonucl. Fusion 1990, p. 855
[9] P. Moroz, N. Hershkowitz and J.T. Tataronis, Proc. 1992 Int. Conf. Plasma Phys., p. 909
[10] C. Litwin, University of Wisconsin report PLR-92-12, to be published
[11] R. Klima, Czech. J. Phys. B 18, 1280 (1968)
[12] N.C. Lee, G.K. Parks, Phys. Fluids 26, 725 (1983)
[13] N. Fisch and C. Litwin, unpublished (1987)
[14] G.L. Chew, M.I. Goldberger and F.E. Low, Proc. Roy. Soc. Lond. A 236, 112, 1956
[15] D.A. D'Ippolito and J.R. Myra, Phys. Fluids 28, 1895 (1985)

MODELING OF RF-ASSISTED HELICITY INJECTION START-UP AND CURRENT DRIVE*

Y. S. Hwang and M. Ono
Princeton University, P. O. Box 451, Princeton, NJ 08543

ABSTRACT

DC-helicity injection has been successfully applied to the tokamak start-up on CDX-U and CCT tokamaks. Furthermore, CDX-U achieved plasma current levels of up to 10 kA from zero without any ohmic drive. Helicity balance and scaling from CDX-U experiments indicate that plasma current scales with the electron temperature as $T_e^{3/2}$ and it is relatively insensitive to the plasma density. RF heating methods such as electron cyclotron and fast wave heating can be utilized to increase driven plasma current by reducing plasma resistivity. Zero-dimensional modeling has been attempted to design non-inductive start-up and current drive scenarios by combining helicity injection and various RF heating schemes. Plasma current can be driven from zero current by helicity injection and equilibrium field control, and then maintained by helicity injection and bootstrap current with the help of appropriate RF heating to reduce plasma resistivity. Non-inductive start-up and current drive for a 1 MA plasma current in a low-aspect-ratio geometry have been predicted with this model. In a reactor regime, 20 MA of plasma current can be driven and maintained non-inductively with a relatively high efficiency as well.

INTRODUCTION

Steady-state or very long-pulse operation has been identified by the reactor designers as desirable for an attractive fusion reactor. Next generation devices, such as TPX and DEMO, are moving toward near steady-state operation. To achieve this operational scenario, non-inductive current drive methods are essential. Non-inductive start-up methods can also serve an important function during the tokamak start-up, in particular, to minimize the wall eddy current that is expected for the toroidally continuous vacuum vessel of these future devices. Various RF current drive and start-up experiments have been performed, but they still require further improvements. Recent results in the CDX-U dc-helicity injection(DC-HI) experiments[1,2] raise a possibility of combined DC-HI/RF heating for the non-inductive tokamak start-up and current drive. This paper reports possible scenarios by using a zero-dimensional model.

DC-HELICITY INJECTION AND SCALING

The high current emissive cathode has been previously used for breakdown and to inject helicity into the plasma. A preionization experiment on CCT has shown that hot electron injection reduces the loop voltage spike by factor of ten.[3] For a new generation tokamak with a toroidally continuous wall, this type of tokamak start-up may be attractive. Moreover, DC-HI using a hot emissive cathode formed tokamak discharges of up to 10 kA plasma current from zero current on CDX-U. The current multiplication factor defined as a ratio of driven current to injected current, a measure of current drive efficiency, remains relatively high at approximately 20, with an applied cathode voltage of about 400 volts. The power efficiency is also found to be quite high, about 30 % of ohmic drive in this experiment. The efficiency is also found to be insensitive to the plasma density, which is an encouraging result. Furthermore, the DC-HI driven plasma current scales strongly with the electron temperature. It is therefore predicted that an application of additional RF heating can increase plasma current through increased plasma temperature(or decreased plasma resistivity). Also, RF heating can generate high β_{pol} plasmas with larger bootstrap current.[4] Hence, the modeling shows that, by combining DC-HI and RF heating, it is possible to achieve completely non-inductive start-up and current drive with a high efficiency.

START-UP AND CURRENT DRIVE MODELING

Tokamak start-up is initiated with a hot emissive cathode as a preionizer. This cathode continues to be a helicity injection source, providing plasma current ramp-up together with an appropriate equilibrium field control. With given helicity injection efficiency, higher plasma current can be obtained by increased electron temperature. RF heating methods are considered for auxiliary heating to help current ramp-up and maintain plasma current in steady-state. The steady-state plasma current is maintained by DC-HI, bootstrap current and/or RF current drive.

To model this scenario, zero-dimensional calculations are performed. Loop voltage by equilibrium field, V_{EF}, is calculated approximately from Shafranov's formula as following:

$$V_{EF} \approx \frac{dB_v}{dt} 2\pi R \cdot 2a = a\mu_o(\beta_{pol} + \frac{l_i - 3}{2} + ln\frac{8R}{a})\frac{dI_p}{dt}$$

where B_v, I_p, l_i, R and a are vertical field, plasma current, plasma inductance, major and minor radii, respectively. During plasma ramp-up, consumed loop voltage, $L_p \frac{dI_p}{dt}$, is subtracted. Here, total plasma inductance is given by $L_p = R\mu_o(ln\frac{8R}{a} - 1.75)$. Effective loop voltage from helicity injection, $V_{HI} = \eta \frac{2\pi R}{\pi a^2} I_p$, is obtained from helicity balance,

$$V_{HI} = \epsilon I_{inj} V_{cath}/I_p$$

where ϵ, I_{inj} and V_{cath} are power efficiency, cathode injection current and cathode voltage, respectively. Here, η is the plasma resistivity, having strong dependency on plasma temperature. The bootstrap currents are calculated from the formula given by Wilson.[4] For RF current drive, ITER formula of figure-of-merit are used to estimate driven currents.

With the given time evolution of electron density, sufficient plasma temperatures for the given time evolution of plasma current are obtained. Here, DC-HI, bootstrap current and/or RF current drive are considered as current sources. Based on the results of CDX-U helicity injection experiment data, cathode injection currents are varied linearly with the electron density within the limit of electron emission capability. Also, toroidal beta β_T is kept less than Troyon limit.

MODELING RESULTS

As an example, start-up and current drive scenarios for a medium size low-aspect-ratio tokamak(R=0.8 m, a=0.57 m and B_T=0.5 T) are considered by using this non-inductive method. DC-helicity injection and the inductive loop voltage from the equilibrium field change initiate plasma current, then DC-HI and bootstrap current maintain a steady-state plasma current after reaching the flat top. To maintain the plasma current level of 1 MA(edge safety factor $q_a = 3$) with DC-HI efficiency at 30 % of ohmic current drive, plasma temperature should reach 1.3 keV, providing $\beta_T \approx 9.6\%$, $\beta_{pol} \approx 0.2$ and a bootstrap current fraction of 12 %. To achieve this plasma temperature, RF heating power of about 5 MW, in addition to the DC-HI power level of 1.2 MW, is required.

With a smaller DC-HI current drive efficiency, 15 % of ohmic current drive, either plasma temperature should be increased or an additional current drive should be applied. To keep under the toroidal beta limit, an additional current drive such as fast wave(FW) current drive is applied. FW current drive is especially advantageous both in wave accessibility and in providing seed currents on the magnetic axis. With 2 MW of FW current drive power, plasma temperature of 1.5 keV is required. In this case, $\beta_T \approx 11.5\%$, $\beta_{pol} \approx 0.24$ and bootstrap current fraction of 15 % are obtained. Figure 1(a) shows the evolution of various plasma current components with the given density and temperature evolution of Fig. 1(b)-(c). Required RF heating power with the same DC-HI power level is similarly determined to be 4 MW by solving a power balance equation for L-mode confinement scaling. Total required heating power is shown as a solid line in Fig. 1(d). Density of up to $0.3 \times 10^{20}/m^3$ can be accessible by electron cyclotron heating(ECH), and higher density is accessible by FW heating.

Another example shows the possibility of having a reactor grade plasma which is started and maintained only by non-inductive methods. Using an ITER-like plasma size with low-aspect-ratio(R=3.2 m, a=2.28 m and B_T=2.5 T), plasma current of 20 MA and edge safety factor of 3 can be driven by non-inductive methods. Plasma temperature of 6 keV and density of up to $1.5 \times 10^{20}/m^3$ can give a bootstrap current fraction of 12.5 % with $\beta_T \approx 11.2\%$ and $\beta_{pol} \approx .23$, when cathode injection power is increased with the plasma size. Higher aspect ratio devices with the same edge safety factor(R=4 m, a=2.28 m, B_T=5 T and

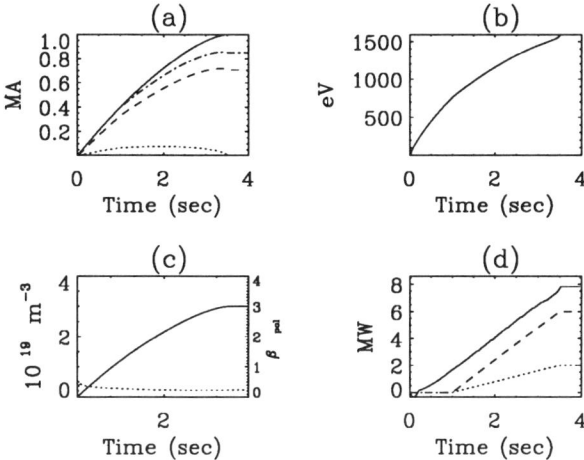

Figure 1: Time evolution of plasma parameters. (a)Total plasma current contributed from the equilibrium field change, helicity injection, bootstrap current and FW current drive, respectively from the bottom trace. (b)Required electron temperature. Ion temperature is assumed to be half the electron temperature. (c) Density as a solid line and β_{pol} as a dotted line. (d) Solid line is required total auxiliary power for ITER-89L scaling. Short dotted line is applied FW current drive and long dotted line is ECH power.

$I_p = 20MA$) can give a higher bootstrap current fraction of 22 %, but much smaller toroidal beta, $\beta_T \approx 5.5$, with higher density of $3.0 \times 10^{20}/m^3$.

SUMMARY

A zero-dimensional model has predicted that plasma current of up to several MA can be driven from zero current by helicity injection and equilibrium field change, and then maintained in steady-state by helicity injection and bootstrap current with the help of appropriate RF heating such as ECH and FW heating to reduce plasma resistivity. Even with lower DC-HI current drive efficiency, similar plasma discharges are predicted with some additional RF current drive power.

* Work supported by U.S. DoE contract No. DE-AC02-76-CHO-3073.

REFERENCES

1. Y. S. Hwang, Ph.D. Thesis, Princeton University, 1993.
2. M. Ono, et al.,in Proceedings of 14th International Conference on Plasma Physics and Controlled Nuclear Fusion Research, IAEA-CN-56/E-2-5(1992).
3. D. S. Darrow, et al., Nucl. Fusion, 1993(to be published).
4. H. R. Wilson, Nucl. Fusion **32**, 257(1992).

THEORY

KINETIC ANALYSIS OF MINORITY GYRORESONANT HEATING: CONVERSION FIELDS IN TOKAMAK GEOMETRY

E. R. Tracy
Physics Department, College of William and Mary, Williamsburg, VA 23185

A. J. Brizard[a], D. R. Cook[b] and Allan N. Kaufman[b]
Lawrence Berkeley Laboratory, Berkeley, CA 94720

ABSTRACT

The explicit analytic theory of gyroresonant minority-ion heating by magnetosonic radiation is generalized from our previous work[1,2]. We include both trapped and passing guiding center orbits. In this paper we focus on the extraction of the conversion fields.. These come in two varieties: a continuum of gyroballistic waves, and a continuum of collective minority-ion Bernstein waves. The extraction of the conversion coefficients is reduced to solving a Fredholm integral equation of the second kind.

INTRODUCTION

We wish to consider the heating of a minority ion species by gyroresonant absorption. For low-field incidence, this can be viewed as a two-stage linear conversion process, as first pointed out in ref.(3). This can be analyzed into the following distinct steps: 1] the incoming magnetosonic wave crosses the resonant layer (where $\omega = \Omega_m$, with ω the magnetosonic wave frequency and Ω_m the minority gyrofrequency). Part of the disturbance continues on as the transmitted magnetosonic wave and part is linearly converted into a continuum of *gyroballistic waves*. 2] the converted gyroballistic waves now propagate freely. These propagate both in x-space, along the guiding center trajectories, and in k-space due to the magnetic field gradients. Eventually these waves come into resonance once again (i.e they satisfy the magnetosonic dispersion relation). This leads to the second conversion: 3] part of the incoming disturbance is linearly converted into an outgoing magnetosonic field (the *reflected* magnetosonic wave) and part continues as a disturbance carried by

---

[a]Present address: Nuclear Eng. Dept., University of California at Berkeley.
[b]Also Physics Dept., University of California at Berkeley.

the gyroballistic continuum. This leads, finally to 4] the gyroballistic continuum now conspires to spin off the collective minority-ion Bernstein wave. This collective wave Landau damps on the continuum before it finally separates. The ultimate absorption profile is composed of two parts: one from the collective Bernstein wave and the remainder from the gyroballistic continuum.

The complete analysis of the 1-D slab case has been carried out (see refs.(1) and (4)). For the 2-D axisymmetric tokamak steps 1]-3] have been carried out for the case of zero perpendicular temperature (cross field drifts are ignored)[2,5]. In this paper we concentrate on the last stage: the calculation of the conversion fields and collective wave spin-off in tokamak geometry. Trapped particle effects are included. We present a solution *in principle*; many details remain to be worked out.

In previous work[1,4] we have shown how to compute the conversion fields for minority gyroresonant heating in slab geometry. We used a generalized form of the Case-van Kampen analysis, as modified by Bateman and Kruskal, which allowed us to develop analytic formulae for the absorption profile. A straightforward extension of these results to tokamak geometries is not possible because the generalized Case-van Kampen modes no longer form an orthonormal set, hence extraction of expansion coefficients for solution of the initial value problem is nontrivial. In the next section we sketch a solution of this problem: by proper choice of the Case-van Kampen modes it is possible to reduce the extraction of the expansion coefficients to finding the solution of a Fredholm integral equation of the second kind. Such equations can be solved formally (by the method of determinants) and various approximate methods of solution are also available[6].

SUMMARY OF ANALYSIS

The free gyroballistic dispersion function is: $D = \omega - \Omega_m(\mathbf{x}) + \mathbf{k}\cdot\mathbf{v}_D(\mathbf{x},\mathbf{l})$. Here, $\mathbf{x}$ and $\mathbf{k}$ are the spatial position and wavevector, ω is the frequency of the original incoming magnetosonic wave, Ω_m is the minority-ion gyrofrequency, $\mathbf{v}_D$ is the drift velocity of the minority ions and $\mathbf{l}$ is a vector of invariants associated with the particle drift motion: $\mathbf{l} = (p_\phi, \mu, h)$ with p_ϕ the toroidal angular momentum, μ the magnetic moment and h the energy. The gyroballistic dispersion relation, $D = 0$, is the Doppler shifted resonance condition. We will for the present restrict ourselves to a particular value of p_ϕ. Hence, in what follows, $\mathbf{l}$ will be a two-dimensional vector. We consider the self-consistent coupling of the gyroballistic waves to the electromagnetic field. This field is driven by the perpendicular current

$$D_A(\mathbf{k})A(\mathbf{x}) = \int d^2 l' g(l') u(\mathbf{x};l');$$

D_A is the magnetosonic wave operator (we will not need its explicit form here), $A(\mathbf{x})$ is the left polarized part of the vector potential, $u(\mathbf{x};l)$ is the left-circular velocity field of the minority ions, and $g(l)$ is the minority-ion distribution function. (We have set the value of various physical constants to unity for clarity.) The gyroballistic waves in turn are driven by the left-polarized electric field:

$$D(\mathbf{x},\mathbf{k};l)u(\mathbf{x};l) = -A(\mathbf{x}).$$

Eliminating the vector potential we get a self-consistent equation for the gyroballistic waves: $D(\mathbf{x},\mathbf{k};l)u(\mathbf{k};l) = -\beta(\mathbf{k})\int d^2l' g(l')u(\mathbf{k};l')$. This equation must be interpreted symbolically. For solution, we must choose a particular representation. If we choose the **x**-representation then in the operator D we have $\mathbf{k} \to -i\nabla_{\mathbf{x}}$ while if we choose the **k**-representation $\mathbf{x} \to i\nabla_{\mathbf{k}}$. We choose the **k**-representation, hence our basic equation is

$$D(i\nabla_{\mathbf{k}},\mathbf{k};l)u(\mathbf{k};l) = -\beta(\mathbf{k})\int d^2l' g(l')u(\mathbf{k};l').$$

Now insert the WKB ansatz: $u(\mathbf{k},l) = \int d^2\xi c(\xi) u_\xi(\mathbf{k},l) e^{-i\xi \cdot (\mathbf{k}-\mathbf{k}_0)}$. The function $c(\xi)$ is the expansion coefficient which depends upon the 'initial conditions', $u(\mathbf{k}_0,l)$, and $\mathbf{k}_0$ is an initial value of **k**. This ansatz leads to the following equations for the slowly varying (in **k**) amplitudes (we suppress the **k**-dependence for clarity):

$$D(\xi;l)u_\xi(l) = -\beta \int d^2l' g(l')u_\xi(l') \equiv -\beta \bar{u}_\xi.$$

This has the solution $u_\xi(l) = F_\xi(l)\delta(D) - \beta \bar{u}_\xi P/D$. Here F is an arbitrary function on the surface $D=0$ and the P symbolizes some prescription for dealing with the pole (e.g. Cauchy's Principal Value). The adjoint eigenfunctions obey

$$D(\xi;l)u^\dagger_\xi(l) = -\beta g(l)\int d^2l' u^\dagger_\xi(l') \equiv -\beta g(l)\bar{u}^\dagger_\xi,$$

with solutions $u^\dagger_\xi(l) = F^\dagger_\xi(l)\delta(D) - \beta g(l)\bar{u}^\dagger_\xi P/D$. In ref.(7) we show that, through proper choice of the functions F and $F^\dagger$, it is possible to insure that the inner product between eigenfunctions can always be cast into the form $<u_{\xi'}|u_{\xi}> \equiv \int d^2 |u^\dagger_{\xi'} \cdot u_\xi = \delta(\xi'-\xi)+K(\xi',\xi)$ where $K(\xi',\xi)$ is a complicated, but in principle known, function of its arguments once the normalizations $\bar{u}$ and $\bar{u}^\dagger$ are chosen. If we now attempt to extract the expansion coefficients, $c(\xi)$, from the 'initial conditions', $u(\mathbf{k}_0, \mathbf{l})$, by projection onto the adjoint eigenfunctions we find:

$$p(\xi) \equiv \int d^2 |u^\dagger_\xi (\mathbf{k}_0; \mathbf{l}) u(\mathbf{k}_0; \mathbf{l}) = c(\xi) + \int d^2 \xi' \, K(\xi', \xi) c(\xi').$$

The unknown we seek in this equation is $c(\xi)$. This is a Fredholm equation of the second kind[6] and can be solved by the method of determinants. Various approximate methods of solution also exist. Once $c(\xi)$ is known the problem is formally solved. Hence this constitutes a solution in principle, yet clearly much work remains to be done to extract physical insight from the solution.

REFERENCES

1. D. R. Cook, A. N. Kaufman, E. R. Tracy and T. Fla, AIP Conf. Proc. **244**, 173 (1991); and to appear in Phys. Lett. **175A** (April 19, 1993).
2. A. N. Kaufman, A. J. Brizard, D. R. Cook, E. R. Tracy and H. Ye, AIP Conf. Proc. **244**, 205 (1991).
3. H. Ye, Ph.D. thesis, Univ. Calif. at Berkeley (1990).
4. D. R. Cook, Ph.D. thesis, Univ. Calif. at Berkeley (1993); D. R. Cook, A. N. Kaufman, E. R. Tracy and T. Fla, to be submitted to Phys. Fl. B.
5. D. R. Cook, A. N. Kaufman, A. J. Brizard, H. Ye and E. R. Tracy, to be submitted to Phys. Fl. B.
6. R. Courant and D. Hilbert, *Methods of Mathematical Physics*, v.1 (John Wiley & Sons, New York, 1989).
7. E. R. Tracy, A. N. Kaufman, A. J. Brizard and D. R. Cook, to be submitted to Phys. Rev. E.

This work was supported by the U.S. Department of Energy under Contract Nos.DE-AC03-76SFOO098 and DE-FG05-84ER53176.

ENHANCED DECAY INSTABILITY AND MODE CONVERSION TO A STRONGLY-DAMPED NONLINEAR WAVE

Lazar Friedland,[a] Allan N. Kaufman,[b] and James J. Morehead[b]

Lawrence Berkeley Laboratory, University of California, Berkeley, CA 94720

When one of the waves in mode conversion or in decay instability is weakly nonlinear, the phase mismatch produced by spatial nonuniformity can be balanced by the nonlinear shift of wavenumber. This produces great enhancement of conversion and unlimited convective growth in the instability.

In the steady state, the spatial evolution of the pair of daughter waves in a *decay instability* or of the two waves in *linear mode conversion* is governed by two coupled equations of the same form. For a medium with *one-dimensional nonuniformity*, the two complex wave amplitudes $A_1(x), A_2(x)$ satisfy [1-7]

$$\mathcal{L}_1 A_1 = \eta A_2, \quad \mathcal{L}_2 A_2 = \sigma \eta^* A_1, \quad \text{with} \quad \mathcal{L}_j \equiv -iV_j \frac{d}{dx} - R_j x, \tag{1}$$

where V_j is the group velocity ($\dot{x} = \partial \omega / \partial k$) of wave j, and R_j is its refraction ($\dot{k} = -\partial \omega / \partial x$). Thus in the eikonal limit $[d/dx \to ik]$, $\mathcal{L} \to k \, \partial \omega / \partial k + x \, \partial \omega / \partial x$, where k and x are measured from the phase-space conversion point. The complex coupling strength is η (proportional to pump amplitude in the decay instability), while $\sigma = +1$ for the stable case (standard linear conversion) and $\sigma = -1$ for the unstable case (decay instability).

The nonuniformity represented by R_j limits the evolution, by restricting the wave coupling to the region where the wave-number mismatch is small. This limitation can be overcome by a weak nonlinearity $\delta \omega = -\beta |A|^2$ of one of the waves (say, wave 2), so that its wave-number shift can balance the mismatch [8,9]. It has been shown that this nonlinear balance is a generic effect, denoted as *autoresonance* [8,9], occurring when the spatial nonuniformity is sufficiently weak. Here we generalize the theory to include damping v_2 of the nonlinear wave, and treat the case that damping dominates convection: $v_2 \gg V_2 d/dx$. Then $\mathcal{L}_2$ becomes

$$\mathcal{L}_2 = -iv_2 - R_2 x - \beta_2 |A_2|^2. \tag{2}$$

The system (1) is now first-order in d/dx and nonlinear.

---

* Work supported by the U.S. Department of Energy under Contract No. DE-AC03-76SF00098.
[a] *Permanent Address: Hebrew University, Jerusalem, Israel.*
[b] *Also Physics Department, University of California Berkeley.*

To solve it we set $A_j = |A_j|\exp i\varphi_j$, $\eta = |\eta|\exp i\phi$, and obtain (after a little algebra)

$$d\ell n\,|A_1|^2 / dx = -2\sigma\left(|\eta|^2 / v_2 V_1\right)\cos^2\theta,$$
$$R_2 x + \left(\beta|\eta|^2 / v_2^2\right)|A_1|^2 \cos^2\theta = -v_2 \tan\theta, \qquad (3)$$

with $\theta \equiv \varphi_2 - \varphi_1 + \phi - \pi/2$. Letting $\xi \equiv (R_2/v_2)x$; $\alpha \equiv 2\sigma|\eta|^2/V_1 R_2$; $J(\xi) \equiv \left(\beta_2|\eta|^2/v_2^3\right)|A_1|^2$:

$$d\ell nJ / d\xi = -\alpha \cos^2\theta(\xi), \qquad (4a)$$
$$\xi = \tan\theta - J\cos^2\theta. \qquad (4b)$$

This system describes the coupled evolution, of the dimensionless action density J of the first wave and of relative phase θ, with dimensionless distance ξ. Eliminating ξ, we obtain

$$\frac{d\ell nJ}{d\theta} = -\alpha\frac{1 + J(\sin 2\theta)\cos^2\theta}{1 - \alpha J\cos^4\theta} \qquad (5)$$

for J vs. θ, in terms of the *single* parameter α, representing the dimensionless coupling strength. In Fig. 1, we display the phase portrait of (5), for $\alpha = 0.9$, J positive, $|\theta| < \pi/2$. We note that the topology is quite different from the linear case ($J \ll 1$).

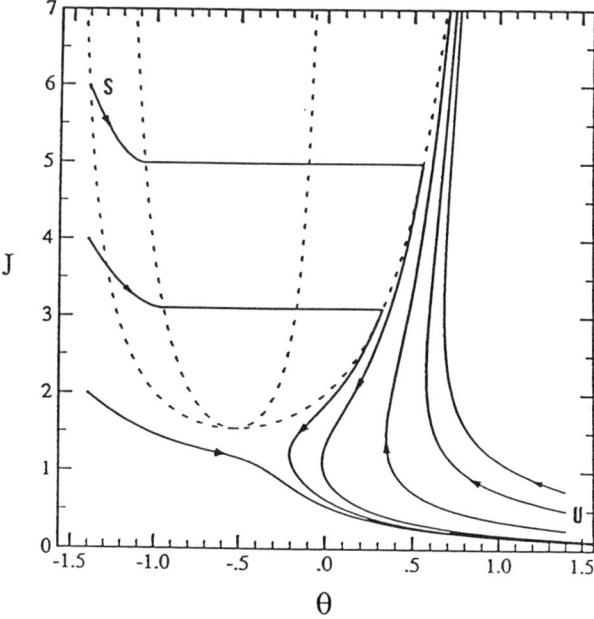

Figure 1. Phase portrait of Eq. (5), for $\alpha = 0.9$: action density J of wave 1 and relative phase θ. The phase space is foliated by the orbits $J(\xi), \theta(\xi)$. The arrows indicate the direction of time.

To develop interpretation, let us first consider the linear limit. Then (5) reads $d\ell nJ/d\theta = -\alpha$, with the solution $J(\theta) = J(0)\exp(-\alpha\theta)$, $\xi = \tan\theta$. (The lowest curve in Fig. 1 has the same topology as the linear limit.) Thus

$$\frac{J(\theta = +\pi/2)}{J(\theta = -\pi/2)} = \frac{J(\xi = +\infty)}{J(\xi = -\infty)} = \exp(-\pi\alpha) \quad (6)$$

yields the standard formula for the linear transmission coefficient. For V_1 positive (without loss of generality), wave 1 is incident from $x = -\infty$. If R_2 is also positive, the wave propagates from $\xi = -\infty$ to $\xi = +\infty$, so (6) states that $J_{final}/J_{initial} = \exp\left[-2\pi\sigma|\eta|^2/|V_1 R_2|\right]$. For stable linear conversion ($\sigma = +1$), this is the usual formula [3], with the Poisson Bracket $|\{\omega_1,\omega_2\}| = |(\partial\omega_1/\partial k)(\partial\omega_2/\partial x)| = |V_1 R_2|$. (For R_2 negative, the wave is incident from $\xi = +\infty$, and the results are the same.) In the unstable case ($\sigma = -1$), we have the standard convective growth [5]: $J_{final}/J_{initial} = \exp\left[2\pi|\eta|^2/|V_1 R_2|\right]$. Note that the *linear* result is *independent* of the damping v_2.

The autoresonant regime corresponds to the *nonlinear* term of (2) balancing the *nonuniform* term. [For β_2 positive (consistent with J positive), this occurs for negative x when R_2 is positive, and for positive x when R_2 is negative, and thus for negative ξ.] In (4b), for large J, we neglect $\tan\theta$, so that $\xi = -J\cos^2\theta$, resulting in $dJ/d\xi = \alpha\xi$, or

$$J(\xi) = J(0) + \alpha\xi^2/2 \quad (\xi < 0). \quad (7)$$

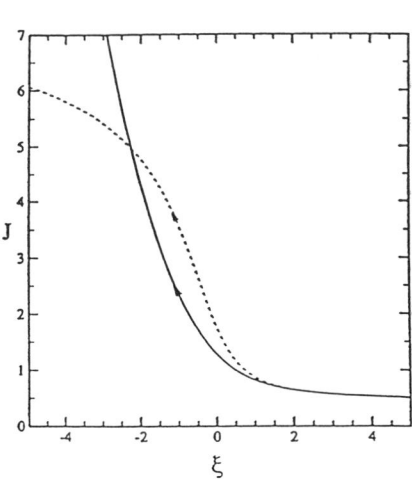

Figure 2. Nonlinear decay instability, autoresonant regime. Wave action of wave 1 grows as ξ^2, in contrast to the linear saturation.

Let us first consider the unstable case ($\sigma = -1$), with R_2 negative, so α is positive. This is illustrated by curve U in Fig. 1 (J vs. θ), and by the solid curve in Fig. 2 (J vs. ξ). Because $d\xi/dt \equiv (d\xi/dx)V_1 \equiv R_2 V_1/v_2$ is negative, the evolution is leftward. The dashed curve in Fig. 2 is the linear result, for which J convectively saturates. The nonlinear autoresonant behavior, on the other hand, demonstrates unbounded algebraic growth, approximated by Eq. (7).

(As J becomes large, one must of course include effects omitted from (1) and (2), such as pump depletion, time dependence, finite V_2, etc.)

The stable case is illustrated by curve S in Fig. 1, and by the solid curve in Fig. 3. (Again, the dashed curve is the linear result.) Here R_2 is positive (for α positive), and the evolution is toward the right. The discontinuity of $dJ/d\xi$ in Fig. 3, and of θ in Fig. 1 occurs when the numerator of (5) vanishes, implying $d\theta/d\xi$ infinite. At this point, θ jumps to the other solution $\theta(J,\xi)$ of (4b). The evolution then enters the autoresonant regime approximated by (7), resulting in *enhanced* mode conversion. In contrast to the linear case, the damping coefficient is now *important*; from $d\xi/dx = R_2/v_2$, we see that J varies more rapidly with x as v_2 decreases.

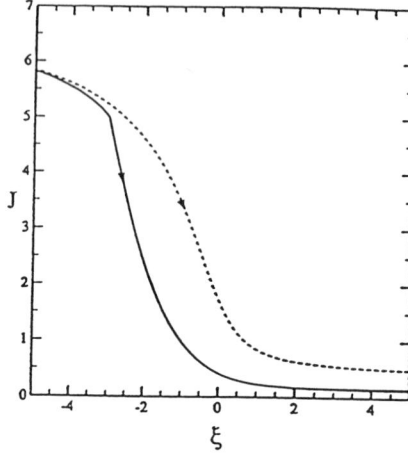

Figure 3. Nonlinear enhancement of mode conversion, in the autoresonant regime.

References

1. L. Friedland, G. Goldner, and A. N. Kaufman, Phys. Rev. Lett. 58, 1392 (1987).
2. L. Friedland and A. N. Kaufman, Phys. Fluids 30, 3050 (1987).
3. A. N. Kaufman and L. Friedland, Phys. Lett. A123, 387 (1987).
4. E. R. Tracy and A. N. Kaufman, Phys. Rev. Lett. 64, 1621 (1990).
5. E. A. Williams, Phys. Rev. Lett. 59, 2709 (1987).
6. R. G. Littlejohn and W. G. Flynn, Phys. Rev. Lett. 70, 1799 (1993).
7. D. R. Cook, W. G. Flynn, J. J. Morehead, and A. N. Kaufman, Phys. Lett. A 174, 53 (1993).
8. L. Friedland, Phys. Fluids B4, 3199 (1992).
9. L. Friedland, Phys. Rev. Lett. 69, 1749 (1992).

RF Heating of the Ionosphere: An Example of Generic Linear Mode Conversion

William G. Flynn and Robert G. Littlejohn
Department of Physics, University of California, Berkeley, California 94720

Abstract

When electomagnetic RF waves are launched into the ionosphere, they may mode convert into electrostatic waves, thereby heating the ionosphere via Landau damping. We investigate this process, primarily to illustrate a recently developed theory of generic linear mode conversion in one dimension. Using a cold fluid slab model, we show that the conversion of electromagnetic waves into electrostatic waves in the ionosphere is a generic process, and may be analyzed using our theory

We have recently identified [1] several generic properties of Hermitian, multi-component, linear wave equations in one dimension, and have derived the general asymptotic solution to any such equation which exhibits these generic properties. Here we show that the standard model [2,3] for linear waves in the earth's ionosphere exhibits the generic properties identified in Ref. [1]. We then show that analysis of this model using the methods described in [1] provides a more unified picture of mode conversion in the ionosphere than previous treatments [3].

Following Mjølhus [3], we assume a slab geometry (described in Fig. 1) and a cold electron plasma with fixed ions. Writing the perturbed electric field as

$$\mathbf{E}(\mathbf{x},t) = \mathbf{E}(z)\exp[i(k_x x + k_y y - \omega t)], \qquad (1)$$

the equation for $\mathbf{E}(z)$ is

$$D\mathbf{E}(z) = \begin{pmatrix} L - N_+ N_- - N_b^2 & N_+^2 & N_+ N_b \\ N_-^2 & R - N_+ N_- - N_b^2 & N_- N_b \\ N_- N_b & N_+ N_b & P - 2N_+ N_- \end{pmatrix} \begin{pmatrix} E_+ \\ E_- \\ E_b \end{pmatrix} = 0. \qquad (2)$$

The vector $\mathbf{N} = c\mathbf{k}/\omega$ is the index of refraction, where $k_z = -i\partial/\partial z$ (for now), and we have expressed D, $\mathbf{E}(z)$, and $\mathbf{N}$ in the basis $\hat{\mathbf{b}} = \mathbf{B}_0/|\mathbf{B}_0|$, $\hat{\mathbf{e}}_\pm = (1/\sqrt{2})[\hat{\mathbf{x}} \pm i(\hat{\mathbf{b}} \times \hat{\mathbf{x}})]$. We also define $Z(z) = \omega_e^2(z)/\omega^2$, $Y = |\Omega_e|/\omega$, so that the usual Stix parameters [4] are $P = 1 - Z$, $L = 1 - Z/(1+Y)$, and $R = 1 - Z/(1-Y)$.

Away from any mode conversion regions, (2) may be solved using standard eikonal theory [5]. Briefly, one first forms the symbol $D(Z, N_z)$ of the the operator D by reinterpreting N_z as a number rather than an operator in D. Then each eigenvalue $\lambda^{(\mu)}(Z, N_z)$ of the symbol $D(Z, N_z)$ determines a polarization or mode (labelled by μ) of the electric field, which is represented in the (Z, N_z) phase space by its dispersion curve $\lambda^{(\mu)}(Z, N_z) = 0$. Such curves are shown in Fig. 2.

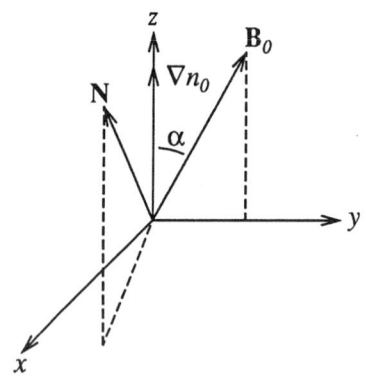

FIG. 1. Geometry for mode conversion in the ionosphere. The unperturbed density $n_0(z)$ varies in the $\hat{z}$ direction. The earth's magnetic field $\mathbf{B}_0$ is taken to be constant and lying in the y-z plane, where it makes an angle α with $\hat{z}$.

Mode conversion occurs when two or more dispersion curves almost cross in phase space or, equivalently, when two eigenvalues of $D(Z, N_z)$ are simultaneously small. In order to cause two eigenvalues of a parameterized Hermitian matrix such as $D(Z, N_z)$ to become small, one must generically vary four parameters. Since our phase space variables (Z, N_z) constitute only two parameters, mode conversion will not generically occur for randomly chosen values of the four additional parameters Y, α, N_x, and N_y which occur in $D(Z, N_z)$ (Fig. 2(a)). (The intersection of the two electromagnetic dispersion curves at $Z = 0$ is nongeneric, but not interesting. It is due to the obvious symmetry which occurs when the plasma vanishes.) However, within the full 6-dimensional parameter space with corrordinates Y, α, $\mathbf{N}$, and Z, we would generically find 2-dimensional subspaces in which two eigenvalues of $D(Z, N_z)$ vanish.

We now show that exactly such spaces exist for the dispersion matrix $D(Z, N_z)$ given in (2). To be definite, we assume that an RF wave of a given frequency is launched from a fixed, ground-based transmitter, which can launch a wave in any desired direction. Thus we treat Y and α as fixed parameters and N_x and N_y as adjustable parameters. For given values of Y and α, we find isolated "critical points" $(Z^c[Y, \alpha], \mathbf{N}^c[Y, \alpha])$ at which two eigenvalues of $D(Z, N_z)$ vanish.

Setting $N_x = 0$ and $N_y \cos \alpha = N_z \sin \alpha$, $N_\pm = 0$ and $D(Z, N_z)$ becomes diagonal. Setting $Z = 1$ causes P to vanish (this is the plasma resonance condition), and then the critical points occur where $N_b^2 = L$ or $N_b^2 = R$. Thus we have

$$N_x^c = 0, \quad N_y^c = A \sin \alpha, \quad N_z^c = A \cos \alpha, \quad Z^c = 1, \qquad (3)$$

where $A = N_b^c$. The possible values for A are

$$N_b^2 = L \Rightarrow A_\pm^L = \pm\sqrt{Y/(Y+1)}, \quad N_b^2 = R \Rightarrow A_\pm^R = \pm\sqrt{Y/(Y-1)}. \qquad (4)$$

Since N_b is real, there are two critical points for $Y < 1$, corresponding to the conversion of a left-polarized wave into a plasma wave and vice versa, while for $Y > 1$ there are four critical points, corresponding to the conversion of both left- and right-polarized waves into plasma waves and vice versa.

If an RF wave of a given polarization (L or R) is launched with $N_x = N_x^c$, $N_y = N_y^c$, then it will be completely transmitted through the plasma resonance layer $Z = 1$, because the dispersion curves of the electromagnetic and plasma waves will exactly cross. A wave launched in a slightly different direction will be partially transmitted, and partially converted into a plasma wave at the resonance layer. In this case, the transmission coefficient T (and conversion coefficient $C = 1 - T$) can be calculated in two steps.

First, one may reduce $D(Z, N_z)$ given in (2) to a 2×2 matrix $\bar{D}(Z, N_z)$, which describes the relevent modes (either L or R, and P) in phase space near the avoided crossing. Using ideas described in Ref. [6], $D(Z, N_z)$ may be reduced simply by eliminating the row and column of $D(Z, N_z)$ which are associated with the non-propagating mode. For example, if we wish to analyze the conversion of an L wave into a P wave, we eliminate the second row and second column of $D(Z, N_z)$. The reduced dispersion matrix may now be analyzed directly using the methods described in Ref. [1]. This calculation is described in Ref. [6]. Writing $T = e^{-2\pi\gamma}$, one finds for $A = A_+^L$,

$$\gamma = (\omega/cZ')(Y/8d)^{1/2}[(N_x - N_x^c)^2 + [2(1+Y)/d](N_y - N_y^c)^2]. \tag{5}$$

Here Z' is dZ/dz evaluated at the avoided crossing, and $d = 1 + (1 + 2Y)\cos^2\alpha$. The result is the same for $A = A_-^L$, and the result for $A = A_\pm^R$ is obtained by substituting $Y \to -Y$ in (5).

Our result (5) agrees with Mjølhus [3]; however, Mjølhus treated $\alpha \ll 1$ as a special case, because certain terms in his reduced dispersion matrix diverged as $\alpha \to 0$. This divergence is due to the fact that he reduced his equations by eliminating E_z rather than E_+ or E_-, as we have done. By eliminating only the non-propagating mode, we avoid this difficulty, and so our formula (5) is valid for all α.

There are, however, certain global aspects of mode conversion in the ionosphere which depend on α. If α is not small, and Y is not too large, then N_y^c for a given critical point (say A_+^L) will be substantially different from the values of N_y^c associated with the other critical points. In this case, for $N_x \approx N_x^c$, $N_y \approx N_y^c$, there will be only one avoided crossing in the phase plane (Fig. 2(b)). If, however, $\alpha \ll 1$, then all values of N_y^c will be small. In this case, for $N_x \approx N_x^c$, $N_y \approx N_y^c$, there will be two avoided crossings if $Y < 1$ (Fig. 2(c)), and four avoided crossings if $Y > 1$ (Fig. 2(d)). Notice, however, that even though all of the avoided crossings occur at the same value of Z, they are all well-separated from each other in phase space (this behaviour is generic). Our result (5) is thus valid at each avoided crossing. However, because transmitted and converted waves may interfere in problems with multiple avoided crossings, one must know their phases as well as there amplitudes. This information can also be obtained using the methods of Ref. [1], but is omitted here for brevity.

This work was supported by the U. S. Deparment of Energy under contract Nos. DE-AC03-76SF00098 and W-7405-Eng-48.

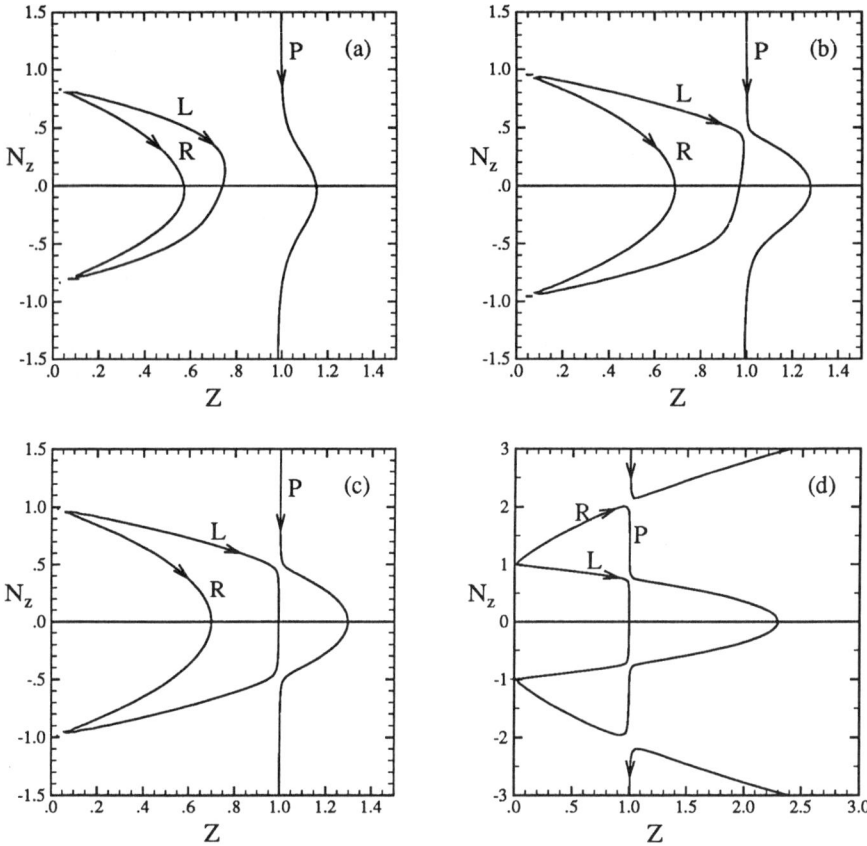

FIG. 2. Dispersion curves for waves in the ionosphere. The modes are labelled L, R, and P, and the arrows indicate the direction of propagation. The individual figures are described in the text.

REFERENCES

[1] W. G. Flynn and R. G. Littlejohn, "Normal Forms for Linear Mode Conversion and Landau-Zener Transitions in One Dimension" (to be submitted to Ann. Phys.)
[2] V. L. Ginzberg, *Propagation of Electromagnetic Waves in Plasma* (Pergamon, Oxford, 1970), Ch. V.
[3] E. Mjølhus, Radio Sci. **25**, 1321(1990).
[4] T. H. Stix *The Theory of Plasma Waves*, (McGraw-Hill, New York, 1962).
[5] R. G. Littlejohn and W. G. Flynn, Phys. Rev. A **44**, 5239(1991).
[6] R. G. Littlejohn and W. G. Flynn, Phys. Rev. Lett. **70**, 1799(1993).

A SIMPLE DERIVATION OF RELATIVISTIC FULL-WAVE EQUATIONS AT ELECTRON CYCLOTRON RESONANCE

D. C. McDonald, R. A. Cairns
Department of Mathematical and Computational Sciences,
University of St Andrews, St Andrews, Fife KY16 9SS, UK.

C. N. Lashmore-Davies
AEA Fusion, Culham Laboratory, Abingdon, Oxfordshire, OX14 3DB, UK (Euratom/UKAEA Fusion Association).

ABSTRACT

When a wave passes through an electron gyroresonance, in a plasma in the presence of a magnetic field gradient, there is a small spread in the resonance due to the electron's Larmor radius. Mathematically this is represented by the inclusion of the so called gyrokinetic term in the resonance condition, Lashmore-Davies and Dendy[1]. The smallness of this term, compared with other effects such as relativistic broadening, suggests that it should be negligible. However, we shall show here, by extending the method of Cairns et al[2] into the relativistic regime, that its inclusion is vital for producing self consistent full-wave equations which describe electron gyroresonance. The method is considerably simpler than those used previously by Maroli et al[3], Petrillo et al[4] and Lampis et al[5] for obtaining similar equations. As an example we include a calculation for the O-Mode passing perpendicularly through the fundamental.

DERIVATION OF THE WAVE EQUATION

To illustrate the technique we consider the O-Mode at perpendicular incidence and simply note that it can readily be extended to oblique incidence and to the X-Mode. We begin by taking the usual uniform plasma expression for the current density,

$$J_z(x) = \int dk\, \sigma(k) E_z(k) \exp(ikx - i\omega t) \qquad (1)$$

Where,
$$\sigma(k) = \sum_{n=-\infty}^{\infty} \int d^3u\, \frac{S_n}{\omega\gamma - n\Omega}$$

And,
$$S_n = 2\pi^{-3/2}\, i\omega_p^2 \varepsilon_0\, J_n(k\rho u_\perp) \exp(ik\rho u_y - in\theta - u^2)$$

We have assumed here a Maxwellian distribution, with thermal velocity u_t.

We now consider an inhomogenous magnetic field. To simplify the algebra, let us consider wave absorption at the fundamental, and assume that in the vicinity of the resonance the gradient scale length is L, so that,

$$\Omega(x) = \omega(1 - x/L) \qquad (2)$$

The most straight forward way to proceed is to insert this expression for Ω into Eq. 1 and then solve to obtain a wave equation. However, the equations formed, by this method, do not conserve energy, because we have not considered the so called gyrokinetic effect. The gyrokinetic effect arises from the fact that Ω must be evaluated not at x, the point at which we are calculating the response of the plasma, but at $x + u_y/\Omega$, the position of the particle guiding centre.

Expanding γ to second order and inserting the corrected Larmor frequency into the resonant denominator, gives,

$$\sigma(k) = \frac{L}{\omega} \frac{S_1}{x + Lu^2/\mu + u_y/\omega}$$

$$+ \sum_{n \neq 1} \int d^3u \, \frac{S_n}{\omega\gamma - n\Omega} \qquad (3)$$

The velocity integration can now be performed, after first applying the following identity to the resonant denominator,

$$\frac{1}{D} = -i \int_\infty^0 dk' \, \exp(iDk') \qquad \text{where, Im } D > 0 \qquad (4)$$

Giving,

$$\sigma = \int_\infty^0 dk' \, \frac{L}{\omega} \omega_p^2 \varepsilon_0 \, \frac{e^{ik'x}}{(1-it)^{5/2}} I_1(\lambda) \exp(-k'^2\rho'^2/4 - \lambda) \qquad (5)$$

Where, $\lambda=k(k+k')\rho'^2/2$, $\rho'=\rho(1-it)^{-1/2}$ and $t=k'L/\mu$. To produce the second order wave equation we expand this to second order in $k\rho$ and $k'\rho$, to give the finished wave equation as,

$$\frac{d}{dx}\left[\left(1 + \frac{1}{2}\frac{\omega_p^2}{\omega^2}F_{7/2}(\mu x/L)\right)\frac{dE_z}{dx}\right] + k_0^2 E_z = 0 \qquad (6)$$

This is the equation derived by Maroli et al[3].

DISCUSSION

By including the gyrokinetic effect into the standard method for calculating the homogenous dielectric tensor, we have produced a technique for deriving full-wave equations describing ECR, which is a considerable simplification on what has gone before. For the O-Mode we have shown it to be in full agreement with the previous work of Maroli et al[3], and the technique can easily be extended to produce wave equations for higher resonances and also to describe the X-Mode.

REFERENCES

1. C. N. Lashmore-Davies and R. O. Dendy, Phys. Fluids 1, 1565 (1989).
2. R. A. Cairns, C. N. Lashmore-Davies, R. O. Dendy, B. M. Harvey, R. J. Hastie and H. Holt, Phys. Fluids B3, 2953 (1991).
3. C. Maroli, V. Petrillo, G. Lampis and F. Engelmann, Plasma Phys. Cont. Fus. 28, 615 (1986).
4. V. Petrillo, G. Lampis and C. Maroli, Plasma Phys. Cont. Fus. 29, 877 (1987).
5. G. Lampis, C. Maroli and V. Petrillo, Plasma Phys. Cont. Fus. 29, 1137 (1987).

ELECTRON CYCLOTRON ABSORPTION AND EMISSION: "VEXATAE QUESTIONES"

M. Bornatici

University of Ferrara, Ferrara, and University of Pavia, Pavia, Italy

F. Engelmann

The NET Team, c/o Max–Planck–Institut für Plasmaphysik, Garching, Germany

ABSTRACT

The salient features of the theory of electron cyclotron absorption and emission for the parameter range for which the radiation spectrum has a line structure are touched upon and contrasted with erroneous results of a recent paper by Arunasalam (Phys. Fluids B 4, 1643 (1992)). In particular, i) the crucial role of the thermal effects on the mode polarization, which in turn affects the absorption profile around the fundamental frequency, is emphasized, showing that using a "Trubnikov O–mode factor" as advocated by Arunasalam is not correct; ii) the profile of the absorption around the fundamental frequency results to be single-humped in all cases, in contrast to Arunasalam's assertion that the absorption profile of the first harmonic O–mode is double–humped; iii) an expression for the local radiation temperature for a bi–Maxwellian electron distribution obtained by Arunasalam and upon which the analysis of the emission spectra of supershot plasmas in TFTR has been based (G. Taylor et al., Proceedings of the Eight Workshop on Electron Cyclotron (EC) Emission and EC Heating, Gut Ising, Germany (1992)) is shown not to be valid.

INTRODUCTION

In an early comprehensive review [1] on propagation, absorption and emission in the electron cyclotron (EC) frequency range, it was pointed out, in particular, that for electron cyclotron absorption (ECA) and emission (ECE) around the fundamental frequency, $\omega \approx \omega_c$, thermal effects are crucial in determining the ordinary (O) mode polarization independently of the value of the electron density; the same is true for the extraordinary (X) mode, in the finite–density regime. These results did show that the earlier Trubnikov–like treatments, which are based on the cold plasma approximation to the mode polarization, needed substantial revision in this frequency range. In this respect, a recent paper by Arunasalam[2] fails in yielding a correct theoretical description of EC absorption and emission. Actually, Arunasalam's paper is an astonishing collection of such egregious errors that a prompt rectification is required.[3]

ELECTRON CYCLOTRON ABSORPTION AND MODE POLARIZATION

To lowest order in the finite–Larmor radius parameter $[(\omega/\omega_c)N_\perp u_\perp]^2$ ($\lesssim 1$), and for propagation at angles for which $N_\parallel^2 \gg \max\left\{\left|\frac{n\omega_c}{\omega} - 1\right|, \frac{T}{mc^2}\right\}$, so that the Doppler effect in the EC resonance is dominant with respect to the relativistic one, the EC absorption around the n-th harmonic is proportional to $|E_{eff}|^2 \equiv |E_x - iE_y + \frac{\omega - n\omega_c}{n\omega_c}\frac{N_\perp}{N_\parallel}E_z|^2$. The evaluation of the effective electric field, E_{eff}, requires a thorough approach.[1,3] In particular, for the first harmonic <u>ordinary mode</u> one gets, for a

Maxwellian plasma,[1,3]

$$E_{eff} = -\left\{ \frac{\left[1-\left(\frac{\omega_p}{\omega_c}\right)^2\right]^{1/2}}{\frac{(1+2\cos^2\vartheta)\sin\vartheta}{1+\cos^2\vartheta}} \right\} \sqrt{2}\left(\frac{T}{mc^2}\right)^{1/2} \frac{1}{Z(\zeta_1)} E_z, \qquad (1)$$

$Z(\zeta_1)$ being the familiar (nonrelativistic) plasma dispersion function, with argument $\zeta_1 = \frac{1}{\sqrt{2}} \frac{\omega - \omega_c}{\omega} \frac{(mc^2/T)^{1/2}}{|N_\parallel|}$. The upper and lower forms in Eq. (1) refer, respectively, to the finite–density regime, i.e. for $\left(\frac{\omega_p}{\omega_c}\right)^2 > 2\left(\frac{T}{mc^2}\right)^{1/2} |N_\parallel|$, for propagation at angles such that $\sin^4\vartheta \gg 4\left[1-\left(\frac{\omega_p}{\omega_c}\right)^2\right]^2 \cos^2\vartheta$, and the tenuous–plasma limit $\left(\frac{\omega_p}{\omega_c}\right)^2 < 2\left(\frac{T}{mc^2}\right)^{1/2}|N_\parallel|$. From Eq. (1) it appears that, for the first harmonic O–mode, the effective electric field, E_{eff}, and, thus, the absorbed power ($\sim |E_{eff}|^2$) is non-vanishing due to <u>thermal effects</u>, independently of the value of the electron density. This is in marked contrast with the situation of the higher harmonics, for which the cold–plasma approximation is generally adequate.[1] Therefore, Arunasalam's result based on the "Trubnikov O–mode factor", peculiar to the cold, tenuous–plasma approximation, (see Eq. (18) and the Appendix of Ref. 2), is incorrect for the first harmonic O–mode.

For the first harmonic <u>extraordinary mode</u> and for the finite–density regime, the effective electric field is again given by Eq. (1) with the curly brackets-factor replaced by $\left[1+\left(\frac{\omega_p}{\omega_c}\right)^2\right]\left[2-\left(\frac{\omega_c}{\omega_p}\right)^2\right]^{1/2}\left(\frac{\omega_c}{\omega_p}\right)^2$.[1]

From the mode polarization (1), the EC absorption profile, i.e. the frequency dependence of the absorption, results to be described by the function

$$Im\left\{-\frac{1}{Z(\zeta_1)}\right\} = \frac{Im\{Z(\zeta_1)\}}{|Z(\zeta_1)|^2} \qquad (2)$$

for both the O–mode and X–mode (in the latter case restricted to the finite-density regime).[1,3] The profile function (2) exhibits a single peak in marked contrast with the (erroneous) double-humped structure found by Arunasalam[2] for the first harmonic O–mode.

For quasi-perpendicular propagation, i.e. for $N_\parallel^2 \lesssim \min\left\{\left|\frac{n\omega_c}{\omega}-1\right|, \frac{T}{mc^2}\right\}$, for which the relativistic effect in the resonance condition has to be retained along with the Doppler effect, the evaluation of the absorption coefficient requires, in general, a numerical approach.[4,5] Such numerical analyses confirm, in particular, the single humped structure of the absorption profile for the first harmonic O–mode at all angles of propagation.

LOCAL EC RADIATION TEMPERATURE FOR A BI-MAXWELLIAN PLASMA

Consider a plasma consisting of two Maxwellian electron populations: the bulk electrons (temperature T_b and density n_b) and the suprathermals (temperature $T_h (> T_b)$) and density $n_h = [\eta/(1-\eta)]n_b (> n_b)$). Noting that Kirchhoff's law is valid for each electron species separately, so that $j_i = N_r^2[\omega^2 T_i/(2\pi)^3 c^2]\alpha_i$, $i = b, h$, j_i and α_i being, respectively, the emission and absorption coefficient of the i-th electron population, the (local) radiation temperature, $T_{rad} \equiv [(2\pi)^3 c^2/\omega^2](j/N_r^2\alpha)$, can be expressed as[6]

$$T_{rad} = \frac{\alpha_b + \frac{T_h}{T_b}\alpha_h}{\alpha_b + \alpha_h} T_b. \qquad (3)$$

Thus, the presence of a suprathermal ($T_h > T_b$) electron population results in a local enhancement of the radiation temperature over the bulk temperature T_b.

For the particular case of perpendicular propagation, for the first harmonic O-mode and second harmonic X-mode, one has[3]

$$\frac{\alpha_{1,h}^{(O)}|_{max}}{\alpha_{1,b}^{(O)}|_{max}} = \frac{\alpha_{2,h}^{(X)}|_{max}}{\alpha_{2,b}^{(X)}|_{max}} = \frac{n_h}{n_b} = \frac{\eta}{1-\eta}, \quad (4)$$

$\alpha_{n,i}^{(O,X)}|_{max}$ denoting the maximum value of the absorption coefficient of the i-th electron population around the n-th harmonic, attained for

$$\frac{mc^2}{T_i}\left(\frac{n\omega_c}{\omega} - 1\right) = \begin{cases} n + \frac{3}{2}, & n \geq 1, \text{ for the O - mode}, \\ n + \frac{1}{2}, & n \geq 2, \text{ for the X - mode}. \end{cases} \quad (5)$$

Inserting Eq. (4) into Eq. (3) yields

$$T_{rad}(N_\parallel = 0) = \frac{1 + \frac{T_h}{T_b}\frac{\eta}{1-\eta}}{1 + \frac{\eta}{1-\eta}}T_b \quad (6)$$

which is Arunasalam's result (62).[2] However, the derivation given here displays that, to obtain Eq. (6), one is referring to different points in space for $i = b$ and $i = h$ if $T_b \neq T_h$, see Eq. (5), which invalidates Eq. (6). In fact, noting that $|d\ln\alpha_i/d\ln\omega| \approx \frac{mc^2}{T_i}$, whereas the frequency shift between the maxima of α_h and α_b is $\Delta\omega/n\omega_c \approx \frac{T_h}{mc^2}$, cf. Eq. (5), it follows that the result (6) is meaningless even as an approximation since it amounts to neglecting terms which are of the order of (or larger than) the terms containing $\eta/(1-\eta)$ which are retained. It has to be noted that the improper result (6) has been used as the basis for a recent interpretation of EC emission spectra of supershot plasmas in TFTR.[7]

EC RADIATION FOR A NONUNIFORMLY MAGNETIZED BI-MAXWELLIAN PLASMA

To evaluate the EC radiation from a plasma in an inhomogeneous magnetic field requires the solution of the equation of radiative transfer.[1]

For a bi-Maxwellian (bulk+suprathermals) distribution function the specific intensity of the EC radiation, i.e. the energy flux density per unit solid angle and frequency, around the n-th harmonic and for propagation perpendicular to the magnetic field, is[8]

$$I^{(i)}(\omega < n\omega_c, N_\parallel = 0) = \frac{\omega^2}{(2\pi)^3 c^2}\left[T_b\left(1 - e^{-\tau_{n,b}^{(i)}}\right) + T_h\left(1 - e^{-\tau_{n,h}^{(i)}}\right)e^{-\tau_{n,b}^{(i)}}\right] \quad (7)$$

for the radiation emitted on the low magnetic field side, whereas

$$I^{(i)}(\omega < n\omega_c, N_\parallel = 0) = \frac{\omega^2}{(2\pi)^3 c^2}\left[T_b\left(1 - e^{-\tau_{n,b}^{(i)}}\right)e^{-\tau_{n,h}^{(i)}} + T_h\left(1 - e^{-\tau_{n,h}^{(i)}}\right)\right] \quad (8)$$

for the high field side. In Eqs. (7) and (8), the index $i(=O,X)$ refers to the mode; $\tau_{n,b}^{(i)}$ and $\tau_{n,h}^{(i)}$ denote, respectively, the optical thickness (for perpendicular propagation) of the bulk and suprathermals. Expressions (7) and (8) are (approximate) solutions of the equation of radiative transfer for a linear spatial profile of the magnetic field, obtained by neglecting the spatial variation of the density and temperature across the resonance of both electron populations.[8]

The salient feature of solutions (7) and (8) is the strong asymmetry of the suprathermal radiation with respect to the direction of (perpendicular) propagation. Simple limiting cases are readily recovered from these relations, e.g.:

i) for observation on the low field side and $\tau_{n,b}^{(i)} \gtrsim 1$, the radiation by the suprathermals is effectively absorbed by the bulk, which counteracts the local suprathermal enhancement predicted by Eq. (3). In particular, for an optically thick bulk plasma, $\tau_{n,b}^{(i)} > 1$, the plasma tends to radiate like a black body at the bulk temperature, i.e. $I^{(i)} \simeq (\omega^2/(2\pi)^3 c^2) T_b$;

ii) for observation on the high field side, the suprathermal radiation is unaffected by the bulk absorption; for parameters such that $\tau_{n,h}^{(i)} < 1 < \tau_{n,b}^{(i)}$, i.e. when the suprathermal population is optically thin, whereas the bulk is optically thick, the intensity of radiation is

$$I^{(i)} \simeq \frac{\omega^2}{(2\pi)^3 c^2}\left(T_b + \tau_{n,h}^{(i)} T_h\right), \qquad (9)$$

the enhancement over the bulk temperature being thus significant for $(T_h >)\tau_{n,h}^{(i)} T_h \gtrsim T_b$.

The analysis given just displays the obvious fact that the local radiation temperature (3) alone is insufficient for determining the emitted radiation, but that the equation of radiative transfer must be used. As a consequence, the interpretation of the observed ECE spectra from a tokamak plasma based only on the evaluation of the local radiation temperature, as done in Ref. 7, is not possible.

REFERENCES

[1] M. Bornatici, R. Cano, O. De Barbieri, and F. Engelmann, Nucl. Fusion **23**, 1153-1257 (1983), and references therein (Review Paper).
[2] V. Arunasalam, Phys. Fluids **B4**, 1743 (1992).
[3] M. Bornatici and F. Engelmann, submitted to Phys. Fluids.
[4] M. Bornatici, U. Ruffina, and E. Westerhof, Comments Plasma Phys. Controlled Fusion **9**, 73 (1985).
[5] I. Fidone, G. Giruzzi, G. Granata, R.L. Meyer, M. Bornatici, and E. Mazzucato, Phys. Fluids **B1**, 1937 (1989).
[6] M. Bornatici, U. Ruffina, and E. Westerhof, Plasma Phys. Controlled Fusion **28**, 629 (1986); in particular, Eq. (27).
[7] G. Taylor, V. Arunasalam, P.C. Efthimion, and B. Grek, Proceedings of the Eight Joint Workshop on Electron Cyclotron Emission and Electron Cyclotron Heating, Gut Ising, Germany (1992).
[8] M. Lontano, R. Pozzoli and E.V. Suvorov, Nuovo Cimento **63B**, 529 (1981).

Electron Cyclotron Power Absorption in Plasmas with Non-Maxwellian Electron Velocity Distributions

Su Yue and Arnold H. Kritz

Lehigh University, Bethlehem, Pennsylvania 18015, USA

Gary R. Smith

Lawrence Livermore National Laboratory
Livermore, California 94551, USA

ABSTRACT

Tokamaks with sufficiently strong supplementary heating develop non-Maxwellian electron velocity distributions. Because the absorption of electron cyclotron power is proportional to $\nabla_\mathbf{v} f$, even small deviations from a Maxwellian distribution can significantly affect power deposition [1]. Following an approach used to study microinstabilities in a plasma with an arbitrary, numerically specified, electron distribution [2], we have developed a computational module to study electron cyclotron power deposition in plasmas that have distributions motivated by those in actual tokamaks. Also, we compare the deposition results obtained using an energy balance approach with those obtained using a Taylor expansion of the dielectric tensor. We illustrate the limitations of the latter approach.

INTRODUCTION

In studies of electron cyclotron heating (ECH) and current drive in tokamaks, a variety of approximations and assumptions enter the calculation of the power deposition. For example, depending on the temperature and cyclotron harmonic, the computation of power deposition may be within non-relativistic [3] or weakly relativistic [4] approximations or be fully relativistic [5][6][7]. Also, the deposition of ECH power in toroidal plasmas is often determined using a dispersion relation based on a Maxwellian electron velocity distribution. In this paper we use the fully relativistic description of the plasma and examine the dependence of ECH absorption on a small population of electrons in motion relative to an isotropic Maxwellian bulk distribution. We also examine the validity of the Taylor expansion approach relative to the energy balance method for deducing the imaginary wave number from the complex elements of the dielectric tensor.

VELOCITY DISTRIBUTION WITH STREAMING ELECTRONS

We consider a plasma equilibrium appropriate for the TPX tokamak and evaluate the effect that a small population of streaming electrons would have on the energy deposition of an ECH wave propagating in this plasma. In particular, we consider a distribution of the form

$$f = (1-\eta)f_b + \eta f_s$$

where f_b and f_s describe the Maxwellian bulk and the shifted-Maxwellian streaming populations [1]. The streaming electrons are taken to have an average momentum p_0 relative to the bulk plasma. Thus the parameters that define the electron velocity distribution are the bulk temperature T_b, the temperature T_s of the streaming electrons, the energy of the streaming electrons, $E_s = m_0 c^2 \{[1+(p_0/m_0 c)^2]^{1/2} - 1\}$, and the fraction η of streaming electrons.

METHODS FOR CALCULATION OF ECH ABSORPTION

In general, the dielectric tensor can obtained by using Vlasov's equation and Maxwell's equations for an arbitrary distribution function. We use a standard relativistic representation of the dielectric tensor that involves a sum over cyclotron harmonics [5]. Once the elements of the dielectric tensor are known, there are two approaches that have been employed to compute the imaginary wave number that then yields information about the local deposition of the electron cyclotron power. One involves a Taylor expansion of the dielectric tensor [3] and the other involves an energy balance approach [8]. In our studies we have found that the Taylor expansion approach is flawed and that energy balance approach must be used to obtain reliable results.

SUMMARY OF RESULTS

The TORCH code was used to trace rays for the TPX plasma, thereby providing the electron density and temperature, parallel and perpendicular wave numbers, and magnetic field along the ray path. By employing an algorithm for deposition in a plasma with an arbitrary distribution function [2], we investigated the dependence of electron cyclotron wave deposition on the population of the streaming electrons. This dependence is illustrated in Fig. 1, where the imaginary wave number k_i is plotted as a function of the magnetic field along the ray. The increase in k_i, as the population of streaming electrons is varied from 0 to 5% of the bulk density, would alter the power deposition profile and, consequently, the current that is driven by the ECH deposition.

The effect of streaming electrons with $p_\| N_\| > 0$ and $p_\| N_\| < 0$ is illustrated in Fig. 2. We consider two cases with $\eta = 5\%$ and $T_s = 10$ keV compared to the case where there are no streaming electrons. When the electrons stream with a velocity corresponding to 20 keV in the positive direction ($p_\| N_\| > 0$), there is an increase in the rate of absorption while in the negative direction ($p_\| N_\| < 0$) there is a decrease in the rate of absorption.

In Fig. 3 we illustrate the dependence of the imaginary wave number on the temperature of the streaming electrons. We find that for $\omega_{ce}/\omega > 0.99$, the increase in k_i, due to streaming electrons, decreases with increasing T_s.

Problems associated with the use of the Taylor expansion approach for calculating k_i are illustrated in Fig. 4. In particular, we note that the Taylor expansion completely fails at the second harmonic for the 10 keV plasma. Also, for $\omega_{ce}/\omega > 1$, the Taylor expansion approach yields a result for k_i which is too large. This failure at the fundamental becomes more severe as the plasma temperature increases.

CONCLUSIONS

We found that streaming electrons may play an important role in electron cyclotron absorption processes in TPX. The population density, current direction, mean momentum, and temperature of streaming electrons can significantly affect absorption.

We explored the relativistic accuracy of the Taylor expansion and energy balance approaches for calculating k_i. We found that the Taylor expansion results at the fundamental were incorrect for $T_e \geq 5$ keV. Furthermore, the Taylor expansion results at the second harmonic were incorrect for $T_e > 1$ keV. For the parameter regimes that we investigated, the weakly relativistic calculations were accurate for $T_e \leq 10$ keV.

ACKNOWLEDGMENTS

This work has been performed by Lehigh University and LLNL and supported by the United States Department of Energy under Contracts DE-FG02-92-ER5-4141 and W-7405-ENG-48.

REFERENCES

[1] M. Bornatici, U. Ruffina, and E. Westerhof, Plasma Phys. and Controlled Fusion **28**, 629 (1986).

[2] Y. Matsuda and Gary R. Smith, J. Comput. Phys. **100**, 229 (1992).

[3] D. B. Batchelor and R. C. Goldfinger, "RAYS: A Geometrical Optics Code for EBT," ORNL/TM-6844 (1980).

[4] I. P. Shkarofsky, Phys. Fluids **9**, 561 and 570 (1966); J. Plasma Phys. **35**, 319 (1986); I. P. Shkarofsky and A. K. Ghosh, Technical Report 232-1, MPB Technologies Inc. (1984).

[5] B. A. Trubnikov, "Electromagnetic Waves in a Relativistic Plasma in a Magnetic Field" in *Plasma Physics and the Problems of Thermonuclear Reactions*, ed. M. A. Leontovich, trans. J. B. Sykes, Pergamon Press, London, Vol. **3**, 122 (1959).

[6] D. B. Batchelor, R. C. Goldfinger and H. Weitzner, Phys. Fluids **27**, 2835 (1984).

[7] E. Mazzucato, I. Fidone, and G. Granata, Phys. Fluids, **30**, 3745 (1987).

[8] M. Bornatici, R. Cano, O. DeBarbieri, and F. Engelmann, Nucl. Fusion **23**, 1153 (1983).

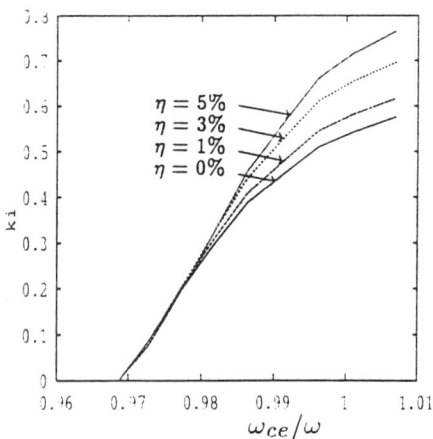

Fig. 1. Imaginary wave number versus magnetic field for η ranging from 0 to 5%. $T_s = 10$ keV, $E_s = 20$ keV, $\omega_{pe}^2/\omega^2 \approx 0.79$, $N_\parallel \approx 0.25$, and $T_b \approx 10$ keV.

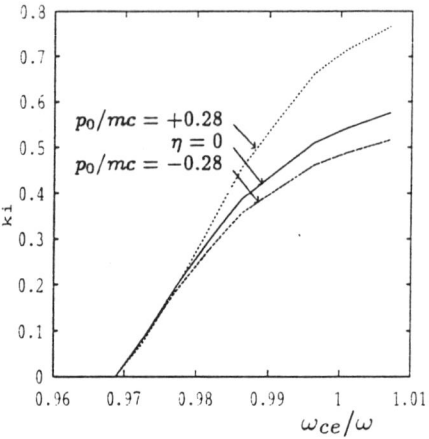

Fig. 2. Imaginary wave number versus magnetic field for $p_\parallel N_\parallel > 0$ and $p_\parallel N_\parallel < 0$ streaming electrons with $\eta = 5\%$ and other parameters as in Fig. 1.

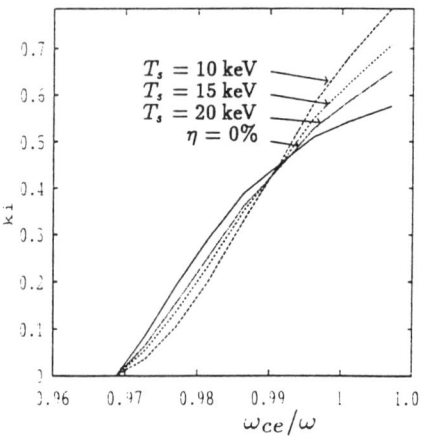

Fig. 3. Imaginary wave number versus magnetic field for T_s ranging from 10 to 20 keV with $\eta = 5\%$ and $E_s = 40$ keV. ω_{pe}^2/ω^2, $N_\parallel$, and T_b are the same as in Fig. 1.

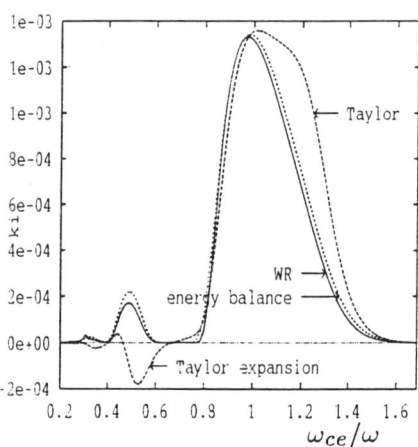

Fig. 4. Comparison of energy balance and Taylor expansion methods for fully relativistic calculation of $n_i = ck_i/\omega$; the weakly relativistic energy balance result is also shown. $\omega_{pe}^2/\omega^2 = 0.1$, $N_\parallel = N_\perp$, and $T_b = 10$ keV.

PERPENDICULAR ION ACCELERATION BY SHORT SCALE-LENGTH RF FIELDS

K.J. Reitzel, G.J. Morales, and V.K. Decyk
Physics Department, University of California, Los Angeles, Los Angeles, CA 90024

OVERVIEW

One of the outstanding challenges in RF-interactions with plasmas is the explanation of the ubiquitous observation of fast ion tails under conditions that do not satisfy a single-particle or a collective resonance, as has been most clearly demonstrated recently in tokamak[1] and space[2] experiments. This analytical, numerical, and simulation study has isolated a basic process that has the potential to explain the universality of ion acceleration. It consists of the interaction between large Larmor orbit ions and oscillating electric fields that have a sharp spatial envelope, as illustrated in Fig. 1. The symmetry breaking provided by the amplitude gradient results in runaway energy absorption at nonresonant frequencies. These basic features suggest that the process may be important to fast wave antenna performance, current drive in the strong-damping regime, alpha-particle relaxation, and dissipation of Alfvén wave filaments. In general, ion acceleration conditions are achieved when plasmas with a high level of density fluctuations are irradiated with RF, since sharp amplitude gradients develop locally.

RESPONSE TO EVANESCENT FIELDS

Using an exponential to represent a typical evanescent RF structure, $E_x(x,t) = E_0 \exp(-\alpha x)\cos(\omega t + \theta)$, we investigate ion acceleration by numerically integrating the equation of motion for various initial RF phases θ. Figure 2 shows the change in perpendicular ion velocity after a gyroperiod. For the electric field amplitude chosen ($p \equiv q\alpha E_0/M\omega^2 = 0.3$, $\nu \equiv \omega/\Omega_i = 10.24$), it is found that the adiabatic invariant is broken. The most dramatic effects occur while the ion passes through the high field side of the structure and a substantial "kick" is apparent. For $p = 0.3$, we see in Fig. 3 that the sign of the kick is essentially phase-independent; an average energy gain of about 20 times the initial value occurs. For a weak electric field, $p = .001$, it is seen from Fig. 3 that the kick is a nearly sinusoidal function of θ and requires an analytic treatment to determine the net energy gain. To this end, we use the conserved canonical momentum to cast the equation of motion as:

$$\frac{d^2}{dt^2} x + \Omega^2 x = \frac{qE_0}{M} \exp(-\alpha x) \cos(\omega t + \theta) \ . \qquad (1)$$

Expanding the particle orbit consistently to second order in the scaled electric field p, the phase averaged energy gain during one gyration through the evanescent structure is obtained from a lengthy, but straightforward calculation

$$\frac{<\Delta(KE_\perp)>}{KE_\perp(0)} = C_\nu(\rho)\, p^2 \ . \qquad (2)$$

The function C_ν is positive definite and overwhelmingly dominated by the modified Bessel function:

$$C_\nu(\rho) \approx \frac{2\pi^2 \nu^6}{\alpha^4 \rho^3} \frac{d}{d\rho} [I_\nu^2(\alpha\rho)] \,, \qquad (3)$$

where ρ is the unperturbed ion gyroradius. Figure 4 shows a comparison of the analytic expression and the results obtained from numerically integrating Eq. (1) for small p. The excellent agreement validates our premise that an evanescent RF field causes irreversible acceleration in both the small and large amplitude regimes.

ANTENNA NEAR-FIELD SIMULATIONS

A 2-1/2-D particle-simulation code has been developed to investigate the nonresonant acceleration of ions at the plasma edge. Figure 5 displays the self-consistent electric field amplitude for $\omega/\Omega_i = 6.4$ generated by an antenna located at $x = 0$. The local power density absorbed by the ions $<\underline{E} \cdot \underline{j}_i>$ indeed verifies that strong ion acceleration develops in the region having a sharp spatial gradient. The slope of the time evolution indicated in Fig. 6 confirms that the ion heating scales as E_o^2. The saturation seen at large times is due to quenching of the electric field by electron Landau damping parallel to the magnetic field.

CONCLUSION

The results of analytical, numerical, and particle simulation studies conclusively demonstrate that a runaway ion acceleration process exists for RF fields that have a sharp spatial envelope. Second order perturbation theory indicates that the broken spatial symmetry of the evanescent structure is responsible for the acceleration calculated, and that there exists no field amplitude threshold. This process may account for the ubiquitous ion heating at nonresonant frequencies observed in RF heating experiments and space plasmas.

ACKNOWLEDGMENT

This work is sponsored by USDOE and ONR.

REFERENCES

1. J.D. Evans, G.J. Morales, and R.J. Taylor, Phys. Rev. Lett., 69, 1528 (1992).

2. P.M. Kintner, et al., Phys. Rev. Lett., 68, 24148 (1992).

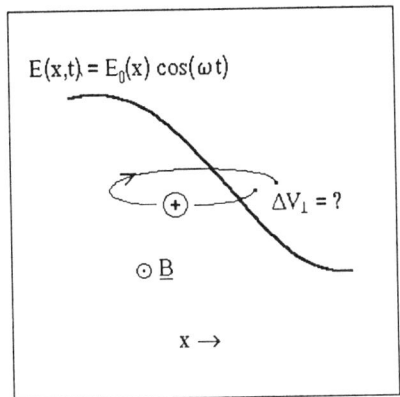

Fig. 1: Characteristic geometry of the interaction.

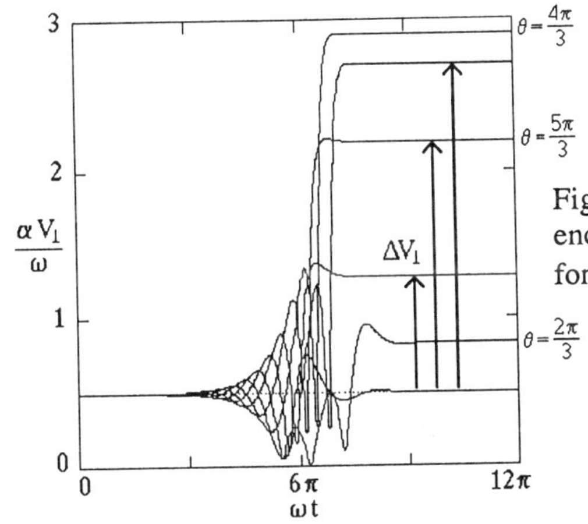

Fig. 2: Velocity kick in a single encounter with the evanescent field for different phases. $p=0.3$, $\nu=10.24$

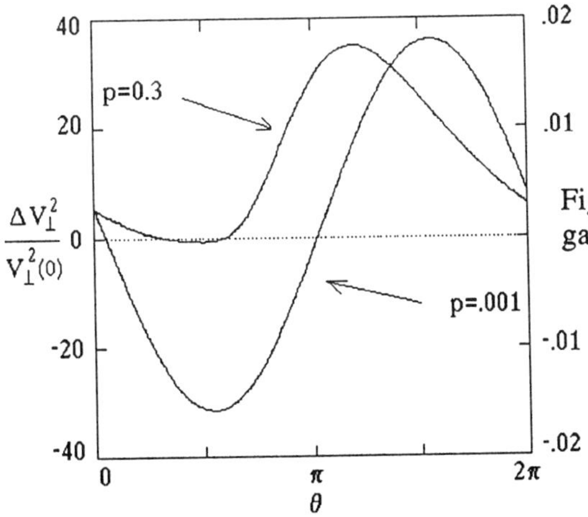

Fig. 3: Phase dependence of energy gained during one gyroperiod.

280 Perpendicular Ion Acceleration

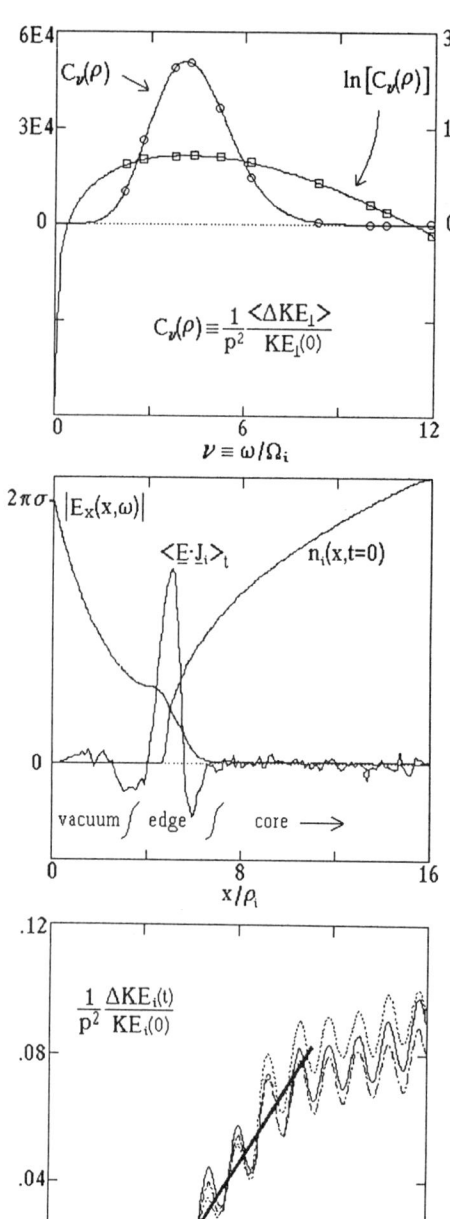

Fig. 4: Theoretical predictions (solid curves) and results from numerical integration for p=10⁻⁵ (open points).

Fig. 5: Spatial profile of power absorbed by ions in a particle simulation. The antenna is located at x=0. ν=6.4

Fig. 6: Simulation shows that ion heating scales as E_0^2. The three curves represent a 70% variation in p. RF turned on at $\omega_{pe}t=100$.

STATISTICAL APPROACH TO LHCD MODELING USING THE WAVE KINETIC EQUATION

K. KUPFER*
General Atomics, San Diego, CA 92186-9784

D. MOREAU and X. LITAUDON
Association EURATOM-CEA sur la Fusion Contrôlée
CE Cadarache, Saint Paul lez Durance, France

ABSTRACT

Recent work has shown that for parameter regimes typical of many present day current drive experiments, the orbits of the launched LH rays are chaotic (in the Hamiltonian sense), so that wave energy diffuses through the stochastic layer and fills the spectral gap.[1] We have analyzed this problem using a statistical approach, by solving the wave kinetic equation for the coarse-grained spectral energy density. An interesting result is that the LH absorption profile is essentially independent of both the total injected power and the level of wave stochastic diffusion.

THE WAVE KINETIC EQUATION

We consider the propagation of short wavelength lower-hybrid waves in a tokamak. The local spectral energy density of the rf field is denoted $U(x, k, t)$, where x is the position vector, k is the wavevector, and $U\,dx\,dk$ is the energy in the six-dimensional volume element $dx\,dk$. For simplicity we assume there is no mode conversion to the fast wave, so that every point in the (x, k) phase space is associated with a unique polarization vector. Also, the background plasma is assumed to be stationary. In this case, U obeys the wave kinetic equation (WKE)

$$\left(\frac{\partial}{\partial t} + \dot{x}\frac{\partial}{\partial x} + \dot{k}\frac{\partial}{\partial k}\right)U + 2\gamma(x,k)U = S(x,k,t) \quad , \tag{1}$$

where γ is the damping rate (electron Landau damping) and S is the rf source. The quantities $\dot{x}$ and $\dot{k}$ are time derivatives of x and k along the ray trajectories induced by the local dispersion relation $D(x, k, \omega) = 0$, where ω is the frequency of the waves. Upon inverting the dispersion relation to obtain $\omega = \Omega(x, k)$, the ray trajectories can be written in Hamiltonian form,

$$\dot{x} = \frac{\partial}{\partial k}\Omega \quad , \qquad \dot{k} = -\frac{\partial}{\partial x}\Omega \quad . \tag{2}$$

A detailed derivation and discussion of the WKE has been given by Ref. 2.

Once the source function is specified, Eq. (1) can be solved by following increments of injected power along ray trajectories in phase space. Appropriate canonical coordinates in tokamak geometry are $(x, k) \to (r, \theta, \phi, k_r, m, n)$, where m and n are the poloidal and toroidal mode numbers. In the cylindrical approximation both θ and ϕ are ignorable coordinates, so that m and n are conserved; in which case the ray trajectories are integrable and the solution of (1) can be written explicitly. We consider the point

---

*Oak Ridge Associated Universities Postdoctoral Program.

source $S = (2\pi)^{-2}\delta(r-r_o)\delta(k_r - k_{ro})$, where r_o is just inside the cutoff near the plasma edge and k_{ro} satisfies the dispersion relation at $r = r_o$. (Here m and n are considered parameters.) In this case, the solution of (1) is

$$U = (2\pi)^{-2}\,\delta[\Omega(r,k_r) - \omega]\,\frac{\cosh[\nu y(r)]}{\sinh\nu}\quad ; \quad \text{for} \quad r_1 \leqslant r \leqslant r_2\ , \qquad (3)$$

where

$$\nu = \int_{r_1}^{r_2} dr\, 2\gamma(r)/v_r(r) \quad \text{and} \quad y(r) = \nu^{-1}\int_{r_1}^{r} dr'\, 2\gamma(r')/v_r(r')\ . \qquad (4)$$

Here $v_r = |\partial\Omega/\partial k_r|$ and the dispersion relation is used to solve for k_r as a function of r. The turning points where $v_r = 0$ are denoted by r_1 and r_2, so that the ray is confined to the region $r_1 \leqslant r \leqslant r_2$, where r_1 is the caustic and r_2 is the cutoff. From (3) we can calculate the absorbed power density, $P(x) = \int dk\, 2\gamma(x,k)U(x,k)$, and the energy density in the rf parallel electric field (which determines the electron quasilinear diffusion coefficient), $\varepsilon_o|E_\shortparallel|^2/2 = \int dk\, \alpha_\shortparallel(x,k)U(x,k)$. Here $\alpha_\shortparallel = 2\,\partial D/\partial\varepsilon_\shortparallel(\omega\partial D/\partial\omega)^{-1}$ follows from the cold plasma dispersion relation, where $\varepsilon_\shortparallel$ is the parallel component of the dielectric tensor.

There are two distinct regimes: (1) the multipass regime (for $\nu \lesssim 1$), when the absorption is peaked at the caustic; and (2) the singlepass regime (for $\nu \gtrsim 5$), when nearly all the power is absorbed before the ray reaches the caustic. Figure 1 shows the radial profiles of P and $|E_\shortparallel|^2$ for a single field harmonic (m,n) in the multipass regime. The solid curve is the solution of the WKE, as given above, and the dashed curve is the corresponding fullwave solution, determined numerically by a cylindrical fullwave code.[3] The figure shows three important results: (1) that the fullwave solution in the multipass regime develops a node structure due to interference between the ingoing going and outgoing waves; (2) the WKE solution averages accurately over this fine scale interference pattern; and (3) the singular behavior of the WKE solution in the vicinity of the turning points (caustic and cutoff) occurs over a very short scale-length and is a negligible effect. We have also compared fullwave and WKE solutions in the singlepass regime, in which case the two methods converge, as expected.

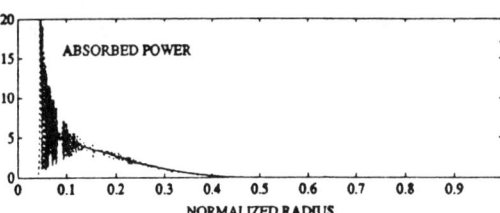

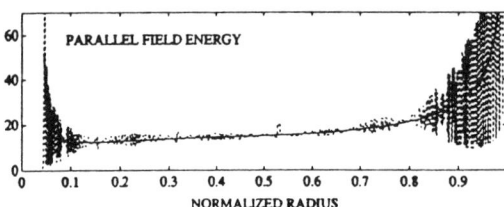

FIG. 1. Radial profiles of the absorbed power (top) and parallel electric field energy (bottom) in the multipass regime ($\nu < 1$) for a single field harmonic ($m = 100, n = 450$) in cylindrical geometry. The solid curve is the WKE solution from Eq. (3) and the dashed curve is the numerically determined fullwave solution. Parameters are typical for LHCD on Tore Supra.

THE WAVE DIFFUSION MODEL

In toroidal geometry the inherent poloidal asymmetry typically causes the formation of a thick stochastic layer in ray phase space.[1] This allows m to upshift (and downshift) as the rays fill the available stochastic phase space, until all the injected power is absorbed. When the spectral gap is large and the electron Landau damping is weak, the rays make many passes through the plasma before being absorbed. In this case, the chaotic dynamics allows the WKE to be approximated as a diffusion equation for the appropriately coarse-grained phase space energy density. The quasilinear treatment of (1) leads to the result that $\langle U \rangle \approx (2\pi)^{-2} \widehat{U}(m) \delta(\Omega_o - \omega)$, where $\Omega_o(r, k_r, m) = \omega$ is the unperturbed (cylindrical) approximation to the dispersion relation, and $\widehat{U}$ obeys the diffusion equation,

$$\tau(m) \frac{\partial}{\partial t} \widehat{U} - \frac{\partial}{\partial m} D \frac{\partial}{\partial m} \widehat{U} + \nu(m)\widehat{U} = \delta(m) P_{\text{in}} \quad . \tag{5}$$

Here $\tau(m) = \int_{r_1}^{r_2} dr |\partial \Omega_o / \partial k_r|^{-1}$ is the radial transit time, D is the diffusion in m per radial transit, and $\nu(m)$ is the damping decrement per transit, as defined in (4) using the unperturbed orbits. Also, P_{in} is the total injected power and we have idealized the LH source as being a delta function spectrum in n and m, with n treated as a parameter.

Equation (5) is solved on the domain $m_a \leqslant m \leqslant m_b$ with a reflecting boundary condition ($\partial \widehat{U}/\partial m = 0$) at m_a and an absorbing boundary condition at m_b. The lower boundary, as discussed in Ref. 1, corresponds to a propagation limit, since the dispersion relation cannot be solved for real k_r if $m < m_a$, where m_a can be calculated in the cylindrical approximation. The upper boundary m_b is somewhat arbitrary because the electron Landau damping becomes very large at large m, so that $\widehat{U}$ decays rapidly. It is sufficient to chose m_b large enough that the power flow across the boundary is negligible. Thus in steady-state (5) yields the conservation relation $P_{\text{in}} = \int dm \, \nu(m) \widehat{U}(m)$.

The effect of the chaotic dynamics on the evolution of the spectral energy density enters (5) through the diffusion coefficient D. Although D can be a function of m, we have found that taking D to be a constant is an adequate approximation. Typical values for D obtained from toroidal ray-tracing codes are in the approximate range 10^3 to 10^4 (assuming a circular plasma cross-section and parameters typical of Tore Supra and JT-60). Since the characteristic timescale for the divergence of neighboring trajectories in the chaotic phase space must be much shorter than the time scale for absorption, one finds that a rough criterion for the validity of the diffusion model is that $\sqrt{D}$ must be smaller than the typical m upshift required for strong electron Landau damping.

Because the damping decrement in (5) depends implicitly on the electron distribution function, one must solve (5) together with the electron Fokker-Planck equation, where the latter includes rf driven quasilinear diffusion. We refer to this coupled nonlinear system as the wave diffusion/Fokker-Planck system, or simply the WD/FP. The situation is entirely analogous to the usual coupled ray-tracing/Fokker-Planck approach (RT/FP), except that in the RT/FP one seeks an exact solution of (1), which requires the use of a toroidal ray-tracing code to calculate the orbits of a large ensemble of rays for many radial transits. We have implemented numerical solutions of the WD/FP using a simple one-dimensional Fokker-Planck model at each radial grid point. A typical steady-state solution is shown in Fig. 2 for JT-60 like parameters with

$n_{eo} = 1.5 \times 10^{19}$ m^{-3}, $T_{eo} = 5$ keV, and a launched $n_{\parallel}$ of 1.5. Good agreement is found between the WD/FP and RT/FP, as long as the rays make many passes before being absorbed, in which case the effect of the chaotic dynamics is well described by (5).

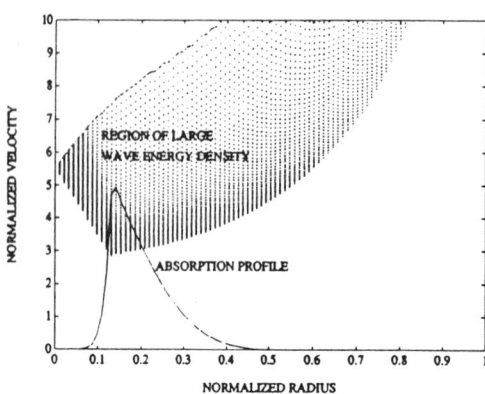

FIG. 2. Typical solution of the WD/FP. The rf energy density in the shaded region is large enough to produce quasilinear flattening of the electron tail. [Here the parallel velocity is normalized to the local electron thermal velocity $\{T_e(r)/m_e\}^{1/2}$.] The profile of the absorbed rf power density is superimposed.

Numerical solutions of the WD/FP over a large range in the parameters D and P_{in} indicate the following simple scaling. Defining $W(m) = \nu \widehat{U}/P_{in}$ as the normalized absorption per unit m, one finds that in steady-state W is essentially independent of $\widehat{U}$ due to strong quasilinear flattening of the electron tail, i.e., $\gamma \propto \partial f_e/\partial v_{\parallel} \propto 1/\widehat{U}$. It follows from (5) that $\widehat{U} \propto P_{in}/D$, so that quasilinear flattening occurs as long as P_{in}/D is larger than some critical value. For typical LHCD parameters P_{in}/D is 100 to 1000 times the critical value, so that the radial profiles of the absorbed power and the local current drive efficiency are independent of substantial variations in either D, or P_{in}.

ACKNOWLEDGMENTS

This work was supported in part by the EURATOM-CEA Association, in part by an appointment to the U.S. Department of Energy Fusion Energy Postdoctoral Research Program administered by Oak Ridge Associated Universities, and in part by U.S. DOE Contract No. DE-AC03-89ER51114.

REFERENCES

1. K. Kupfer and D. Moreau, Nucl. Fusion **32**, 1845 (1992).
2. S.W. McDonald, Phys. Reports **158**, 337 (1988).
3. D. Moreau et al., Nucl. Fusion **30**, 97 (1990).

FULL-WAVE THEORY OF A QUASI-OPTICAL LAUNCHING SYSTEM FOR LOWER-HYBRID WAVES: PRELIMINARY RESULTS

G. Cincotti, F. Gori, M. Santarsiero, R. Serrecchia
Dipartimento di Fisica. Università "La Sapienza" di Roma. Italy

F. Frezza, G. Schettini
Dipartimento di Ingegneria Elettronica.Università "La Sapienza" di Roma. Italy

F. Santini
Associazione EURATOM-ENEA sulla Fusione. C.R.E. Frascati. Italy

ABSTRACT

Numerical studies on the use of an advanced launcher to couple lower-hybrid waves to a plasma, for current drive in tokamaks, are currently under development. The study of the coupling has been carried out in a rigorous way, through the solution of the scattering from cylinders with parallel axes in the presence of a plane of discontinuity for electromagnetic constants. We present the general features of the proposed method together with preliminary results on launched spectra and coupled power.

1. INTRODUCTION

When a beam of radiofrequency radiation is to be injected into a plasma for current drive purposes at the lower-hybrid frequency[1], one is faced with the problem that a propagating plane wave is completely reflected by the plasma. The solution is to use a coupling mechanism via slow waves. This is generally accomplished by a sophisticated arrangement of waveguides[2]. An alternative approach that has been proposed not long ago[3] is to produce slow waves by scattering at a grating made of, e.g., conducting cylinders.

A rigorous analysis of such a system is not easy because it entails the solution of a heavy scattering problem. The main difficulty arises from the presence of the plane reflecting surface. Indeed, solutions of the scattering problem from cylindrical structures in homogeneous media are available[4,5]. In the presence of plane interfaces only solutions are known that hold in the limit of wires or perfectly reflecting surfaces[6]. It is not possible to apply these approximations to the present case, because the radii of the cylinders are comparable with the operating wavelength and the plasma surface is not a perfectly reflecting one. It is possible to solve this problem in a general way by using the plane wave expansion of cylindrical functions[7].

In the next paragraph the general solution is presented together with a method that holds in the case of a constant density plasma and leads to remarkable numerical simplifications. Numerical results will be given in the third section, while the last one is dedicated to future developments and conclusions.

2. N CYLINDERS IN FRONT OF A PLANE PLASMA SURFACE

A. The general solution.

Owing to the various geometrical features of the interacting waves and bodies, the imposition of the right boundary conditions is not a trivial task. To solve this problem it is customary to expand the diffracted field in terms of cylindrical functions, i.e., the product of a Hankel function of integer order H_n times a sinusoidal angular factor ($\exp\{in\vartheta\}$). Since the reflection properties of a plane of discontinuity for electromagnetic constants are generally known for incident plane waves[2,4], in order to get the rigorous solution it is essential the use of the analytic plane wave expansion of the above-mentioned cylindrical functions[7].

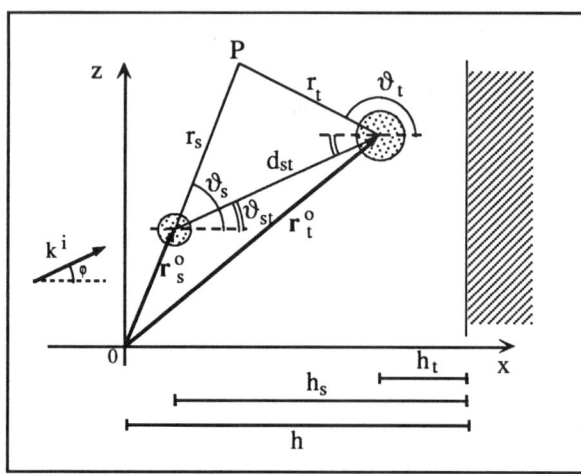

Fig.1. Geometry and notations used in this paper

The problem consists in calculating the coefficients of the expansion of the field diffracted from every cylinder, in the presence of a plane surface with reflection coefficient $\Gamma(n_{//})$, in terms of the cylindrical functions.

Such coefficients can be determined imposing the right electromagnetic boundary conditions on the conducting cylinders; to this aim it is convenient to express the field in terms of cylindrical functions centered on the different cylinders.

In this preliminary study we consider a plane wave with wavevector $\mathbf{k}^i$ as the incident field. The linear polarization with the magnetic vector parallel to the axes of the cylinders has been chosen to properly launch a lower-hybrid slow wave.

The magnetic field $\mathcal{H}_{tot}$ can be expressed as the sum of the following fields:

$\mathcal{H}_i$: field of the incident plane wave;

$\mathcal{H}_r$: field due to the reflection of $\mathcal{H}_i$ from the plane surface;

$\mathcal{H}_d$: field diffracted from the cylinders;

$\mathcal{H}_{dr}$: field due to the reflection of $\mathcal{H}_d$ from the plane surface.

The $\mathcal{H}_i$ and $\mathcal{H}_r$ fields can be expanded in terms of Bessel functions[4] J_n, while the diffracted field can be expressed as a sum of cylindrical functions with unknown coefficients c_{sn} (s=1,...,N where N represents the total number of cylinders). The notations used throughout the paper are that shown in fig. 1.

By using the plane wave expansion of the cylindrical functions[7]

$$\mathcal{H}_n(kr)\,e^{in\vartheta}\Big|_{x=h} = \int_{-\infty}^{\infty} F_n(n_{//},kh)\,e^{ikn_{//}z}\,dn_{//} \tag{1}$$

we obtain the field $\mathcal{H}_d$ in a form which allows one to evaluate $\mathcal{H}_{dr}$ by means of the reflection coefficient.

Imposing the vanishing of the tangential component of the electric field on the cylindrical surfaces, we find the following linear system:

$$i^m\, e^{ik^i_{//}z^0_t}\left[e^{ik^i_\perp x^0_t}\,e^{-im\varphi} + \Gamma(k^i_{//})e^{2ik^i_\perp h}\,e^{-ik^i_\perp x^0_t}\,e^{im(\varphi-\pi)}\right] +$$

$$+ \sum_{s=1}^{N}\sum_{n}\left\{\left[H_{n-m}(kd_{st})e^{i(n-m)\vartheta_{st}} + i^m A^{st}_{nm}\right](1-\delta_{s,t}) + G_n(ka_t)\delta_{s,t}\delta_{m,n}\right\} \times$$

$$\times\, i^n\, e^{-in\varphi}\, c_{sn} = 0 \tag{2}$$

where a_t is the radius of the tth cylinder, $\delta_{i,j}$ is the Kronecker symbol,

$$A^{st}_{nm} = \int \Gamma(n_{//})F_n(n_{//},kh_s)\,e^{ik_{//}(z^0_t-z^0_s)}\,e^{ik_\perp h_t}\,e^{-im(\pi-\alpha)}\,dn_{//}, \tag{3}$$

$\alpha = \sin^{-1}(n_{//})$, and $G_n(\xi) \equiv \dfrac{J'_n(\xi)}{H'_n(\xi)}$.

The solution of such system leads to the evaluation of the unknown coefficients c_{sn}; therefore the total magnetic field $\mathcal{H}_{tot}$ is fully determined.

B. A remarkable case: the constant density plasma.

In this case the reflection coefficient Γ becomes independent from $n_{//}$ and the previous formulas for reflected fields take a simpler form. In particular, it is possible to avoid the evaluation of the integral (3) in the linear system (2). It can be shown that the field $\mathcal{H}_{dr}$ is proportional to that due to an arrangement of cylinders specularly placed beyond the plasma surface (image method):

$$\mathcal{H}_{dr}(x_t,z_t) = \Gamma\, \mathcal{H}_d(2h_t - x_t, z_t)\ . \tag{4}$$

3. NUMERICAL RESULTS

The method outlined in the previous section has been applied to the evaluation of the diffracted field, the launched spectrum, and the coupled power in different experimental configurations.

As an example we consider the layout shown in fig.2, where an alignment of N identical cylinders in front of a plasma, having density $n_0=2n_c$, is sketched. The shape of the coupled power spectra are reported in fig.3 and fig.4, for N=5 and N=20,

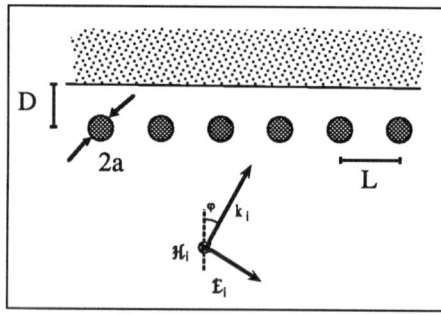

Fig.2. A single layer quasi-optical grill

respectively. In these examples an operating frequency of 8 GHz has been chosen. In both cases the ratio between coupled and incident power is about 25%, whereas the directivity varies from 83% (N=5) to 90% (N=20). We stress that such values have not been optimized using the available free parameters. Up to now the solution for a non-homogeneous plasma has only been performed for a single cylinder.

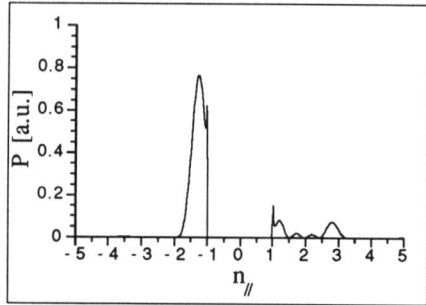 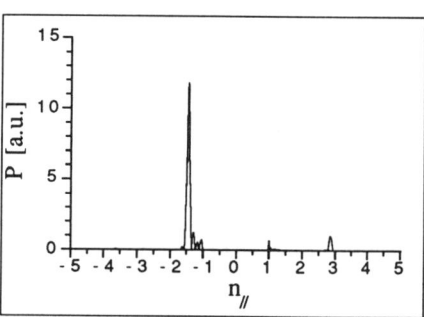

fig.3. Coupled power spectrum   fig.4. Coupled power spectrum
(N=5; kL=2.9; ka=0.85; kD=1.1; φ=45°)   (N=20; kL=2.9; ka=0.85; kD=1.1; φ=45°)

4. CONCLUSIONS

The results of this work, even if in a preliminary way, seem to indicate a value of the coupled power of a single array of cylinders higher then that reported in ref.3, thus suggesting the possibility of working with lower electric fields in a double array cavity. We remark that the double array configuration can be studied as a particular case of that outlined in sec.2.

The use of rods shaped in a different way is in progress. A more realistic modeling needs the use of incident fields different from plane waves. We have therefore planned the study of the diffraction of an incident gaussian beam from a quasi-optical grill; such a beam can actually be transmitted from a radiofrequency generator to the grill in the form, e.g., of a HE_{11} mode of a corrugated waveguide.

REFERENCES

1. R.A.Cairns, *Radiofrequency Heating of Plasmas*, Adam Hilger, Bristol 1991
2. M. Brambilla, Nucl. Fusion **16**, 47 (1976)
3. M.I. Petelin, E.V. Suvorov, Sov. Tech. Phys. Lett. **15**, 882 (1989)
4. C.A.Balanis, *Advanced Engineering Electromagnetics*, J.Wiley & So., NY 1989
5. H.A.Ragheb and M.Hamid, Int.J.Electronics **59**, 407 (1985)
6. J.R. Wait, Can. Journal of Physics **32**, 571 (1954)
7. G. Cincotti, F. Gori, M. Santarsiero, F. Frezza, F. Furnò, G. Schettini, Optics Comm. **95**, 192 (1993)

THE EFFECT OF POLOIDAL ANTENNA WIDTH ON LOWER-HYBRID WAVE PROPAGATION

R A Cairns
University of St Andrews, St Andrews, Fife, KY16 9SS, UK

V Fuchs
Centre Canadien de Fusion Magnétique, Varennes, Québec, Canada J3X 1S1

ABSTRACT

In simulations of lower hybrid heating and current drive in tokamaks, an important part of the calculation is the determination of the ray paths from the antenna to the central region of the plasma. The role of the parallel wavenumber spectrum and the need to launch a set of rays which cover it adequately is well known. However, the antenna also has a finite poloidal extent and a corresponding poloidal wavenumber spectrum, which will contribute to the spreading of wave energy within the tokamak and affect the absorption and current profiles. We describe a technique for estimating the spatial width of the beam produced by a finite width antenna, taking account of both the poloidal spread in launch position and the spectral width. The method uses standard ray tracing methods and estimates the beam width from data on three rays.

INTRODUCTION

Simulations of radio frequency current drive and heating in a tokamak (as described in, for example, references 1-3) involve ray tracing to determine the way in which energy is carried from the antenna to the central region of the plasma. A recent study by Bizarro and Moreau[4] discusses the effect of stochasticity of ray paths and emphasises the need for a sufficiently large number of rays to cover any stochastic region of the parallel spectrum densely. An antenna element also has a finite poloidal extent, and a corresponding poloidal wavenumber spectrum. The effect of this finite width has been studied by Pereverzev[5], using a WKB method in which the effect of diffraction on the beam width can be studied. Here we develop an entirely different technique for studying the same problem. The basic theory of our method is much simpler than that developed by Pereverzev and, rather than developing a new set of equations to describe wave propagation, makes use of standard ray tracing codes which are readily available.

ESTIMATE OF THE SPATIAL BEAM WIDTH

Our method is based on the observation that the diffraction of a beam can be calculated by following an ensemble of rays spread across the source, with the power in each weighted according to the spatial profile of the power produced by the source. Also there needs to be a number of rays launched from each point on the source, with a spread in initial wavenumbers and the power weighted according to the spectrum of the source. For simplicity we take the wave to be launched from an antenna with a Gaussian spatial profile

$$\exp\left(-\frac{1}{2}\theta_0^2/\varphi^2\right)$$

Here θ_0 is the poloidal angle, with the subscript zero denoting that it refers to the launching point of the wave. Without a subscript it will refer to the poloidal angle following a ray. The angle φ is a constant, giving a measure of the extent of the antenna. If this width is small then we can represent the corresponding poloidal wavelength spectrum by the continuous distribution

$$\exp\left(-\frac{1}{2}m_0^2\,\varphi^2\right).$$

Now suppose that at a subsequent point on the ray path the central ray of the beam is at the radial position r. A ray launched with non-zero m_0, θ_0 will have a radial displacement which can be approximated by

$$\Delta_r = \left(\frac{\partial r}{\partial \theta_0}\right)\theta_0 + \left(\frac{\partial r}{\partial m_0}\right)m_0,$$

the partial derivatives being evaluated at the central ray. Turning this round, we see that a ray can arrive at the radial position displaced by Δr from the central ray if it is launched from θ_0 with a value of m_0 given by

$$m_0 = \frac{1}{\left(\frac{\partial r}{\partial m_0}\right)}\left(\Delta r - \left(\frac{\partial r}{\partial \theta_0}\right)\theta_0\right).$$

The power carried by such a ray will be weighted according to both θ_0 and m_0 as discussed above, to give a weighting

$$\exp\left\{-\frac{1}{2}\frac{\theta_0^2}{\varphi^2} - \frac{1}{2}\varphi^2\left[\Delta r - \left(\frac{\partial r}{\partial \theta_0}\right)\theta_0\right]^2 \Big/ \left(\frac{\partial r}{\partial m_0}\right)^2\right\}.$$

Integrating this over θ_0 we obtain the combined effect of all the rays which can arrive at the position $r + \Delta r$. This integral can be done straightforwardly, with a result proportional to $e^{-\frac{1}{2}r^2/\sigma_r^2}$, where

$$\sigma_r = \sqrt{\varphi^2\left(\frac{\partial r}{\partial \theta_0}\right)^2 + \frac{1}{\varphi^2}\left(\frac{\partial r}{\partial m_0}\right)^2}.$$

In a similar way, the angular width σ_θ is given by

$$\sigma_\theta = \sqrt{\varphi^2 \left(\frac{\partial \theta}{\partial \theta_0}\right)^2 + \frac{1}{\varphi^2}\left(\frac{\partial \theta}{\partial m_0}\right)^2}$$

The unknown quantities which appear in these formulae are the partial derivatives which depend, of course, on the behaviour of the rays. However, they can be estimated from the behaviour of just three rays. The first of these is the central ray ($\theta_0 = m_0 = 0$) while the second and third are launched with $m_0 \neq 0$ and $\theta_0 \neq 0$ respectively. From the angular and radial beam widths, a single total width is easily calculated.

This theory assumes a smooth mapping of the source onto the beam at subsequent positions, a condition which is violated near a radial reflection where the rays behave as in Fig. 1. In effect the beam is folded over in this region, and a reasonable way to measure the true width is to assume that the part of the beam which our theory assumes to be outside the plasma is reflected in the boundary so as to lie inside the plasma. Once the reflection region is passed, the rays are again mapped smoothly from the source and there is no further problem.

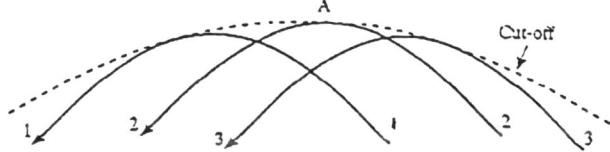

Fig. 1: Ray behaviour at a reflection

The spectral width of radiation reaching a given point can be calculated in an analogous way, but we shall not give details of this part of the calculation here.

SOME ILLUSTRATIVE RESULTS

We have investigated some representative cases for a tokamak with the parameters of ITER and an antenna with $\varphi = 0.23$. In these calculations the rays were followed until 99% of the central ray was absorbed. Fig. 2 shows results correponding to the value 1.98 for the parallel wavenumber and launched, as shown by the figure, from a position above the equatorial plane on th outside of the torus. The three rays shown in the figure are those used in the width calculation. The variation in width, predicted by the theory, is shown in Fig.3. There is a broad peak around toroidal angles 5-7, corresponding to the reflection of the beam at the top of Fig.2. The dip in the middle is related to the crossing of the rays and is obviously spurious. It should be filled in, though, since it is very narrow, it will probably not matter much in the application of this theory to absorption calculations. The beam is then refocussed towards the second reflection point and diverges again as it goes towards the centre and is absorbed.

For comparison, Fig4 shows the evolution of an ensemble of rays. Five bundles are launched from points evenly spread across the antenna width, each containing a range of components with poloidal wavenumber varying from 0 to 14. The width is in reasonable agreement with our predictions.

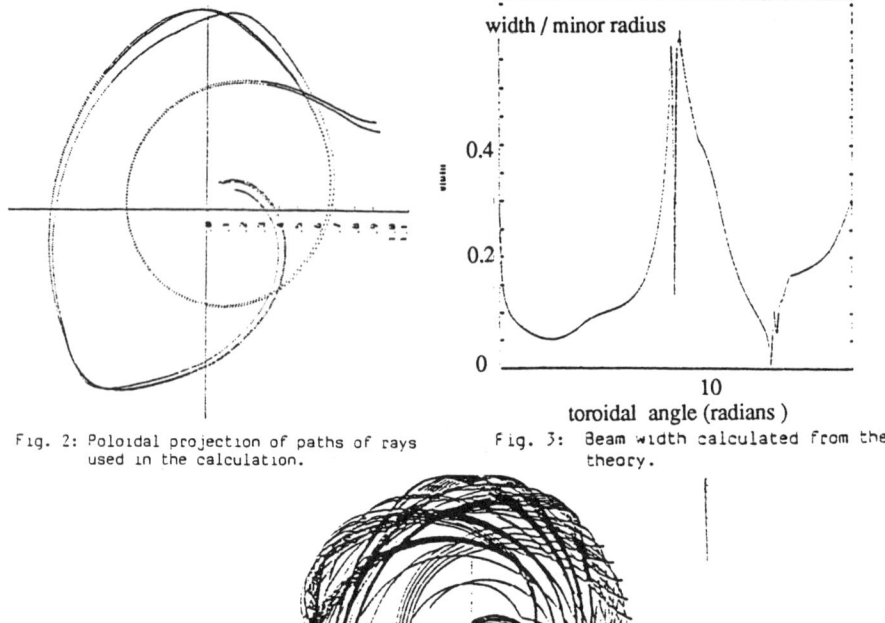

Fig. 2: Poloidal projection of paths of rays used in the calculation.

Fig. 3: Beam width calculated from the theory.

Fig. 4: Poloidal projection of an ensemble of rays.

CONCLUSIONS

The ray paths shown in Fig. 4, which all have the same toroidal wavenumber, show the importance of taking the poloidal width of the antenna into account. To launch an ensemble of rays covering both the toroidal spectrum and the effects of the spatial and spectral spread in the poloidal direction would require a very large number of rays, and our intention is to develop the method presented here so that the poloidal effects can be incorporated into lower hybrid simulation codes without expending a great deal of computer time on ray tracing.

REFERENCES

1. P. T. Bonoli and R. Englade, Phys. Fluids **29**, 2937 (1986).
2. P.T. Bonoli, M. Porkolab, Y.Takase and S.F. Knowlton, Nucl. Fusion **28**, 991 (1988).
3. V. Fuchs, I.P. Shkarofsky, R.A. Cairns and P.T Bonoli, Nucl. Fusion **29**, 1497, (1989).
4. J.P. Bizarro and D. Moreau, submitted to Phys. Fluids.
5. G.V. Pereverzev, Nucl. Fusion, **32**, 1091 (1992).

INTERACTION OF ICRF WAVES WITH LOWER-HYBRID DRIVEN SUPRATHERMAL ELECTRONS

Abhay K. Ram, Abraham Bers
Plasma Fusion Center, M.I.T., Cambridge, Massachusetts, USA
Vladimir Fuchs
Centre Canadien de Fusion Magnetique, Varennes, Québec, Canada
Robert W. Harvey
General Atomics, San Diego, California, USA

ABSTRACT

We determine the conditions for which the interaction of mode converted ion-Bernstein waves (IBW) with the energetic electron tails created by lower hybrid waves (LHW) can lead to an enhancement in the current drive efficiency. This may help explain the "synergy" results obtained on JET.

INTRODUCTION

The study reported here was motivated by JET experiments in which it was observed that the efficiency of LH current drive (CD) was significantly enhanced in the presence of waves in the ion-cyclotron range of frequencies (ICRF) typically used for minority ICR heating.[1] The ICRF spectrum, generated by a monopole antenna configuration, was symmetric in $k_\parallel$ and, thus, incapable of generating a net current by itself. Consequently, any enhancement in the CD efficiency is due to the interaction of ICRF waves with the asymmetric LHCD electron distribution function. Two aspects of the experimental conditions required for the enhancement play an important role in our studies. First, the ICRF spectrum, excited in the monopole configuration, ensures sufficient ICRF power at low values of $k_\parallel$'s. Second, the existence of ion cyclotron and ion-ion hybrid resonaces inside the plasma near the center ensures the existence of a mode conversion layer inside the plasma. Thus, the incident ICRF power, propagated by fast Alfvén waves (FAW), can, in principle, mode convert to ion-Bernstein waves (IBW's). This mode conversion is efficient for small $k_\parallel$'s and for small minority concentrations.[2]

Modeling studies and experiments clearly show that the LH energy and momentum deposition onto the electrons is typically localized and off-center. In order to enhance the current drive efficiency ICRF waves have to interact with energetic electrons on the same flux surfaces where LHW's are Landau damped and generate current. Consequently, in our effort to arrive at an understanding of the JET observations, we simplify our analysis to a single flux surface where the LH absorption is a maximum. This allows us to isolate the physically relevant mechanisms that are responsible for the enhancement in CD efficiency. Since the experimental conditions permit mode conversion, we study the effect of both FAW's and IBW's on LH generated electron distribution functions.

TWO-DIMENSIONAL VELOCITY SPACE FOKKER-PLANCK STUDIES

The relativistic evolution of the flux surface averaged, and gyro-angle averaged, electron distribution function in a uniform d.c. magnetic field, and acted upon by RF waves with frequencies ω below the electron cyclotron frequency ω_{ce}, is given by the Fokker-Planck equation:

$$\frac{\partial}{\partial t} f_o = \frac{\partial}{\partial p_\|} D^{LH} \frac{\partial}{\partial p_\|} f_o + \frac{\partial}{\partial p_\|} D^{FW/IB} \frac{\partial}{\partial p_\|} f_o + \left(\frac{\partial}{\partial t} f_o\right)_{collisional} \quad (1)$$

where the D's are the appropriate quasilinear diffusion coefficients and $p_\|$ is the component of the electron momentum along the magnetic field. Following the usual procedures,[3] we express the D's in the following convenient forms:

$$D^{LH} = \pi e^2 \sum_{k_\|} \left\{ |E_{kz}|^2 \delta\left(k_\| - \frac{\omega_{LH}}{v_\|}\right) D_o^{LH} \right\} \quad (2a)$$

$$D^{FW/IB} = \pi e^2 \sum_{k_\|} \left\{ |E_{ky}|^2 \delta\left(k_\| - \frac{\omega_{IC}}{v_\|}\right) D_o^{FW/IB} \right\} \quad (2b)$$

where

$$D_o^{LH} = \frac{1}{|v_\||} \left[\left\{ J_0 - \frac{v_\perp}{v_\|} J_1 \mathrm{Im}\left(\frac{E_{ky}}{E_{kz}}\right) \right\}^2 + \left\{ \frac{v_\perp}{v_\|} J_1 \mathrm{Re}\left(\frac{E_{ky}}{E_{kz}}\right) \right\}^2 \right] \quad (3a)$$

$$D_o^{FW/IB} = \frac{1}{|v_\||} \left[\left\{ \frac{v_\perp}{v_\|} J_1 + J_0 \mathrm{Im}\left(\frac{E_{kz}}{E_{ky}}\right) \right\}^2 + \left\{ J_0 \mathrm{Re}\left(\frac{E_{kz}}{E_{ky}}\right) \right\}^2 \right] \quad (3b)$$

e is the electron charge, v's are the velocities (the component perpendicular to the magnetic field having the subscript $\perp$), ω_{LH} and ω_{IC} are the frequencies of the LHW's and ICRF waves, respectively, J_0 and J_1 are Bessel functions with argument $k_\perp v_\perp / \omega_{ce}$, $\vec{E} = \sum \vec{E}_k \cos(k_\perp x + k_\| z - \omega t)$ is the electric field with the sum extending over the range of $k_\|$'s excited for each wave, and $k_\perp$ as well as the polarizations being determined from the full hot Maxwellian plasma dispersion tensor. The purpose of expressing the diffusion coefficients in different forms for the LHW's and FAW/IBW's is that, while the expression in Eqs. (3a,3b) depend only on the local plasma properties and electric field polarizations, the electric field amplitudes multiplying D_o in Eqs. (2a,2b) can be related to the incident power density, for the appropriate waves.[4] These diffusion coefficients are then used to find the steady-state solution of Eq. (1) using the numerical code CQL3D.[5]

For typical JET-type parameters, we find from Eqs. (2a,2b,3a,3b):

$$\frac{D^{FW}}{D^{LH}} = \frac{|E_{ky}|^2 D_o^{FW} \delta\left(k_\| - \frac{\omega_{IC}}{v_\|}\right)}{|E_{kz}|^2 D_o^{LH} \delta\left(k_\| - \frac{\omega_{LH}}{v_\|}\right)} \approx \frac{1}{4}(k_\perp^{FW} \rho_e)^2 \left(\frac{v_\perp}{v_\|}\right)^2 \frac{\omega_{LH}}{\omega_{IC}} \ll 1 \quad (4)$$

where ρ_e is the electron Larmor radius, i.e. the FAW diffusion coefficient is very small compared to that of the LHW. On the basis of this comparison, even if the FAW deposited its energy on electrons on the same flux surface as the LHW's, one would not expect the FAW to significantly modify the LH generated electron distribution function in JET. Indeed, numerical solutions of Eq. (1) show this to be the case. Therefore, as a next step, we investigate the effect of mode converted IBW's on LHCD.

In order to study the interaction of IBW's with LHCD generated electron tails two issues, that could seriously limit the efficiency of this interaction, need to be resolved. The first issue has to do with the high (parallel) phase velocities of IBW's at mode conversion. Since low $k_\parallel$'s undergo mode conversion, the IBW $k_\parallel$-spectrum lies below the LH spectrum. Thus, the IBW's and LHW's are acting on different parts of the electron distribution function. The second issue has to do with the amount of mode converted power to the IBW's. Mode conversion calculations based upon the Fuchs-Bers model [2] show that, for JET-type parameters with a hydrogen minority in a deuterium plasma, less than 25% of the incident FAW power, depending on the minority concentration, is mode converted to IBW's. (The table below gives the mode conversion coefficient evaluated at $k_\parallel = 1$ m^{-1} for different minority concentrations η, the ratio of the hydrogen density to the electron density.)

Table I Mode conversion coefficient C as function of minority concentration η

η	0.1	0.05	0.03	0.01
C(%)	0.1	2.5	12	25

However, both of these problems are satisfactorily resolved by the dramatic effect of toroidicity on the propagation of IBW's.[6] Both $|k_\parallel|$'s and electric field amplitudes are significantly enhanced along the IBW rays away from mode conversion. The upshifted $k_\parallel$'s allow the IBW's to interact with the energetic electron tails and the stronger electric field increases the diffusion coefficient. The electric field amplitudes can increase by factors ranging from 3 to about 10. The variations in $k_\parallel$'s and electric field amplitudes typically occur over short radial distances of propagation of IBW's.[6] An additional benefit of the toroidal effect on IBW's is that, with the mode conversion layer centrally located in the plasma, $k_\parallel$'s are upshifting as the IBW's propagate towards the flux surface of maximum LHW absorption. By accounting for the increase in the electric field amplitude in the diffusion coefficient, numerical solutions of Eq. (1) show that the IBW's indeed enhance the current drive efficiency.

RESULTS

Our numerical examples are for JET-type parameters. On the flux surface under consideration, we take an electron density n_e of 2×10^{19} m^{-3}, a hydrogen-deuterium plasma with a 3% hydogen minority, $\nu_{LH} = \omega_{LH}/2\pi = 3.7$ GHz,

$\nu_{IC} = \omega_{IC}/2\pi = 48$ MHz, and the local toroidal magnetic field of 3.6 Tesla. We normalize the diffusion coefficients to the collisional diffusion coefficient and express them in the form $D/D_c = \alpha D_o$ where $D_c = m_e v_{te} \nu_o$ with m_e and v_{te} being the electron mass and thermal velocity, respectively, and ν_o being the electron-electron collision frequency. For LHW's we find from LHCD simulations that the flux-surface averaged $\alpha_{LH} = 2.0$ and for IBW's, combining all the information discussed above, we find that $\alpha_{IB} = 0.1$. For the case of only lower hybrid waves in a plasma where the temperature of all the species is 1.4 keV, we find that the current drive efficiency is $\gamma_{LH} = n_e RI/P_d = 0.24 \times 10^{20}$ A/W m^2, where $R = 3$m is the major radius, I is the current and P_d is the power dissipated. The lower hybrid spectrum is taken to extend from $3.5v_{te}$ (electron Landau damping limit) to $10.6v_{te}$ (corresponding to the $n_\parallel = 1.8$ of the incident LH spectrum). When ICRF waves are used in typical JET scenarios of LH current drive the bulk electrons and ions are also heated. We take account of this heating by assuming that the temperature of all the species has increased to 4 keV. The lower hybrid spectrum extends from $3.5v_{te}$ to $6.3v_{te}$ (corresponding to $n_\parallel = 1.8$) and the IBW spectrum extends from $5.7v_{te}$ to very near the speed of light. The calculated current drive efficiency is found to be $\gamma_{LH+IB} = 0.43 \times 10^{20}$ A/W m^2. This corresponds to approximately a 75% increment in the current drive efficiency when compared to the case of LHW's only. This is comparable to the observed increases in the current drive efficiencies on JET.

In conclusion, our studies show that an increase in the current drive efficiency is likely to occur when IBW's interact with the energetic electron tails. The effect of FAW's on LH current drive efficiency is negligible.

This work is supported in part by DOE Grant No. DE-FG02-91ER-54109 and by DOE Contract No. W-7405-ENG-48. The CCFM is supported in part by Atomic Energy of Canada Ltd., Hydro-Quebec, and Institut National de la Recherche Scientifique.

REFERENCES

1. C. Gormezano, M. Brusati, A. Ekedahl, P. Froissard, J. Jacquinot, and F. Rimini, Proceedings of the IAEA Technical Meeting on Fast Wave Current in Reactor Scale Tokamaks (Synergy and Complementarity with LHCD and ECRH), Arles, France, September 23-25, 1991. Eds. D. Moreau, A. Bécoulet, and Y. Peysson, p. 244.
2. V. Fuchs and A. Bers, Phys. Fluids 31, 3702 (1988).
3. C.F. Kennel and F. Engelmann, Phys. Fluids, 9, 2377 (1966); I. Lerche, Phys. Fluids, 11, 1720 (1968).
4. A. Bers, Plasma Physics - Les Houches 1972, Eds. C. DeWitt and J. Peyraud (Gordon and Breach, N.Y., 1975), p. 113.
5. M.G. McCoy, G.D. Kerbel, R.W. Harvey, AIP Conf. Proc. 159, 77 (1987); R.W. Harvey, M.G. McCoy, G.D. Kerbel, AIP Conf. Proc. 159, 49 (1987).
6. A.K. Ram and A. Bers, Phys. Fluids B3, 1059 (1991).

Selfconsistent modeling of RF heating of fast particle populations and beams

D. Van Eester

Laboratorium voor Plasmafysica - Laboratoire de Physique des Plasmas
Associatie 'Euratom - Belgische Staat' - Association 'Euratom - Etat Belge'
Koninklijke Militaire School - Ecole Royale Militaire
Brussels - Belgium

ABSTRACT

A model is presented that studies the RF heating of tokamak plasmas in a fully selfconsistent way. It involves the simultaneous description of the temporal evolution of the heated species' velocity distribution function, the electric field responsible for the acceleration of the particles and the modification of background ion and electron temperature profiles resulting from collisional relaxation of the heated particles on the thermalized background. The nonmaxwellian character of the heated species is fully accounted for in the wave equation. As examples RF heating of both very energetic beams and minorities is discussed.

INTRODUCTION

Current-day tokamak plasmas contain particle populations that are highly nonmaxwellian. To study the RF heating of such populations in a selfconsistent way the Fokker-Planck and wave equations, respectively providing the velocity distribution function f of the heated particles and the RF electric field E, have to be solved simultaneously: f should be computed accounting for the actual E, and E as well as the local power absorption should be evaluated for the nonmaxwellian f. Moreover, to account for the power transfer from the heated species to the thermalized background the proper transport equations have to be solved as well. A coupled wave equation/Fokker-Planck/heat transport solver is presented in which f is the solution of either the steady state or the time dependent Fokker-Planck equation and in which a hat function representation which allows to find the dielectric tensor for an arbitrary distribution function is adopted in the wave equation. It permits to study the temporal evolution of the plasma when tails are created by the RF, when combined RF/NBI is applied or in the presence of fusion produced α-particles.

THE PHYSICAL MODEL

The adopted wave equation allows to study waves in anisotropic plasmas [1]. For all species the full dielectric tensor is considered i.e. finite Larmor radius corrections to all orders are retained. It is supposed that both the electrons and the majority ions are characterized by a bimaxwellian velocity distribution function with a parallel drift velocity. The dielectric tensor for such species is well known [2]. When species have a nonmaxwellian distribution function f the corresponding dielectric tensor $\bar{\bar{\mathcal{K}}}$ is computed using a piecewise linear 'hat' function representation of f [3]. The starting point for the computation is Ichimaru's expression of the dielectric tensor $\bar{\bar{\mathcal{K}}}$ valid for nearly uniform plasmas [4]; the effect of particle trapping has been neglected. It is assumed that the velocity distribution f does not depend on the gyrophase ϕ and that it is known on a grid of perpendicular and parallel points $(v_{\perp i}, v_{//k})$ that is sufficiently refined to guarantee that a linear interpolation gives an accurate approximation of the value of f at points in between those of the grid. The contributions from the elementary surfaces S_{ik} to the total integral can be computed using Taylor series expansions around the mid points of S_{ik}. The result is of the form

$$\bar{\bar{\mathcal{R}}} = [1- \sum_\alpha \left(\frac{\omega_{p\alpha}}{\omega}\right)^2]\bar{\bar{I}} - 2\pi \sum_\alpha \left(\frac{\omega_{p\alpha}}{\omega}\right)^2 \sum_{n=-\infty}^{\infty} \sum_{j=0}^{\infty} \varepsilon_\alpha{}^j \sum_{m=0}^{j} \sum_{\ell=0}^{j} a_{mj\ell} \sum_i \sum_k \bar{\bar{\mathcal{M}}}(n,j,m,l,...)$$

in which $\omega_{p\alpha}$ is the plasma frequency, Ω_α the cyclotron frequency, ω the antenna frequency, $\varepsilon_\alpha \equiv \frac{k_\perp \Delta_{\perp i}}{4\Omega_\alpha}$, $k_\perp$ the perpendicular wave number, $a_{mj\ell} = \frac{(-1)^m j!}{(j-m)!m!(j-\ell)!\ell!}$ and $\Delta_{\perp i} = v_{\perp i+1} - v_{\perp i}$; ε_α is small for a sufficiently fine grid. The last two sums are on all elementary areas S_{ik} of the grid. The elements of the tensor $\bar{\bar{\mathcal{M}}}$ in the above involve simple integrals over S_{ik} which can be evaluated analytically.

The nonmaxwellian distribution provided on a grid is obtained through the integration of the steady-state or time dependent Fokker-Planck equation expressed in terms of the parallel and perpendicular velocity components [5]. To allow relatively large time steps to be taken a Cranck-Nicholson [6] scheme is adopted.

Power transferred from the heated species to the majority ions and the electrons through Coulomb collisions modifies the temperature profile of the latter species. To account for the temporal evolution of the RF heating in a fully selfconsistent way the wave equation should be solved using the appropriate temperature profiles. The time evolution of the temperature profile is modeled by solving the heat transport equation for the majority and electrons.

EXAMPLES OF SELFCONSISTENT HEATING MODELING

As a first example the heating of a small minority of H in a D plasma of JET is studied. The following plasma parameters are taken: central electron density $N_{eo}=5 \, 10^{19} m^{-3}$, central electron temperature $T_{eo}=15 keV$, central ion temperature $T_{Do}=10 keV$, minor radius $a_p=1.25m$, major radius $R_0=3.09m$, frequency $f=53 MHz$, magnetic field $B_0=3.45T$, parallel wave number $k_{//}=8m^{-1}$. The density and temperature profiles are parabolic and parabolic squared, respectively. The minority concentration is 1% and a total incoming RF power flux of $15 MW/m^2$ is assumed. The cyclotron damping layer is assumed to lie 0.2m away from the equatorial plane. Figure 1(a) shows the temporal evolution of the average parallel and perpendicular minority H energy as well as the RF and collision relaxation power densities P_c of the minority population at $x=R-R_0=0.5m$. In absence of RF a steady state maxwellian distribution is observed. When the RF field is switched on after 300 milliseconds the RF power power density P_{RF} very rapidly increases to a value of $1.9 \, 10^5 W/m^3$ in 50ms and then starts decreasing on a slower time scale. After 500ms i.e. roughly half the Spitzer slowing down time ($\tau_s \approx 1000 ms$ for the present example) the steady state is reached. Under stationary conditions the incoming RF power is exactly balanced by that passed on to the other species by collisional relaxation ($P_{RF}=P_c$). Because the minority concentration is small a very energetic minority tail with perpendicular energy $<\mathcal{E}_\perp>=900 keV$ is formed and most of the RF power is passed on to the electrons ($P_c \approx P_{ce}$). It is interesting to note that cyclotron heating also provokes diffusion in the parallel direction: the asymptotic $m<v_{//}>^2$ is 28keV; the asymptotic $<\mathcal{E}_{//}>$ is 92keV. On top of Fig.1(a) the fraction of the minority absorption in the total absorbed power is written down for 3 different times. When the RF is just switched on the minority is still maxwellian and absorbs about one third of the incoming RF power. As the perpendicular tail develops it absorbs gradually more power until the minority distribution reaches its stationary value. In the final steady state the fraction of the minority absorption has doubled ($P_H/P_{TOT} = 0.63$ after 900ms). The related power deposition profiles are shown in Fig.1(b). The perpendicular tail formation is reflected by the gradually more pronounced broad power deposition profile. Note that the tail is only formed at one side of the minority cyclotron layer since the electric field at the other side is too small to significantly heat the particles there.

To study the effect of RF power on very energetic NB injected particles a nonzero source term is included in the Fokker-Planck equation. Because both the RF and collision operators conserve the number of particles a sink term then has to be introduced in the model as well. We assume that the beam particles are lost after 200ms. Figure 2 shows the time evolution of the beam in absence of RF and for $x=-0.4$m at different times after the start of NBI injection. The plasma parameters are the same as in the previous example except that the H minority is replaced by a D beam with $N_{beam}/N_{Dmajority}=0.15$ in the steady state regime. Because minority heating is the dominant heating mechanism for the chosen JET parameters, artificially removing the minority permits to highlight the beam heating physics. The beam particles are injected at an angle of $45°$ with respect to the local magnetic field and initially have an energy of 120keV. Because they did not have the time to diffuse away at small times particles are only found in the neighbourhood of the source. For larger times the collisional relaxation has set in and a maxwellian distribution function of thermalized particles is gradually formed. The steady state is reached after about 500ms. This is larger than the confinement time (200ms) but significantly shorter than the slowing down time (2500ms).

When RF is applied to the beam, the usual very energetic tail is formed. Figure 3 plots the asymptotic power deposition profiles for a given perpendicular RF electric field $E_+=1.5$kV/m and $E_-=4.5$kV/m. For comparison, the power deposition profile at the outset of the heating and corresponding to the last distribution function in Fig. 2 is shown in the same figure as well. The fraction of the incoming RF power absorbed by the beam is 37%; 36% goes to the majority ions and the rest is absorbed by the electrons. Due to the absorption of the RF power the wave amplitude is significantly lower at $x=-0.4$ than it is at $x=0.8$m. When the beam tail is formed this effect is enhanced and results in the majority deuterium cyclotron layer being screened as well. When the steady state is reached the beam particles absorb 70% of the incoming power. The fraction of the power going to the majority has dropped to 18%. The corresponding RF power densities are 0.27MW/m^3 and $P_{RF}=0.90$MW/m^3 at the high and low field side edges of the integration interval respectively. Typical average perpendicular energies of the beam tail particles in the damping region are of order 100-200keV.

CONCLUSION

A model was presented to study RF heating in a fully selfconsistent way. It involves a wave equation that allows to account for the nonmaxwellian distribution functions, a Fokker-Planck code that accounts for NBI and RF power sources and a heat transport model that includes Coulomb relaxation and direct heating effects.

REFERENCES

1. D. Van Eester, Proc. Europhys. Top. Conf. on RF Heating and CD of Fusion Devices, Brussels, Europhysics Conf. Abstracts 16E, 129 (1992).
2. T.H. Stix, The Theory of Plasma Waves (Mc Graw Hill, New York, 1965), p. 190.

3. D. Van Eester, Plasma Phys. Contr. Fusion **35**, 441 (1993).
4. S. Ichimaru, Basic Principles of Plasma Physics. A Statistical Approach (Benjamin Inc., Reading, 1973]), p. 51.
5. A.M. Messiaen et al., Plasma Phys. Contr. Fusion **35A**, 15 (1992).
6. C.F.F. Karney, Comp. Phys. Reports **4**, 183 (1986).

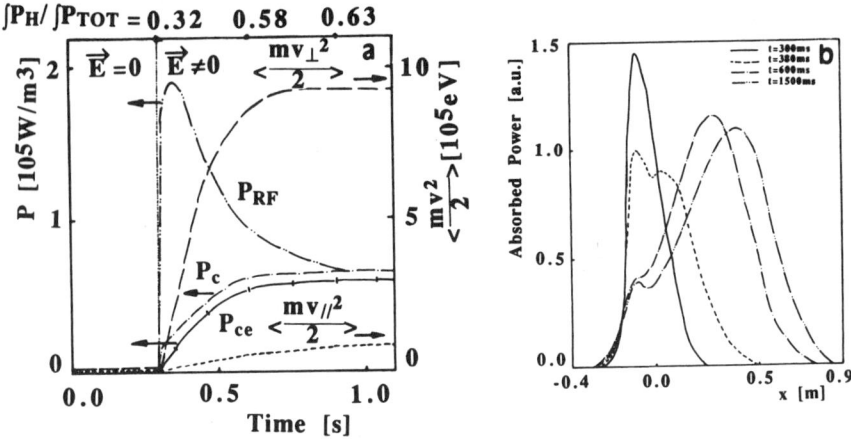

Fig.1 Typical temporal evolution of minority related quantities during RF heating (a) and associated minority power deposition profile at four different times (b).

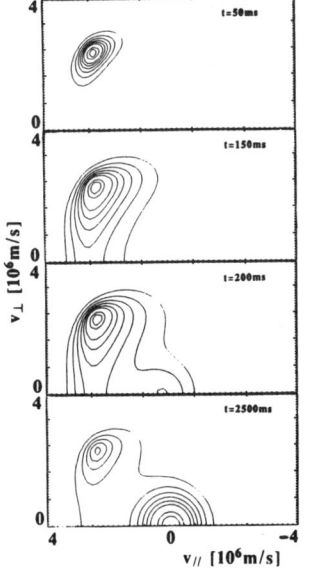

Fig.2 Temporal evolution of the beam distribution function with no RF at $x = -0.4$ m.

Fig.3 Steady state RF power deposition profile for combined RF+NBI heating. The deposition at the outset of RF is shown for comparison.

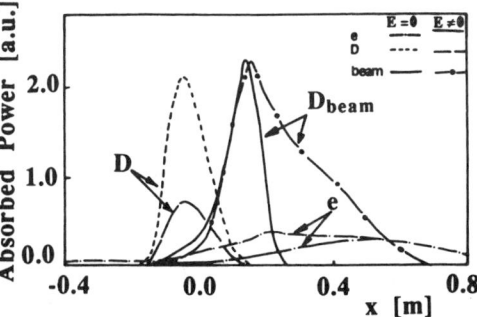

GLOBAL WAVE MODELING OF ELECTRON INTERACTIONS WITH FAST MAGNETOSONIC WAVES*

E. F. Jaeger, D. B. Batchelor, and M. Murakami

Oak Ridge National Laboratory, Oak Ridge, TN 37831-8071

INTRODUCTION

Electron interactions with fast magnetosonic waves are of interest for both direct electron heating and fast-wave current drive (FWCD) in tokamaks. Here we apply the full-wave ICRF code PICES[1] to examples of both of these applications. To realistically account for the actual D-shaped magnetic geometry of present-day tokamaks, PICES is interfaced with the 3-D MHD equilibrium code VMEC.[2] Likewise, to correctly model the real toroidal structure of both source and image currents in ICRF current drive antennas, PICES is interfaced with the 2-D recessed antenna impedance code RANT.[3] Both current drive and electron heating by fast waves can be strongly altered through modification of the $k_\parallel$-spectrum by the poloidal magnetic field. A poloidal mode expansion in PICES allows such variations in $k_\parallel$ to be included correctly. In this paper, comparisons are made to observations of the direct electron heating profile on TFTR[4] and to the FWCD efficiency on DIII-D.[5] We also extrapolate to make predictions for future tokamaks such as TPX and ITER.

THE PICES FULL-WAVE CODE

The PICES (Poloidal Ion Cyclotron Expansion Solution) code was designed specifically to treat electron interactions with fast magnetosonic waves. Poloidal and toroidal mode expansions are used to solve the reduced-order[6] wave equation in full toroidal geometry using general flux coordinates:

$$-\nabla \times \nabla \times \vec{E} + \frac{\omega^2}{c^2}\left(\vec{E} + \frac{i}{\omega\epsilon_0}\vec{J}_p\right) = -i\omega\mu_0 \vec{J}_{ext} \qquad (1)$$

In Eq. (1) $\vec{J}_{ext}$ represents the antenna current, and $\vec{J}_p$ is the plasma current as determined by the constitutive relation $\vec{J} = \overleftrightarrow{\sigma} \cdot \vec{E}$, which is expressed as a sum over poloidal and toroidal harmonics of the electric field.[7] Because the plasma conductivity $\overleftrightarrow{\sigma}$ is expanded to second order in Larmor radius, only harmonic resonances up to second order are retained to lowest order. However, higher-order resonances are still included in the wave solution and power absorption by adding additional terms to σ_{xx} and σ_{xy} that are obtained from a selective series expansion of the modified Bessel function, I_ℓ, for small argument.

The calculation includes a full (nonperturbative) solution for $E_\parallel$ and uses the reduced-order form of the plasma conductivity tensor[6] to eliminate numerical problems associated with resolution of very short wavelength ion Bernstein waves.[7] By solving the wave equation in two spatial dimensions (ρ, ϑ), PICES calculates the wave electric field, $\vec{E}$, the wave plasma current, $\vec{J}_p$, and the power absorbed per unit length. The poloidal mode expansion in ϑ allows variations in $k_\parallel$ to be included correctly, while the ρ dimension is treated by finite differences. Variations in the toroidal dimension ζ are included by summing toroidal harmonics weighted by a particular antenna spectrum. Toroidal harmonics are uncoupled because of the azimuthal symmetry and can therefore be calculated individually. Poloidal

---

* Research sponsored by the Office of Fusion Energy, U.S. Department of Energy, under contract DE-AC05-84OR21400 with Martin Marietta Energy Systems, Inc.

harmonics, on the other hand, are tightly coupled by the toroidal $(1/R)$ dependence of $\vec{B}$ as well as by the noncircularity of the magnetic flux surfaces and must therefore be calculated simultaneously.

DIRECT ELECTRON HEATING ON TFTR

Recently, direct electron heating experiments have been carried out by Murakami et al.[4] on NBI-heated supershot deuterium plasmas in TFTR. Modulation of the rf power allows direct measurement of the electron power deposition profile, q_e. Experimental results in Fig. 1(a) show that q_e is strongly peaked in the plasma core, and the total integrated power, P_e, is about 72% of the applied power, $P_{rf} \sim 1.65$ MW. The magnitude and profile shape agree well with those predicted by PICES in Fig. 1(b). For hydrogen plasmas, PICES predicts a very different heating profile due to Alfvén wave heating of electrons in a narrow band located at the shear Alfvén resonance.[8]

FAST-WAVE CURRENT DRIVE ON DIII-D

In addition to providing an alternative to minority ion cyclotron heating (but without the associated ion tails), direct electron heating provides the basis for FWCD in tokamaks. The first evidence of this process in a large tokamak has been reported by Pinsker et al.[5] on DIII-D, where about 25% of the total plasma current was sustained by fast waves launched from a directionally phased four-strap antenna array. Figure 2 (from Ref. 5) shows the observed dependence of current drive efficiency on electron temperature for co-FWCD in $B_0 = 1$ T discharges. Circles and squares indicate normal and ECH preheated discharges, respectively. For $Z_{\text{eff}} = 3.5$, the efficiency predicted by PICES (shown by the solid and dashed lines for co- and counter-FWCD, respectively) agrees quite well with the observed values.

FUTURE EXPERIMENTS ON TPX AND ITER

A good model for electron interactions with fast waves should not only describe present observations but should also make definite predictions about future observations. In this section, we consider predictions for future experiments on TPX and ITER. Figure 3 shows a frequency scan for TPX parameters. The toroidal mode number is varied with frequency to match the wave's phase velocity to the electron thermal velocity at plasma center. Peaks in the current drive efficiency γ (solid line) occur at frequencies $f \sim 45$ MHz and 75 MHz, where parasitic ion absorption by deuterium (P_D), hydrogen (P_H), and neutral beam deuterium (P_{beam}) are minimum. The remaining fraction which is absorbed by electrons (P_e) is maximum at these same frequencies. More detailed calculations at 45 MHz suggest that about half of the total required plasma current in TPX can be sustained by FWCD, the remaining fraction being provided by the bootstrap current. As expected, the resulting current profile is centrally peaked and gives a self-consistent q-profile that is less than unity on axis.

Such steady-state operating points do not appear to be practical in the latest designs for ITER-R, where average wall loading (W) is 1 MW/m^2 and plasma current (I_p) is about 25 MA. However, if the critical electron temperature gradient model for power balance holds, there appear to be some low current scenarios $(I_p \sim 5$ MA and $q_{95} \sim 14)$ where steady-state operation may be practical.[9] These cases exhibit reduced wall loading $W \sim 0.2$ MW/m^2 with $Q \sim 10$. An example is shown in Fig. 4(a) where 34 MW of fast-wave power at 21 MHz drives about 2 MA of current with 3 MA of bootstrap current. The equilibrium plasma assumes $T_{e0} = 33$ keV with $n_{e0} = 0.7 \times 10^{20}$ m^{-3}. In this case the current drive efficiency $\gamma \sim 0.29$ A/W·m^2 is limited only by the $k_\parallel$-upshift, and the resulting current profile is quite broad. An alternate scenario in Fig. 4(b) at 69 MHz gives a more centrally peaked current profile with $\gamma \sim 0.31$ A/W·m^2, but in this case, one-third of the power is lost to tritium ions at the second harmonic resonance. This parasitic absorption increases if a tritium tail develops. Less than 1% of the rf power is lost to fast alpha absorption.

SUMMARY AND CONCLUSIONS

The full-wave ICRF code PICES appears to be a good model for electron interactions with fast waves. By this we mean that it describes a large class of present observations on TFTR and DIII-D with only a few adjustable parameters. In addition, it makes definite predictions about observations on future tokamaks such as TPX and ITER. There are still problems which require work, however. A better current drive calculation is needed to calculate electron fluxes directly from a solution to the quasilinear equation rather than from fits to such solutions, as is now the case. Likewise, a self-consistent ion tail calculation would eliminate much of the present uncertainty involving parasitic ion losses.

REFERENCES

1. E. F. Jaeger, D. B. Batchelor, and D. C. Stallings, Nucl. Fusion **33** (1993).
2. S. P. Hirshman, and D. K. Lee, Comput. Phys. Commun. **31**, 161 (1986).
3. D. B. Batchelor et al., Fusion Technology **21**, 1214 (1992).
4. M. Murakami et al., Paper to be presented at the 10th Topical Conference on Radio Frequency Power in Plasmas.
5. R. I. Pinsker et al., IAEA-CN-56/E-2-3 (1992).
6. D. N. Smithe, P. L. Colestock, R. J. Kashuba, and T. Kammash, Nucl. Fusion **27**, 1319 (1987).
7. M. Brambilla, and T. Krucken, Nucl. Fusion **28**, 1813 (1988).
8. R. Majeski, personal communication, December 14, 1992.
9. W. M. Nevins, personal communication, January 18, 1993.

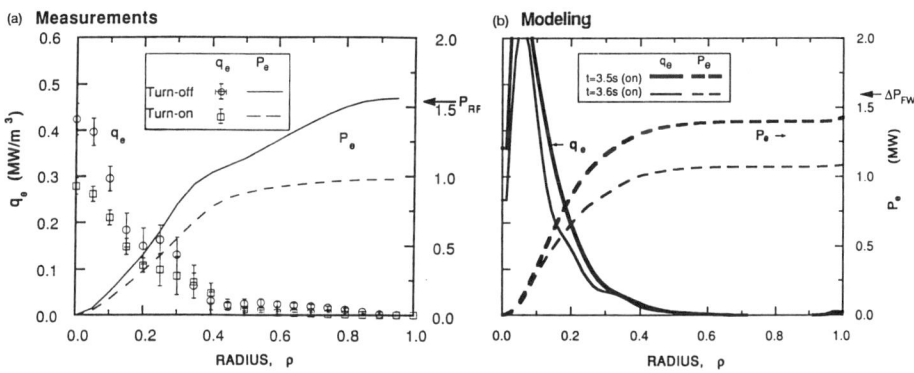

Fig. 1. Direct electron heating in TFTR.

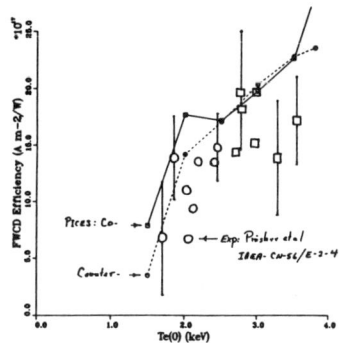

Fig. 2. FWCD efficiency vs electron temperature in DIII-D. Experiment: Pinsker et al., IAEA-CN-56/E-2-4.

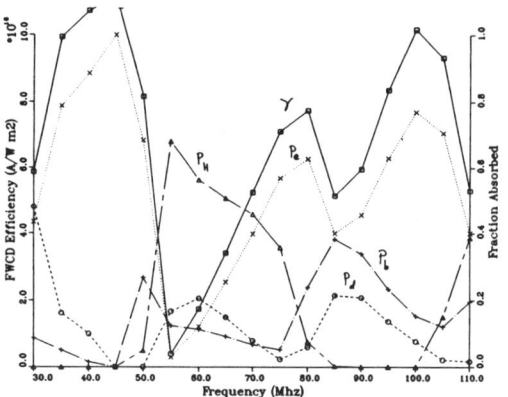

Fig. 3. Frequency scan in TPX for $V_{ph} = V_{th}$.

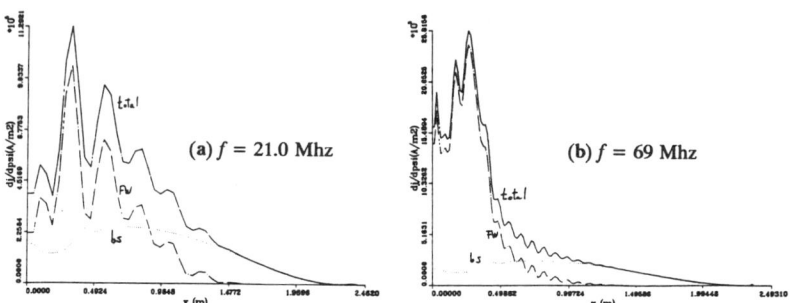

Fig. 4. Current profiles in ITER-R (reduced wall loading $W = 0.2$ MW/m² and $Q \simeq 10$). (a) $I_{bs} = 3.16$ MA, $I_{fw} = 2.16$; (b) $I_{bs} = 3.16$ MA, $I_{fw} = 2.28$.

THE EFFICIENCY OF FAST WAVE CURRENT DRIVE FOR A WEAKLY RELATIVISTIC PLASMA

S.C. CHIU and Y.R. LIN-LIU
General Atomics, San Diego, CA 92186-9784

C.F.F. KARNEY
Princeton Plasma Physics Laboratory, Princeton, NJ 08543

ABSTRACT

Current drive by fast waves (FWCD) is an important candidate for steady-state operation of tokamaks. Major experiments using this scheme are being carried out on DIII–D. There has been considerable study of the theoretical efficiency of FWCD.[1-7] In Refs. 4 and 5, the nonrelativistic efficiency of FWCD at arbitrary frequencies was studied. For DIII–D parameters, the results can be considerably different from the Landau and Alfvén limits. At the high temperatures of reactors and DIII–D upgrade, relativistic effects become important. In this paper, the relativistic FWCD efficiency for arbitrary frequencies is studied. Assuming that the plasma is weakly relativistic, i.e., T_e/mc^2 is small, an analytic expression for FWCD is obtained for high resonant energies ($u_{\rm ph}/u_{T_e} \gg 1$). Comparisons with the results from a numerical code ADJ[7] and the nonrelativistic results[5] shall be made and analytical fits in the whole range of velocities shall be presented.

I. INTRODUCTION

Current drive by fast waves (FWCD) is an important candidate for steady-state tokamak experiments. For this reason, there are major experiments being carried out or planned in DIII–D, Torre-Supra, JET, and other tokamaks. There was also considerable theoretical work.[1] Fisch and Karney[2] and Ehst and Karney[3] have obtained empirical formulae for the Landau and Alfvén limits. The authors[4,5] recently obtained a unified nonrelativistic formula applicable to both limits and in intermediate regimes. This is especially useful for intermediate frequencies such as in DIII–D. As the electron temperature increases to above 10 keV, relativistic effects are expected to be increasingly important. One of the authors[6,7] made fully relativistic studies of the current drive efficiency in the Landau and Alfvén limits. This paper extends the work of Ref. 5 for arbitrary frequencies to include relativistic effects and extends the work of Ref. 6 to intermediate regimes. Analytic expressions for the efficiency are obtained that agree well with the numerical adjoint code ADJ. We find that in the Alfvén limit the nonrelativistic efficiency formula is quite accurate even at energies when relativistic effects are separately important in absorption and current density.

II. WEAKLY RELATIVISTIC EFFICIENCY OF FWCD

For all practical purposes, the assumption that $\lambda^{-1} = T_e/mc^2 \ll 1$ is valid, i.e., the plasma is weakly relativistic. We shall assume this in the following; nevertheless, the resonant electron energy can still be fully relativistic.

We express momentum per unit mass W in units of $V_{T_e} = (T_e/mc)^{1/2}$, also the efficiency $J_{\scriptscriptstyle \parallel}/P_D$ in units of $2e/m\nu_0 V_{T_e}$ where ν_0 is the electron collision frequency. Then

$$\eta = J_{\scriptscriptstyle \parallel}/P_D = G(W_p)/D_A(W_p) \quad , \tag{1}$$

where W_p is the resonant momentum per unit mass. For a Maxwellian distribution, the absorption function $D_A(W_p)$ can be expressed in terms of elementary functions. For large λ (the weakly relativistic approximation),

$$D_A(W_p) \simeq \left[\frac{2}{\gamma_p^3} + \frac{\alpha_R - 1}{\gamma_p}\left(1 + \frac{1}{\gamma_p}\right) + \frac{\alpha_2}{\gamma_p}\right] - \frac{1}{\lambda\gamma_p^2}\left[\frac{2(\alpha_R - 1)}{\gamma_p} + \alpha_2\right]$$

$$+ \frac{1}{\lambda^2\gamma_p^3}\left[\frac{6}{\gamma_p^2} + \frac{6(\alpha_R - 1)}{\gamma_p} + 2\alpha_R\right] + O(\lambda^{-3}) \quad , \tag{2}$$

where

$$\alpha_R + i\alpha_I = -\frac{c^2\Omega_e}{V_{T_e}^2\omega}\frac{D(\epsilon_{33}^R - i\epsilon_{33}^I)}{|\epsilon_{33}|^2(S - n_{\scriptscriptstyle \parallel}^2)} \quad , \tag{3}$$

S, D, ϵ_{33} are plasma dielectric functions, W_p is the resonant momentum, and $\alpha_2 = (\alpha_R - 1)^2 + \alpha_I^2$. One can derive analytic expressions for $G(W_p)$ in the high $[G_H(W_p)]$ and low $[G_L(W_p)]$ energy limits. In the high energy limit, $G_H(W_p)$ is fairly complicated, although it is in terms of elementary functions and can be efficiently calculated. We only write down the leading order term in a W_p^{-1} series:

$$G_H(W_p) \simeq W_p^2 \, a_1(\gamma_p) \left[\frac{2}{\gamma_p^3} + \frac{2(\alpha_R - 1)}{\gamma_p^2} + \frac{\alpha_2}{\gamma_p}\right] \quad , \tag{4}$$

where γ_p is the relativistic-γ of resonant electrons,

$$a_1(\gamma_p) \approx \frac{4}{\gamma_p^2(Z + 1 + 4\kappa_1\gamma_p)} \quad , \tag{5}$$

where $\kappa_1 = K_1(\lambda)/K_2(\lambda)$, K_i are modified Bessel functions. On the other hand, the low energy limit is given by

$$G_L(W_p) = \frac{1}{W_p}\left[\frac{1}{4}I_4(\infty) + (\alpha_R - 1)\bar{I}_2(\infty) + \alpha_2 I_0(\infty)\right] \quad , \tag{6}$$

where

$$I_n(\infty) = \int dW \frac{2\pi V_{T_e}^3 f_M}{n_0} \frac{\chi_s}{\gamma^2} W^n \quad , \tag{7}$$

χ_s being the Spitzer function, f_M the Maxwellian. $I_n(\infty)$ are functions of Z and λ and can be tabulated or approximated by simple functions. We can obtain a smoothly joint function of $G(W_p)$ as follows:

$$G(W_p) = e^{\lambda(\gamma_p - 1)} G_L(W_p) + \left[1 - e^{-\lambda(\gamma_p - 1)}\right] G_H(W_p)$$

$$+ \frac{W_p \, e^{-\lambda(\gamma_p - 1)}}{\gamma_p(Z + 1 + \kappa_1\gamma_p)} \left\{\frac{A + BZ}{Z + 1} + \frac{C\,\alpha_R}{Z + 4} W_p^2\right\} \quad , \tag{8}$$

where $A = 11.5\alpha_2 + 23\alpha_R - 1.5$, $B = 7.5\alpha_2 + 15\alpha_R + 4.5$, and $C = 8$. There are weak temperature dependences in A, B, C which can be neglected for parameters of interest. The exponential factors are suggested by the background Maxwellian.

The analytic expression $\eta_A = J_\parallel/P_D$ is compared with the numerical evaluation by the adjoint code ADJ, and η_{NR} for the non-relativistic limit of η. In the relativistic calculations, the bulk temperature is taken to be 25 keV. Figure 1 is a comparison of η_A with results from ADJ for various values of Z. The agreement is quite good. Figure 2 compares the relativistic calculations for the Alfvén ($\alpha_R = \alpha_I = 0$) and Landau ($\alpha_R \gg 1$) limits at 25 keV with non-relativistic calculations. The striking characteristic is that the non-relativistic η_{NR} appears quite accurate in the Alfvén limit even at relativistic energies ($\gamma_p \lesssim \sqrt{2}$), while in the Landau limit the divergence between η_{NR} and η (relativistic) occurs at much lower energies (~ 100 keV). The result at the Alfvén limit is surprising because relativity is supposed to have an effect at $\gamma = \sqrt{2}$. In a closer look at the absorption (Fig. 3), we find that the relativistic effect is actually quite prominent at lower energies. Apparently, when $\alpha_R = \alpha_I = 0$, the relativistic decrease in $j_\parallel$ and in P_D are similar so that the ratio is rather insensitive to λ, the bulk temperature. The accuracy of Eq. (5) becomes worse as γ_p becomes larger than about $\sqrt{2}$. This can be improved with a more accurate function. Results will be presented in a more detailed paper.

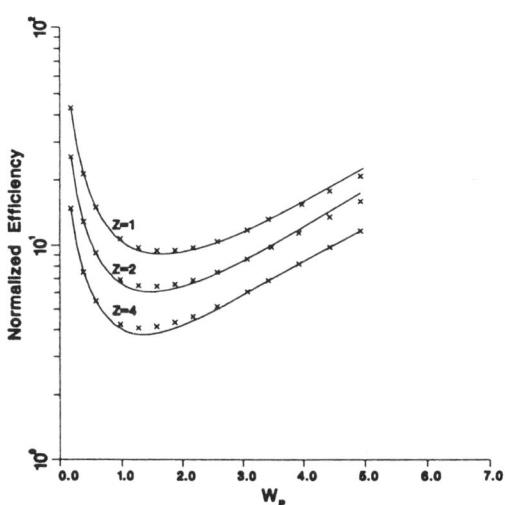

FIG. 1. Normalized J/P versus resonant momentum for $1/\lambda = T_e/mc^2 = 0.05$ for $Z = 1, 2, 4$. Solid lines are from ADJ code; crosses are from the analytic expression.

III. CONCLUSIONS

A general weakly relativistic theory of current drive is presented for parallel acceleration. The analytic expression for efficiency is found to be in good agreement with numerical results from the ADJ code. An interesting result is that the cold efficiency formula is quite accurate up to 0.5 MeV for the Alfvén limit but less so in the Landau limit. From absorption, it is found that even in the Alfvén limit, relativistic effects set in at lower energies, but this is compensated by a similar effect on $j_\parallel$ so that the ratio is not much changed.

IV. ACKNOWLEDGMENT

This is a report of work supported by U.S. DOE Contracts DE-AC03-89ER51114 and DE-AC02-76CH03073.

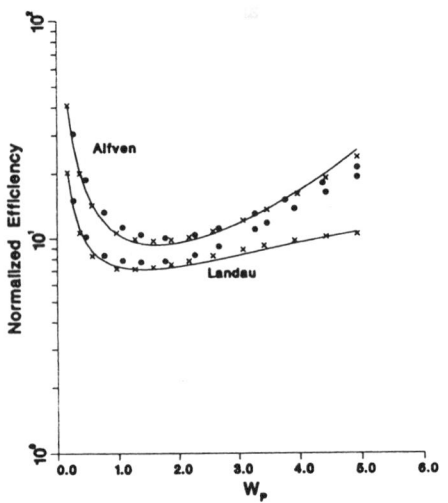

FIG. 2. Comparison between relativistic and non-relativistic J/P for $Z = 1$ at the Landau and Alfvén limits. Solid lines are relativistic ADJ results at $Z = 1$ and $\lambda^{-1} = 0.05$; crosses are from corresponding analytic expression; solid circles are non-relativistic results.

V. REFERENCES

1. N.J. Fisch, Rev. Mod. Phys. **59**, 175 (1987).
2. N.J. Fisch and C.F.F. Karney, Phys. Fluids **24**, 27 (1981).
3. D.A. Ehst and C.F.F. Karney, Nucl. Fusion **31**, 1933 (1991).
4. S.C. Chiu, V. Chan, R.W. Harvey, and M. Porkolab, Nucl. Fusion **29**, 2175 (1989).
5. S.C. Chiu, C.F.F. Karney, R.W. Harvey, and T.K. Mau, in *Proc. Europhysics Topical Conference on Radiofrequency Heating and Current Drive of Fusion Devices*, Brussels (1992), pp. 173.
6. C.F.F. Karney and N.J. Fisch, Phys. Fluids **28**, 116 (1985).
7. C.F.F. Karney, N.J. Fisch, and A.H. Reiman, in *Proc. 8th Topical Conference on RF Power in Plasmas*, Irvine (1989) (American Institute of Physics), p. 430.

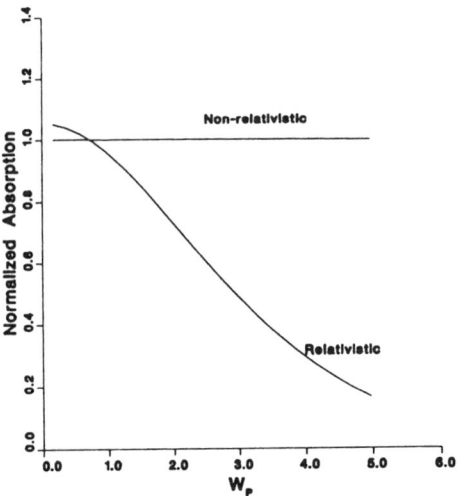

FIG. 3. Normalized absorption D_A versus W_p at the Alfvén limit.

SURFACE WAVES IN NONUNIFORM PLASMAS AND THEIR ABSORPTION AT THE LOCALIZATION OF SURFACE PLASMONS*

Yu.M. Aliev

Lebedev Institute, Russian Academy of Sciences, 117924 Moscow, Russia

J. Berndt, H. Schlüter

Experimental Physics II, Ruhr–University, 4630 Bochum, Germany

A. Shivarova

Faculty of Physics, Sofia–University, 1126 Sofia, Bulgaria

ABSTRACT

Plasmas produced by travelling surface waves are connected with density nonuniformities both in radial direction and in axial direction. Radial density profiles may lead to plasmon excitation close to the discharge tube walls and thus increased energy transfer to the plasma. Here the effect of axial plasma inhomogeneity is considered. In the geometrical optics approach — usually employed for models of this type of plasmas — at the end of the discharge resonant behaviour is expected, though the validity of approximation fails. The resonance is investigated on the basis of a second–order one–dimensional differential equation for the distribution of the electromagnetic surface wave field along a plane plasma–vacuum interface, assuming an axially linear density profile as often observed. Enhancement of electric field strength, storage of surface charge density and strong absorption can be expected in the region of spatial localization of surface plasmons. The strength of the resonance is noticeably affected by collisions.

I. INTRODUCTION

It is well known that electromagnetic surface waves (SWs) can propagate along the surface formed by the discontinuity of refractive indices. The possibility of sustaining gas discharges by this type of wave has given rise to a wide range of applications, e.g.[1] and continued theoretical studies. The effects of radial plasma density profiles on SW propagation has been previously treated, mostly by numerical calculations[2,3]. Recently it has been shown[4] that radial nonuniformity in SW plasmas leads to plasmon generation and associated increase of power transfer to the plasma, although the central plasma frequency ω_p of SW discharges is considerably larger than the wave frequency ω: A sufficient radial density drop is possible for large enough (electron – neutral) collision frequencies ν, at high ω still satisfying the demand $\nu/\omega < 1$. However, it should be stressed that the conditions are more important and easier to fulfill towards the end of the discharge column

---

*Support by the National Foundation for Scientific Research in Bulgaria (project F27), Alexander von Humboldt Foundation and Deutsche Forschungsgemeinschaft (exchange program, SFB 191) is gratefully acknowledged.

310 Surface Waves in Nonuniform Plasmas

in view of axial plasma density decrease.

The origin of axial nonuniformity from an interplay of dispersion and power balance has been investigated repeatedly, leading to essentially linear axial density profiles. The influence of axial inhomogeneity on dispersion is commonly treated by using local dispersions relations (geometrical optics approach). On this basis at the end of the discharge a ("bounded system") resonance is predicted (at $(\omega_p/\omega)^2 \approx 2$) which could lead to considerable power storage and to power transfer to the plasma. However, fast changes of the wave number k render questionable the validity of the geometrical optics approach. Below the resonance is considered for a model situation also concerning the influence of collisions.

II. BOUNDARY VALUE PROBLEM

The essential features may be studied at the system of a plane plasma–vacuum interface, the SW travelling in z-direction. The plasma permittivity

$$\varepsilon = \varepsilon(z) = 1 - \{\omega_p^2(z)/[\omega(\omega + i\nu)]\} \tag{1}$$

is based on linear hydrodynamics (cold electrons), now for simplicity of analysis assumed independent of x (thus neglecting radial plasma density profiles). Using Maxwell's equations, the magnetic field strength (B_y) is governed by:

$$\varepsilon \frac{\partial}{\partial z}\left(\frac{1}{\varepsilon}\frac{\partial B}{\partial z}\right) + \frac{\partial^2 B}{\partial x^2} + \frac{\omega^2}{c^2}\varepsilon B = 0 \tag{2}$$

At the boundary ($x = 0$), the conditions here to be statisfied (p referring to plasma, v to vacuum) are:

$$B^p(x=0, z) = B^v(x=0, z) = B_0(z) \tag{3}$$

$$\varepsilon^{-1} \left[\partial B^p(x,z)/\partial x\right]\big|_{x=0} = \left[\partial B^v(x,z)/\partial x\right]\big|_{x=0} \tag{4}$$

For the homogeneous case ($\varepsilon' = \frac{d\varepsilon}{dz} = 0$), (2) can be solved by ($x \leq 0$)

$$B = B_0(z)\exp(a_p \cdot x) \tag{5}$$

with a_p being constant, leading to (at the interface $x = 0$):

$$\frac{d^2 B_0}{dz^2} + \frac{\omega^2}{c^2}\frac{\varepsilon}{1+\varepsilon} B_0 = 0 \tag{6}$$

The coefficient at B_0 defines a wave number (k^2), and solutions are $\sim \exp(ikz)$. For the inhomogeneous case, a description may be tried by using the boundary equation (6) with $\varepsilon = \varepsilon(z)$. This is valid under the assumption that the influence of axial inhomogeneity on the transverse distribution of the SW field is small, equivalent to a geometrical–optics–type description concerning the x–direction only. Then it is consistent to neglect the term $-\frac{\varepsilon'}{\varepsilon}\frac{dB_0}{dz}$, justified for wavelengths small compared to the density scale length $\left(\frac{d\ln|\varepsilon|}{dz}\right)^{-1}$, usually fulfilled.

With a linear density profile $\omega_p^2(z) \approx (1 - z/L)$ (and $z = 0$ chosen for the "resonance" point $\mathrm{Re}(\varepsilon_p) = -1$), now (6) leads to an equation with singular turning point, often employed in ionospheric and cyclotron heating problems[5],

$$\frac{d^2 B_0}{dz^2} + \frac{\omega^2}{c^2} B_0 - \frac{\omega^2}{c^2} \frac{L}{2z} B_0 = 0, \qquad (7)$$

into which collision terms can easily be incorporated. The solutions are Whittaker functions, their quasi–periodic behaviour (for $z \leq 0$) replacing that of the $\exp(ikz)$–functions of the homogeneous case or $\exp(i \int k dz)$–functions of the geometrical optics approach and determine the behaviour towards and beyond the resonance point. If the second term of (7) is neglected, Hankel functions are obtained instead. On the basis of this treatment absence of noticeable reflection can be concluded analytically as well as strong peaks of the (modulus of) electric field components $E_x = (-i/\varepsilon)(c/\omega)(dB_0/dz)$ and $E_z = (-iB_0)/\sqrt{-(1+\varepsilon)}$.

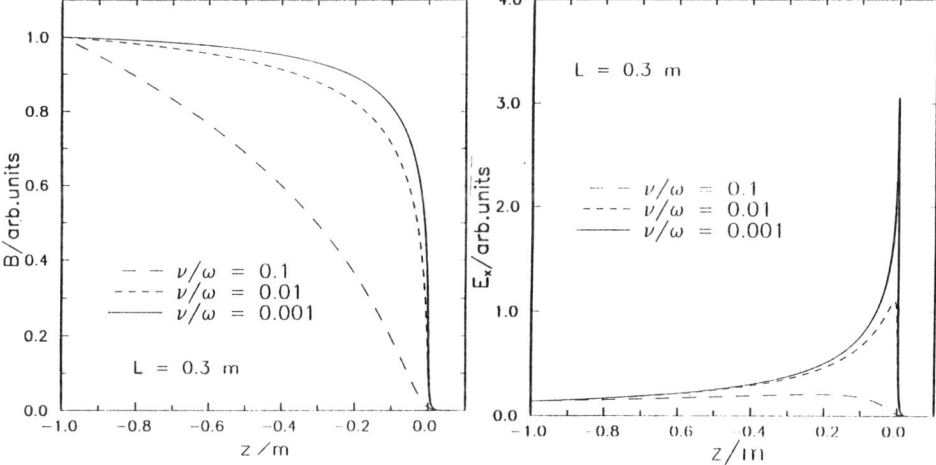

Figure 1: Modulus B_0 for 2,45 GHz.   Figure 2: As in fig. 1, but modulus E_x.

III. NUMERICAL RESULTS

For $\omega = 2\pi \cdot 2.45$ GHz, used in most experiments, the typical behaviour of B_0 is visualized in fig. 1: The modulus of B_0 is shown for a scale length of $L = 0.3\,m$. In fig. 2 the resultant modulus of E_x is depicted for the same conditions. The results for E_z look similar, with modulus E_z larger than modulus E_x. In (geometrical optics) approaches collisions are usually neglected concerning the dispersion[6], their influence being weak away from the resonance point indeed, whereas there is of cause a strong effect on the resonance. In fig. 3 the effect of enlarged inhomogeneity with $L = 0.15\,m$ is demonstrated for E_x. Fig. 4 depicts the behaviour for varied frequencies ω with $L = 0.30\,m$ and $\nu/\omega = 0.01$ with modulus E_z shown in this case. The curves are normalized to the equal modulus B_0 at $z = -1\,m$. The numerical (complex oscillatory) solutions are obtained by

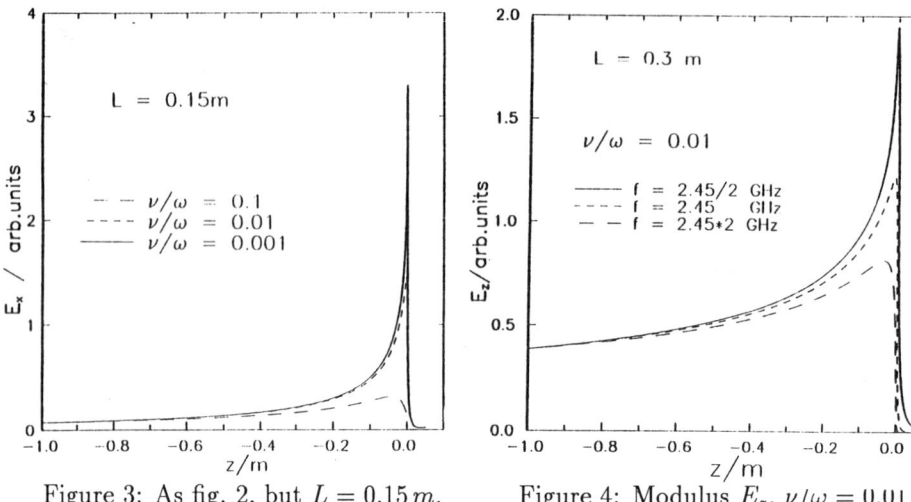

Figure 3: As fig. 2, but $L = 0.15\,m$. Figure 4: Modulus E_z, $\nu/\omega = 0.01$.

backward integration starting beyond the resonance point at positive z. For the cases considered, no indication of reflection appeared in the numerical calculations.

IV. SUMMARY

The results demonstrate the possibility of strong surface charge density, power storage and absorption at the end of the SW sustained discharge where the spatial localization of "surface plasmons" occurs. The large field strengths encountered can be expected to influence the discharge behaviour. Strong collisional damping reduces these effects, though in experimental situations, low enough ν/ω are realizable to show the resonance effect. Moreover, in the case of larger ν/ω (with ν/ω still $\ll 1$), as mentioned, the likelyhood of plasmon generation associated with radial density profiles grows. The extreme sharpness and strength of resonance due to axial nonuniformity for rather low values of ν/ω may be limited by effects not contained in the model presented. Such effects deserve further attention, in particular concerning improved description of transverse field distribution in the presence of axial density inhomogeneity.

REFERENCES

[1] M. Moisan and Z. Zakrzewski, Rev. Sci. Instrum. **58**, 1895 (1987).
[2] R. Darchicourt, S. Pasquieres, C. Boisse–Laporte, P. Leprince and J. Marec, J. Phys. D.: Appl. Phys. **21**, 293 (1988).
[3] M. Zethoff and U. Kortshagen, J. Phys. D.: Appl. Phys. **25**, 1574 (1992).
[4] Yu.M. Aliev, V.Yu. Bychenkov, A.V. Maximov and H. Schlüter, Plasma Sources Sci. Technol. **1**, 126 (1992).
[5] T.H. Stix, The Propagation of Electromagnetic Waves in Plasmas (Pergamon Press, 1964).
[6] E. Benova, I. Zhelyazkov, Physica Scripta **43**, 68 (1991).

ANTENNA DESIGN
AND RF TECHNOLOGY

HIGH-POWER AND LONG PULSE CAPABILITY OF THE LH SYSTEMS ON TORE SUPRA AND TEST BED FACILITY

G. REY, M. GONICHE, G. BERGER-BY, P. BIBET, J.P. BIZARRO, J.J. CAPITAIN,
J. CARRASCO, Y. DEMERS*, D. GUILHEM, G.T. HOANG, X. LITAUDON, R. MAGNE,
D. MOREAU, Y. PEYSSON, M. SEKI**, J. SCHLOSSER, G. TONON.

Association EURATOM-CEA sur la Fusion Contrôlée - CE Cadarache
F-13108 Saint Paul Lez Durance CEDEX, FRANCE

* Tokamak de Varennes (CCFM) 1804 Montée Ste Julie Varennes, Quebec, Canada JBX.

** JAERI, Naka-Machi, Naka-Gun, IBARAKI-KEN, JAPAN.

ABSTRACT

During the 92 TORE SUPRA experimental campaign, 48 % of the plasma shots has been devoted to lower hybrid experiments. 75 shots with a mean pulse duration of 27 s (10 to 60 s [1]) have been performed using the LH system with a mean RF power of 2.8 MW corresponding to an incident power density at the launcher mouth of 17 MW/m^2. More than 40 RF shots - 1 to 5 s long - have been also achieved with injected power between 0.6 to 0.94 of the maximum available power (6.8 MW). With a typical launcher-limiter distance of 3 cm, the thermal load on the carbon guard of the launcher is lowered to less than 1 % of the total injected power (3 MW).

On test bed facility, modules of antenna made of Dispersion Strengthened Copper (D.S.C.) have been successfully tested in collaboration with the CCFM (Canada) and JAERI (Japan). Super long shots up to 100 mn have been achieved with transmitted power density of 50 MW/m^2.

For next step device, as TORE SUPRA CONTINU, D.S.C. material and RF power density of 25 MW/m^2 are relevant options for high temperature quasi continuous operating conditions.

1. THE LH SYSTEM OF TORE SUPRA

The RF power is supplied by 16-500 kW-ThCSF klystrons working at 3.7 GHz in quasi continuous regime [2]. Each klystron feeds 2 superposed modules of the antennae via a 25 m long partly oversized transmission waveguide. The launcher is composed of 16 multijunctions [3,4] made of Cu-Zr modules allowing working temperature up to 400°C. The $N_{//}$ spectrum of the antenna is centered at 1.9 and can be changed from 1.3 to 2.5 by changing the phase of the wave between adjacent klystrons. All the facing components surrounding the antennae are covered with carbon tiles bolted or brazed on water cooled structure.

2. HIGH POWER CAPABILITY AND LONG PULSE OPERATION

2.1 Conditioning is mainly based on baking at temperature of 150-200°C for working temperature at 120-150°C and by sequences of 800-10 ms long RF pulses with a duty cycle of 1/10 in vacuo between plasma shots. In fig. 1 is plotted the averaged injected power for RF pulse longer than 1 s against the cumulated shots on plasma for the 2 grills : 4 MW are performed with the 2 grills after 40 plasma shots. In darked spots is plotted the injected power for RF pulse length exceeding 8 s and corresponding to almost breakdown free shots required for the experimental campaigns. The associated power density of 20 MW/m^2 can be considered as a reliable working power density easily obtained even after opening session.

2.2 The 92 experimental campaign. In fig. 2, are collected more than 120 RF shots with pulse duration of 1 to 62 s and RF power above 2 MW. 75 shots have been devoted to long pulse

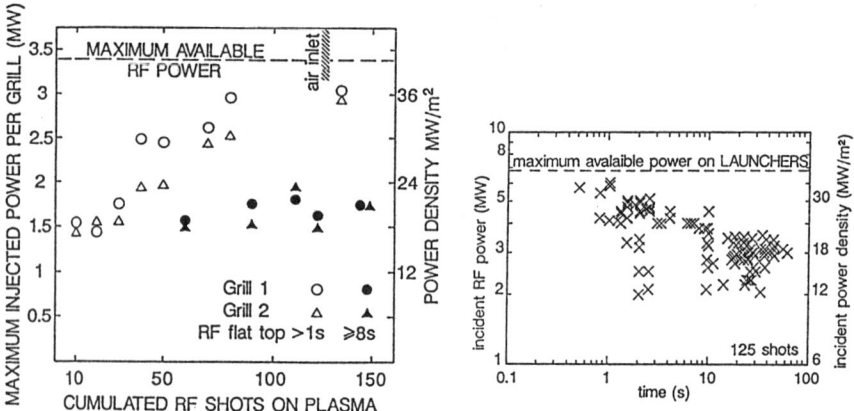

Fig. 1. Maximum injected power per grill versus the cumulated RF shots on plasma.

Fig. 2. LH performances.

experiments, the averaged time duration and injected power are respectively 27 s and 2.8 MW corresponding to an incident power density of 17 MW/m^2 ; limitation of the injected RF energy is related to exhaust power capability of the plasma facing components. More than 40 shots - 1 to 5 s long - have been also performed with RF power above 2/3 of the maximum available power (6.8 MW).

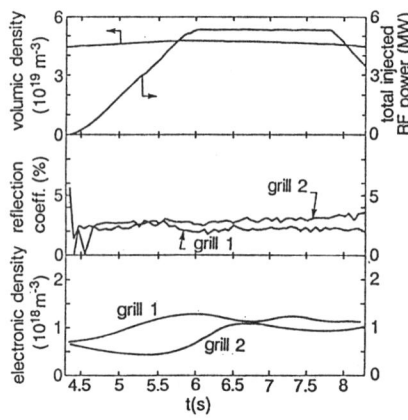

Fig. 3. High power capability

Fig. 3 illustrates the high power capability of a 5.2 MW - 2 s long breakdown free pulse operated at high density (volume averaged density of 4.5 10^{19} m^{-3}) : the reflection coefficient of the 2 launchers lies constant at values of 2 to 3 % when the electronic density at the grill mouth, measured with Langmuir probes, is in the range of 0.5 to 1.5 10^{18} m^{-3}. From swan code simulation associated to the measured reflextion coefficient, the electric fields in reduced waveguides are deduced : peak power density exceeds 80 MW/m^2 for an incident power density of 35 MW/m^2.

The main plasma and RF parameters of the record shot of 60 s flat top 1 MA are reported in fig. 4. The RF energy injected in the torus reaches 180 MJ, very weak outgassing is observed during the 62 s - 3 MW of RF power corresponding to a power density of 18 MW/m^2. From the 3.10^{-4} mPa pressure increase observed and the 40 m^3 s^{-1} measured pumping speed, outgassing rate of 2 10^{-4} Pam s^{-1} can be deduced.

On fig 5 is plotted the heat load on the carbon guard referred to be the total power (in the range of 3 MW) : values below 1 % are obtained when the grill-limiter distance is greater than 3 cm. The experimental points are well fitted by an exponential low with an e-folding length λ_Q = 1 cm.

For a typical 3 cm grill-plasma distance and 3.2 MW power, the surface temperature increase of the carbon guard limiter saturates at 300°C with a time constant of 2-3 s in good agreement with simulations from a 2D thermohydraulic code (CASTEM 2000) and flux of 3.5 MW/m² at the leading edge.

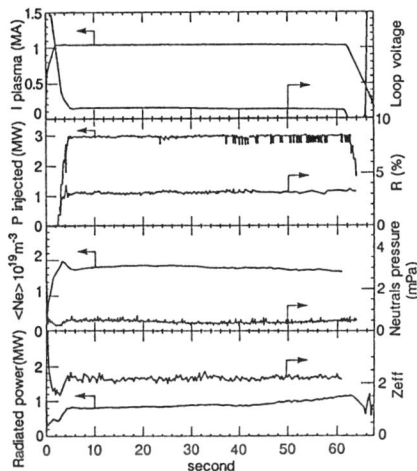

Fig. 4. 1 mm RF discharge

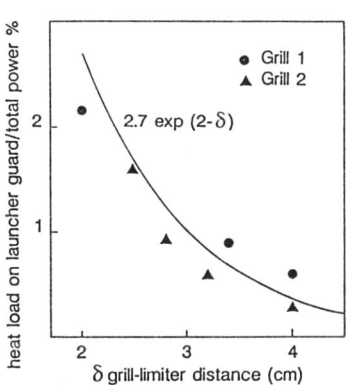

Fig. 5. Heat load on launcher guard.

3. HIGH POWER AND LONG PULSE TESTS ON NEW MODULES OF ANTENNA

3.1 New multijunction modules made of Dispersion Strengthened Copper (D.S.C.) - namely the Glidcop AL15 - OFHC copper with 0.15 % of alumina powder - have been tested successfully on the RF test bed in association with TdV [5] (Canada) and JAERI [6] (Japan) laboratories. Such material are very promising for new devices due to their excellent thermal and electrical properties (90 % of IACS) associated to a high yield strength at high temperature (160 MPa et 400°C).

3.2 Conditioning and outgassing properties. The conditioning procedures based on high temperature (300 to 450°C) baking are listed at the bottom of fig. 6. High RF peak power density up to 200 MW/m² have been easily performed : 10 to 30 RF shots were needed to obtain breakdown free pulses for both modules operating as well in travelling waves in the JAERI module as in standing waves in the TdV multijunction. As indicated on fig. 6, the outgassing rate is decreased by 2 orders of magnitude by the RF conditioning while 1 order of magnitude is still gained after a 450°C baking. Slight effect the outgassing is observed when baking is performed with H_2 pressure (up to 300 Pa) to reduce copper oxide and the multipactor associated to the large S.E.E. coefficient of oxides.

On fig. 7 is plotted the outgassing rate of DSC modules against temperature obtained during RF power injection. The outgassing rate scales exponently with the temperature the module with 1 order of magnitude increase every 120°C. Further increase of the outgassing are observed above 400°C, such an effect may be related to the baking temperature performed only up to 450°C which is lower than the operating temperature of the module.

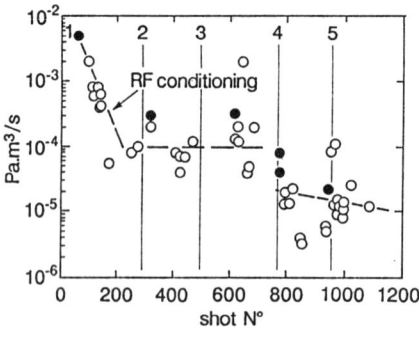

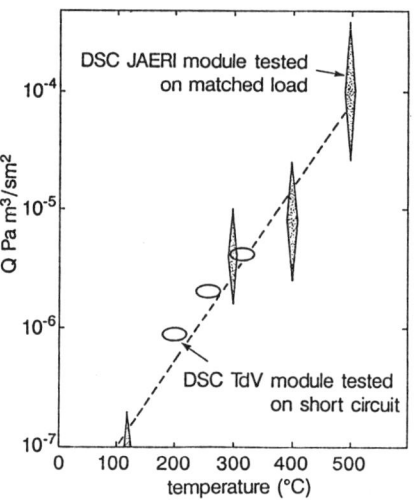

• first shot o following shots

1-2 baking 300°C - 60h
3 baking 300°C - 50h+10h with H_2 at 10Pa
4 baking 450°C - 60h
5 baking 450°C - 50h+10h with D_2 at 10Pa

Fig. 6. Outgassing flux at T = 500°C.

Fig. 7. Outgassing rate of DSC of module

3.3 Long pulse capability. Super long shots up to 100 mn time duration at power density of 50 MW/m² have been successfully achieved (fig. 8). Despite an air cooling system connected at the top and the bottom of the module, steady state of the temperature and outgassing is not obtained. With water cooling, constant outgassing rate at the level of 10^{-7} Pa m s^{-1} has been achieved with an equilibrium temperature of 100°C.

4. CONCLUSION

For TORE SUPRA CONTINU, 20 MW of LH waves are required to sustain 2 MA at volume averaged density of $5\ 10^{19}$ m^{-3} 32 × 700 kW cw klystrons at 3.7 Ghz will be needed to supply 4 × 5 MW launchers. As for next step devices, DSC material and operating RF power density below 30 MW/m², will be relevant options allowing in quasi continuous and high temperature regime, reliable working conditions.

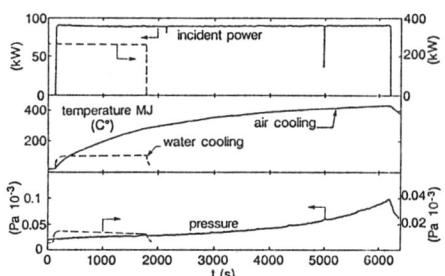

Fig. 8. In plain line : 100 mn long pulse with air cooling in dotted line a 3 mn long pulse with water cooling at 150 MW/m² of power density.

REFERENCES

1. Equipe TORE SUPRA pres. by G. TONON. Plasma Phys. Control Fusion 35 (1993) A105-A122.
2. R. MAGNE et al. (1988) Proc. 15th Symp. on Fus. Tech. Utrech. Vol. 1, p. 524.
3. G. REY et al. (1988) Proc. 15th Symp. on Fus. Tech. Utrech. Vol. 1, p. 514.
4. D. MOREAU, Proc. Int. Conf. on Cont. Fus. and Plasma Heat. (1984) Lausanne Vol. 1, p. 216.
5. Y. DEMERS et al. Europhys. Top. Conf. RF Heat. Cur. Drive of Fus. Devices. Brussels July 92.
6. M. GONICHE, M. SEKI. Rapport to be published.

CHARACTERISTICS OF A LARGE MULTIJUNCTION LAUNCHER FOR HIGH-POWER LHCD EXPERIMENTS ON JT-60U

M. Seki, Y. Ikeda, K. Ushigusa, O. Naito,
T. Kondoh, S. W. Wolfe, and T. Imai

Japan Atomic Energy Research Institute, Naka Fusion Research Establishment
Naka-machi, Naka-gun, Ibaraki-ken, 311-01, Japan

ABSTRACT

This paper presents overview of a large multijunction launcher for JT-60U. The launcher is featured by the multijunction module with the oversized taper waveguide, in order to simplify structure of the launcher. This launcher allows high performances of current drive and current profile control by using very sharp and highly directive spectrum. Initial result of coupling property is also described. A good coupling was observed at a power level of ~0.8 MW with plasma-launcher distance of <14 cm.

INTRODUCTION

Since the steady state operation is an important issue for a tokamak reactor, the establishment of non-inductive current drive is a key item. The lower hybrid current drive (LHCD) will be an useful method in a next generation tokamak. Therefore the extended data base of LHCD is necessary for design of the future launcher. And the main objectives of LHCD experiments for JT-60U are to drive plasma current efficiently and to control current profile in high plasma current (I_p = 3-4MA) and high density ($n_e > 3 \times 10^{19} m^{-3}$) regions. In order to perform these experiments successfully, a large multijunction launcher (CD-2) was constructed in addition to the 3-divided multijunction launcher (CD-1') for injection of high rf power up to ~10MW into JT-60U plasma.

In this paper, the necessity of the simple structure for the launcher is explained. The design of the CD-2 launcher composed of the simple multijunction modules is described. The first result of coupling property is also presented.

DEVELOPMENT OF THE SIMPLE MULTIJUNCTION MODULE

To perform current drive with high efficiency, many waveguides of the launcher are needed for launching sharp and directive spectrum. On the other hand, number of transmission line should be reduced from a point of view of system engineering. To satisfy both requirements, the launcher with E-plane multijunction technique was proposed. Because this method enables to increase the effective waveguides in spite of the same transmission lines. The launcher with 3-divided multijunction module was developed in the JT-60 and the current drive experiments by using the launcher (CD-1) showed good performance with high efficiency.

The CD-1 launcher, however, proved validity in LHCD experiments, this launcher is not sufficient for a future launcher. Because 12-18 waveguides are required to improve the spectrum and to reduce power density at grill mouth. For the purpose an 18-divided module, in which an rf power was split at 3-junction points as 1 -> 3 -> 6 -> 18, was fabricated. This 18-divided launcher showed good rf property. However unfortunately its structure was very complicated, so that it was hard to construct, very long, and high cost. It was recognized that the simplification of launcher structure was critical problem for adaptability to the next LHCD system. Therefore a simple multijunction module must be developed.

To realize a simple structure of a launcher, a multijunction module composed of 12-divided waveguides and over-sized taper waveguide was employed as shown in Fig. 1. The main issue in the R&D of this new type multijunction module was to investigate the effect of higher modes in the taper waveguide on the rf property. According to the results reported previously [1], the higher mode effect was negligible in the case of the reflection coefficient less than 10% in the sub-waveguides. Under this coupling condition, the reflection coefficient in a transmission line was deduced to be less than 5-10%. When the reflection coefficient exceeded ~10%, the phase shift by ~40deg. occurred in sub-waveguides. In other words, the low reflection coefficient is the key point whether this simple multijunction module is available or not. It should be pointed out that if a good coupling is observed with the CD-2 launcher, the high performances in LHCD experiments are expected.

DESIGN OF A LARGE MULTIJUNCTION LAUNCHER

In order to improve current drive efficiency, the CD-2 launcher was designed to radiate very sharp and high directive spectrum, which demanded many waveguides at the grill mouth. The 48 waveguides lined in the toroidal direction were considered through the optimization of the launching spectrum under the conditions of RF port size and the engineering limits. This spectrum will allow high current drive efficiency by 20-30% compared to that of CD-1' launcher. As mentioned above, simplification of launcher structure is important. The CD-2 launcher consists of 4x4 multijunction modules and its structure is very simple and compact due to adoption of over-sized taper waveguide in the module. The CD-2 launcher is shown in the photograph 1. To obtain good coupling, a launcher position is adjustable in a stroke of 60mm shot by shot. A fast moving probe will be available to measure density and temperature in front of the grill mouth for studying coupling mechanism. Moreover a gas puffing in RF port will operate to control coupling property actively by regulation of a neutral gas pressure. This idea is based on the analysis of a distant coupling phenomena of CD-1' launcher as reported in the reference [2]. The CD-2 launcher has two acceleration monitors to estimate the electromagnetic force on the launcher when a plasma disruption occurs. The design of the structure, especially supporting parts, will strongly depend on the estimation of the torque. These results will be described in near future. Photograph 2 shows the grill mouth of the CD-2 launcher in the RF port. The gas puffing port is a rectangular pipe equipped on the left side of the RF port. The fast moving probe is installed on the lower side of the launcher.

INITIAL RESULTS OF COUPLING PROPERTY

Before high power experiments in JT-60U, the CD-2 launcher was pre-baked at a temperature of 250°C during two weeks in the test stand and short rf pulses of 100msec were injected into every 2-module at power level of 200-400kW. The monitoring system for rf measurements, temperature measurements, CCD-TV monitor were prepared in JT-60U LHCD system and inter-lock systems were also arranged. At first the total system performance was checked by using small rf power with adjustment of inter-lock level to protect the grill mouth, vacuum window, and klystron. The injection power into vacuum chamber of JT-60U was gradually increased with monitoring temperatures of the launcher by thermo-couples and infrared TV, finally injection power reached up to 500kW-20msec and/or 200kW-2sec in each multijunction module. The reflection coefficients in vacuum loading were ~15,

80, and 10% for the frequencies of 1.74, 2, and 2.23GHz, respectively. The vacuum conditionings have been performed successfully.

The coupling experiment has started by using JT-60U plasma. The CD-2 launcher was set behind the first wall, and was gradually pushed to plasma. To obtain good coupling, plasma position was closed to the launcher head. The input power up to 1MW was injected into plasma after aging, and the coupling property was investigated at a power level of 800kW. According to the results shown in Fig. 2, the coupling property strongly depends on the distance (δ) between plasma surface and launcher head as expected. The reflection coefficient rapidly decreases when δ is less than the critical length of ~14cm. This improvement of coupling suggests that the density at the grill mouth exceeds around the cut-off density (~5×10^{10} cm^{-3}), even though the peripheral density is not measured by a probe. The calculation shows the same coupling property to experimental results.

The incident and reflective power are sketched in Fig. 3 for 16 modules in the shot of Ip=1.8MA, Bt=3T, ne~1.7×10^{19} m^{-3} with injection power of ~0.8MW. As known from the figure the columns 1-3 show the reflection coefficient less than several percents. On the other hand the reflection coefficient of column 4 is relatively high up to 50%, it is main reason why the δ of the column 4 is larger (~9cm) than others (~6cm) as shown in Fig. 4. It suggests that after optimization of plasma position in front of the grill mouth, the multijunction module with oversized taper waveguide will be available. And the CD-2 launcher is expected to perform current drive and current profile control efficiently by using high power with low reflection coefficient.

SUMMARY

A large multijunction launcher in JT-60U was newly constructed to excite very sharp and highly directive wave spectrum. This launcher is featured by adoption of an oversized-taper waveguide in a multijunction module. The multijunction module enables to make the structure of launching system simple. The launcher composed of 16 modules (4 columns x 4 rows) has 48 waveguides in toroidal direction and then totally 192 waveguides. An improvement of current drive efficiency of this launcher is expected more than 20% comparing to previous experiments with 3-divided multijunction launcher in JT-60U. The coupling experiments at a power level of ~1MW have been started since February 1993. The coupling of reflection coefficient less than several percents was observed at the distance between plasma and the launcher head of 6cm. This result suggests that the CD-2 launcher composed of simple multijunction launcher is feasible. And the high performance on current drive and current profile control are expected.

ACKNOWLEDGMENTS

The authors would like to express our appreciation to the members of the JT-60U.

REFERENCES

1. Y. Ikeda, M. Seki, et al., 14th Symposium on Fusion Engineering, San Diego, September 30 - October 3, 1991
2. Y. Ikeda, M. Seki, et al., 17th Symposium on Fusion Technology, Rome, September 14-18, 1992

Photo. 1. Overview of the CD-2 launcher.

Photo. 2. The gas puffing port and the fast moving probe.

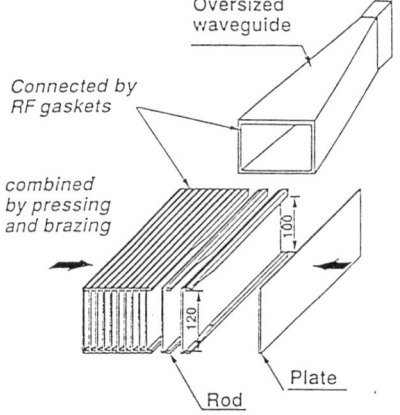

Fig. 1. Structure of the new multijunction module.

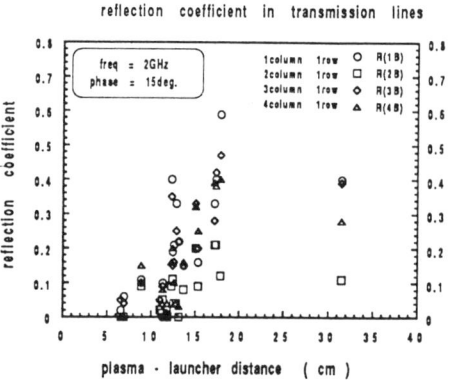

Fig. 2. Reflection coefficient vs the plasma launcher distance.

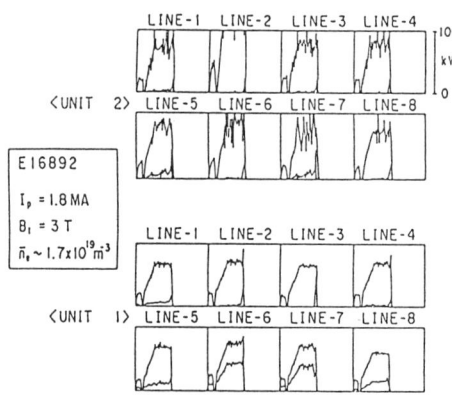

Fig. 3. Typical rf wave form in a good coupling.

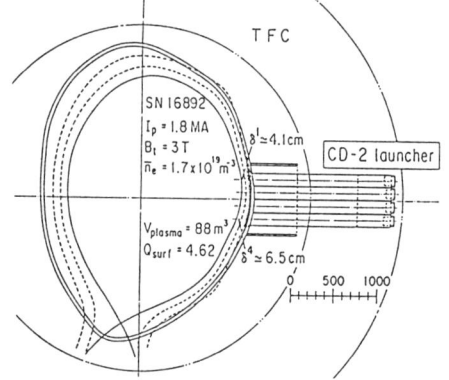

Fig. 4. Plasma launcher distance in a good coupling.

COMBLINE ANTENNAS FOR LAUNCHING TRAVELING FAST WAVES

C.P. Moeller, R.W. Gould,* D.A. Phelps, and R.I. Pinsker

General Atomics, San Diego, CA 92186-9784

ABSTRACT

The combline structure shows promise for launching traveling fast magnetosonic waves with adjustable $n_\parallel$ ($3 \leqslant n_\parallel \leqslant 6$) for current drive.[1] In this paper, the dispersion and damping properties of the combline antenna with and without a Faraday shield are given. The addition of a Faraday shield which eliminates the electrostatic coupling between current straps as well as between the straps and plasma offers the advantage of eliminating the need for the lumped capacitors which are otherwise required with this structure. The results of vacuum dispersion and damping measurements on a low power model antenna are also given.

INTRODUCTION

For current drive, it is necessary to launch a unidirectional wave, and desirable to vary the launched $n_\parallel$ in the range 3 to 6 during a discharge while presenting a matched load to the generator, even as the plasma position and edge density vary. These additional requirements are much more easily achieved by a radiating slow wave structure, fed at one end, than by the array of individually fed current straps presently used for launching fast waves. Such a structure is inherently a traveling wave device, in which the mutual coupling between radiating elements is part of the wave propagation, in contrast to individually fed radiating elements, for which the mutual coupling leads to unequal loading of the elements. A single point feed also eliminates the need for most of the external matching networks the individually fed current straps require. The input will appear matched to the external transmission line if the structure is long enough to radiate all of the incident power, if the output is terminated in a matched load, or if the structure is made part of a resonant ring. With a matched input, for the same total power, the voltage at the vacuum feedthrough may be lower than that at the feedthroughs of the individually fed straps with their high VSWRs.

A slow wave structure which has elements closely resembling the current straps of present fast wave antennas is the combline,[2] which is shown schematically in Fig. 1 configured as an antenna. Except for being open at the front, this structure is identical to commercially available bandpass combline filters. As with any bandpass filter, as the frequency varies within the passband, the phase shift from element to element ranges from 0 radians at the lower cutoff to π radians at the upper cutoff, although as will be seen, the range from $\pi/4$ to $\pi/2$ is the most desirable with regard to ohmic loss and radiated wavenumber spectral purity. The two disadvantages of the combline antenna as shown in Fig. 1 are the lack of a Faraday shield, and the lumped capacitance which is required at the end of each current strap. Although the required capacitance is small, the required area is inconveniently large. Both these drawbacks can be overcome by configuring a Faraday shield as shown in Fig. 2, so as to eliminate capacitive coupling between current straps, as will be discussed in the next section.

---

*Permanent Address: California Institute of Technology, Pasadena, CA 91125.

VACUUM DISPERSION PROPERTIES

The vacuum properties will be analyzed with the plasma replaced by a conducting wall, with the surface impedance of the plasma treated as a perturbation on the vacuum solution, in a extension of an approach used by Golant.[3] We can regard the array of current straps and the conducting walls as a section of multi-conductor transmission line, which is governed by the equations:

and
$$\frac{\partial V_r}{\partial y} = -i\omega \sum_s L_{rs} I_s , \quad (1)$$

$$\frac{\partial I_r}{\partial y} = -i\omega \sum_s C_{rs} V_s , \quad (2a)$$

where y is the coordinate along a conductor, $V_r(y)$ and $I_r(y)$ are the voltage and current on the rth conductor at y, and L_{rs} and C_{rs} are the mutual inductance and capacitance per unit length, respectively, between the rth and s^{th} conductors. With an ideal Faraday shield, which does not affect inductances but completely shields each bar electrostatically, Eq. (2a) is replaced by

$$\frac{\partial I_r}{\partial y} = -i\omega C_0 V_r , \quad (2b)$$

where C_0 is the capacitance per unit length of each bar to its shield. With all the conductors grounded at $y = 0$, and assuming an infinite array of identical elements, we can let $V_r = V^0 \sin(\beta y) \exp(-ir\theta)$ and $I_r = I^0 \cos(\beta y) \exp(-ir\theta)$, where θ is the phase shift from element to element to be determined and β is the propagation constant along y. From (1) and (2a), respectively, we then obtain $V^0/I^0 = -i(\omega/\beta)\sum_s L_{0s} \exp(-is\theta) \equiv -i(\omega/\beta)L(\theta)$ and $I^0/V^0 = i(\omega/\beta)\sum_s C_{0s} \exp(-is\theta) \equiv i(\omega/\beta)C(\theta)$ where the $r = 0$ element is typical in the infinite array. With a shield the second equation is replaced by $I^0/V^0 = i(\omega/\beta)C_0$. Without the shield, $\beta/\omega = [L(\theta)C(\theta)]^{1/2} = 1/c$ for a TEM wave. With the shield, $\beta/\omega = [L(\theta)C_0]^{1/2}$. The dispersion relations are determined by the boundary conditions at $y = \ell$: $I_r/V_r = i\omega C_e$, C_e the lumped capacitance, without a shield and $I_r = 0$ and $\beta\ell = \pi/2$ with a shield. The corresponding dispersion relations are respectively

and
$$L(\theta)C_e = 1/[c\omega \tan(\omega\ell/c)] , \quad (3a)$$

$$L(\theta)C_0 = 1/(2\ell\omega/\pi)^2 . \quad (3b)$$

It is apparent that if $C_e = 0$ in (3a), the only solution is $\omega\ell/c = \pi/2$, giving zero pass band width and zero group velocity. With an electrostatic shield, corresponding to (3b), the lumped capacitance is unnecessary.

The total electromagnetic field energy in a unit cell without a shield is $W_{\text{cell}} = (1/4)V_p^2\{\ell/[c^2 L(\theta) \sin^2(k_0\ell)] + C_e\}$, while with the shield it is $W_{\text{cell}} = (1/8)V_p^2\ell\{C_0 + (\pi/2)^2/[\omega^2\ell^2 L(\theta)]\}$, where V_p is the peak voltage at the end of the strap. The power flow along the structure is then just $P_0 = W_{\text{cell}} d\omega/d\theta$.

We have solved for $C(\theta)$ using a variational technique for the geometry shown in Fig. 1. $L(\theta)$ is then obtained from $L(\theta) = 1/[C(\theta)c^2]$. With an ideal Faraday shield, $L(\theta)$ is presumed not to change, while we can calculate C_0 approximately using the same variational technique. An example of the dispersion properties is shown in Fig. 3, where, referring to Fig. 1, $w = 5$ cm, $t = 2.5$ cm, $s = 5.4$ cm, $\ell = 31.25$ cm, and $d_2 = 15$ cm to a conducting front wall. For case A, $d_1 = 5$ cm, $\ell = 31.25$ cm, and

and $C_e = 16$ pF; for case B, $d_1 = 2.5$ cm and $C_e = 20$ pF; case C is similar to A except $C_e = 0$, $\ell = 40$ cm, and there is a Faraday shield for which $C_0 = 96$ pF/m. Corresponding values of $V_p/(P_0)^{1/2}$ at the passband center, P_0 the input power, are $28\,V/W^{1/2}$ for case A, $32\,V/W^{1/2}$ for case B, and $25\,V/W^{1/2}$ for case C.

Regarding the spatial spectrum, the k_z values of the fields in the structure must satisfy $k_z = (\theta/p) + (2\pi N/p)$, where p is the period and N is an integer. The most troublesome harmonic has $N = -1$. By making p small, $k_{z(-1)}/k_{z(0)}$ can be made so large that the evanescent region can filter out the harmonic. In our example, $3 \leqslant |k_{z(-1)}/k_{z(0)}| \leqslant 7$.

PLASMA DAMPING

We have evaluated the plasma damping by determining $\text{Im}\,n_z$. n_z can be written in a Taylor series as

$$n_z(\rho_p) = n_{z0} + \left.\frac{dn_z}{d\rho}\right|_{\substack{\rho=\rho_r \\ u=u_r}} (\rho_p - \rho_r) + \cdots, \qquad (4)$$

where n_{z0} is the vacuum n_z evaluated above with a conducting wall at u_0, ρ_r is the surface impedance at u_r with this wall present, while ρ_p is the surface impedance at u_r with a plasma replacing this wall (see Fig. 4). The ρ's are normalized to 377 ohm, and $u = k_0 x$.

The derivative in (4), $dn_z/d\rho$, can be evaluated from dn_{z0}/du_0, the change in n_{z0} due to the movement of the front conducting wall, which we evaluated numerically. The position u_0 of the wall can be chosen so that $\text{Re}\rho_p \approx \rho_r$, minimizing the importance of higher terms of the series. Assuming a 5 cm vacuum region($u_r/k_0 = 5$ cm), which is large in view of the radiated k_z, the distance along the structure for power to be reduced by $1/e$ is shown in Fig. 5 for two idealized DIII–D plasma profiles as a function of n_z. For purpose of comparison, the effective series resistance at the bottom of the strap necessary to reduce the power in the structure by $1/e$ in a meter is ≈ 0.57 ohm for case A and ≈ 0.24 ohm for case B. Even with this small loading resistance, the incident power is efficiently radiated, demonstrating the advantages of a slow wave structure. The predicted vacuum ohmic damping is in comparison only $\approx 1\%$/m for case A and 2%/m for case B.

Preliminary damping measurements using resistive films on a low power model have given the result that the measured damping is somewhat stronger than that calculated by the above model, so the curves of Fig. 5 are probably rather conservative.

CONCLUSIONS

We have described a periodic structure for efficiently launching traveling fast waves of high $n_\parallel$. The $n_\parallel$ can be varied over a wide range with a moderate frequency change. The shielded version of this structure requires no loading capacitors and has lower peak voltages than the corresponding unshielded structure.

REFERENCES

1. Nathaniel J. Fisch and Charles F.F. Karney, Phys. Fluids **24**, 27 (1981).
2. C.P. Moeller, S.C. Chiu, and D.A. Phelps, in *Proc. Europhys. Top. Conf. on RF Heating and Current Drive of Fusion Devices 1992*, Brussels, Vol. 16E, p. 53.
3. V.E. Golant, Sov. Phys. Tech. Phys. **16**, 1980 (1972).

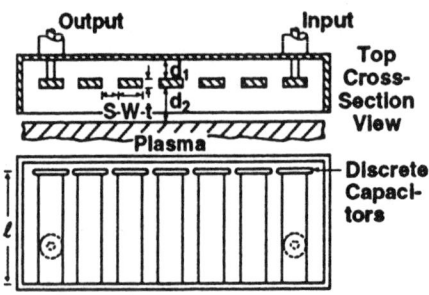

FIG. 1. Schematic of unshielded combline antenna.

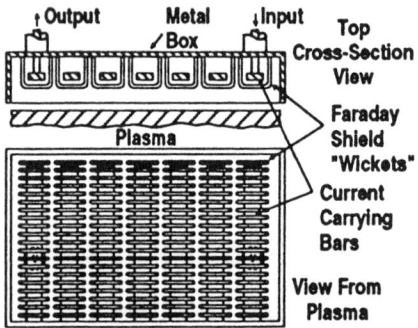

FIG. 2. Schematic of shielded combline antenna.

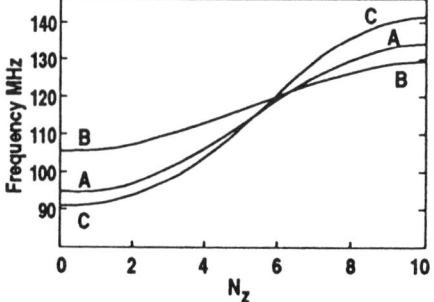

FIG. 3. Vacuum dispersion of the unshielded (cases A and B) and shielded (case C) combline antennas. A has 5 cm backplane spacing, compared to 2.5 cm for B. C is similar to A except for the shield. Other dimensions are given in the text.

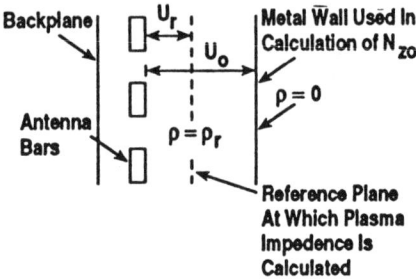

FIG. 4. Geometry of the loading calculation.

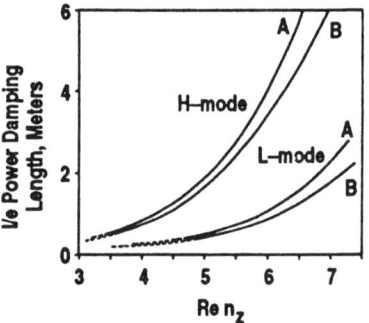

FIG. 5. Damping length for two idealized DIII–D plasma profiles for cases A and B.

FOLDED WAVEGUIDE DESIGNS FOR TOKAMAKS*

D. J. Hoffman, T. S. Bigelow, C. H. Fogelman, J. J. Yugo,
J. B. O. Caughman, W. L. Gardner, and M. D. Carter
Oak Ridge National Laboratory, Oak Ridge, TN 37831-8071

P. H. Probert
University of Wisconsin, Madison, WI, 53707-1687

E. Barbato
ENEA, Frascati Research Centre, Frascati, Italy

ABSTRACT

The folded waveguide (FWG) has been tested[1] to the megawatt level in RFTF and shows great promise for tokamak use. It has three primary advantages: low electric field (anywhere) per unit power coupled to the plasma, strong structural capabilities, and better spectral content than loops. A tokamak test is now needed. Potential candidates include C-Mod at 80 MHz and FTU at 433 MHz. The waveguide test on the first machine will be directed at conventional ion cyclotron heating, while the test on the latter will be directed at direct electron heating. In addition, a variation[2] of the folded waveguide is proposed to be tested on Phaedrus-T. In this paper, we discuss the advantages of the waveguide, the design layout, some of the potential physics programs, and how these programs may have an impact on its potential use in ITER.

BACKGROUND

The FWG has been described in a number of papers. Figure 1 shows the base configuration. Generally, it is characterized by folding a fundamental-mode waveguide back and forth upon itself so that the total width of the structure is $1/n$ times the width of the fundamental waveguide, where n is the number of folds. In the original configuration,[3] the FWG was energized as a resonant cavity, thereby rendering either end of the waveguide at a current maximum. Alternate apertures on the plasma face were closed on the waveguide to keep the poloidal orientation of the antenna's electric field in the same direction, like conventional loops. Since the antenna face is at a voltage minimum, the shorting of alternate faces was only a minor perturbation on the cavity resonance. Because of this resonance, the antenna was originally conceived as a single-frequency antenna, unless it could be designed with a movable backwall or with tunable inductive slugs.

This configuration was tested on RFTF to very high power levels (1000 kW unloaded), despite its high Q (≈ 1800). The antenna has low internal and plasma-facing electric fields because it is intrinsically a low impedance device and because the plasma-facing surface is at a voltage minimum. Tests showed that the FWG had capacitive electric fields that were two orders of magnitude below that of conventional loops.[1] Thus, the FWG may be attractive for tokamaks from a voltage, impurity, and Faraday shield standpoint (i.e., no shields needed).

Although the antenna has ultra-high power capability, additional measurements and calculations were needed to estimate how the antenna would couple to the plasma. Detailed 3-D measurements showed that the field structure was like that of a conventional loop except that the FWG had "bumps" in the toroidal rf magnetic field at

---

*Research sponsored by the Office of Fusion Energy, U.S. Department of Energy, under contract DE-AC05-84OR21400 with Martin Marietta Energy Systems, Inc.

the apertures of the polarization plate and that this rf field experienced minimal sign reversal when measured along the toroidal span of the antenna. Since the bumps evanesce away from the antenna in a short distance (not unlike the bumps that result from Faraday shields, etc., in conventional loops), analysis indicated that the first effect would have minimal impact on the plasma. The second, illustrated in Fig. 2, is probably an advantage because it facilitates directional spectra that may be needed for large machines like ITER.

Given that the radiating field structure is akin to conventional loops, coupling measurements can be made in the following manner: (1) total flux linkage to the plasma per unit internal electric field; (2) change in Q when resistive material is put in front of the antenna; (3) actual plasma coupling measurements in the (sparse) RFTF plasma; and (4) 3-D plasma calculations. Although the first three measurements do not properly treat the effects of different spectra on a hot tokamak plasma, they can be used to approximate relative loading if compared to a loop of similar size. Given a loop of identical toroidal and poloidal dimensions energized at 600 A rms and a waveguide operating at an internal electric field of 25 kV/cm (realized on RFTF tests in plasma), the FWG was seen to have an advantage in power capability over the loop of about 13 times.

The plasma calculation involved inferring every current on the antenna face plate that generated the rf magnetic fields that had been measured. Using the Mantis code, Carter et al.[4] could then infer the coupling to a magnetized warm tokamak plasma. Again, the result is consistent with the three previously measured coupling parameters. This analysis revealed that the FWG has a very broad current channel poloidally on the face plate, similar to Faraday shield currents, the difference being that the Faraday shield currents are effectively canceled at the back of the Faraday shield element, thus canceling its "coupling," while the current path in the FWG is closed deep into the waveguide, thus generating large rf flux in the plasma.

The realization that this current channel could be generated by structures other than the FWG has led to other related antennas,[2] primarily a folded TEM waveguide or stacked striplines, to achieve the same current path with relatively low electric fields in and on the antenna structure. Although the use of a TEM waveguide makes the spectra revert to more conventional loops, these waveguide variations may restore the possibility of wide or low-frequency range to the FWG concept.

TOKAMAK TESTS

To date, three tokamaks plan to test important aspects of the FWG. All three tests will make important contributions to the feasibility of using ceramic-free ICRF antennas for ITER. A summary of each follows.

C-MOD. C-Mod uses conventional ion heating scenarios at 80 MHz. Since the RFTF tests were at 80 MHz, C-Mod represents the least extrapolation from the FWG data base. The waveguide for C-Mod will be designed for 2-MW, 3-s capability; plans are to test at it 4 MW if initial tests are successful. Since this represents the first ultra-high-power test in a tokamak, the antenna will be uncooled. A single waveguide will be installed through the C-Mod port. Pending results from the first antenna on C-Mod, a "monopole" or "dipole" B_4C-coated polarization plate[1] will be added. Design, coupling analyses, and heating analyses will start immediately, with a high-power test planned on RFTF in FY 1995 and on C-Mod in FY 1995 and 1996. This experiment will form the basis of the high-power ion cyclotron heating data base.

Phaedrus-T. The mission of the Phaedrus-T tokamak is to perform low-frequency (ion cyclotron frequency or less), low-phase-velocity (phase velocity less than electron thermal speed) current drive. Two scenarios to accomplish this are (1) to launch fast waves at $f < f_{ci}$ which have the necessary k_z to be mode-converted to a kinetic Alfvén wave at a resonant surface, which will then Landau damp on electrons, or (2) to employ a two-ion plasma and launch fast waves which will be mode-converted to ion Bernstein waves at the ion-ion hybrid resonant layer and then also be Landau damped on electrons. Both scenarios, especially the second, are most straightforward if the fast wave is launched from the high field side. In Phaedrus-T this requirement leads us to seek fast-wave launchers in which the voltage can be kept low, so that problems with the feed lines can be minimized. Also, in the case of scenario 1, the antenna must be efficient at driving low loading resistances such as 0.1 or 0.2 Ω. Thus, we are led to the stacked stripline (SSL) antenna, shown in Fig. 3. This antenna, now under construction, is designed to be 5 cm thick so as to fit in the narrow space between the Phaedrus-T vacuum chamber and the plasma on the high field side. The feed lines consist of a separate wide copper foil and ground return for each element in the stack (a total of 10 feed lines) in a Kapton-insulated sandwich. The maximum rf voltage, in a worst-case loading with 500 kW of power, is less than 200 V. This experiment will establish the data base on FWG variations (and possible frequency extension) and on current drive arrays for tokamaks.

Frascati Tokamak Upgade. The Frascati Tokamak Upgrade (FTU) is a high field, high-density (7 T, $n \approx 10^{14}$ cm^{-3}) tokamak with $\approx 8 \times 40$ cm^2 ports. Megawatt power is available at 433 MHz, and thus a conventional folded waveguide can convey the power to the tokamak. At this frequency and field, the fast wave can be used for direct electron heating, analogous to the direct electron heating on DIII-D. Analysis by Barbato et al.[5] indicates that the waveguide can couple to electrons with >10% single pass absorption for electron temperatures greater than 3.5 keV (which are achievable with the lower hybrid system). Central power deposition was predicted, and the loading was seen to be sufficient for >2 kW/cm^2 for typical lower-hybrid-heated FTU plasmas. This experiment will form the basis of direct electron heating data base.

SUMMARY

Tests of the FWG indicate very high power capability. A possible ultimate use of the ceramic-free antenna is on the ITER system. In order to establish the bases for considering use on ITER, three major experiments will use the waveguide to explore three important facets of its potential. In particular, C-Mod can establish its use in conventional ion heating modes, Phaedrus-T in current drive arrays, and FTU in the direct electron heating mode.

REFERENCES

1. G. Haste et al., "The folded waveguide: a high-frequency rf launcher," submitted to Fusion Eng. and Design (1992).
2. W. Gardner et al., "Comparison of the folded stripline and stacked stripline concepts to the folded waveguide launcher," 10th Topical Conf. on RF Power in Plasmas, Boston, April 1–3, 1993.
3. T. Owens, IEEE Trans., Plasma Sci. PS-**14**, 934–946 (1986).
4. M. D. Carter et al., Proc. 9th Topical Conf. on RF Power in Plasmas, Charleston, 1991 (AIP, New York, 1992), p. 164.
5. E. Barbato et al., "Proposal of a fast-wave experiment at 433 MHz on FTU by a folded waveguide launcher," to be published in Proc. of the 20th EPS Conf., Lisboa, Portugal, July 26-30, 1993.

330 Folded Waveguide Designs for Tokamaks

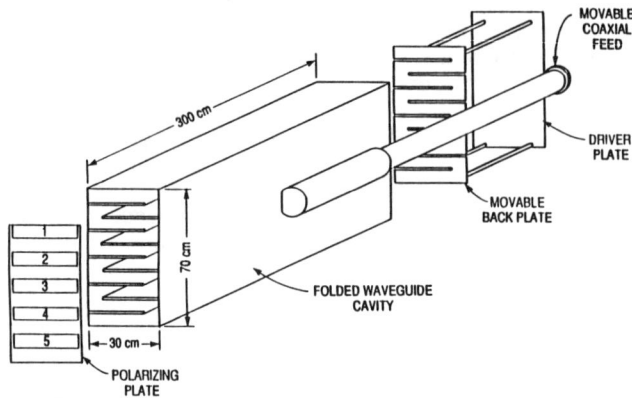

Fig 1. The folded waveguide with polarizer plate. This waveguide was tested on RFTF at ~ 1000 kW.

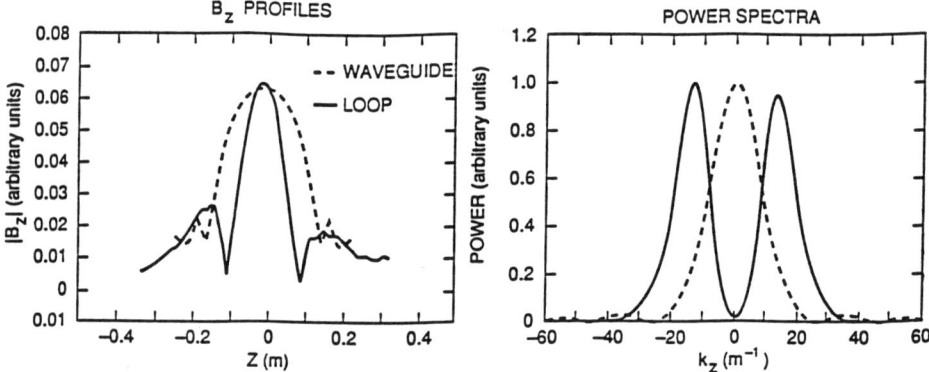

Fig. 2. Measurements of the toroidal rf magnetic field for a conventional loop and folded waveguide. The graph on the right is the spectra corresponding to the above measurements.

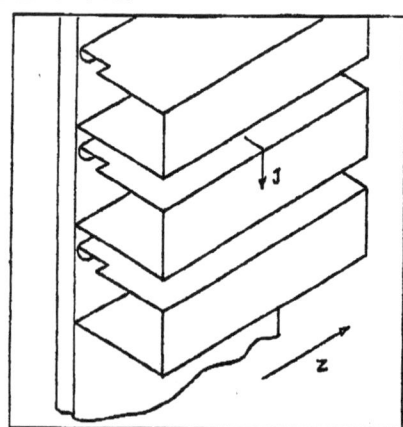

Fig. 3. The approximate current sheet for the FWG or variations such as the stacked stripline or folded stripline antenna.[2]

COMPARISON OF THE FOLDED STRIPLINE AND STACKED STRIPLINE CONCEPTS TO THE FOLDED WAVEGUIDE LAUNCHER*

W. L. Gardner, J. B. O. Caughman, and D. J. Hoffman
Oak Ridge National Laboratory, Oak Ridge, TN 37831

P. H. Probert
University of Wisconsin, Madison, WI 53706

ABSTRACT

Two new concepts are being developed as possible upgrades to the folded waveguide launcher. The folded stripline is a folded waveguide with an additional conductor positioned inside. The term *stripline* refers to the resemblance of the design to microwave microstrip line. The conductor provides support for TEM mode propagation, which eliminates cutoff and the nonlinear frequency dependence of the waveguide impedance and phase velocity. A natural extension to the folded stripline is the stacked stripline, which comprises several stacked, independent TEM waveguides. Initial measurements indicate that both concepts have better magnetic flux coupling than the folded waveguide.

INTRODUCTION

The folded waveguide cavity launcher, as depicted in Fig. 1, has been analyzed and tested for the last several years.[1-5] The main advantages it offers over currently deployed loop-based designs are a higher coupled magnetic field (power) to the plasma and a low-output electrostatic field component that nearly eliminates the need for a Faraday shield. Its main disadvantages seem to be the large port size required for support of the waveguide fundamental mode, given reasonable power density and voltage handling limits, and the nonlinear frequency dependence of the waveguide impedance and phase velocity, which makes the launcher very sensitive to tuning. These disadvantages have led to two concepts based on supporting a TEM mode by introducing extra conductors to the basic cavity design.

DESCRIPTION

The concept known as the folded stripline merely adds a folded inner conductor to the basic folded waveguide structure, as shown in Fig. 1, and evolved from initially centering the inner conductor. The folded stripline behaves more like a two-conductor stripline to produce better coupling to the current-carrying straps. The current straps replace the polarizer plates to help channel current to create the correct magnetic field vector. The inner conductor is supported by the back cavity wall of the half-wave resonator. The preferred method of coupling rf power to the cavity is through the center conductor of a coax cable connected to the inner conductor of the cavity. Tuning and matching can be done either externally to the cavity with standard coaxial tuning

---

*Research managed by the Office of Fusion Energy, U.S. Department of Energy, under contract DE-AC05-84OR21400 with Martin Marietta Energy Systems, Inc.

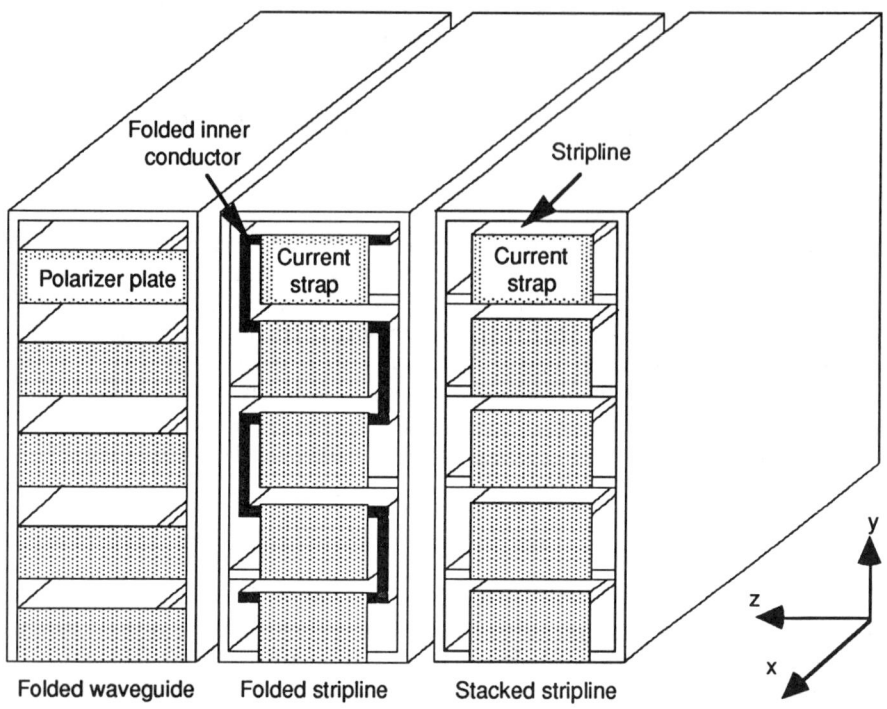

Fig. 1. A side-by-side comparison of the design features of the three resonant cavity concepts.

elements, by a combination of adjustments to cavity length and position of the coaxial line coupler along the inner conductor (tap point adjustment), or by a combination of these. Experience with the folded waveguide suggests that external tuning and matching are preferable as a practical matter.

The concept known as the stacked stripline takes the previous concept a step further by using multiple stripline resonators and current straps, as shown in Fig. 1. Note that each resonator is independent of the others except at the output, where mutual coupling can be significant. This means that it is very important to tune each line separately to the same resonant frequency with the other lines shorted and to match each to the same impedance tap point prior to operating them in a ganged mode. It is also assumed that the input power is divided from a single source, so it is equally important that all of the inputs have the same phase relationship.

RESULTS

We constructed same-size, small-scale cavities of each concept to ascertain how they compare with each other. Each had outside dimensions of 6.0 × 14.0 × 61.3 cm. The output end of each was configured as shown in Fig. 1. All couplers were designed with a sliding tap point to allow freedom of travel in the long dimension for impedance matching. The input signal for the stacked stripline was split using an

ANZAC DS-312, four-way splitter and equal line lengths to individual stripline connectors. The fifth resonator was undriven but was tuned like the others to provide the best output coupling. Tuning for the stacked stripline was done by flexing the inner conductor of each stripline resonator to slightly change the characteristic impedance along the line until they all resonated at the same frequency. Measurements were taken with an HP 8753C Network Analyzer.

The measurement of importance for this preliminary comparison was the relative magnetic field coupled to a loop probe at the output end of the launcher. The loop probe, 2 mm in diameter, was positioned to intercept the magnetic flux vector in the z-direction (Fig.1). A comparison of scans of relative signal strengths in the y-direction is shown in Fig. 2. Both of the stripline concepts seem to have better magnetic flux coupling along the vertical centerline than the folded waveguide. The folded waveguide, in turn, has significantly more flux linkage than loop-type designs. The stacked stripline has the best coupling, though its profile is significantly more peaked. This peaking is primarily due to the nonoptimized geometry for multiply coupled resonators, and the fact that only four of the five resonators are being driven.

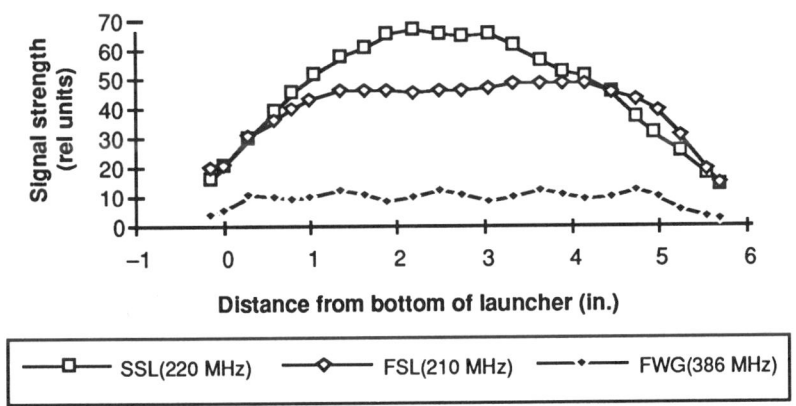

Fig. 2. Vertical scans of the coupled magnetic flux in the z-direction for the stacked stripline (SSL), folded stripline (FSL), and folded waveguide (FWG). Cavity resonant frequencies are also shown.

Horizontal scans were also made for the striplines. A representative scan for the stacked stripline is shown in Fig. 3 and is very similar in form to those taken for the folded stripline case. The structure is very reminiscent of that observed for loop-type launchers, where there are large return currents in the side walls. These currents can be adjusted by slotting the side walls as is done for loop-type launchers.

The features of both stripline designs suggest that they are hybrids of the folded waveguide and loop-type geometries. The key features that make them attractive are the low electric field at the output and the high magnetic flux coupling, which is aided by the fact that several short current straps carry the current rather than the one or two long ones used in loop-type designs—i.e., there is very small current droop because the ratio of the physical length to quarter wavelength is small.

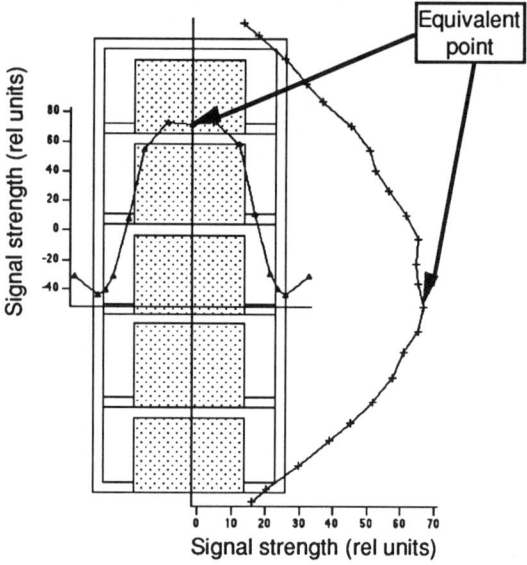

Fig. 3. Vertical and horizontal scans of the stacked stripline coupled magnetic flux. Signal strength for the scan crossing point is also indicated.

REFERENCES

1. T. L. Owens, IEEE Trans., Plasma Sci., PS-**14** (6), 934 (1986).
2. T. L. Owens et al., AIP Conf. Proc. 159: Appl. of RF Power to Plasmas, 298 (1987).
3. G. R. Haste and D. J. Hoffman, AIP Conf. Proc. 190: Appl. of RF Power to Plasmas, 266 (1989).
4. F. W. Baity et al., AIP Conf. Proc. 244: RF Power in Plasmas, 298 (1991).
5. G. R. Haste et al., Fus. Engr. & Dsgn., accepted for publication.

THREE-DIMENSIONAL EFFECTS FOR RADIO FREQUENCY ANTENNA MODELING*

M. D. Carter, D. B. Batchelor, and D. C. Stallings

Oak Ridge National Laboratory, Oak Ridge, TN 37831-8071

ABSTRACT

Electromagnetic field calculations for radio frequency (rf) antennas in two dimensions (2-D) neglect finite antenna length effects as well as the feeders leading to the main current strap. The 2-D calculations predict that the return currents in the sidewalls of the antenna structure depend strongly on the plasma parameters, but this prediction is suspect because of experimental evidence. To study the validity of the 2-D approximation, the Multiple Antenna Implementation System (MAntIS) has been used to perform three-dimensional (3-D) modeling of the power spectrum, plasma loading, and inductance for a relevant loop antenna design. Effects on antenna performance caused by feeders to the main current strap and conducting sidewalls are considered. The modeling shows that the feeders affect the launched power spectrum in an indirect way by forcing the driven rf current to return in the antenna structure rather than the plasma, as in the 2-D model. It has also been found that poloidal dependencies in the plasma impedance matrix can reduce the loading predicted from that predicted in the 2-D model. For some plasma parameters, the combined 3-D effects can lead to a reduction in the predicted loading by as much as a factor of 2 from that given by the 2-D model, even with end-effect corrections for the 2-D model.

INTRODUCTION

The MAntIS code[1] solves Maxwell's equations in vacuum for a 3-D periodic slab geometry. The slab is bounded on one side by a plasma and on the other by a conducting wall. The plasma boundary is represented by a generalized impedance matrix in Fourier space. Both directions that are tangential to the plasma (toroidal, or z, and poloidal, or y), are Fourier-analyzed. The direction perpendicular to the plasma (radial, or x) is solved for analytically using prescribed current elements.

Each current element consists of a filament and two feeder legs at each end of the filament. The filaments carry currents in the y and z directions at constant radial location. The feeders carry currents in the x direction in a way that analytically maintains continuity in the current at the ends of each filament. The current elements can be rotated arbitrarily about the x-axis, and the filaments may have a prescribed finite divergence (charge) along the current path.

MODELING

The impedance matrix for the results of this paper was generated by the ORION1-D code.[2] A cold plasma was considered with the density profile shown in Fig. 1. Other pertinent plasma parameters are the central magnetic field, $B_0 = 4$ T; the frequency, $f = \omega/(2\pi) = 40$ MHz; the minor radius, $a = 0.53$ m; and the major radius, $R = 2.25$ m. An absorber was used so that no power was reflected from the high field side of the slab model for a tokamak.

The rf antenna to be modeled consisted of two antenna modules placed side by side in the toroidal direction with septa in between and at each side. The dimensions of the antenna and the filaments used to model the rf currents in MAntIS are shown in Fig. 2. Two filaments were used to simulate the peaking of current near the edges of each main strap. All filaments for the main straps were kept in phase with one another (frequently referred to as monopole phasing). The filaments used to describe

---

* Research sponsored by the Office of Fusion Energy, U.S. Department of Energy, under contract DE-AC05-84OR21400 with Martin Marietta Energy Systems, Inc.

**The submitted manuscript has been authorized by a contractor of the U.S. Government under contract DE-AC05-84OR21400. Accordingly, the U.S. Government retains a nonexclusive, royalty-free license to publish or reproduce the published form of this contribution, or allow others to do so, for U.S. Government purposes.

the currents in the septa were kept constant throughout all runs and were phased in time roughly π radians behind the current in the main straps. The distribution of the currents on the septa was determined by visual inspection of the calculated rf magnetic field at $y = 0$ with a gap between the plasma and main straps, d, of 0.025 m. This distribution was relatively insensitive to d.

The amplitude and phase of the filaments for the main strap was determined by integrating the z component of the rf magnetic field, B_z, at $y = 0$, $z = 0$ along the septa between $x = 0$ and $x = 0.3$ m. The current amplitude was adjusted until the real part of the integral of B_z over the septa was zero, and then the phase was adjusted to make the imaginary part of the integral of B_z over the septa also go to zero. The norm used to determine zero was the amplitude of the integral of B_z between the plasma and the septa.

RESULTS

MAntIS was benchmarked in the 2-D limit by using an antenna with a length equal to $2\pi a$ ($m = 0$ limit) and adjusting the main strap current as described above. A uniform current distribution was specified along all of the current elements. The solution required a current amplitude in all four main strap filaments of 3.56 A, with a time phase lead of 0.22 rad. The current returned in the conducting backwall was 0.148 A and in the septa was 2 A. The difference in the main strap and return current amplitudes of roughly 1.4 A is frequently observed in the 2-D limit and is often referred to as "current robbing" by the plasma. The results for both the real and imaginary power were within 10% of the results found by the 2-D RANT code.[3] Comparable values for the current robbing and phase shift were also observed in the 2-D RANT code. This agreement is rather good, considering the filament model used to describe the septa and the fact that the filaments in the septa were not adjusted from the values found for the 0.9-m-long antenna.

By adding the total currents in the septa, backwall, and main straps, we studied the current-robbing effect for main straps with poloidal extent of 0.9 m. The contrast between this 3-D modeling and the 2-D results is shown in Fig. 3. As shown in Fig. 3, the current-robbing phenomenon observed in the 2-D limit does not occur at an appreciable level when finite length antennas with feeders are considered because the feeders provide a low inductance path for the return currents. This result shows that the current patterns on 3-D antennas are far less sensitive to changes in plasma parameters than 2-D models generally indicate. Also, for the 3-D modeling, the phase lead on main straps was very small for 0.9-m-long antenna, with the largest phase lead equal to $\pi + 0.054$ rad for $d = 0.015$ m.

The robust nature of the current pattern on 3-D antennas results in a launched power spectrum that is depressed for small values of the toroidal wave number, $k_z = n/R$, relative to predictions by 2-D models. Figure 4 shows the power spectra per mode for the same values of d using the 2-D and 3-D models. A rule of thumb for the validity of the 2-D approximation can be based upon the value of the return current in the conducting walls; for a monopole configuration, the ratio of total to driven current in the 2-D model must remain close to $(1 + i0)$ for the 2-D model to represent the toroidal mode spectrum accurately.

For the plasma parameters shown in Fig. 1, there is an additional reduction in plasma loading caused by 3-D effects. This reduction occurs because a significant portion of the poloidal antenna spectrum is cut off by the plasma. The combination of relatively small a and a low plasma edge density causes the width of the launched toroidal power spectrum to be reduced at moderate poloidal mode numbers.

CONCLUSIONS

Three-dimensional effects can have a substantial influence on plasma loading predictions for some plasma parameters and antenna configurations. Two effects that are fully 3-D can be isolated. First, the loading can be reduced for monopole antenna phasing because of robust return currents in the sidewalls of the antenna structure. These currents are robust because feeders and conducting sidewalls provide a low inductance path that prevents the plasma from significantly affecting their distribution. The return currents reduce the loading by depressing the launched power spectrum at low toroidal mode numbers. Second, the loading can be reduced if the plasma does not allow propagation of the power spectrum at moderate poloidal mode numbers. This condition typically occurs for devices with small minor radius and low edge density. Overall, for the plasma parameters chosen here, 3-D predictions for antenna loading can be reduced by more than a factor of 2 from 2-D results, even with end effect corrections for the 2-D results.

REFERENCES

1. M. D. Carter, D. B. Batchelor, and E. F. Jaeger, ORNL/TM-12264, February 1993.
2. E. F. Jaeger, D. B. Batchelor, and H. Weitzner, Nucl. Fusion **28**, 53 (1988).
3. D. B. Batchelor, M. D. Carter, R. H. Goulding, et al., Fusion Technol. **21**, 1214 (1992).

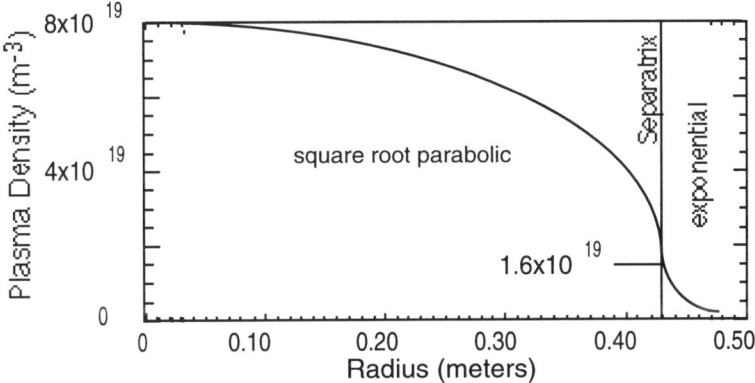

Fig. 1. Plasma density profile used in all modeling. The e-folding length is 0.02 m.

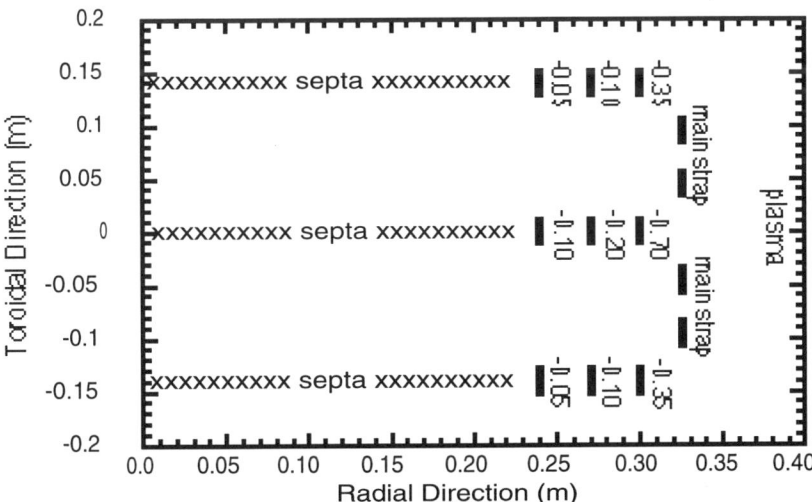

Fig. 2. Schematic showing location of current filaments used to model the antenna. Numbers indicate the amplitude of the current in each filament. Amplitudes for septa were determined for a case with plasma 0.025 m from main straps and held constant for all runs. Magnetic flux through center septum was made zero by adjusting the amplitude and phase of the two filaments used to model the main strap.

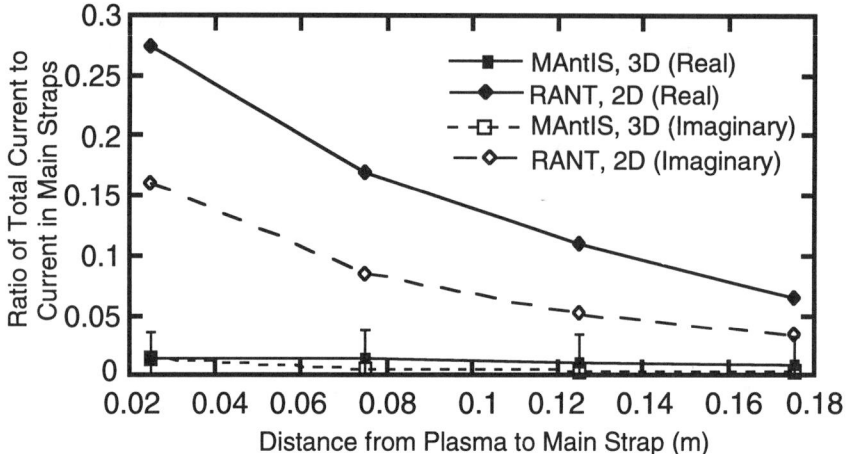

Fig. 3. For finite length antennas, the feeders to the main strap and the sidewalls of the antenna provide a low inductance path for the return currents. This path prevents the current from being "robbed" by the plasma, as happens in the 2-D (m = 0) limit. The MAntIS data is for main straps that have 0.9 m-length in the poloidal direction and uniform current along the strap. Errors in the MAntIS calculations occur because a finite number of poloidal modes are used to calculate the currents in the conducting backwall.

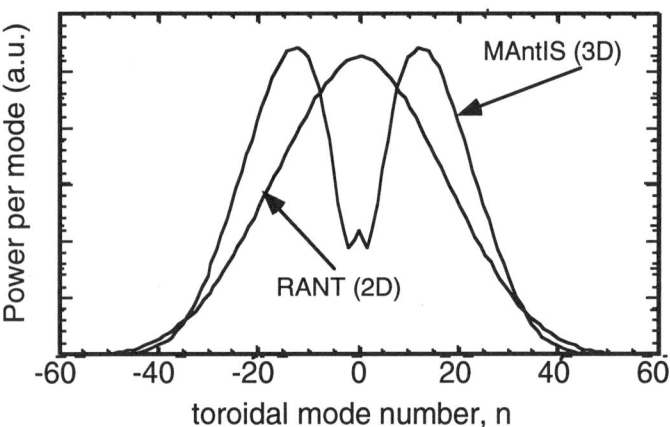

Fig. 4. An average over the poloidal mode launched power spectrum for the 3-D MAntIS calculation shows a depression in the toroidal mode spectrum near n = 0 caused by return currents in the sidewalls of the antenna. The main straps in the 3-D calculation are 0.9 m long with d = 0.025 m. The 2-D calculation by the RANT code shows the effects of current robbing by the plasma and does not demonstrate the depression near n = 0.

CONVERSION OF THE FOUR-STRAP ARRAY IN DIII–D TO A TUNABLE TRAVELING WAVE ANTENNA

D.A. PHELPS, C.P. MOELLER, C.C. PETTY, and R.I. PINSKER
General Atomics, San Diego, CA 92186-9784

P.M. RYAN, R.H. GOULDING, and D.J. HOFFMAN
Oak Ridge National Laboratory, Oak Ridge, TN 37830

ABSTRACT

The *in situ* conversion of the two-standing wave resonator driven four-strap array in DIII–D to a new type of traveling wave antenna (TWA) is introduced. First observations are reported on (1) broad frequency range tunability; (2) narrow frequency band phase control; (3) single pass transmission loss and modification caused by resistive and reactive sheet simulators of plasma loading; and (4) recirculating power resonance and gain in a traveling wave resonator (TWR) driven TWA.

1. INTRODUCTION

To maximize fast wave current drive (FWCD) in the ion cyclotron range of frequencies (ICRF), the rf radiation spectrum must track an optimum phase velocity[1] that increases with electron heating[2] and changes with L–H and other mode transitions, or ELMs. Broadband frequency and phase tunability is needed to access the desired FWCD, electron heating, and ion cyclotron resonance heating scenarios. A transmitter/tuner/antenna system that meets these needs, but also switches rapidly between scenarios, is the ICRF enabling technology challenge for the future.

The combline TWA[3–7] exhibits phase control between $(0, \pi)$ within a preselected cutoff frequency passband, and thereby suits its principal mission — FWCD at the highest frequency commensurate with transmitter power, FWCD efficiency, and avoidance of ion cyclotron resonance. In-band rapid switching between electron heating during startup, and FWCD during L and H–mode, is another accessible scenario.[8] To access ion heating scenarios, however, requires an in-vacuum tunable capacitive load, a change in the tune during a vent, or sets of comblines tuned to preselected frequencies.[8] This latter solution is viable for next generation tokamaks, such as ITER[4] and Ignitor.[5,6] Issues for TWAs in general are the lack of physics and technology validation at high rf power, and/or with plasma loading, sheath rectification and impurity generation.[9] To perform near term research on as-built antennas, including broad phase and frequency tunability, consider the externally tuned TWA[6] described herein.

2. THE EXTERNALLY TUNED TWR DRIVEN TWA

The TWR driven TWA circuit in Fig. 1 utilizes an as-built boxed-in four-strap array with septa and Faraday shield, such as the DIII–D four-strap array mock-up at ORNL (the Mock-up), or the B4C shielded four-strap array now in DIII–D (B4C Prototype). Each strap is made into a transmission line cavity, and tuned for standing wave resonance between the strap-short and an external coax-short using a line stretcher (or tuning capacitor).[6,10] Power is critically coupled into or out of just the end strap-cavities. Interior strap-cavities are powered only by mutual coupling. Power not radiated or absorbed in the walls is recirculated in a TWR, using a 3 dB hybrid.

© 1994 American Institute of Physics

Parameters based on a resolution of theory with experiment are given in Table I. Note that (1) $Z_0 = 42/93$ Ω implies with/without a Faraday shield (for example); (2) k_{mn} are the TWA coupling factors; (3) $k_{m,m\pm1} = 39\%$ without septa (as preferred); and (4) $R_p = k_{23}X_L \gtrsim 3$ Ω is the critical load resistance (where $kQ = 1$). Notably, L and H–mode loads in DIII–D[11,12] are less than this R_p as desired.

Recently, toroidal wave number (k_{tor}) spectra for a 1 m TWA were simulated. Remarkably, nearly 100% of the power can be absorbed by the plasma without significant change in the k_{tor} spectrum relative to no absorption, except well above the optimum k_{tor} for FWCD. Since higher k_{tor} are more evanescent,[11–13] there will be much less change in the launched fast wave spectrum.

3. EXPERIMENTAL RESULTS

Single-pass observations of the S-parameters at TWA ports P1 and P2 were made without the TWR in Fig. 1. The magnitude and phase of reflection (S_{11}, S_{22}) and transmission (S_{21}, S_{12}) parameters were measured. Transmission reciprocity was observed. The observed 2 MHz passband was predicted by theory,[3] using the parameters in Table I. The observed smooth phase variation, about $\pi/2$ at midband, is consistent with a theory for unidirectional traveling wave (versus a standing wave). By lengthening the strap-cavities, the passband was down-shifted from about 61 to 63 MHz to about 48 to 50 MHz, as predicted.[3] Following this procedure, wider band tuning (e.g., 30 to 120 MHz) can be achieved.

Frequency responses for the Mock-up are displayed in Fig. 2. Filter-like passbands are predicted by applying combline theory[3] to a TWA with half-wave resonant strap-cavities.[6] There are no spurious passbands below the first upper filter passband at about 140 MHz, as expected.[3] The highest transmission efficiency occurred for no loading, and is less than unity due to ~10% reflection from P1 and P2, wall absorption, and/or free space radiation. Tuners were not used to reduce these reflections, in order to study TWA passband responses. The family of passband profiles in Fig. 2 are parametric in (a) resistive (carbon filled plastic), or (b) reactive (aluminum) sheet dummy loads of larger area than the Faraday shield. The load was changed by changing the shield-sheet gap. The observed passband attenuation and slight down-shift in Fig. 2(a) compares with expectation for a RLC resonant circuit.[6] As a potential calibration technique, each external short was replaced by a 2.7 ohm resistor load. The observed attenuation was slightly greater than that for the 1.5 cm gap. This suggests that plasma relevant 1 to 3 Ω/strap loads[11,12] can be achieved with plasma relevant 2 to 8 cm shield-sheet gaps. A family of profiles for shield-metal sheet spacings are shown in Fig. 2(b). The slight increase in frequency with smaller gaps is consistent with reduced antenna inductance caused by the metal sheet. This compares with driven four-strap array measurements, wherein 3% to 5% reduction in inductance was observed in the presence of plasma.[12] These results suggest a recipe for establishing a complex plasma surface impedance[6]: (1) set the metal wall gap for the right reactance; (2) interstice the resistive sheet to produce the right resistance; and (3) iterate as necessary.

Figure 3 exhibits the transmission passband modifications introduced when the TWA is driven by a TWR, as typified in Fig. 1. At the observed maximum TWR resonance (about 62 MHz), the roundtrip electrical length at midband (i.e., for 90 deg phasing/strap) was a little under 2 m, which is consistent with prediction.[6] Furthermore, the observed maximum gain of 1.8:1 is consistent with prediction[6] for a 0.7 roundtrip transmission efficiency and the 3 dB hybrid coupler.

3. CONCLUSIONS AND RECOMMENDATIONS

The reversible conversion of the B4C prototype in DIII–D to an externally tuned TWA has been discussed. Several critical enabling technologies have been demonstrated:
1. Tunability over 30 to 120 MHz using a line stretcher in each strap cavity.
2. Smooth phase change between $(0, \pi)$ cutoffs within a 2 MHz transmitter bandwidth.
3. Single pass transmission loss and passband modification caused by resistive and reactive sheet simulators of plasma loading.
4. Recirculating power gain in a traveling wave resonator (TWR) driven TWA.

Recommendations for future externally tuned TWA research tasks are:
1. Modify or delete the as-built septa and shields to improve phase control, stability, directionality and coupling to plasma.
2. Demonstrate the *in situ* conversion of a driven four-strap array to a TWA, as recently demonstrated with the Mock-up. Note in Fig. 4(a) that only two SWR locations need to be shorted. One tuned line inputs power, the other extracts power. A 3 dB hybrid coupler can be used to recirculate power.
3. Conduct near term 200 W experiments in DIII–D. Investigate how TWAs respond to plasma transitions and fluctuations in reactive and resistive loading.

Recommendations for alternative TWA concepts are:
1. Develop the present 9 to 27 strap "thin" combline TWA concepts[5,6] for DIII–D. These comblines can fit in and between ports and tile surface to vessel wall recesses.
2. Study the hybrid TWA concept,[13] that mixes features of the combline elements, as sketched in Fig. 4(b). This takes advantage of combline enhancements near a preselected frequency for FWCD, while retaining full conventional capability and broadband access to driven FWCD and ion/electron heating scenarios.
3. Design TWAs for two-strap center grounded arrays like those in TFTR and Alcator, and four Jet-type strap arrays like those being fabricated for DIII–D and JET.

ACKNOWLEDGMENT

This is a report of work sponsored in part by the U.S. Department of Energy under Contract Nos. DE-AC03-89ER51114 and DE-AC05-84OR21400.

REFERENCES

1. M. Porkolab, Proc. AIP Conf. RF Power in Plasmas 244, p. 197 (1991).
2. J.S. deGrassie et al., Bull. Am. Phys. Soc. **37**, 1513 (1992).
3. C.P. Moeller et al., Proc. Euro. Top. Conf. RF Heating (1992) p. 53.
4. D.A. Phelps et al., Bull. Am. Phys. Soc. **37**, 1513 (1992).
5. C.P. Moeller et al., 6th Wrkshp. on RF Antenna Design & Phys., Boulder (1993).
6. D.A. Phelps et al., *ibid.* (1993).
7. C.P. Moeller, to be presented at this conference (April 1–3, 1993).
8. R.L. Freeman, GA, private communication (December 1992).
9. J. Jacquinot et al., private communication, *ibid.*, Ref. 6 (1993).
10. F. Durodie, 16th Symp. on Fusion Technol., London (1990).
11. M.J. Mayberry et al., to be submitted to Nucl. Fusion, GA-A19565 (1989).
12. R.H. Goulding et al., *ibid.*, Ref. 1 (1991).
13. D.J. Hoffman, ORNL, private communication (March 16, 1993).

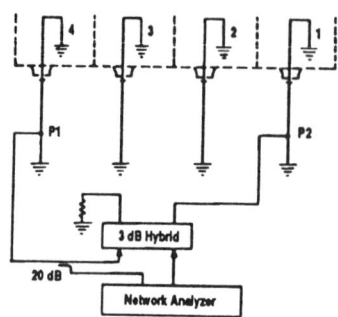

FIG. 1. A TWR driven TWA.

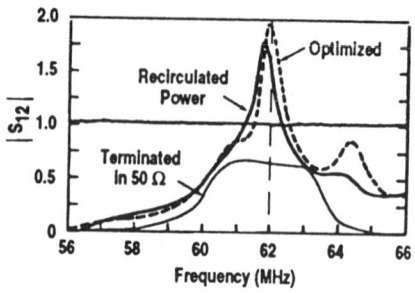

FIG. 3. Observed TWR–TWA circuit gain.

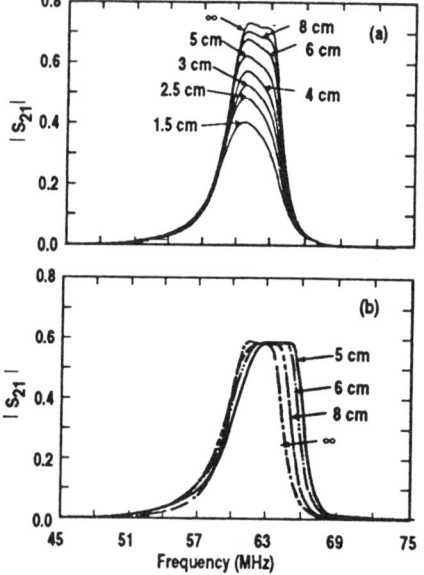

FIG. 2. TWA $|S_{21}|$ profiles parametric in (a) resistive and (b) reactive sheet loading.

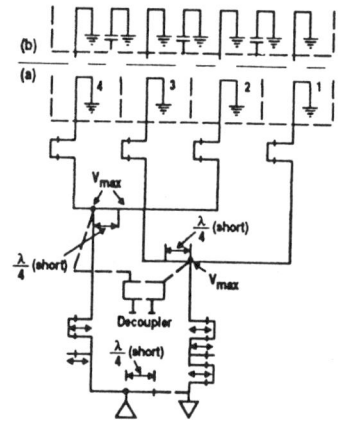

FIG. 4. Conversion of the driven four-strap array in DIII–D to (a) an externally tuned TWA; (b) a hybrid thereof with combline TWA elements.

Table I. Parameters for the DIII–D Prototype and Mock-up

Antenna	ℓ_p (m)	v/c	Z_0 (Ω)	R' (Ω/m)	$k_{12}=k_{34}$	k_{23}	$k_{13}=k_{14}$	k_{14} (%)
Strap Proto.	0.45	0.52	42/93	—/0.015	9.5	7.4	1.6	0.4
Strap Mock-up	0.45	0.52	35/75	0.12/0.015	7.0	6.5	0.7	0.2
Strip Proto.	0.34	1	32	0.005				
Strip Mock-up	0.38	1	32	0.005				
Coax Proto.	*	1	30	0.086	*Strap 1=0.75; 2=0.82; 3=0.80;			
Coax Mock-up	0.13	1	40	0.02	and 4=0.75 m			

DESIGN OF LONG-PULSE FAST WAVE CURRENT DRIVE ANTENNAS FOR DIII-D*

F. W. Baity, D. B. Batchelor, K. C. Bills, C. H. Fogelman, E. F. Jaeger, J. L. Ping,
B. W. Riemer, P. M. Ryan, D. C. Stallings, D. J. Taylor, J. J. Yugo
Oak Ridge National Laboratory, Oak Ridge, Tennessee 37831-8071, USA

ABSTRACT

Two new long-pulse fast wave current drive (FWCD) antennas will be installed on DIII-D in early 1994. These antennas will increase the available FWCD power from 2 MW to 6 MW for pulse lengths of up to 2 s, and to 4 MW for up to 10 s. Power for the new antennas is from two ASDEX-type 30- to 120-MHz transmitters. When operated at 90° phasing into a low-density plasma ($\sim 4 \times 10^{19}$ m^{-3}) with hot electrons (~ 10 keV), these two new antennas are predicted to drive approximately 1 MA of plasma current.

ANTENNA DESCRIPTION

The new antennas are designed for installation at the 0° and 180° toroidal locations in midplane ports originally housing movable limiters. The basic specifications of the antennas are as follows:

Power	(initial operation)	2 MW
	(ultimate)	4 MW
Frequency		30–120 MHz
Phasing	(between adjacent elements)	0°, ±90°, 180°
Pulse length		10 s

The opening in the vessel wall is approximately 1-m wide by 0.5-m tall, but the port tapers rapidly to a nearly square cross section at the port flange. The width of a practical antenna array which can be fitted into this port is 0.7 m. Each array comprises four elements in the toroidal direction. The arrays are modular: the elements are identical except for the coaxial feeders and can be installed or removed individually, simplifying maintenance. The 10-s pulse length dictated that all components, with the exception of the Faraday shield, be water cooled. In order to retain modularity, there are four separate water feeds. The arrays are supported by horizontal mounting plates in the port, with accommodation for thermal growth relative to the DIII-D vessel.

The Faraday shield consists of a single layer of 13-mm-diameter nickel-plated molybdenum rods with a 100-µm-thick plasma sprayed coating of boron carbide on the plasma-facing side. The rods are inclined at a 12° angle to match the pitch of the local magnetic field at the shield location. The rods are mounted to the antenna housing individually by thin Inconel strips to allow for differential thermal expansion of the Faraday shield relative to the antenna housing. The length of the Inconel strips was chosen to provide the desired magnetic coupling between adjacent antenna elements.

Figure 1 is a front view of the antenna array, showing the four elements with the feed lines extending to the vacuum feedthroughs at the port cover flange.

---

*Research sponsored by the Office of Fusion Energy, U.S. Department of Energy, under contract DE-AC05-84OR21400 with Martin Marietta Energy Systems, Inc.

ANTENNA SPECTRUM AND CURRENT DRIVE EFFICIENCY

The constraints posed by the port configuration dictated that the arrays have exactly four elements. The directionality of the launched waves is influenced chiefly by the fact that there are four array elements, the radial position of the sidewall return currents, and the distance to the cutoff layer in the plasma. The vacuum spectrum 2 cm in front of the antenna is shown in Fig. 2 for two locations of the antenna sidewalls, one flush with the current strap and one recessed 3.4 cm, corresponding to the location chosen for the antennas.

The antenna dimensions were input to the RANT code to calculate wave amplitudes at the plasma. The output from RANT was then used as input to PICES[1] for calculation of the current drive efficiency. The PICES results at 120 MHz for a plasma with $B_t = 2$ T, $n_e = 4 \times 10^{19}$ m^{-3}, $T_e = 10$ keV, and $T_i = 10$ keV and for an antenna to plasma gap of 2 cm is shown in Fig. 3 as a function of phasing between adjacent elements. For 4 MW (total for two antennas) and 90° phasing, the driven current is expected to be 1 MA.

CURRENT STRAP CONFIGURATION

At the upper frequency of operation of 120 MHz, a 42-cm high antenna with a phase velocity of 0.6c (due to the Faraday shield) is longer than a quarter wavelength. Since there is room for only one coaxial feeder per element, the current strap is divided into two poloidal segments. 3-D modeling of the rf magnetic field amplitudes was conducted to compare the double strap configuration to a single strap at 60 and 120 MHz. The poloidal distributions of the integrated toroidal magnetic flux are shown in Fig. 4. The droop at the ends is due both to finite wavelength effects and to the effect of the ends of the antenna housing. In terms of the total flux at the plasma boundary, the double strap is superior at 120 MHz and the single strap is superior at 60 MHz. However, the double strap reduces the voltage between the antenna and Faraday shield relative to the single strap, so the double strap is nearly equivalent to the single strap at 60 MHz.

The maximum power capability of the antenna with a peak voltage of 30 kV appearing anywhere in the antenna structure is shown in Fig. 5 as a function of antenna loading resistance for 60, 90, and 120 MHz. The value plotted is the total power for the array with the 30-kV peak voltage occurring on any element of the array at 90° phasing.

The wide frequency range of operation causes voltage and current maxima to occur throughout the transmission line. In particular, the vacuum feedthrough is near a voltage minimum at 120 MHz, but near a voltage maximum at 60 MHz. Thus, all components must be designed to handle high voltages. Based on superior voltage standoff in feedthrough tests compared to copper, silver, and gold, nickel plating was selected for all components in vacuum.

ANTENNA PHASING AND IMPEDANCE MATCHING

The external transmission line system will be of the same design as that used successfully on the first-generation FWCD antenna on DIII-D. The circuit is shown schematically in Fig. 6. The loops connecting elements 1 and 3 and elements 2 and 4 provide fixed phasing of either 0° or 180° depending on the total lengths of the loops. The five tuning elements (ignoring the hybrid couplers for the moment) provide the desired phasing between elements 1 and 2 as well as matching the impedance to 50Ω.

The hybrid coupler between the resonant loop tees is adjusted to cancel the mutual coupling between elements in the antenna structure, with the result that the two stub tuners are set to the same lengths (the impedances looking into the resonant loop tees are identical), and the interline phasing is set by adjusting the phase shifter in the matched line section. The use of the hybrid coupler is primarily needed when each antenna is fed by two transmitters and provides equal loads to both transmitters, thus maximizing the total input power to the antenna.

Recent low-power tests on the existing FWCD array demonstrated the principle of operation of the hybrid coupler. With the hybrid coupler located at a high voltage point in the unmatched line sections, it was sufficient to set the stubs connected to the hybrid under vacuum loading conditions and leave them fixed under plasma loading conditions.

SUMMARY

Two new four-element FWCD arrays are being readied for installation on DIII-D at the beginning of 1994. Each array will be capable of 4-MW operation for 10 s in the frequency range of 30 to 120 MHz. At full power and with optimum plasma conditions, the antennas should be able to drive 1 MA of plasma current. Extensive rf modeling was performed on the antenna design to ensure the achievement of the antenna specifications throughout the operating parameter space.

REFERENCE

1. E. F. Jaeger and D. B. Batchelor, Full-Wave Calculation of Fast Wave Current Drive in Tokamaks Including $k_{\parallel}$ Variations, in: Fast Wave Current Drive in Reactor Scale Tokamaks, ed. D. Moreau, A. Bécoulet, Y. Peysson, Proceedings of the IAEA Technical Committee Meeting, 87-96 (Arles, France, 1991).

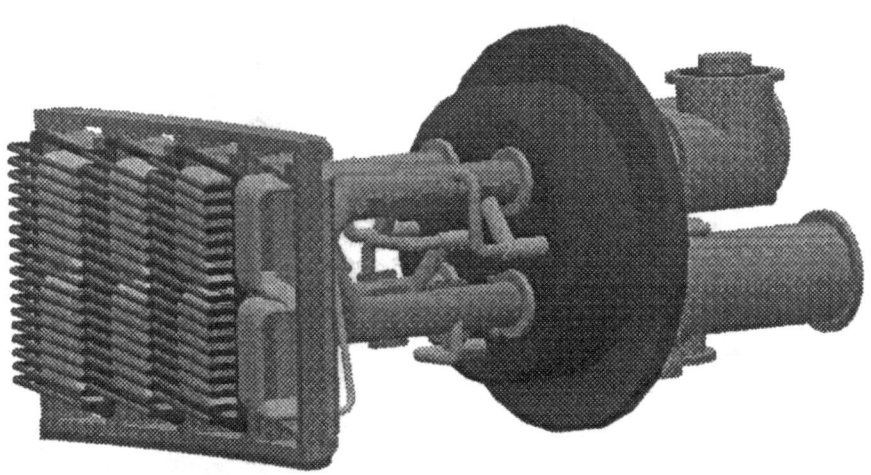

Fig. 1. Front view of the long-pulse FWCD antenna.

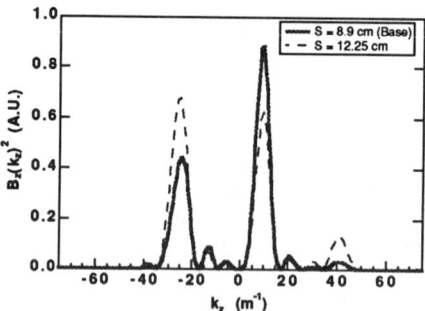

Fig. 2. Vacuum $B_z(k_z)^2$ spectra 2 cm from the antenna for 90° relative phasing with baseline and flush septa/sidewall lengths.

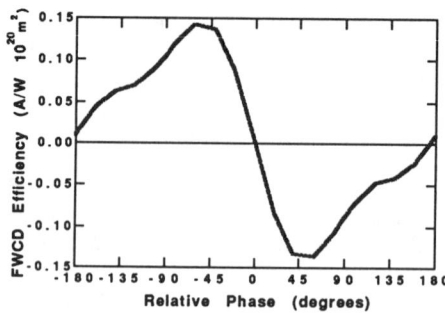

Fig. 3. PICES results at 120 MHz for a plasma with $B_t = 2$ T, $n_e = 4\times10^{19}$ m^{-3}, $T_e = 10$ keV, and $T_i = 10$ keV and for an antenna to plasma gap of 2 cm.

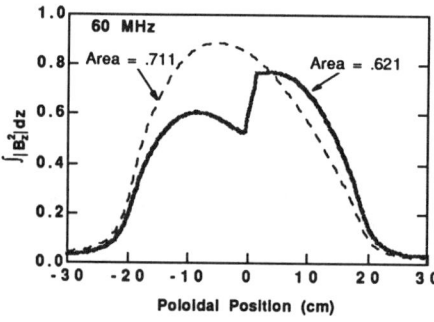

Fig. 4. Net $\int |B_z^2| dz$ distribution 2.5 cm in front of the current strap for single and double strap geometry at 60 and 120 MHz.

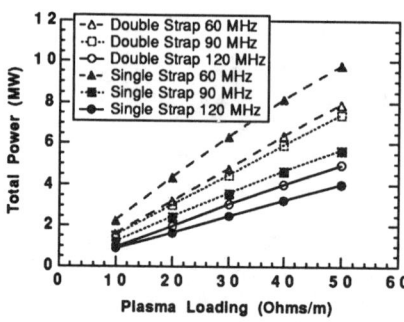

Fig. 5. Power comparison between single and double strap designs for 30 kV peak line voltage limit for the four strap array at 90° relative phasing (end effects included).

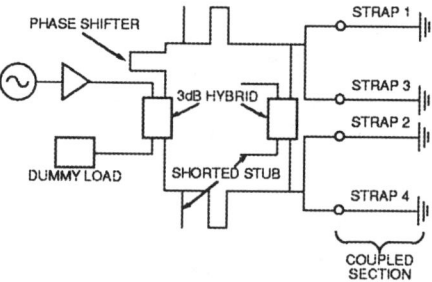

Fig. 6. Phasing and impedance matching circuit for the new antennas during initial operations. Ultimately a second transmitter will replace the first 3dB hybrid junction and dummy load.

EFFECT ON ANTENNA STRUCTURE OF HIGH-POWER RF DURING PLASMA OPERATION*

G. Haste, C. E. Thomas, A. Fadnek, and M. Carter
Oak Ridge National Laboratory, Oak Ridge, TN 37830-8071

B. Beaumont, A. Becoulet, H. Kuus, and B. Saoutic
Centre d'Etudes Nucléaires de Cadarache, 13108 Saint-Paul-lez-Durance, France

INTRODUCTION

High-power, long-pulse operation on the Tore Supra tokamak results in considerable stress on the plasma-facing components. The ICH antennas must deliver high-power rf (up to 4 MW per antenna) in this environment. The antenna structure is therefore subjected to the power flux resulting from the interaction between rf and the edge plasma. The structure's response during operation is described, as is the condition of the antenna after prolonged use.

TORE SUPRA ANTENNAS

ICH power is launched from two side-by-side resonant double loops[1] per antenna. Bumper limiters are located along the sides and at the top and bottom. The Faraday shield has evolved through two versions, as indicated in Fig. 1. In the first version, two tiers of water-cooled tubes pass through an uncooled septum between the two current straps. Carbon tiles are brazed to the outer surfaces of the outer tier of tubes. In the second version, water on one side flows through a single tier of tubes into a plenum on the septum, and out through tubes on the other side. A coating of boron carbide covers the outer surfaces of the water-cooled tubes and the cooled septum.

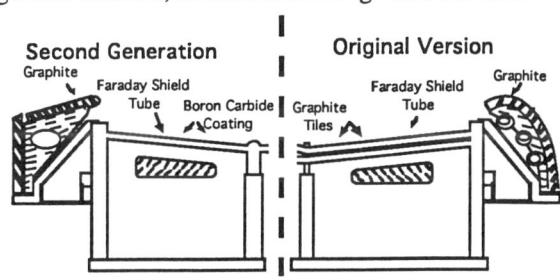

Fig. 1. Evolution of Tore Supra Faraday shields.

Each strap of each antenna is independently powered by a separate transmitter. The phase between straps may be specified. There are six transmitters, each of 2-MW capability, so 4 MW per antenna is available. Although 4 MW has by now been reached for one antenna, the operation described here ranged between 2 and 2.5 MW per antenna.

ANTENNA RESPONSE DURING PLASMA OPERATION

IR imaging, using a camera which viewed the antennas through endoscopes, permitted a time history of the temperatures of the front face of the antenna structure. For these experiments, the antenna was 2 cm beyond the shadow of the limiter. For

---

*Research sponsored by the Office of Fusion Energy, U.S. Department of Energy, under contract DE-AC05-84OR21400 with Martin Marietta Energy Systems, Inc.

that reason, very little temperature increase was seen for the structure, except during and following the rf pulse. Noise pickup during rf prevented a reliable determination of the antenna temperatures, so the bulk of the information applies to the post-pulse period.

Briefly stated, rf leads to localized heating. The hot spots correspond to concentrated power flux and/or lack of cooling. As an example of the latter, the uncooled inconel septum in the first version got hot enough (>600°C) to limit operation. There was a large increase in the nickel concentration when using that antenna and an increase in the radiated power consistent with the observed heavy impurity increase. That antenna was operated only with the straps out of phase (–180°). Conversely, when the antenna with the cooled septum was operated with a –180° phase difference, there was little or no temperature increase for the septum. However, as shown in Fig. 2, even the cooled septum could become hot when using other phase differences.

Often, the hottest spot on the antenna face was on one of the two diagonally opposite corners, those where the magnetic field lines just graze the corner (considering the "q-tilting" of the field lines). In general, the lower of those would be the hottest, but as shown in Fig. 3, phase differences other than 180° could cause the upper corner to be about as hot. The heating in the upper corner was enough to debraze the carbon tile there, with the result that the corner tile fell off during one shot.

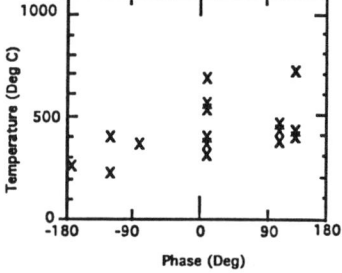

Fig. 2. Maximum septum temperatures as phase is varied.

CONDITION OF ANTENNA FACE

After one of the antennas was removed, damage was observed to the tile on the bottom corner, such that part of the tile was missing. Why the part was missing is not clear, but in the light of experience with the upper corner (on another antenna), thermal shock is a serious contender. After the part had broken off, the area had clearly been eroded by the plasma.

Three of the carbon tiles on one of the side bumpers (on the "electron drift" side) had been broken, but that bumper had suffered a mechanical jolt, so that may have been the cause for the tile breakage seen. The sharp edges still present on the breaks indicated negligible plasma erosion after the fractures.

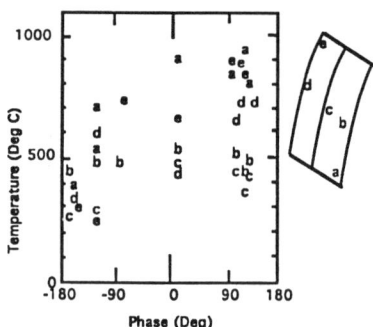

Fig. 3. Maximum temperatures on antenna face as phase is varied.

A band of discoloration extends along this side bumper, about where the maximum heating is expected. That location is at the maximum of the product of the exponentially decaying heat flux times the cosine of the angle between the normal to the tile surface and the magnetic field direction. Very small-scale damage occurred to the tiles along that band. There are sharp edges where one tile overlaps the next, and areas of a few square millimeters were broken at those edges.

Localized pitting, consisting of a large number of small pits in an area of about 4 × 20 mm, was seen at a few random spots on the antenna face. The cause is unknown.

About 20 of the carbon tiles were missing from the water-cooled tubes of the Faraday shield, predominantly in the area where intense heating was seen. Thermal shock is suspected for their departure.

CORNER HEATING

Although corner tiles are not cooled as well as the rest of the bumper tiles, the high temperatures on the corners seem to indicate localized heating, as opposed to poor cooling. Enhanced heating at the corners is somewhat surprising because the corners are well below (~6 cm) the ends of the current straps. The maximum current, and therefore the maximum rf B fields, occur at the center of the antenna, about 30 cm from the corners. The corners seem to be remote from the strong field regions; nevertheless, some of the antenna power is being coupled there, with a high power density.

One possible mechanism for concentrating power at the corners is the electromagnetic excitation of the plasma sheath.[2] Electric fields (electromagnetic rather that electrostatic) can heat electrons via Fermi acceleration. As a result, a large positive sheath potential is formed. Ions, accelerated through the sheath potential, will deposit their energy on the material surface—in this case, the carbon tile.

The driving force for this mechanism is an electric field parallel to the static magnetic field. Image currents in the antenna structure give rise to such electric fields, particularly at the bottom and top. The principal image currents flow vertically in the sides of the structure, beside the current strap, but at the top and bottom of the sides, these currents must turn and flow horizontally.

Measurements of the rf magnetic field distributions can be used to estimate the parallel electric field near the corner, for vacuum conditions. The value of the electric field in a plasma is not apparent because of the large dielectric constant and the flux compression expected for a plasma in close proximity. Additional work will be required to determine whether this mechanism, which gives qualitative agreement with the observations, is of the right magnitude to explain the effects.

SUMMARY

The ICH antennas in Tore Supra have operated at high power for long pulses. In the first version of the Faraday shield, the septum became very hot and was probably the source for nickel impurity in the plasma. Septum heating was greatly alleviated in the second version.

The bumper limiters had several broken tiles, some of which were known to be due to mechanical causes, some to thermal shock. The tiles subjected to the most thermal stress were at the corners. The reason for the intense heating there is not known with certainty, but electromagnetic excitation of the sheath is a possible cause.

A new Faraday shield is now being designed which will take advantage of the lessons from the first two versions; this shield will have a cooled septum, a single tier of tubes, and boron carbide coating. The bumper limiters will be repaired, but their design will not be altered.

REFERENCES

1. F. W. Baity et al., "Compact loop launcher design study for Tore Supra," Proc. 13th European Conf. on Controlled Fusion and Plasma Heating (Schliersee, April 14–18, 1986), Vol. 10c, Part II, pp. 161.
2. M. D. Carter, D. B. Batchelor, and E. F. Jaeger, "Electromotive Excitation of a Plasma Sheath," Phys. Fluids B **4** (5), May 1992, p. 1081.

POWER COMPENSATORS FOR PHASED OPERATION OF ANTENNA ARRAYS ON JET AND DIII-D*

R. H. Goulding, D. J. Hoffman, and P. M. Ryan
Oak Ridge National Laboratory, Oak Ridge, TN, 37831-8071

G. Bosia, M. Bures, D. Start, and T. Wade
Jet Joint Undertaking, Abingdon, OXON, OX14 3EA (U.K.)

C. C. Petty and R. I. Pinsker
General Atomics, San Diego, CA 92138

ABSTRACT

A system has been designed to allow operation of the JET A2 antenna arrays with phasing between elements of other than 0 or π at increased power. The system uses a hybrid coupler to transfer power in the direction opposite to that occurring at the antenna elements.[1] This results in nearly equal currents and voltages on the feed lines and antenna structures, with nearly equal input powers at the rf generators. The direction and magnitude of the power transfer is controlled by adjustable tuning elements mounted on two of the four coupler ports. Much insight has been gained on optimization of the circuit design to maximize bandwidth (~10 MHz for the present JET design with 20 MHz a possibility) and power-handling capability. The concept has been tested initially on the DIII-D tokamak, with a compensator installed in the feed circuit for the four-element FWCD antenna array. In this application it acts to equalize input impedances, and will eventually be used to allow operation with multiple transmitters.

THE JET POWER COMPENSATOR SYSTEM

The basic design of the JET power compensator system (shown in schematic in Fig. 1) has been described previously.[2] Each of the four antenna elements is driven by a separate generator, with the phase and magnitude of the current on each controlled by feedback circuits. At phases other than 0 or π, power flows between each pair of elements due to mutual inductance, primarily through nearest neighbors. The power flowing into each inner element from the nearest element leading in phase is offset to some extent by that flowing out to the other

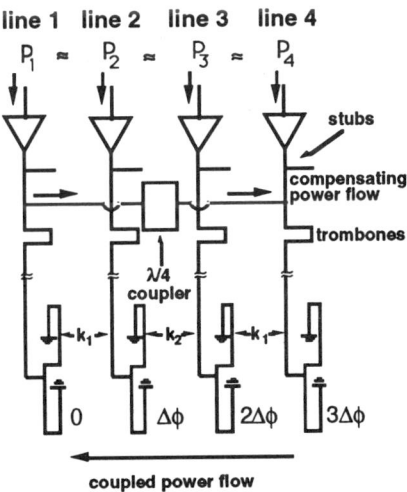

Fig. 1. Schematic of the JET power compensator system.

---

*Research sponsored by the Office of Fusion Energy, U.S. Department of Energy, under contract DE-AC05-84OR21400 with Martin Marietta Energy Systems, Inc.

neighboring element. The balance is exact for identical elements with $k_2 = k_1$ (where $k = M'/L'$, the ratio of the mutual to self-inductance per unit length). There is a net power flow from the leading to the lagging outer element, but by transferring power externally in the opposite direction, it is possible to produce equal maximum current and voltage amplitudes on feed lines and antenna elements with equal input powers. This power transfer also balances the impedances on the outer lines. In the JET design, this is accomplished using a $\lambda/4$ coupler connected across lines 1 and 4 at a location between the tuning stubs and the phase shifters ("trombones"). The direction and magnitude of the transferred power is controlled by adjustable reactances terminating two of the coupler ports, labeled Y_2 and Y_4 in Fig. 2.

POWER COMPENSATOR DESIGN CONSIDERATIONS AND PERFORMANCE

The currents I_c flowing through the coupler defined at the tee junctions, or "coupler tees" (shown in Fig. 2), are given by the expression $I_{ci} = Y_{ii}V_i + Y_{ij}V_j$, where the Y_{ij}'s are the admittance matrix elements for the two-port network consisting of the coupler and connecting lines. The power transferred through the coupler is given by the expression $P_{ci} = 1/2\ Y_{14}\ |V_1|\ |V_4|\ \sin \phi_{ij}$, where ϕ_{ij} is the phase of V_j relative to V_i. The value of Y_{14} and the other Y_{ij}'s are determined by the values chosen for Y_2 and Y_4 (ref. 2).

Two important design parameters which affect power handling with a power compensator installed are the

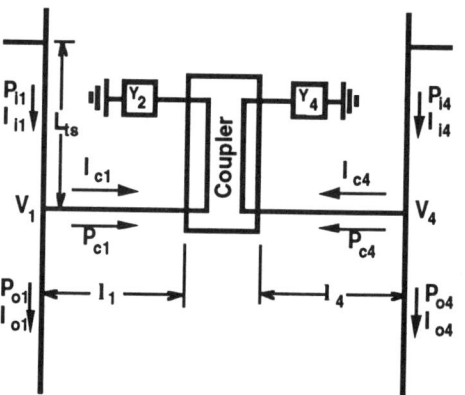

Fig. 2. Schematic of coupler installed between outer feed lines.

stub-tee separation (L_{ts}) and the tee-coupler separation (l_1 and l_4). In order to minimize the currents I_c and I_i, and thus the voltages on the coupler and feed lines, it is desirable to keep V_1 and V_4 small. This can be done at any frequency by making $L_{ts} = 0$. If this is not possible, as is true in the JET case, then this can be done by choosing $L_{ts} = (2n+1)\lambda/8$ for a selected midband frequency. If L_{ts} is located too close to an impedance minimum, a large value of Y_{14} is required to achieve the required magnitude of the transferred power. The corresponding values of Y_{11} and Y_{44} then become very large, causing large circulating currents (and high voltages) between the coupler and tuning stubs, even though V_1 and V_4 are small. It has also been found that the optimum length for l_1 and l_4 is again an odd multiple of $\lambda/8$. For JET, the selected values $L_{ts} = 2.41$ m and $l_1 = l_4 = 4.25$ m result in an operating band extending approximately from 40 MHz to 50 MHz.

During the upcoming phase of JET operation, the two pairs of antennas will be separated toroidally such that the gap between straps 2 and 3 is ~35% greater than that between 1 and 2 (and 3 and 4); k_2/k_1 is then expected to be in the range 1/3 to 1/2. Assuming that one generator is producing maximum power, and the output of the others is reduced to compensate for inductive power transfer at the antennas, the ratio of the actual input power to the total available power for a relative phasing of $\pi/2$ and equal currents in the straps is given by $P/P_{max} = 1/(1+k_1 Q)$ without a compensator across the outer lines, and $P/P_{max} = 1/[1+ (k_1 - k_2) Q]$ with a compensator for equal currents at the strap grounds. Q is defined as $\omega L'/R'$ where R' is the resistance per unit length of the antenna elements. For the A2 antennas, at $f = 41.9$ MHz, with $R' = 9.2$ Ω/m, the value of Q is ~8, and assuming $k_1 = 0.1$, $k_2 = 0.05$ gives a 30% improvement in P/P_{max} with the coupler present. This is reduced from a 75% improvement with $k_1 = k_2$. Furthermore, if the currents and voltages are allowed to become unbalanced, with the powers adjusted to keep the maximum voltage on each feed line below 30 kV, the increase in input power made possible by the compensator is reduced to ~15% for $k_2/k_1 = 0.5$. This is because the reduction in k_2 causes the voltage on the inner element, lagging in phase relative to its outer neighbor, to increase to a value comparable to that of the lagging outer line without a compensator.

The impedance balancing function of the compensator is of significant help in allowing operation at low values of loading, even with $k_2 \neq k_1$. Figures 3a and 3b are plots of the effective values of R_c as a function of the actual value of R_c for a case in which the currents and voltages are allowed to unbalance so as to achieve maximum possible input power. R_c is defined as the resistance calculated at an impedance minimum in the unmatched region of an outer feed line, ignoring antenna coupling and feed line losses. $R_{c\ eff}$ is the corresponding resistance value on each of the four feed lines with coupling taken into account. These calculations were performed using models described elsewhere.[1] The input powers are again calculated so that $V = 30$ kV is not exceeded on any line and $P_{in} \leq 2$ MW on all lines. For this calculation, $f = 41.9$ MHz, $k_1 = 0.1$, $k_2 = 0.05$, and $\Delta\phi = \pi/2$. In Fig. 3a, it can be seen that the values of $R_{c\ eff}$ on lines 1 and 4 are nearly balanced by the compensator, but that line 3 is shifted to a lower loading value. In Fig. 3b, the compensator is not present, and the effective loadings on lines 1 and 4 become severely unbalanced. For a given value of $R_c \leq 3$ Ω, the minimum value of $R_{c\ eff}$ is 1 Ω or more higher with the compensator than without it. The value of R_c at which $R_{c\ eff}$ on any line goes to zero and operation is strictly impossible (the "Yagi" point), is 0.8 Ω with the compensator and 1.8 Ω without it.

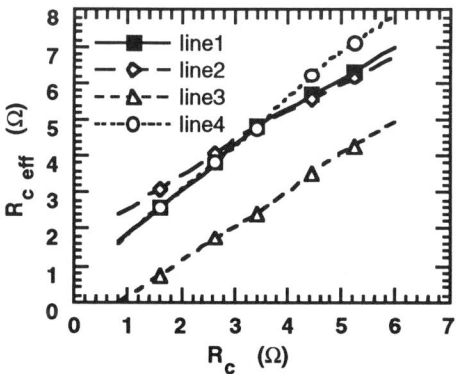
Fig. 3a. $R_{c\ eff}$ with compensator.

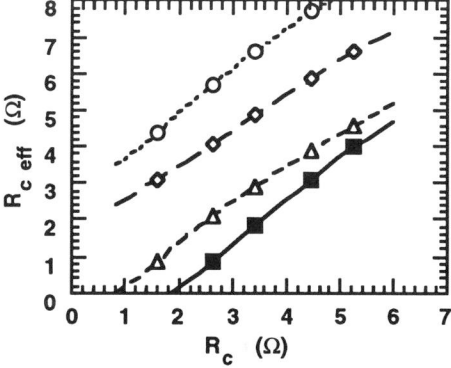
Fig. 3b $R_{c\ eff}$ without compensator.

THE DIII-D POWER COMPENSATOR

Initial phased operation of the DIII-D FWCD array using a power compensator has begun. A small (3-in.-diam ports) $\lambda/4$ coupler was used, limiting maximum input power at this point to ~250 kW. In the feed circuit of the DIII-D FWCD antenna array (Fig. 4), pairs of antenna elements are connected by external resonant loops instead of each being driven individually. Power from a single generator is fed through a power splitter and matching networks to the two loops. In this configuration, a power compensator can be used to produce equal currents on all antennas with equal input powers to the resonant loops, regardless of the ratio k_2/k_1.

In the present DIII-D design, the coupler tees are located at high impedance points located $\lambda/2$ from the resonant loops. At this location, the required value of Y_{12} is independent of resistive plasma loading and antenna phasing, and the antennas are truly decoupled, in that the impedances on the generator side of the coupler tees are independent of phasing. Initial operation in vacuum has confirmed the phase insensitivity. Figure 5 is a Smith chart showing the reflection coefficient measured on the generator side of the hybrid power splitter. The phase is rotated through 2π by moving only the phase shifter on the generator side of the right-hand shorted stub. A very good match is maintained at all phases without any retuning, indicating a cancellation of nearly all of the coupling occurring between the two resonant loops.

The one drawback of the high impedance location for the coupler tees is that, as previously mentioned, the voltages in the coupler are high. Other locations are being considered for future use.

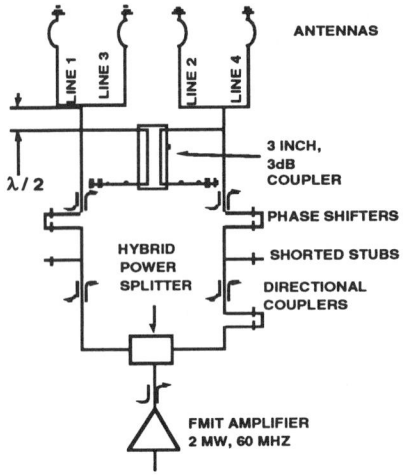

Fig. 4. Schematic of DIII-D antenna feed system with power compensator.

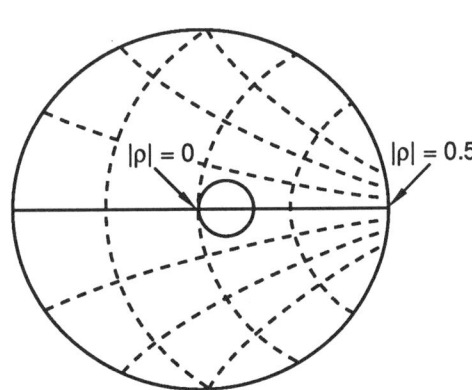

Fig. 5. Reflection coefficient at power splitter input for phases between 0 and 2π.

REFERENCES

1. G. Bosia and J. Jacquinot, in Proc. IAEA Tech. Committee Meeting on Fast Wave Current Drive in Reactor Scale Tokamaks, Arles, France, Sept. 23–25, 1991.
2. R. H. Goulding et al., Proc. 17th Symposium on Fusion Technology, Rome, Sept. 11–15, 1992.

Electrical Characterization of the JET A_2 Antenna: Comparison of Model with Measurements *

P. M. Ryan and R. H. Goulding,
Oak Ridge National Laboratory, Oak Ridge, TN 37831-8071

V. Bhatnagar, A. Kaye, and T. Wade
JET Joint Undertaking, Abingdon, U.K.

Introduction

The JET experiment is replacing its previous (A_1) antennas with upgraded designs (A_2) for its upcoming "pumped diverter" operation.[1] These antennas are more directional than the previous two-strap A_1 antennas when operated as a phased array. The frequency range is 23 to 57 MHz, with particular interest in frequencies around 32, 42, and 48 MHz for various experimental scenarios. Figure 1 shows a full four-element array; note that the power for both the inner and outer straps is introduced through ports located behind the outer straps. A full-scale low power "flat" mockup was tested at JET; strap lengths were adjusted to give balanced operation with resonance at 42 MHz. A second mockup module, differing only slightly from the original, was subsequently fabricated and both modules were sent to ORNL for additional measurements and to test the operation of the power compensator circuit.[2]

There are benefits to using a transmission line model to characterize coupled antenna systems, primarily in the ease of incorporating the antennas into the overall analysis of the transmission, tuning, and matching system. The characteristics of the array under arbitrary phasing are also needed for the design, analysis, and control of the power compensator. There are aspects of the JET A_2 antenna geometry that differ considerably from previously modeled cases. As can be seen in Fig.1, each transmission line feeds two poloidally-stacked straps connected in parallel. The parallel straps present different electrical loads at the match point due to geometrical differences. Currents in one section of the strap influence other sections of the same strap as well as in neighboring straps due to internal inductive coupling. The lengths of the inner and outer straps differ; moreover, the inner straps are fed from ports located behind the outer straps, resulting in increased coupling between the inner and outer straps due to the long feed lines and in greater disparity between the electrical loads presented at the inner and outer feed ports. The present effort is to determine whether a more general coupled transmission line model can characterize the array response with sufficient accuracy for the purpose of design and analysis.

Electrical Characteristics

Each antenna is divided into a number of sections of arbitrary length, each of which is characterized by self-inductance (L'), self-capacitance (C'), mutual inductance (M'), and mutual capacitance per unit length (K'), as calculated from a 2D magnetostatic code[3], and by resistance per unit length (R') which may be ohmic losses or plasma loading impedance. The electrical characteristics are calculated using a transmission line model which uses an iterative technique to solve the coupled telegraphy equations:

---

*Research sponsored by the Office of Fusion Energy, U.S. Department of Energy, under contract DE-AC05-84OR21400 with Martin Marietta Energy Systems, Inc.

© 1994 American Institute of Physics

$$\frac{\partial^2 I_i}{\partial x_i^2} = j\omega C_i' \left(j\omega L_i' + R_i'\right) I_i(x_i)$$

$$+ \sum_{m=1}^{N} \left(-\omega^2 C_i' M_{im}' + j\omega s_{im} K_{im}' \left(j\omega L_m' + R_m'\right)\right) I_m(x_{im}) \exp(-j\omega d_{im}/c)$$

For straps i and m, the direction of increasing x_m to direction of increasing x_i is represented by s_{im} and the distance between x_i and x_m is d_{im}. The effect of the Faraday shield on the L' and C' of the radiating sections of the strap was calculated with a 3-D magnetostatic model[4]. The four port impedances are calculated for four linearly independent phasings (eigenmodes) and transformed to scattering matrix representation using standard techniques.

Figure 2 shows a typical geometry used to determine the coupling matrixes for the radiating sections of the straps. Scattering parameter measurements on the mockups were made with a network analyzer connected between two ports at a time; the two unused ports were terminated with 50 ohm resistances. The connections for the measurements are not what will be used on JET; installation of a current probe reduced the diameter of the feedline and a 30 Ω to 50 Ω, flange-to-Type N adapter was used in place of the standard JET 30 Ω feedthrough. The two mockup modules were inclined 6.8° to the horizontal and the two interior straps were spaced 52.2 cm apart, while the distance between straps in each module is fixed at 40.2 cm. This corresponds to the antenna orientation for a radial location designed to accommodate large cross-sectional plasmas.

The transfer of power from strap 1 to strap 4 during phased operation depends primarily on the interstrap transmission coefficients S_{12}, S_{23}, and S_{34}. Figure 3 shows good agreement between the measured and calculated values of $|S_{11}|$, $|S_{12}|$, and $|S_{22}|$ vs. frequency between 23 and 57 MHz, when resistive straps representing a plasma load of approximately 21 Ω /m are mounted on one mockup module. However, it appears that $|S_{12}|$, while peaked near the nominal resonance frequency of 42 MHz, departs from the model by beginning to increase again at the higher frequencies.

This effect is more apparent when lossless straps are mounted on both mockup modules, as in Figure 4. Here the measured $|S_{12}|$ and $|S_{34}|$ differ somewhat in their frequency response, but both show a shift in their peaks to higher frequencies while the model remains peaked near the antenna resonant frequency. The coupling between inner straps, $|S_{23}|$, shows frequency response that agrees with the model although the magnitude is low. Good agreement can be obtained by reducing the appropriate mutual inductance by 40%. The calculated mutual inductance between the straps is high due to a conducting ground plane immediately behind the modules, which will be present when mounted on JET but not in the measurements.

Magnetic Field Measurements

The toroidal magnetic field (B_z) distribution for one mockup module has been measured at a radial location 1.5 cm in front of the Faraday shield with a magnetic loop probe mounted on an automatically scanning coordinate measuring machine for three different frequencies - 25, 35, and 57 MHz. The poloidal field distributions calculated with the 3-D magnetostatic model (constant strap current) have been weighted by the current distributions obtained from the transmission line code and are compared with the measurements at 35 MHz in Fig. 5. The bumps in the measured field for the outer straps are caused by the hollow circular rings inserted into the strap to allow access to the mounting fixtures behind the strap (see Fig. 1); these structures were not modeled.

Conclusions and Discussion

The coupled transmission line model agrees well enough with measurements to be a useful tool for design and analysis, particularly when the radiating sections are loaded. However, the power transmission coefficients between straps in the same module (S_{12}, S_{34}) are greater than predicted at the higher frequencies (50-60 MHz), and are somewhat different for two modules that are almost identical. They also show a greater sensitivity to the spacing between modules than is predicted by the model. In addition, a 50 Ω match (-24 dB) at 16 MHz is now observed with the resistive straps (Fig. 3a) that was not seen previously. Additional tests seem warranted, including inspecting the strap connections, minimizing room reflections, and measuring the characterisitics of the modified feedthroughs and adapters.

References

[1] R. Lobel et al, Proc. 16th SOFT, London, U.K., 1990, 1104-1108
[2] R. H. Goulding et al, Proc. 17th SOFT, Rome, Italy, Sept. 14–18, 1992.
[3] G. L. Chen et al, AIP Conf. Proc. **159**, 382 (Kissimee, FL 1987)
[4] P. M. Ryan et al, *Fusion Engineering and Design* **12**, 37 (1990)

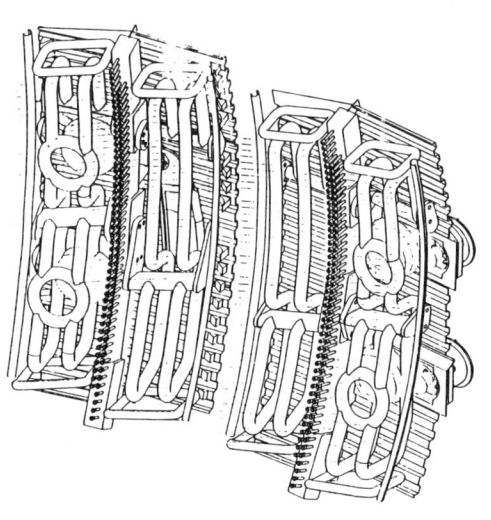

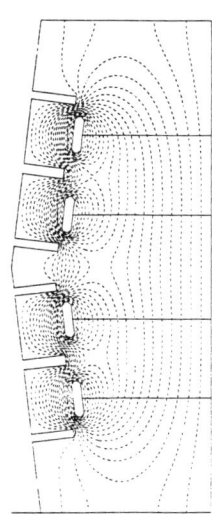

Fig 1. JET A2 Antenna Array

Fig 2. 2D calculation of coupling parameters.

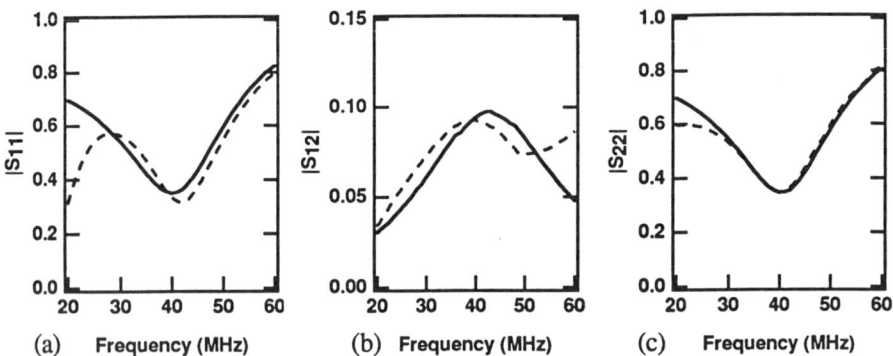

Fig 3. Measured (dashed) and calculated (solid) S-parameters for resistive straps (a) S_{11}, (b) S_{12}, (c) S_{22}

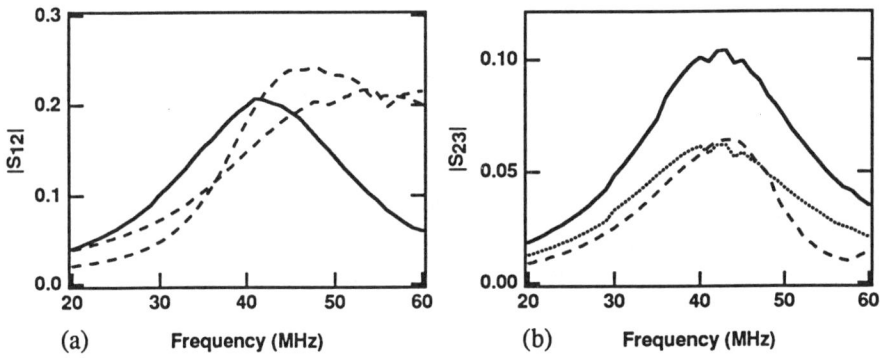

Fig 4. Measured (dashed) and calculated (solid) S-parameters for lossless straps.
(a) Slight differences between modules are seen in S_{12} and S_{34} (dashed)
(b) Dotted curve is S_{23} calculated by using 0.6 times the calculated M_{23}'.

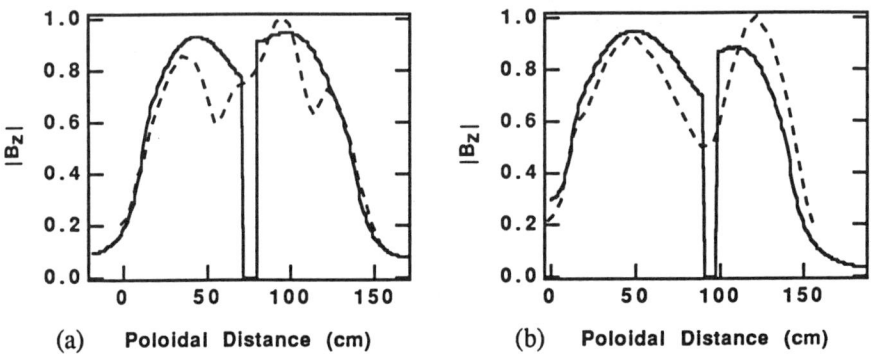

Fig 5. Comparison of calculated (solid) and measured (dashed) B_z profile along the centerline of the strap at 35 MHz for (a) outer straps and (b) inner straps.

SELF-CONSISTENT 3-D ICRH ANTENNA MODELING WITH PLASMA

Y. L. Ho, W. Grossmann, A. Drobot
SAIC–McLean, VA

M. D. Carter, P. M. Ryan, D. B. Batchelor
ORNL

ABSTRACT

A new solver has been developed for the ARGUS electromagnetic field solver package to analyze the ICRH antenna coupling problem. The key advantages are that complex 3-D structures, e.g., Faraday screen, side walls..., are represented realistically, and that all currents flowing in the structures are solved self-consistently. Plasma response is included as a surface impedance in Fourier-space while the field solution around the antenna (satisfying $\nabla \times (\nabla \times E) = (\omega/c^2 E)$ is computed in real space. The plasma impedance matrix is calculated separately, e.g., analytically or from a full-wave code such as ORION.

INTRODUCTION

Generally, theoretical modeling and simulation of ICRF heating and current drive phenomenology have been carried out by decomposing the problem into two separately analyzed parts: one that concentrates on the absorption and other such effects that take place inside the plasma, and one that deals primarily with the electromagnetic coupling of the antenna to the plasma. The first kind of analysis, usually making use of full wave codes[1] and ray-tracing codes,[2] requires that the toroidal and poloidal antenna power spectra be known *a priori*. This antenna power spectra can be obtained by solving the antenna plasma coupling problem. Among the many approaches here, Lehrman and Colestock[3] and Theilhaber and Jacquinot[4] have treated the current flow in the antenna current straps in the presence of plasma self-consistently. However, their treatment tend to be limited to idealized antenna structures. To address this limitation, we have developed a new frequency-domain solver with self-consistent plasma response for the ARGUS simulation code capable of modeling highly complex 3-D antenna structures.

THE ARGUS SIMULATION CODE

The simulation code ARGUS is a general-purpose field solver designed to handle problems involving complex geometrical structures. Its key advantage is the sharing of realistic geometric representation amongst disparate solvers, thus permitting different aspects of a problem to be analyzed together. Currently, five field solvers are applicable to the antenna modeling problem: an electrostatic field solver, an electromagnetic normal mode solver, an electromagnetic time-domain solver, an electromagnetic frequency-domain solver with plasma response, and a particle-in-cell module for self-consistent treatment of particle and field interaction. Our prior work with ARGUS in modeling ICRF antennas relied primarily on the time-domain solver.[5] These work demonstrated the ability of ARGUS to model the effect of complex 3-D structures, e.g., side-slits, on the antenna vacuum field pattern.

NEW ARGUS FREQUENCY DOMAIN SOLVER WITH PLASMA RESPONSE

To include plasma response, it is advantageous to work in the frequency-domain. Here, we describe briefly the essential details of our present implementation. The new solver scheme requires two levels of iteration: an inner loop and an outer loop. The inner loop solves the Helmholtz equation

$$\nabla \times (\nabla \times E) = (\omega/c)^2 E$$

in vacuum with complex internal boundaries given the tangential components of the electric field at the boundaries of the computational domain. Here, E is the electric field, ω is the antenna frequency, and c is the speed of light in MKS units. An iterative conjugate gradient solver is used for this loop.

The outer loop is where we introduce the plasma response. We have chosen to model the plasma as a surface impedance,

$$E_T(\omega, k_\theta, k_z) = Z(\omega, k_\theta, k_z) B_T(\omega, k_\theta, k_z)$$

where E_T and B_T denotes the tangential electric and magnetic field components, Z is the plasma impedance matrix, k_z is the toroidal mode number, and k_θ is the poloidal mode number. The elements in Z can be evaluated from a variety of plasma models, i.e., a full wave code, a slab code, or analytical solutions. At each iteration, the new E_T at the plasma-vacuum interface is evaluated from B_T from the previous iteration. This new E_T forms the boundary condition for the inner loop iterations. Fourier transforms are used to communicate between real and Fourier space representation of E_T and B_T. As a detail, we also move the components in B_T that depends on E_T to the left-hand-side in the above equation for E_T to enhance implicitness.

To demonstrate the feasibility of our numerical scheme, we have used a simple uniform density cold plasma slab model with plane wave solutions to calculate Z. Figure 1 describes a test problem configured with a plasma interface at some standoff distance from a mockup of the DIII-D fast wave antenna, operating in π-phasing. For a 31.8 Mhz case with toroidal field strength, $B = 1\,T$, edge plasma density, $n = 4 \times 10^{19} m^{-3}$, and the plasma positioned 2.5 cm in front of the Faraday shield, we find the actual total antenna impedance, measured at the feed point (not the usual per unit length loading) to be:

$$Z = (3.55 - i\,32)\,\Omega.$$

where Ω is given in Ohms. The antenna power spectral distribution over the full antenna face associated with the above simulation, as a function of distance from the bottom of the antenna, is shown in Figure 2.

As a further example, we show in Figure 3 a series of three antennas differing only in the structures of the Faraday bars: one with no bars (Figure 3a); one has three bars covering only the radial current feeder (Figure 3b); and one with six bars (Figure 3c). The plasma is positioned at 4 cm in front of the antenna bay. The current strap is 11.2 cm wide in a 22.6 cm wide bay and recessed 2.8 cm behind the bay front. For these cases, the plasma impedance matrix is obtained from the ORION full-wave slab code.[1] The plasma is highly absorbent and is representative of the SST experimental condition proposed at PPPL. In Figure 4, we compare the resultant antenna current distribution for the three cases, plotted as the poloidal magnetic field measured in front he of current straps at the toroidal symmetry plane versus antenna height. The result clearly show the slow down of the phase velocity as the number of Faraday bars increased, and the discreet bar effects causing the bumps in the curve for the six bar case. These calculations partially illustrates the utility of the ARGUS simulation code: the ability to predict antenna performance allowing the testing of different ideas.

PRESENT WORK

Currently, we are performing detailed benchmark of ARGUS against the 2-D RANT and the new 3-D MAntIS code[6] developed at ORNL. The comparison with the 2-D code will help us discern 3-D effects not present in 2-D models. In the future, we plan to enhance the convergence speed of the solver, increase its ease of use, and apply it to actual designs. On the research front, we plan to examine particle orbits and add sheath effects to the model.

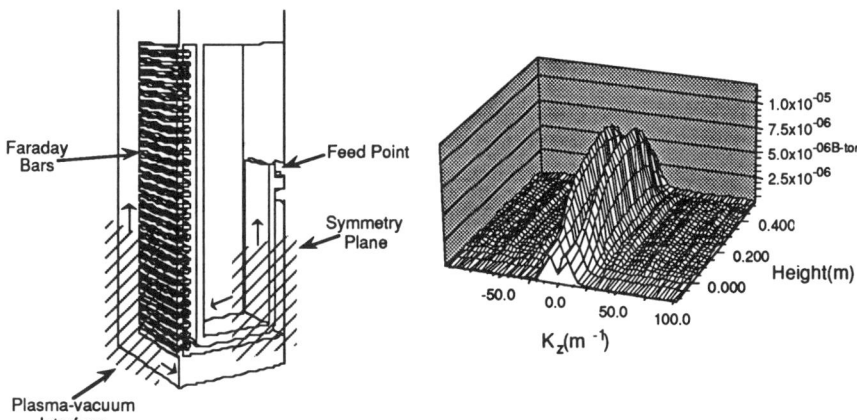

Figure 1. Cutaway view of a single strap of the mockup DIII-D ICRH antenna. Center septum not shown.

Figure 2. Toroidal field spectra launched by the antenna shown in Figure 1 as a function of distance from the bottom of the antenna. This case has density, n 4×10^{19} m^{-3}, toroidal field $B = 1$ T, and plasma positioned 2.5 cm in front of the Faraday shields.

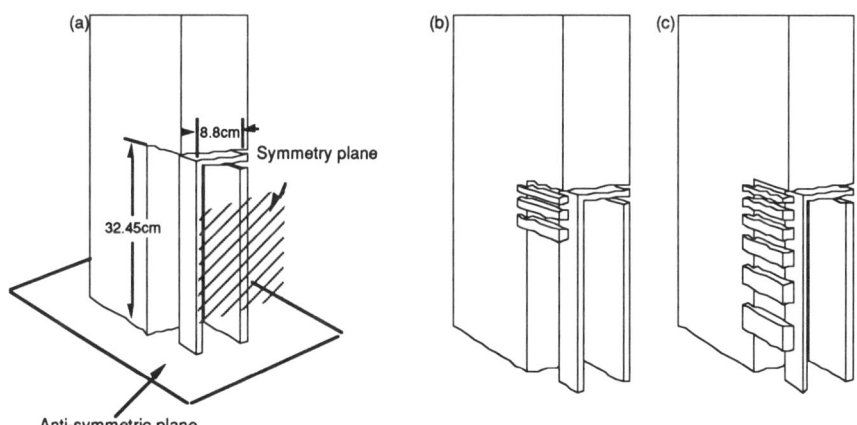

Figure 3. Three identical antennas with different Faraday bar structures used for comparision: (a) no bars; (b) 3 bars; (c) six bars. These are single strap antennas. In each case, only one quarter of the antenna is shown.

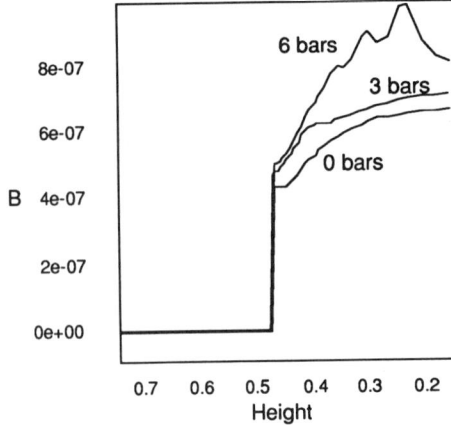

Figure 4. Comparison of the sharp current distribution as a function of antenna height (measured from the center short) for the three antennas shown in Figure 3.

ACKNOWLEDGMENT

This work was supported by the United States Department of Energy.

REFERENCES

1. See, for example, E. F. Jaeger, D. B. Batchelor, M. D. Carter, and H. Weitzner, *Nucl. Fusion* **30**, 505 (1990).

2. See, for example, V. P. Bhatnagar, R. Koch, P. Geilfus, R. Kirkpatrick, and R. R. Weynants, *Nucl. Fusion* **24**, 955, (1984).

3. I. Lehrman and P. L. Colestock, *IEEE Trans. Plasma Sci.* **PS-15**, 285 (1987).

4. K. Theilhabar and J. Jacquinot, *Nucl. Fusion* **24**, 541 (1984).

5. M. Kress, Y. L. Ho, W. Grossmann, A. Drobot, D. B. Batchelor, P. M. Ryan, and M. Carter, in *Radio Frequency Power in Plasmas*. AIP Conference Proceedings 244, edited by D.B. Batchelor (AIP, New York, 1992), p. 213.

6. M. D. Carter et al., *Bull. Am. Phys.* **37**, 1604 (1992).

A "3-D" ICRF ANTENNA COUPLING MODEL FOR HEATING AND CURRENT DRIVE APPLICATIONS IN TOKAMAKS

M. H. Bettenhausen and J. E. Scharer
University of Wisconsin-Madison, Madison, WI 53706

ABSTRACT

Analysis and computer simulation is presented for investigating coupling of Ion Cyclotron Range of Frequency (ICRF) waves to tokamak plasmas using phased coil antenna arrays. The model accounts for 3-D antenna and feeder current effects, an antenna cavity which is finite in all three dimensions and warm plasma effects. The antenna radiation resistance and the spectrum of the ICRF power coupled to the plasma are calculated from the model. The effect of the three-dimensional cavity on the parallel wavenumber spectrum of the coupled power is investigated. Emphasis is on coupling of the fast wave but the model includes parallel electric fields for consideration of slow or ion Bernstein mode excitation. The model also permits studying the effects of variations of the angle between the Faraday shield bars and the tokamak axis. Radiation resistance and spectrum results from the model are presented.

INTRODUCTION

Accurate modeling of ICRF antennas is of interest to support operation and understanding of existing ICRF experiments and to aid in the design of future ICRF systems. The model presented here is an extension to the models developed by Brambilla[1] and Chiu et al[2]. The model incorporates a finite length antenna in a three-dimensional recessed cavity and includes parallel electric field and warm plasma effects.

DESCRIPTION OF THE MODEL

A rectangular coordinate system is assumed with x, y and z corresponding to the radial, poloidal and toroidal directions, respectively. Figure 1 shows the geometry used and defines various parameters. The current distribution on the antenna straps is assumed to be sinusoidal in the poloidal direction with a phase velocity of ω/k^* on the straps which face the plasma. The current distribution on the radial feeders is taken to be uniform.

The electric fields excited by the assumed antenna current distribution are calculated by solving Maxwell's equations in four separate regions: in the plasma, $x > 0$, and in the three vacuum regions as labeled in Fig. 1. Boundary conditions are applied between adjacent regions to produce a unique solution.

© 1994 American Institute of Physics

364 A "3-D" ICRF Antenna Coupling Model

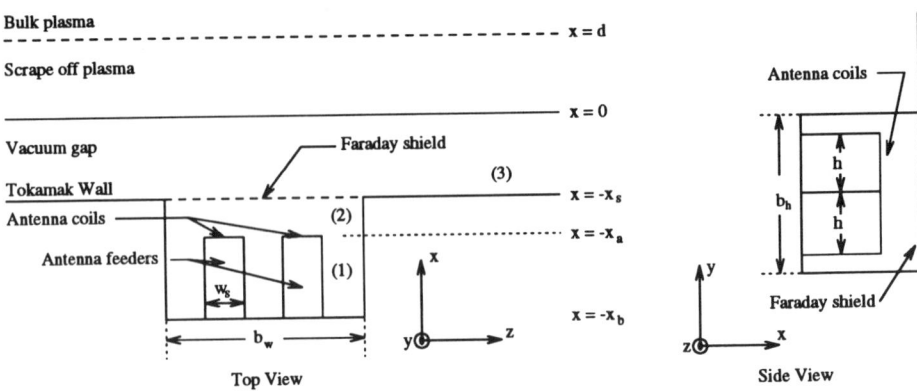

Figure 1: Antenna model geometry.

The plasma wave fields are calculated using the model developed by Brambilla[1]. This is a warm plasma slab model which includes the dominant finite Larmor radius terms to second order in $(k_\perp \rho)^2$. These results are used to calculate a plasma surface impedance relating the electric and magnetic fields at the vacuum-plasma interface.

The boundary conditions at the Faraday shield assume that the shield acts as a perfect conductor parallel to the sheild bars and as a perfect insulator perpendicular to the bars. The tokamak and cavity walls are also assumed to be perfectly conducting so that the tangential electric fields are zero on the walls. A jump condition is applied at the antenna current elements such that $\hat{x} \times (\vec{B}_> - \vec{B}_<) = \mu_0 \vec{J}_s$ where $\vec{J}_s$ is the surface current on the antenna and the less than and greater than subscripts denote the valued of $\vec{B}$ for $x < -x_a$ and $x > -x_a$. Finally, the electric field components E_y and E_z are continuous at the antenna.

RESULTS

The computer code WICS (Wisconsin Ion cyclotron Coupling Solver) is used to calculate the antenna radiation resistance and parallel wavenumber spectrum of the coupled power based on the model discussed above.

Table 1 presents values of radiation resistance calculated with WICS for the dimensions of the TFTR Bay M antenna and TFTR supershot plasma conditions (shot no. 67911). The currents on the two antenna straps are phased 180 degrees apart for these cases. The results demonstrate that as the cavity height is increased the radiation resistance increases. This is likely due to smaller return currents on the cavity walls. Similarly, as the length of the antenna is increased the radiation resistance per meter antenna length increases provided the ends of the antenna are far enough from cavity walls. This result is due to the decreased importance of end effects due to the radial feeders and the finite antenna length.

R_A (Ω)	b_h (cm)	h (cm)	R_A/m (Ω/m)
4.1	81	38	5.4
8.7	550	38	11.4
11.8	550	50	11.8
20.7	550	100	10.4

Table 1: Variation of radiation resistance with antenna and cavity height.

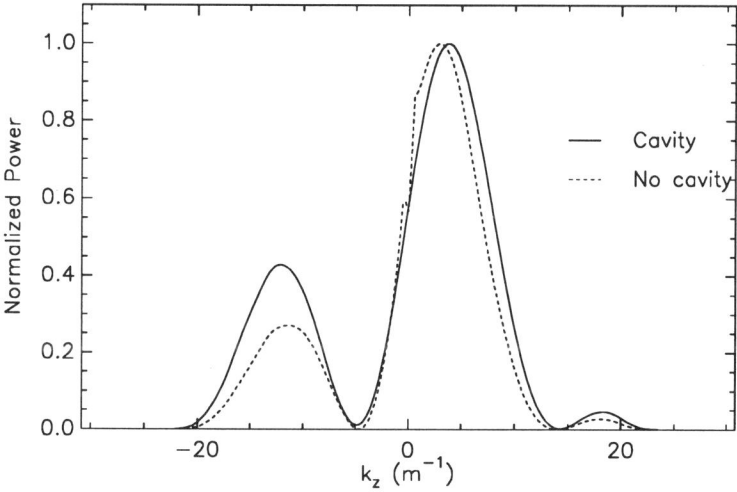

Figure 2: Wavenumber spectrum for coupled power with $(0, \pi/2)$ phasing.

Figure 2 shows the parallel wavenumber spectrum of the coupled power spectrum for the TFTR Bay M antenna for a 90 degree phase difference between currents on the straps. The spectrum has been normalized to give a maximum of 1.0 for each case. The presence of the cavity increases the percentage of the coupled power in the secondary lobe of the spectrum at negative k_z. Figure 3 shows a similar result for in-phase excitation of the antenna straps. The differences in the spectrum shown in these figures are primarily due to a reduction in the power coupled to low k_z rather than an increase in the power coupled to higher k_z. This is consistent with the reduction in radiation resistance due to the cavity as discussed above.

DISCUSSION AND FUTURE PLANS

The model presented here is being used to study the effects of changes in antenna and cavity dimensions and in the plasma parameters on antenna perfor-

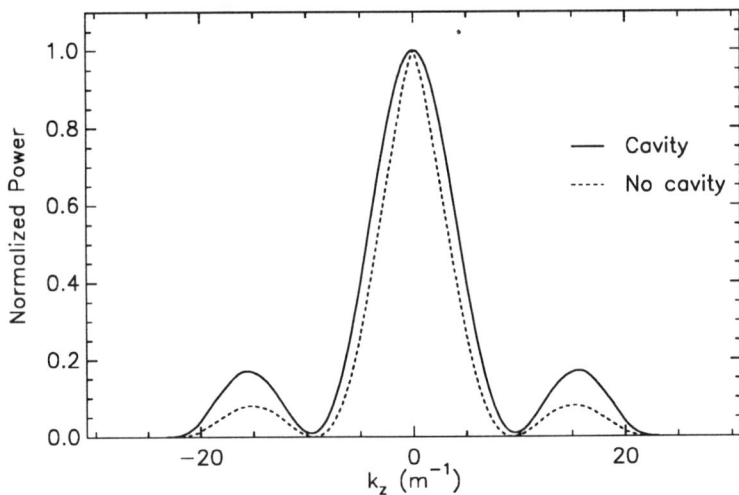

Figure 3: Wavenumber spectrum for coupled power with $(0,0)$ phasing.

mance. Particular emphasis will be placed on the plasma density profile. These studies will investigate ICRF antenna performance issues on TFTR, the Joint European Torus (JET) and antenna design issues for the Tokamak Plasma Experiment (TPX). The model is particularly well suited for consideration of the importance of the alignment of the Faraday shield with the tokamak magnetic field and the importance of the relative heights of the antenna straps and the cavity.

ACKNOWLEDGEMENTS

The authors thank D. Batchelor, M. Carter and and D. Stallings for useful discussions and helpful suggestions and for the use of the 2D antenna coupling code RANT for benchmarking our code. The authors also thank M. Brambilla for providing a copy of his computer code FELICE which helped us develop a better understanding of ICRF coupling physics while developing our code and R. Majeski for helpful discussions and information about the TFTR antennas and plasma conditions.

This work supported by U.S. DOE Grant Nos. DE-FG02-86ER53218 and DE-FG02-93ER54196 and Princeton Contract S-03587-G.

REFERENCES

1. M. Brambilla, Nuclear Fusion, 28, 549 (1988).

2. S.C. Chiu, M.J. Mayberry and W.D. Bard, Nuclear Fusion, 30, 2551 (1990).

Ion Cyclotron Range of Frequencies (ICRF) Heating of Fast Ions in Fusion Plasmas[*]

J. E. Scharer, N. T. Lam and R. S. Sund
University of Wisconsin, Madison
and
O. Sauter, General Atomics

Abstract

We consider the problem of fundamental and second harmonic deuterium and fast tritium beam heating for ITER and TFTR plasmas in the presence of a population of fast alpha particles. For the fundamental deuterium heating case, we justify the replacement of a 0.1-0.4 % fast alpha slowing down population by an absorption equivalent Maxwellian of T_α= 600-800 keV. The absorption, transmission and reflection coefficients are calculated using XWAVE[1], a code which treats full wave effects and perpendicular magnetic field gradients from a fundamental definition of power conservation. A range of D/T ratios and deuterium tail temperatures are examined for a range of alpha particle concentrations. We find that the region near the two-ion D-T hybrid resonance can provide a substantial enhancement of the alpha particle absorbed power and wave reflection. We also have used XWAVE and the SEMAL code[2] to describe second and higher harmonic heating of the deuterium, tritium and alpha particles for the case of the TFTR supershot and second harmonic minority tritium beam scenarios. We find that a significant (> 5%) alpha particle heating occurs for the second harmonic deuterium heating case and that the single pass tritium second harmonic heating is very small.

I. Introduction

We examine the effect of small concentrations of fast ions and fast alpha particles produced in fusion plasmas on ion cyclotron frequency range (ICRF) wave absorption and wave reflection. We examine the fundamental and second harmonic deuterium and minority second harmonic tritium heating scenarios. We consider the fast alpha particle slowing down distribution and show that its absorption can be approximated by a Maxwellian chosen such that the number of resonant particles in the absorption region matches that of a slowing down distribution.

We consider the effects of a small concentration of alpha particles on ICRF wave propagation and absorption for an ITER plasma at the fundamental deuterium cyclotron resonance. We then consider the cases of second harmonic deuterium for the TFTR supershot case and minority second harmonic beam tritium heating for TFTR parameters. We are interested in the fundamental deuterium regime where the initial slowing down alpha distribution can have large gyroradii such that $(k_\perp \rho_{sd})^2 \sim 1$ for the fast wave but the Maxwellian equivalent thermal velocity for the alphas satisfies $(k_\perp \rho_{at})^2 \ll 1$. For the fundamental and second harmonic regimes we utilize the XWAVE code which includes perpendicular gradients in the magnetic field and second order gyroradius effects for cases where $(k_\perp \rho_{at})^2 \ll 1$. For the second harmonic regime where the third harmonic for tritium and alpha particle higher harmonics are included we utilize the SEMAL code. This code is based on a finite-element solution of the plasma wave equation and can take into account an arbitrary number of cyclotron harmonics.

II. Applications to ITER and TFTR

We consider the case of fundamental deuterium and alpha particle heating in a tritium plasma with ITER parameters. We assume the plasma to have $n_e = 9 \times 10^{13}/cm^3$, B=4.85 T, R_0=6 m, n_D/n_T=0.5, $T_e=T_T$=10-15 keV and T_D=10-20 keV with a corresponding fast alpha particle concentration of $n_{\alpha f}$=0.1-0.4 % n_e. The two ion hybrid resonance lies at x=-0.7 m towards the high field side from the fundamental resonance on the axis. The corresponding left-hand circularly polarized component peaks in a localized region near the two ion hybrid resonance. The Maxwellian equivalent alpha particle distribution is chosen to have the same number of resonant particles in this region as the slowing down distribution. The $|E_{left}|$ peak extends over a comparatively narrow interval near the two-ion hybrid resonance, and most of the power that goes to the alphas is absorbed in that narrow region. Consequently, we choose the number of resonant particles at x_p, the location of the $|E_{left}|$ peak, to be the basis of equivalency between two distributions. (Recall, $x_p \approx x_{ii}$ = the location of the two-ion hybrid resonance). We denote the fast alpha particle slowing down distribution by g_{sd} and the Maxwellian by g_M. Thus, $G_M[v_{res}(x_p)] = G_{sd}[v_{res}(x_p)]$ where $G_j(v_\parallel) = \int_0^\infty g_j(v_\perp, v_\parallel) v_\perp dv_\perp$ and $v_{res}(x) = [\omega - \Omega_\alpha(x)]/k_\parallel$.

An explicit expression is obtained by substituting $g_{sd}(v_\perp, v_\parallel) =$

$$\frac{3 n_{\alpha f}}{4\pi \ln[1+(v_0/v_c)^3]} \frac{H(v_0-v)}{(v_\perp^2+v_\parallel^2)^{3/2} + v_c^3},$$

where $v_0 = 1.3 \times 10^7$ m/s is the alpha birth velocity, v_c is the critical velocity, and H is the Heaviside step function. The criterion then becomes,

$$\frac{\exp(-r^2/t^2)}{t} \leq \frac{\sqrt{\pi}}{2 \ln[1+b^3]} \left\{ 0.5 \ln\left[\frac{(r+1)^2}{(b+1)^2} \frac{b^2-b+1}{r^2-r+1}\right] + \sqrt{3} \tan^{-1}\left(\frac{2b-1}{\sqrt{3}}\right) - \sqrt{3} \tan^{-1}\left(\frac{2r-1}{\sqrt{3}}\right) \right\}$$

where $t = \sqrt{2T_\alpha/m_\alpha}/v_c$, $b = v_0/v_c$, and $r = |v_{res}|/v_c$. For $k_\parallel = 6$ m^{-1}, a solution of the above equation yields $t = 1.24$, i.e., a Maxwellian temperature $T_\alpha \approx 600$ keV. Therefore, we have taken $T_\alpha = 600$ keV for the following calculations.

Table I shows the total power absorption for each species for $k_\parallel = 6$ m^{-1} for the case of the lowest temperature plasmas noted above. Note that a substantial fraction of the incident power can be absorbed by the fast alphas when the absorption by the deuterium and electrons is not complete. Figure 1 shows the total absorbed power density over the region where it is significant.

Table I.

$n_{\alpha f}/n_e$ (%)	D (%)	α (%)	e (%)	R (%)
0.1	36	9.9	18	36
0.2	34	18	17	31
0.3	32	25	16	26
0.4	31	31	15	22

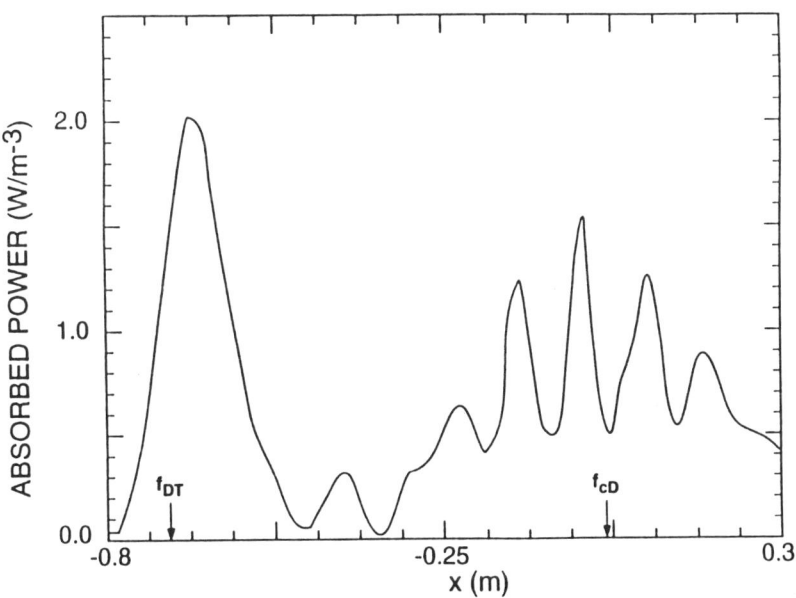

Figure 1. ITER fundamental deuterium heating

Figure 2 corresponds to a TFTR supershot with $n_{eo} = 9 \times 10^{13}$/cm, $T_e = 10$ keV, $T_H = 50$ keV, $T_D = T_T = 20$ keV, $n_D/n_T = 1$, $n_H/n_e = 2\%$, $n_\alpha/n_e = 0.25\%$ @ $T_\alpha = 800$ keV. The wave frequency corresponds to $\omega = 2\omega_{CD0}$ on axis at $B_0 = 4.5T$ utilizing the SEMAL code. Note the significant alpha particle heating on the low field side (9%) as well as comparable hydrogen and deuterium heating.

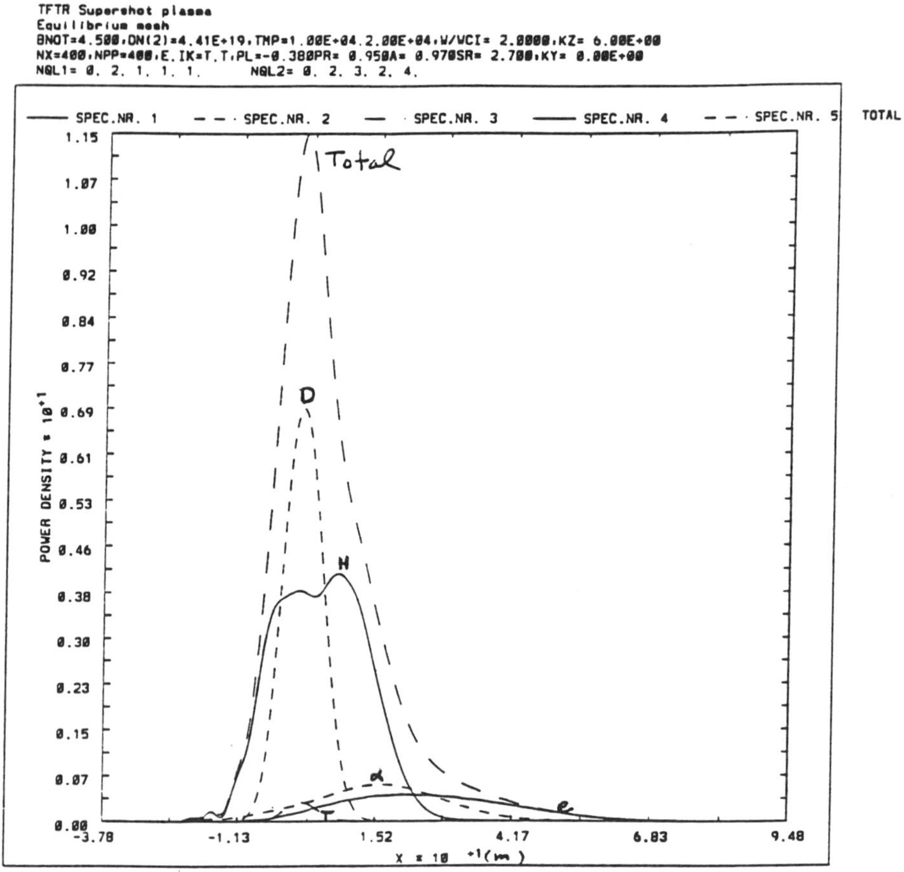

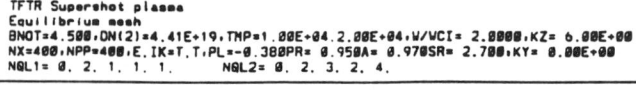

Figure 2. TFTR Supershot at $\omega = 2\omega_{CD0}$

[1]R. S. Sund and J. E. Scharer, Physics of Fluids B3, 1326(1991)
[2]O. Sauter and J. Vaclavik, Nuclear Fusion 32, 1455(1992)

Acknowledgments
*We appreciate the assistance of C. Phillips, R. Majeski and R. Wilson in determining the TFTR ICRF experimental parameters. This research is supported by U.S. DOE Grant Nos. DE-FG02-86ER53218 and DE-FG02-83ER54196.

DESIGN OF THE ION CYCLOTRON SYSTEM FOR TPX*

D. W. Swain, S. Shipley, J. Yugo, R. Goulding, D. Batchelor, D. Stallings
Oak Ridge National Laboratory, Oak Ridge, TN 37831-8071

E. Fredd
Princeton Plasma Physics Laboratory

INTRODUCTION

The TPX experiment[1] will operate for very long pulse times (≥ 1000 s) and will require current drive of several different types to explore the advanced physics operating modes as one of its main missions. Fast wave current drive (FWCD) using ion cyclotron waves in the 40–80 MHz range will be used as one of the main current-drive mechanisms. For initial operation, 8 MW of rf will be supplied, along with 8 MW of neutral beams and 1.5 MW of lower hybrid power.

The ion cyclotron (IC) system is a major part of the TPX heating and current drive system. The IC system must
- supply 8 MW of power through two main horizontal ports;
- be upgradable to provide up to 12 MW of rf power through two ports;
- operate for 1000-s pulses every 75 min;
- drive current using FWCD with high reliability;
- be bakeable to 350°C for cleaning; and
- incorporate shielding to attenuate the neutron and gamma flux from DD operation so that hands-on maintenance can be performed exterior to the vacuum vessel.

The system will consist of four modified FMIT power units (now at PPPL) that will be upgraded to deliver 2 MW each to the plasma. Two antennas, each with six current straps, will be located in adjacent ports. A sophisticated matching system is needed to provide experimental flexibility and reliability.

DESIGN CHOICES

For operation at a toroidal field of 4 T, frequencies that could be used for efficient FWCD (within the transmitter frequency range) are around 45 MHz or 77 MHz, where there are no significant ion resonances in the plasma region, as shown in Fig. 1. Ion heating (either H minority or second-harmonic D) can be done near 61 MHz.

The requirement for efficient FWCD has led to the choice of a 12-strap IC system, with six straps in each port. An alternate design of four straps per port was studied but resulted in inefficient utilization of the IC power for current drive. Figure 2

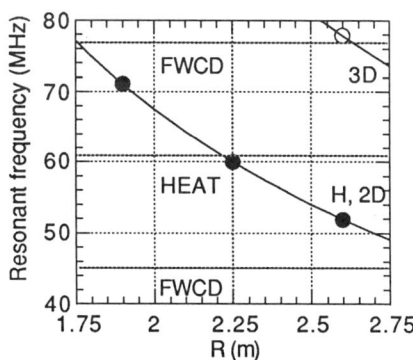

Fig. 1. Resonant frequencies vs R.

*Research sponsored by the Office of Fusion Energy, U.S. Department of Energy, under contract DE-AC05-84OR21400 with Martin Marietta Energy Systems, Inc.

shows the normalized rf power spectra launched into a typical TPX plasma ($<n>$ ≈ 0.5×10^{20} m^{-3}, T_e ≈ 5 keV) vs n_z (= c/v_{phase}) by two four-strap antennas and by two six-strap antennas. For both cases, the relative phasing between current straps was chosen to put the peak in the FWCD spectrum at an n_z value corresponding to a toroidal phase velocity for the wave of $v_{phase} = 1.4$ v_{Te} (i.e., $n_{zpeak} = 4.8$). The four-strap spectrum at 77 MHz has a secondary peak at n_z ≈ −12 with a substantial amount of power, whereas the six-strap system does not.

We define f_{good} as the fraction of the total launched power that is in the region corresponding to efficient FWCD (i.e., the region with $v_{Te} \leq v_{phase} \leq 3v_{Te}$). Analyzing spectra such as those in Fig. 2 yields the results shown in Fig. 3, where f_{good} is plotted vs frequency for the four-strap and six-strap antenna designs. While the relative difference in f_{good} is small at 40 MHz, the six-strap configuration is significantly more efficient at higher frequency.

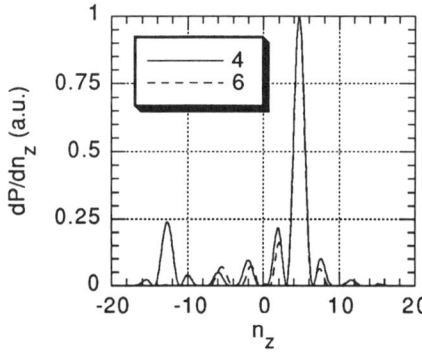

Fig. 2. Spectra at 77 MHz.

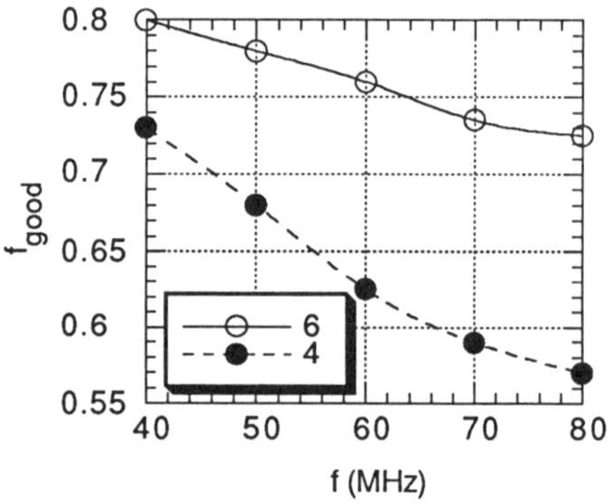

Fig. 3. Fraction of launched power in the region of efficient FWCD.

ANTENNA DESIGN

A drawing of a single six-strap FWCD antenna is shown in Fig. 4.[2] Each strap is grounded at the center and connected to coax lines at both ends. The coax lines traverse a large box in the port containing liquid shielding material that thermalizes and attenuates the neutrons from the plasma. Each coax goes through two right-angle bends to prevent neutron streaming, then through a ceramic vacuum feedthrough and thence through the port flange cover to the tokamak exterior.

A Faraday shield assembly (not shown in Fig. 4) covers the current straps. The shields are made of a single layer of 1.5-cm-diam Cu-coated Inconel tubes tilted at an angle of 12° relative to the horizontal (to align approximately with the local magnetic field), covered by a plasma-sprayed layer of B_4C that is 0.1 to 0.3 mm thick. The Faraday shield is about 30% optically transparent.

The Faraday shield tubes (and all other elements of the antenna) are water-cooled so that they can operate for the 1000-s pulses in the ambient heat loads. For the upgraded power case with 24 MW of beams, 18 MW of IC, and 3 MW of lower hybrid, the heat loads can be up to 180 W/cm^2 on some Faraday shield tubes due to ripple-trapped neutral beam ions, 50 W/cm^2 from plasma radiated power, and 3 W/cm^2 from rf losses.

TUNING AND MATCHING SYSTEM

The tuning and matching system for three straps, which is powered by one transmitter, is shown in Fig. 5. Each current strap is part of a resonant loop that consists of the strap, the coax inside the antenna structure, and coax external to the tokamak connected between the top and bottom feeds to complete the loop. The loops are connected to a transmission system that contains matching components (a line stretcher and stub tuner for each loop), and then to a three-way power splitter. There is a pre-matching section (line stretcher and stub tuner), which already exists at PPPL, between the splitter and the transmitter. This circuit is replicated four times for the total of four transmitters and 12 current straps.

An analysis of the response of this circuit has been carried out using a coupled-circuit, lossy transmission line model. Table I shows the results of the calculations. For a resonant loop side length of 8.8 m, the loops are resonant at three frequencies in the FMIT tuning range, shown in the leftmost column. The loading resistances, R', calculated for a typical plasma density profile expected in TPX ($<n> = 0.8 \times 10^{20}$ m^{-3}, steep edge density profile) using the RANT full-wave plasma code,[3] are shown in the second column. The total power to the plasma can be computed from the formula

$$P_{total} = N_{straps} R' h F_{avg} I_0^2 / 2 \;,$$

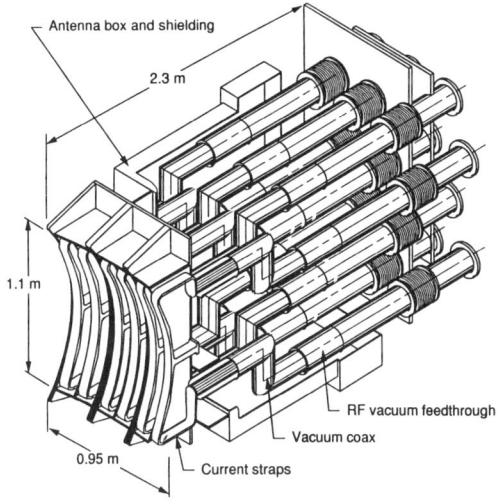

Fig. 4. Perspective view of the IC antenna.

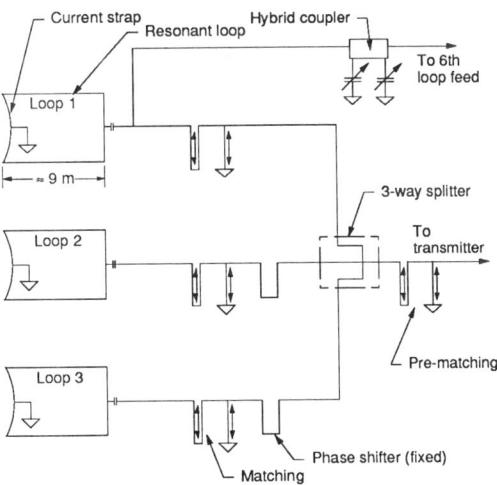

Fig. 5. Tuning and matching circuit driven by one transmitter.

where h is the length of the current strap. F_{avg} is a quantity < 1 that accounts for the fact that the current flowing in the strap decreases from its peak value, I_0, at the strap center (due to the finite wavelength of the current relative to the strap length); it also includes the effect of decreased rf fields near the end of the strap.[4]

The value of I_0 was obtained from the circuit model calculations and was determined by requiring that the peak voltage, V_{max}, anywhere in the rf circuit equal 30 kV.

This value of V_{max} was chosen to ensure reliable operation of the IC system; it has been achieved on many operating tokamak experiments.

The last column of the table shows the total power that can be delivered to the plasma by the two-port, 12-strap array when the straps

Table I. IC system power results

f (MHz)	R' (Ω/m)	F_{avg}	I_0 (kA)	P_{total} (MW)
45	9.1	0.73	0.67	19.7
61	10.7	0.63	0.59	15.7
77	11.5	0.54	0.55	12.4

are phased to drive current (i.e., to launch a spectrum such as that shown in Fig. 2). The results indicate that the system is easily capable of delivering 12 MW to the plasma at the lower two frequencies; at 77 MHz, the system only barely meets the 12-MW requirement.

The current straps in each antenna are coupled by mutual inductance; the value of the coupling coefficients, k_{ij}, is determined by the geometry of the current straps and the septa between straps. For the baseline design, the nearest-neighbor coupling can be as much as 10%. This can cause a significant change in the power required to drive each strap. Table II shows the results of a coupled-circuit calculation at 77 MHz with $R' = 12.4$ Ω/m (the high-frequency case of Table I), a phase angle between straps of 90°, a total power to the plasma of 8 MW, and a requirement that the currents in the six straps be equal (to achieve an optimum FWCD spectrum). For this case, the first strap requires negative power, while the sixth strap needs more than twice the average strap power of 1.33 MW. In order to compensate for the difference in power between straps caused by the interstrap coupling, a hybrid coupler (shown in Fig. 5) will be used to couple power from the first strap to the last strap in the six-element antenna, as is being implemented on DIII-D and JET.[5] Calculations indicate that this should allow all the transmitters to see equal power loads, which will optimize the power-delivery capability of the rf system.

Table II. Coupled circuit results

Strap	Power (MW)
1	−0.04
2	1.32
3	1.33
4	1.34
5	1.32
6	2.73

REFERENCES

1. W Reiersen et al., "Conceptual Design Overview", TPX Doc. 91-930319-PPPL/WReiersen-02, March, 1993 (unpublished).
2. D. Swain et al., "System Design Description: Ion Cyclotron System", TPX Doc. 23-930319-ORNL/DSwain-01, March, 1993 (unpublished).
3. G. Neilson et al., "TPX Physics Design Description", Sec. 7.4, TPX Doc. 93-930319-PPPL/GNeilson-01, March, 1993 (unpublished).
4. P. M. Ryan, private communication.
5. R. H. Goulding et al., "A Power Compensation System for Phased Operation of the JET Antenna Array", in Fusion Technology 1992, Proceedings of the 17th Symposium on Heating and Current Drive, Rome, 1992 (to be published).

DESIGN AND COUPLING CHARACTERISTICS OF LOWER-HYBRID LAUNCHER FOR TPX

A.E. Hubbard and M. Porkolab
MIT Plasma Fusion Center, 175 Albany St., Cambridge, MA 02139

S. Bernabei and N. Greenough
Princeton Plasma Physics Laboratory, Princeton NJ 08543

P. Goranson, D. Swain and J. Yugo
Oak Ridge National Laboratory, Oak Ridge, TN 37831

ABSTRACT

The physics and mechanical design of the LHCD launcher for the proposed TPX experiment is presented. The main role of this system is current drive and current profile control, requiring a flexible and well defined spectrum. The launcher features 32 independently phasable guides in each of 4 rows. Coupling calculations indicate that low reflection coefficients can be achieved over the whole range of phasing by adjusting the launcher position. Good directivity is predicted over a wide range of densities. The mechanical design of the launcher is complicated by the high expected thermal loads and radiation fluxes. A design which incorporates these requirements is outlined.

I. INTRODUCTION

The role of the Lower Hybrid system for TPX is to provide auxiliary current drive, to supplement the anticipated high bootstrap current fraction and enable pulses of up to 1000 seconds, as well as to modify the current profile in such a way as to access attractive advanced tokamak regimes. The design requirements for the launcher have been established after extensive modelling of various scenarios using the ACCOME code[1]. It is required to launch 1.5 MW of power at a frequency of 3.7 GHz, with a central $N_\parallel$ variable over a range of at least 2.0-3.0. In order to achieve reliable operation during the long pulse operation, a relatively conservative limit of 25 MW/m^2 has been set.

In order to meet these requirements, the launcher will consist of 128 guides with a height of 76 mm and 8 mm spacing, arranged in four horizontal arrays of 32 guides each. Each of the 32 guides in an array will be fed independently, allowing continuously variable phasing and simplifying the launcher design. Coupling studies have been carried out to predict the launched spectra and their dependence on edge conditions. A significant factor in the mechanical design is that, due to the long pulses and high duty cycle of TPX, it is expected that integrated neutron fluxes will be much higher than encountered on existing experiments. Shielding and bends in the guides are thus required. The proposed construction is outlined.

II. COUPLING STUDIES

It is well known that the coupling of LH waves into a plasma depends primarily on the electron density profile near the mouth of the launcher, with the optimum density of order $n_c N_\parallel^2$, where n_c is the critical density corresponding to the plasma frequency. At F = 3.7 GHz, and given the relatively high $N_\parallel$ required for TPX, this implies operation with densities of order $10^{18} m^{-3}$ at the grill mouth. The motivation for the coupling studies has been to quantify the effect of the edge density and grill position, as well as to predict the launched spectra and their directivity.

A Brambilla wave coupling code[2] was used which calculates the spectra for a range of phasings between adjacent guides. The $N_\parallel$ of the central peak varies from 1.7 (approximately the lowest accessible value) for $\Phi = 60°$ to 5.0 (a symmetric spectrum) at 180°. Fig 1 displays a typical spectrum for 90° phasing, which shows that a peak in the negative direction as well as small peaks at high $N_\parallel$ also exist.

Figure 2(a) shows the overall reflection coefficient R as a function of n_e (log scale) for each phasing. It is seen that the optimum densities lie in the range 1-5 x 10^{18} m^{-3}, increasing with $N_\parallel$. We find further that R rises rapidly for densities both above and below the optimal value. For example, at 90° phasing $R < 5\%$ for $n_e = (1-4) \times 10^{18} m^{-3}$. By making an empirical fit to such plots and expressing the reflection coefficient as a function of position, it is calculated that reflection will be in this acceptably low range as long as the grill is held within one scrape-off length of the optimum position. The launcher is designed with the capability of moving over a 5 cm range during a shot, on a time scale of order 1 second, and should be able to match expected variations in edge conditions.

The directivity is defined for the purposes of this study as the power in the main positive peak ($1 < N_\parallel < 6.0$) divided by the total launched power. Figure 2(b) shows that, as expected, the directivity is highest for lower values of Φ and decreases to $< 50\%$ for 180 degree phasing. The practical $N_\parallel$ range of the system for reasonable current drive is then approximately 1.7 - 3.4. The density dependence of the directivity turns out to be weak; in particular, it does not deteriorate at densities well below optimum although the reflected power increases. We interpret this as meaning that much of the power reflected under non-optimal conditions is at higher $N_\parallel$.

By comparing the optimum density range predicted by the Brambilla code with edge simulations for TPX, it is predicted that the optimal position for the grill mouth will be typically 2 cm outside the separatrix. A protection limiter is located at this radius, toroidally separated from the launcher. Heat flux calculations show that as long as the grill is kept slightly behind this limiter, convective loading is not a serious problem. Radiative heat fluxes, up to 40 W/cm² for the full power phase of TPX, dominate and are an important factor in the choice of materials for the grill.

III. MECHANICAL DESIGN

A conceptual mechanical design for the launcher is illustrated in Figure 3.

The 'grille' section which faces the plasma is fabricated of dispersion-strengthened copper (GlidCop), coated with B_4C to reduce surface erosion. Each of the four horizontal arrays will be actively cooled on the top and bottom surfaces, while the 76 x 2 mm septa will be cooled by conduction. Thermal calculations show the maximum temperature rise will be 120 C. Other materials considered, such as molybdenum, titanium or berylium, had excessive temperature rises and/or thermal stresses due to the high heat load from the plasma and 1000 second pulse length.

The front array section immediately behind the grille serves to transition the waveguides from 4 arrays of 32 elements to eight rows of 16 input elements. A staggered arrangement allows for cooling of the input vacuum windows on all surfaces. This section will be fabricated of stacked plates of copper or GlidCop, into which the guides are machined before the assembly is furnace brazed. The bends in the guides also serve a critical role in preventing both neutron streaming down the guides and metallizing of the windows. Tanks both in front of and behind the front array, which will be filled with materials such as Dowtherm A, provide additional radiation shielding in order to meet the requirement for hands-on maintenance of components behind the launcher.

The window assembly, based on a successful PBX-M design, uses titanium alloy plates closely matched to the thermal expansion coefficient of the alumina vacuum seal. The components behind this assembly, which include power splitters and transformers to match the eight standard input waveguides to the narrow guides of the launcher, will be pressurized for easier power handling. Drive screws and keyways allow for radial movement of the entire launcher, as discussed above.

IV. CONCLUSIONS

A lower hybrid launcher has been designed for TPX which will be capable of launching a well controlled spectrum over a range of $N_\parallel$ of 1.7 to 3.4, and should enable the current drive and profile modification required for various advanced scenarios. Coupling studies show low reflections as long as the grill is maintained within one scrape-off length of the optimum density, typically 2 cm from the plasma edge. Directivity is good over an even wider density range. The grill consists of 4 arrays of 32 guides each and is capaple of 1000 second operation with thermal loads of 40 W/cm^2. Radiation shielding is an integral part of the design.

REFERENCES

[1] M. Porkolab, P.T. Bonoli, M. Fenstermacher and J. Ramos, Paper C19, this conference.

[2] M. Brambilla, Nucl. Fus. **16**, p. 47 (1976).

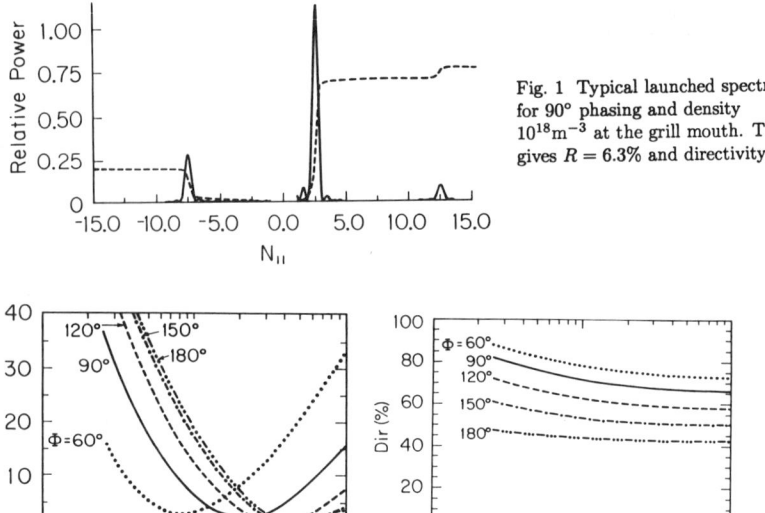

Fig. 1 Typical launched spectrum for 90° phasing and density 10^{18}m^{-3} at the grill mouth. This gives $R = 6.3\%$ and directivity 71%.

Fig. 2 Reflection coefficient (a) and directivity (b) vs edge density (log scale) for various phases.

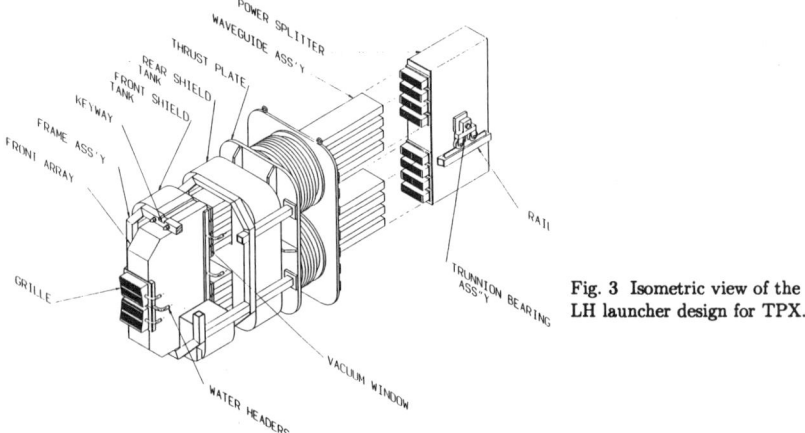

Fig. 3 Isometric view of the LH launcher design for TPX.

ARC DETECTION AND PROTECTION IN HIGH-POWER ANTENNA SYSTEMS[*]

J. B. O. Caughman, D. J. Hoffman
Oak Ridge National Laboratory, Oak Ridge, TN 37831-8071

ABSTRACT

In an effort to establish a criteria for determining the presence of an arc in an antenna system, the change in the input reflection coefficient caused by an arc of arbitrary impedance and location has been calculated for two representative antenna systems. The antenna systems that have been studied are a resonant double loop antenna system (Tore Supra, TFTR) and a coupled four-strap antenna system (DIII-D). The amount of power that is coupled into an arc as a function of the reflection coefficient and arc impedance has been calculated. Results of the calculation have shown that up to fifty percent of the forward power can be coupled to an arc when the measured reflection coefficient is 0.5 and up to ten percent of the forward power can be coupled to an arc when the reflection coefficient is 0.1, depending on the impedance and location of the arc and the particular antenna system. The calculations have shown that changes in the magnitude of the reflection coefficient caused by an arc are often accompanied by changes in voltage ratios on the same current strap or between different straps and by changes in the phase angles of the voltages and/or the reflection coefficient. Recommendations for the use of voltage magnitude and phase detectors as a diagnostic for arc detection and protection are made.

INTRODUCTION

In most ICRF systems, the only determining factor for removing power from the antenna is the magnitude of the reflection coefficient at the input of the antenna system. Power is removed from the antenna if the reflection coefficient exceeds a predetermined value. The value of the coefficient chosen as the "trip" level is somewhat arbitrary and varies from one site to another. Since an undetected arc in the system can cause severe damage, a better method for protecting the antenna system is needed, especially as power levels and system complexity increase. The purpose of this study is to determine how much power can be coupled to an arc as a function of a measurable parameter (reflection coefficient) for an antenna system. The antenna system is electrically modeled both with and without an arc present in the system. Additional tests for an arc condition (voltage ratios, phase angles, etc.) are evaluated to establish additional criteria for setting trip levels to protect rf/antenna systems.

CALCULATION

The antenna systems are simulated using a lossy transmission line model of the feed lines, the matching hardware (stubs and phase shifters), and the antenna structure, including the current strap, the tuning capacitors, etc. Two antenna systems are modeled. One is a resonant double loop antenna structure, such as that used on TFTR and Tore Supra[1] (see Fig. 1a), and the other is a coupled four strap antenna structure with 90° phasing, similar to that used on DIII-D[2] (see Fig. 1b).

The effects of an arc on the system are determined by placing an impedance to ground at various locations in the antenna system, and then calculating the change in the input reflection coefficient of the system, the voltage and current distributions in the system, and the amount of power that is coupled to the arc. The system is in a matched condition before the arc is placed in the system. The arc impedance is modeled as a series resistance and inductance. The magnitude of the impedance is varied, as well as the ratio of the reactive to the resistive component of the impedance. For some of the calculations discussed below, the inductive

---

[*] Research sponsored by the Office of Fusion Energy, U.S. Department of Energy, under contract DE-AC05-84OR21400 with Martin Marietta Energy Systems, Inc.

component of the arc impedance is held constant, and the value of the resistive component is varied. The value of the inductance is determined by using a round wire approximation of the arc, with the radius of the arc being on the same order as the skin depth [3], giving an inductance of ~0.5 nH/cm.

RESULTS

The effects of an arc on several antenna parameters, as a function of the position of an arc on the antenna current strap, are shown in Fig. 2. These results are for a resonant double loop antenna (Tore Supra) with a plasma loading of 6 Ω/m and a driving frequency of 57 MHz. The position of the arc is referenced from the tap point on the current strap. The arc was assumed to have a characteristic length of 10 cm, giving a reactive impedance of ~2 Ω. The real part of the arc impedance was varied from being 10 times the reactive value (solid line) to 0.1 times the reactive value (dashed line).

The reflection coefficient at the input of the antenna can vary from close to zero up to 1.0, depending on the location of the arc, as shown in Fig. 2a. For an arc occurring close to the tuning capacitors, the magnitude of the reflection coefficient is between 0.9 and 1.0, and depends on the ratio of the real to reactive impedance of the arc. The reflection coefficient magnitude is smallest at the voltage minimum on the current strap, which is also the impedance minimum on the strap. At this point the magnitude of the reflection coefficient is rather insensitive to the presence of the arc, unless the arc impedance is less than the strap impedance. However, the change in the value of the phase angle of the reflection coefficient is sensitive to the arc and experiences a phase shift near this point, as shown in Fig. 2b.

The ratio of power coupled to the arc over the total forward power on the antenna as a function of arc position on the current strap is shown in Fig. 2c. For all cases shown, the amount of power that is coupled directly to the arc has a peak on either side of the voltage minimum on the current strap. The peak in the coupled power to the arc occurs for the high resistance arc that is located near the tap point. For this case, almost 50% of the forward power is coupled to the arc. The magnitude of the reflection coefficient at this point is ~0.5. For the lower resistance arc, a peak of roughly 5% of the forward power is coupled to the arc for a reflection coefficient of ~0.5.

The ratio of the voltage at the capacitors as function of the arc position on the current strap is shown in Fig. 2d. Without an arc on the antenna, the ratio of the voltages on the two tuning capacitors is near unity. With an arc, the ratio of the voltages can change from close to 1 to close to 0. The voltage ratio change is most dramatic in the lower arc resistance (and reactance) cases and is most noticeable if the arc occurs away from the tap point. For an arc occurring at the tap point, the capacitor voltage ratio does not change compared to a no arc situation, but the ratio of the capacitor voltage over the system input voltage does change, as shown in Fig. 2e. With no arc on the antenna, the ratio of the capacitor voltage to the input voltage is ~4.6. This ratio changes, however, when an arc occurs on the antenna, especially if the arc is not at the voltage minimum on the current strap.

The effects of an arc on antenna parameters in an multi-strap system are shown in Fig. 3 for the case of an arc occurring at a voltage minimum in the feed line for strap 2 (refer to Fig. 1b). The results are for a system operating at a frequency of 60 MHz and matched to a plasma load of 2 Ω/strap, with 90° phasing between the straps, before the arc was added. The magnitude of the reflection coefficient at the system input is shown as a function of the magnitude of the impedance of the arc. As with the resonant loop antenna case, the reflection coefficient increases as the impedance of the arc decreases, reaching a maximum value of ~0.15 for the parameters shown. The amount of power coupled to the arc as a function of the input reflection coefficient is shown in Fig. 3b. More than 7% of the forward power can be coupled to an arc, even when the reflection coefficient is 0.1.

DISCUSSION AND CONCLUSIONS

As shown in Figs. 2 and 3, a significant amount of power (up to 50% of the forward power) can be coupled to an arc, even when the reflection coefficient is below 0.5. If an arc occurs at a point in the antenna system that causes little change in the reflection coefficient measured at the input to the system, such as at a voltage minimum (current maximum), a great deal of power can be coupled into the arc without the system operator knowing it. Even if only 1% of a forward power of 2 MW is coupled to an arc, the results from 20kW over a surface area of an arc (a few mm^2) can be very damaging. Therefore, using only the level of the reflection coefficient is not enough to protect the antenna. These calculations show that the presence of an arc can cause a change in the ratio of voltages on the strap (or between straps for multi-strap antennas) or a change in the phase angle of the reflection coefficient. This additional information can be measured with passive probes and detectors and can be used in conjunction with the reflection coefficient data to better protect the antenna system by determining when to remove power from the antenna. For future systems, much better protection is needed as remote maintenance requirements for confinement devices force the the antenna systems to be highly reliable.

REFERENCES

1. D. J. Hoffman, et al., Proceedings of the Seventh Topical Conference on Applications of RF Power to Plasmas, Kissimmee, FL, 1987, p.302.

2. R. H. Goulding, et al., Proceedings of the Ninth Topical Conference on Applications of RF Power to Plasmas, Charleston, SC, 1991, p.287.

3. S. Ramo, J. R. Whinnery, and T. van Duzer, Fields and Waves in Communication Electronics (Wiley, New York, 1965).

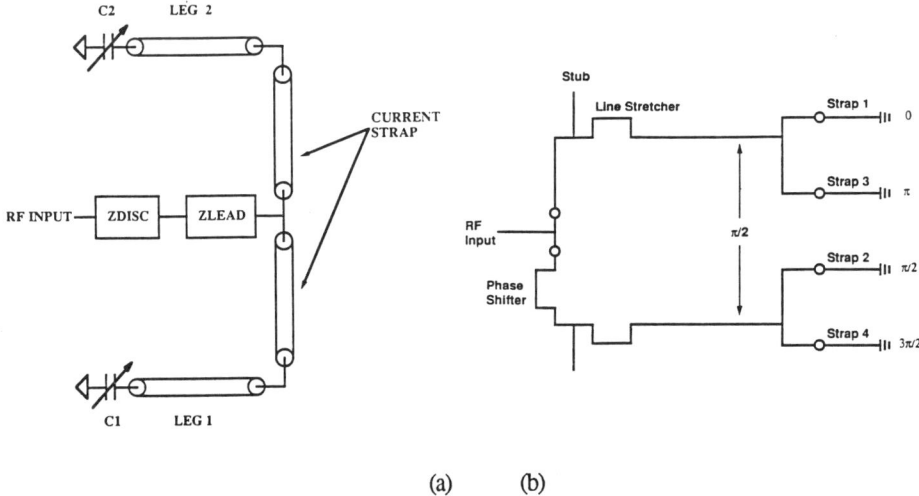

Fig. 1 Antenna systems: a) resonant double loop antenna, b) coupled four strap antenna.

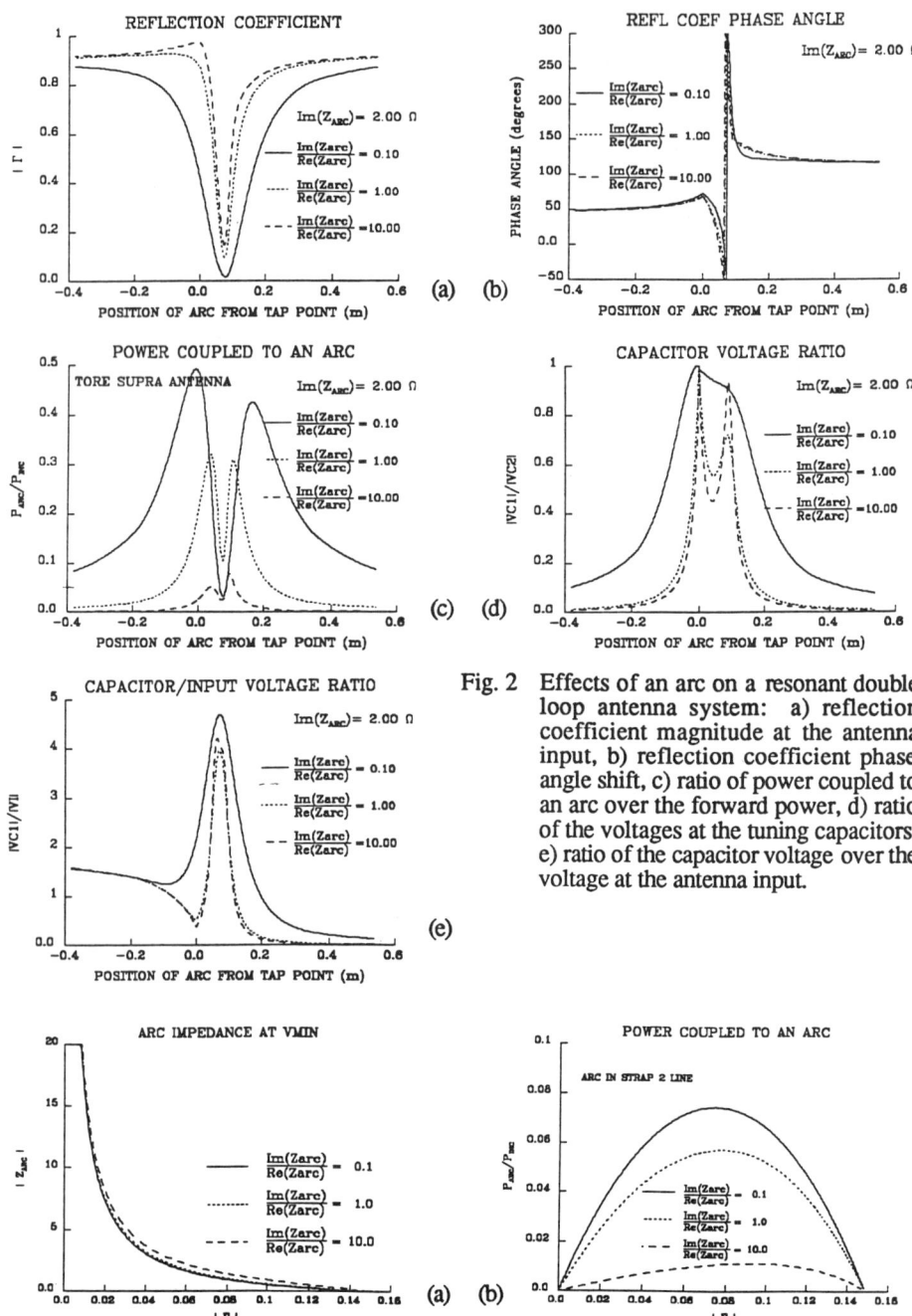

Fig. 2 Effects of an arc on a resonant double loop antenna system: a) reflection coefficient magnitude at the antenna input, b) reflection coefficient phase angle shift, c) ratio of power coupled to an arc over the forward power, d) ratio of the voltages at the tuning capacitors, e) ratio of the capacitor voltage over the voltage at the antenna input.

Fig. 3 Effects of an arc at a voltage minimum on strap 2 of a multi-strap system: a) magnitude of the arc impedance as a function of the input reflection coefficient, b) ratio of the power coupled to an arc over the forward power as a function of the input reflection coefficient.

ICRF Coil for the IDEAL Plasma

R.W. Motley, R. Majeski, S.A. Cohen, M. Diesso, and J.R. Wilson
Princeton University, Princeton, NJ 08544

Abstract

We describe an ICRH coil to drive the plasma in the proposed IDEAL device,[1] a linear plasma machine designed to study the physics and engineering problems of the ITER divertor. In initial operation, 2 MW of CW power at ~ 40 MHz will be applied to a hydrogen plasma via four 0.75-m long multiple saddle coils that excite ICRF slow waves. The waves propagate to a 30 % magnetic beach, where they undergo cyclotron absorption. At full heating power the power flow out the ends of IDEAL is designed to equal that in the ITER divertor. Coil loading and the radial distribution of the E^+ and E^- RF fields have been calculated with the ANTENA Code.[2]

Introduction

The divertor plates of the ITER Tokamak will be subject to intense bombardment by plasma, hot neutrals, and radiation. The power flow along B will be in the range from 200 to 1000 MW/m². There is an acute need for a facility to check materials for the ITER divertor and to test innovative concepts like the gas divertor. IDEAL is designed to deliver 200 MW/m² initially and up to 5 times this level in later operation.

The basic design of IDEAL[1] springs from work on ion-cyclotron-heated mirror machines[3-6], in which cyclotron waves excited in a high field region ($\Omega = \omega/\omega_{ci} < 1$) propagated along the field lines to a magnetic beach, where the wave power was absorbed by cyclotron damping near $\Omega \approx 1$. Typical plasma conditions were P_{rf} = 100-500 kW, $n_e \leq 6 \times 10^{13}$ cm^{-3}, T_e = 10-70 eV, and T_i = 50-400 eV, although lower densities were more common because of the desire for higher ion temperatures.

The Proposed IDEAL Plasma

IDEAL employs the same elements as the ICRF-driven mirror devices, except for the absence of mirror coils. In the central cell, the plasma will be 8 cm in diam. and 25 m long, driven initially by 2 MW of RF power at 40 MHz in the central, high-field (3.5 T) cell, and stabilized by quadrupole fields at each end. The two-dimensional B2 Code[8] applied to the IDEAL geometry predicts $n_e \approx 10^{14}$ cm^{-3}, $T_e \approx 50$ eV, $T_i \approx 170$ eV in the central cell, and near the end sheath 10X the density and 1/10 the temperature, if the input power flows directly to the ions.

The ICRF Antenna

The criteria for the slow-wave coil operating into high density are that the individual out-of-phase coils be spaced about one diameter apart, that the E^+ component of the wave (rotating in the ion direction) be dominant and non zero at r=0, and that the coil loading peak in the density range above 10^{13} cm^{-3}. The saddle coils developed by Shvets[9] to heat plasma on the URAGON torsatron appear to satisfy these requirements. The saddle coils are basically a pair of Stix coils[10] rotated by 90 degrees and placed on opposite sides of the plasma column, exciting the m=+1 mode rather than m=0.

The current paths and the field directions in one period of a saddle coil encircling a plasma are shown in fig. 1. Current loops on opposite side of the column are in phase, giving rise to an alternating field at right angles to the column. Adjacent loops are out of phase. Current is fed to the coil at the midpoint. On each side of the feed are 5 loops grounded at the ends, where the water cooling is provided. The loops consist of four 1/4" water-cooled Cu tubes wound on a radius of 6 cm and are separated axially by 7.5 cm. Such a configuration produces $k_z \approx 0.4$ cm^{-1} or $\lambda_z \approx 15$ cm, appropriate for efficient coupling at high density. Four antennas are planned, two on each side of the gas box.

The antenna will not use Faraday shields. Instead, we will use antenna limiters, possibly insulated to reduced RF-generated electric fields. There will be a ground return for the RF current near the wall.

The ANTENA Code

The McVey ANTENA Code[1] analyzes the coupling of magnetosonic waves to a cylindrical plasma column. The code neglects waves with radial wavenumbers $k \rho_i \sim 1$, so that Alfven resonance effects, important for electron heating, are not properly treated. We neglect reflection of waves at the beach.

ANTENA modeling predicts that the saddle coil will launch a propagating slow wave with fields predominantly in the E^+ (left hand) mode, as shown in fig. 2, with a broad peak near r = 0. The right-handed mode (E^-) and E_z both peak near the plasma edge and are much smaller than E^+. Both E^+ and E^- field profiles are ponderomotively stabilizing.[4,11] The spectrum (fig. 3) shows that $k \sim 0.4$ cm^{-1}, appropriate for a slow wave.

Coupling curves for fixed excitation frequency and plasma profile (parabolic, with width 6 cm, cut off at 4 cm) are shown in fig. 4. These curves show that the operating density in the IDEAL central cell should be variable in the range 10^{13}-10^{14} cm^{-3} by changing the operating frequency between (0.6-0.9) Ω. We would require 80 A rms for 2.5 MW at the peak of the resonance and an antenna voltage to ground of 11.4 kV (1" by 1/4" conductor) at the feed point. The power density at the plasma surface, using 4 antennas, would be ~ 200 W/cm^2.

Electron Heating

Direct electron heating is desirable to more properly simulate ITER's SOL. This may be achieved by operating one of the antennas at a somewhat higher frequency (Ω = 0.9-0.95, rather than 0.7) in order to heat electrons directly via the localized shear Alfven wave.[12] This and other electron heating scenarios, e.g., Ohmic heating, are currently under consideration.

Startup

It would be advantageous to initiate the discharge with the same slow wave antenna. However, it may be difficult to meet the wavenumber matching conditions over a wide range of density. There are a variety of strategies for handling this problem.

One solution involves using a short-pulse, auxiliary plasma source for startup. Such plasmas can be formed by electron-cyclotron heating, lower-hybrid-wave excitation, or arc plasma formed between two end electrodes. Another possibility is to use the ICRH antenna to form a hollow plasma using the m=1 surface wave[13], which has no density cutoff.

References

1. S.A. Cohen, Journal of Fusion Energy 10, 327 (1991).
2. B. McVey, MIT Report PFC/RR-84-12, (1984).
3. W. M. Hooke et al., Phys. Fluids 4, 1131 (1961).
4. J.R. Ferron et al., Phys. Rev. Lett. 51, 1955 (1983).
5. S.N. Golovato et al., Applications of RF Power to Plasmas, 7th Topical Proceedings (1987) p. 254.
6. Y. Yasaka et al., op. cit., p. 246.
7. W.M. Hooke et al., Phys. Fluids 5, 864 (1962).
8. B.J. Braams, Computational Studies in Tokamak Equilibrium, Thesis, University of Utrecht (1986).
9. O.M. Shvets et al., Heating in Toroidal Plasmas, Proc. 4th International Symposium, Rome (1984), Vol. 1, p. 513.
10. T. H. Stix, Phys. Rev. 106, 1146 (1957).
11. S.N. Golovato et al., Phys. Fluids B1, 851 (1989).
12. K. Appert et al., Phys. Fluids 27, 432 (1984).
13. F.J. Paoloni, Phys. Fluids 18, 640 (1975).

386 ICRF Coil for the IDEAL Plasma

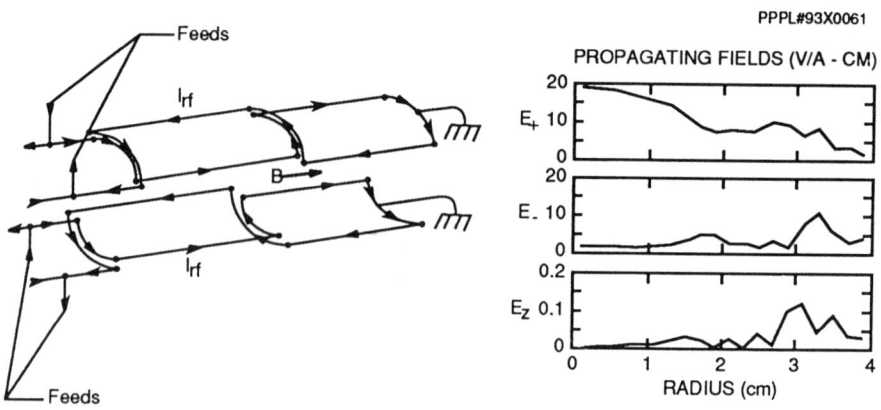

Fig. 1. Winding patterns for one cycle of the saddle coil.

Fig. 2. Propagating fields for the 6-cm coil @ $n_e = 5 \times 10^{13}$, $\Omega = 0.75$.

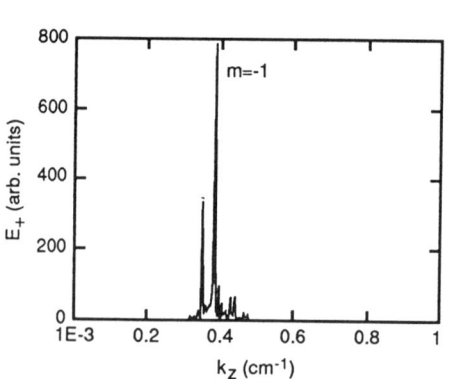

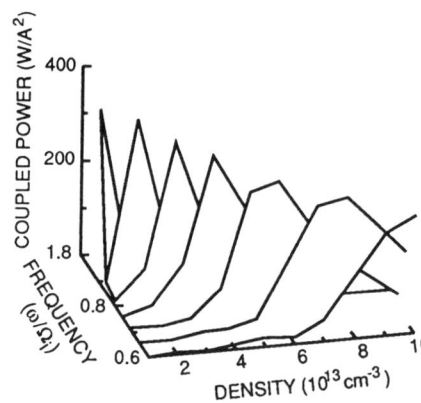

Fig. 3. Spectrum of E^+ field, $m = -1$.

Fig. 4. Slow wave loading curves.

Antenna Conditioning with Insulating Antenna Tiles in Phaedrus-T [†]

T. Intrator, P. Probert, M. Doczy, D. Diebold, and D. Brouchous

University of Wisconsin, Dep't. NEEP, Madison, WI

Abstract

In the course of our Alfven wave heating and current drive experiments several different two and four strap antennas have been installed in Phaedrus-T. The motivation focussing the redesign of the antenna into a four stap design was to enable travelling wave phasing, and to reduce the $k_\parallel \approx 0$ component of the wavenumber spectrum, and consequent edge power deposition. The latest modifications to the 4 strap antenna have dramatically improved its behavior, and enabled us to suppress its RF power induced impurity generation. The remaining gas reflux fueling is significant and is not local to the antenna.

I. Introduction

In our present experiments we launch evanescent wave fields below the ion cyclotron frequency to Landau damp wave power to the electrons at an Alfven resonance inside the plasma. An RF induced density rise[1] has hindered our Alfven wave experiments[2]. Impurity generation can alter the Z_{eff} of the plasma, as well as any Alfven resonance surface by creating spurious ion-ion hybrid and ion cyclotron resonances in the edge. We have devoted some effort to identify and reduce sources of impurities and gas fueling (H, D) reflux during RF operation.

After revising a four strap antenna configuration and our Boronization procedures, we have succeeded in suppressing our impurity generation problem. We still observe a remaining density rise due to gas reflux fueling, not impurity generation.

White light imaging and visible and VUV spectroscopy was used to track the impurity generation and gas reflux. Imaging identifies the source location while spectroscopy can be used to identify the impurity species.

II. Antenna Designs - 2 Strap

The Phaedrus-T fast wave antenna arrays have been revised from two strap to four strap designs in the last two years. The final version of the two strap antenna had Boron Nitride (BN) guard limiters on the outer toroidal edges of the antenna, that interrupted the field lines between antenna and conducting tokamak poloidal limiters. The antenna straps were covered with BN tiles, and there was no Faraday Shield. The Scrape Off Layer (SOL) sheath effects were greatly reduced, compared with the previous Faraday shielded antenna with no insulating limiters.[2]

III. Antenna Designs - 4 Strap I

To facilitate Alfven wave Current Drive experiments, a four strap antenna (4 strap I) was installed in July 1992. Compared to a 2 strap antenna, a 4 strap array is better capable of travelling wave phasing, and will reduce the $k_\parallel \approx 0$ component of the wavenumber spectrum that damps strongly near the SOL and edge region. A BN and AlN60 (BN + Mullite composite) enclosure surrounded the 4 antenna straps, and was intended to reduce the plasma density and parasitic coupling between the antenna straps. The impurity generation, gas reflux and parasitic loading turned out to be substantially worse than the two strap design.

The tiles forming the insulating enclosure were radially recessed approximately 0.1-0.3 cm from the antenna BN guard limiters, and lit up during tokamak discharges. Figure 1 shows the effects of plasma scrubbed Boronized coating on the plasma facing tiles. The elliptic lighter regions correspond to the toroidal projection of the poloidally curved circular section guard limiters onto the flat enclosure tiles. The plasma flows toroidally along a flux surface, scrapes off on the guard limiter edge, and also diffuses radially. The radial scrape off diffusion per unit of toroidal distance coresponds to a 4° angle of attack.

IV. Antenna Designs - 4 Strap II

An antenna redesign (4 strap II) was installed at the end of 1992. We deleted the BN/AlN60 enclosure to reduce total tile area and the plasma facing area of tiling, substituted 6 smaller cover BN tiles per antenna strap, and increased the setback of the antenna strap tile radius from the BN septa.

The visible emission is shown as a surface plot in figure 2, and was observed primarily on the cracks between between the tiles. On the other hand, the antenna loading resistance increased, rather than decreased, leading us to conclude that the antenna coupling to the plasma was not dependent on the impurity generation at the antenna. In spite of this, the overall machine conditioning, including impurity and gas generation was greatly improved.

V. RF Induced Impurity Generation - SPRED data

SPRED data in figure 3 show the evolution of the impurity signatures as the antenna designs have changed. The 2 Strap antenna SPRED data is dominated by He, with some C, O lines, while the 4 Strap I antenna generates a 'dirtier' plasma. The improved 4 Strap II shows reduced continuum (from 200-400 Å from many unresolved Iron lines), with O^{VI}, C^{IV}, O^V as the remaining dominating lines.

VI. Wall and Antenna Conditioning

After experimenting with Boronization techniques, we settled on a 3 point gas feed for the Trimethylboron (TMB) gas feedstock, with two locations near the antenna, one near the main carbon limiter. A schedule of several 250 +° C bakeouts of antenna and limiters after every Up To Air and of 2 Boronizations per week allowed us to reduce the impurity generation to a record low level. Since gas puffing needed to be reduced by 50% during the RF pulse, gas reflux fueling still remained a problem. Discharges in Helium showed no RF induced density rise, evidence that the reflux is due to the chemistry of H,D and limiters.

Figure 4 compares the time history of the line density where the gas puffing was zeroed out both for an ohmic and ohmic + RF shot. Without gas puffing, the particle pump out rate ν_p^{ohm} depends on the balance between fueling, pumping, and confinement. Neglecting confinement changes, the RF fueling rate is approximately $\nu_p^{RF} = \nu_p^{ohm+RF} - \nu_p^{ohm}$. Table I shows a comparison of the ν_p^{RF} for various antenna designs and Boronization procedures.

Gas reflux and associated fueling are not local to the antenna region. Reticon camera images like figure 3 were filmed both with and without an H_α filter to show respectively the hydrogen and impurity visible emissions. The H_α signal at the antenna (and thus antenna localized gas reflux) has neither the time dependence of the RF pulse, nor the spatial shape of the antenna. Fueling has been reduced substantially from uncontrollable to a controllable level, but still remains a technical obstacle to our experimental program.

VII. Conclusions

We have succeeded in substantially reducing the measurable impurity signals, deduced from data both at the antenna location, and from global diagnostics.

The antenna loading does not seem to correlate with the variations in antenna generated impurities, but it has been reduced to the point where parasitic loading is not dominating our experiment.

For our next antenna revision 4 Strap III, we have removed the antenna tiles to ascertain whether insulating guard limiters are enough to suppress SOL and impurity effects, or whether full strap tiling is necessary. For comparison, the TEXTOR antenna has graphite (conducting) guard limiters, no Faraday Shield (FS), and bare antenna straps recessed 3 cm from the limiter radius. From an impurity and SOL modification standpoint the TEXTOR antenna with no FS was shown to be no worse than their antenna with FS.[3] On the other hand, our antenna designs with insulating limiters show small SOL effects[2].

† Supported by DOE-APP Grant DE-FG02-88ER53264

References

[1] R. Behn et al, Plasma Phys, **26**, 173 (1984); GA Collins, F. Hoffmann, B.Joye, R. Keller, A. Lietti, JB Lister, A Pochelon, Phys Fluid, **29** 2260 (1986).

[2] R Majeski et al, *Alfven Wave Experiments in the Phaedrus-T Tokamak*, to be published, Phys Fluids (1993); R. Majeski, et al, *The Phaedrus-T Antenna System*, to be published, Fusion Engin and Design (1993).

[3] R. Van Nieuwenhove et al, Nucl Fusion, **32**, 1913 (1992)

$\nu_p^{ohm}(sec^{-1})$	$\nu_p^{RF+ohm}(sec^{-1})$	$\nu_p^{RF}(sec^{-1})$	Comments
-50	+8	+58	2 Strap, 1 spot TMB
-4.8	+29	+34	4 Strap I, before TMB
-29	+8.7	+38	4 Strap I, 3× pressure
-20	+20	+40	4 Strap I, 3× time
-100	-22	+78	4 Strap II, 1 spot TMB
-43	-19	+24	4 Strap II, 3 spot TMB
-55	-15	+40	4 Strap III, 2 spot TMB

Table I. Particle pump out rates ($\nu_p < 0$), and reflux rates ($\nu_p > 0$), for one e-folding of line density from data like figure 4.

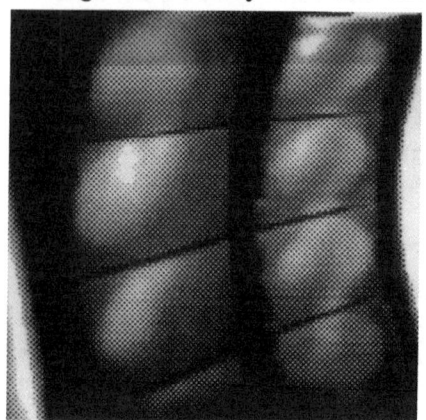

Fig 1. Reticon image of plasma scrubbed regions on the 4 Strap I BN enclosure tiles, showing only center 2 columns, and 3 curved poloidal guard septa.

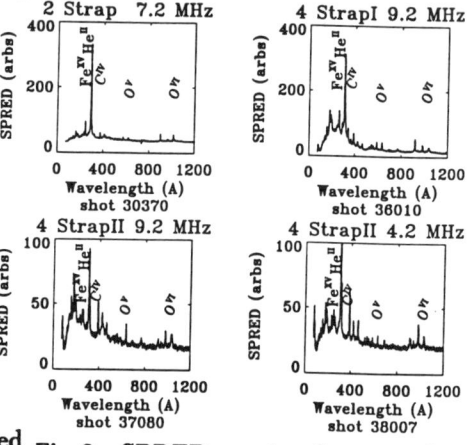

Fig 2. SPRED spectra for several antenna configurations, at ≈ 100 kW of RF power. For 4 Strap II, the Fe continuum is smaller than the 2 Strap antenna, dwarfed by 4 Strap I, and improves for $f_{RF} = 4.2$ MHz.

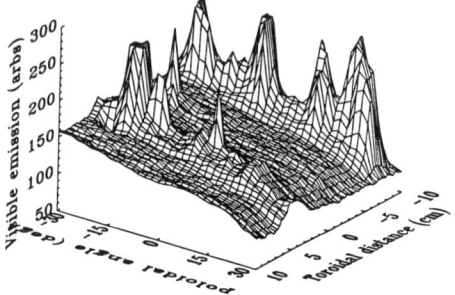

Fig 3. Surface contours of visible emission from the plasma face 4 Strap II, around the center two of four straps. Emission peaks at the cracks between the antenna strap tiles, and increases near the high voltage feeds.

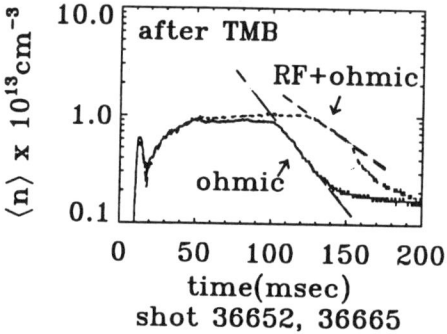

shot 36652, 36665

Fig 4. Averaged density inferred from interferometer line density vs time, showing the particle pumpout.

A PRELIMINARY ENGINEERING ASSESSMENT OF THE ITER CDA ECH LAUNCHER*

T. S. Bigelow and D. W. Swain
Oak Ridge National Laboratory, Oak Ridge, TN 37922

M. Sawan
University of Wisconsin, Madison, WI 53706

ABSTRACT

A preliminary engineering study of the ITER electron cyclotron heating (ECH) launcher configuration proposed by the ITER Conceptual Design Activity (CDA) team[1] has been performed to assess its survivability in the ITER nuclear environment. Potential problem areas are with the vacuum windows, the plasma-facing mirrors, and some of the other high-power waveguide components that are untested in a reactor environment. The study indicates that the CDA design is quite robust, since the mirror power density is relatively low and the windows are well shielded. Although the CDA ECH system is unlikely to be built as proposed, most analysis techniques developed to study this system will apply to future ITER ECH system configurations. The vacuum window is likely to be the most difficult launcher component to develop. Design for a proposed resonant ring for high-power testing of windows using existing lower-power gyrotrons is presented.

INTRODUCTION

ECH was proposed for use on ITER primarily as a tool for plasma disruption suppression by locally controlling the plasma current profile near the $q = 2$ surface. A power level of 20 MW at a frequency of 120 GHz is required for this task. Power is generated by 28 gyrotrons with 1-MW cw capability and is transmitted to the tokamak through corrugated waveguide quasi-optical transmission lines. Local current drive at the moving $q = 2$ surface is possible using a fixed-frequency power source by varying the toroidal launch angle with a rotating mirror. To avoid locating a moving mirror in a plasma-facing nuclear environment, the CDA launcher design uses additional oversized mirrors between the moving mirror and the plasma. There has been some criticism of the CDA launcher design due to the large mirror cross-section and the relatively low power density, which violate one of the traditional claims for ECH system capabilities.

The inherent separation between ECH launcher components and the plasma is one of the principal advantages of ECH in a reactor environment. The potential problems that do exist for an ECH launcher on ITER are primarily with the mirrors and windows, as summarized in Table 1.

LAUNCHER MIRRORS

The proposed CDA launcher combines 28 Gaussian beams launched from a 4 × 7 array of corrugated waveguides that are located >14 m from the plasma edge and around several bends, as shown in Fig. 1. Power is beamed toward the plasma by a series of four mirrors. Mirrors M1 and M4 are ellipsoidal to provide focusing. Mirror M2 is flat and has a ±5° rotation capability that is motorized for high-speed remote operation. The 28 beams have arbitrary phasing so that they propagate incoherently through the launcher. Because of adjacent launch locations and beam diffraction, the beams partially overlap at the plasma edge. Gaussian beam analysis through the launcher indicates that an individual beam will expand by a factor of 2.64 from the input waveguide to the plasma edge, where the beam waist is ~75 mm. The last mirror, M4, is a large (1860 × 2660 mm) ellipsoidal-shaped surface. All mirrors require a surface accuracy of <0.1 mm to provide maximum launched beam quality. A potentially significant problem in the ITER launcher is distortion of mirror surfaces caused by non-uniform thermal expansion from non-uniform power deposition and by surface erosion from neutral particle flux. As indicated in Table 1, there are several sources of power deposition on the mirrors, especially in the plasma-facing mirror M4. Calculations show that microwave loss is the largest power flux into the mirror

---

* Research managed by the Office of Fusion Energy, U.S. Department of Energy, under contract DE-AC05-84OR21400

© 1994 American Institute of Physics

Table 1 ITER environmental effects on the ECH launcher components

ITER environmental factor	Launcher component affected	Potential complications
High frequency, high microwave power density, and long pulse operation	Mirrors	High and uneven power deposition; surface distortion
	Windows	Thermal stress from power absorption; complicated window design, possibly cryogenic or distributed
Fast and thermal neutrons	Mirrors	Power deposition; activation; material property degradation
	Windows	Material properties degradation; higher loss tangent; etc.
Plasma radiation	Mirrors	Power deposition
α, β, γ flux	Mirrors	Power deposition
	Windows	Material properties degradation; higher loss tangent
Neutral particle flux	Mirrors	Surface erosion
Radioactive dust	Windows and mirrors	Contamination

surface. A reduction in particle and plasma radiation flux by a factor of ~200 is provided by the large distance to the plasma and beam duct shielding. Microwave power deposition ranges from 3 W/cm^2 on M4 to 115 W/cm^2 on the rotating mirror M2. Power non-uniformity occurs as a result of the multiple-beam configuration and beam motion as the launch angle is varied. The non-uniformity problem is solved by providing sufficient coolant flow to maintain

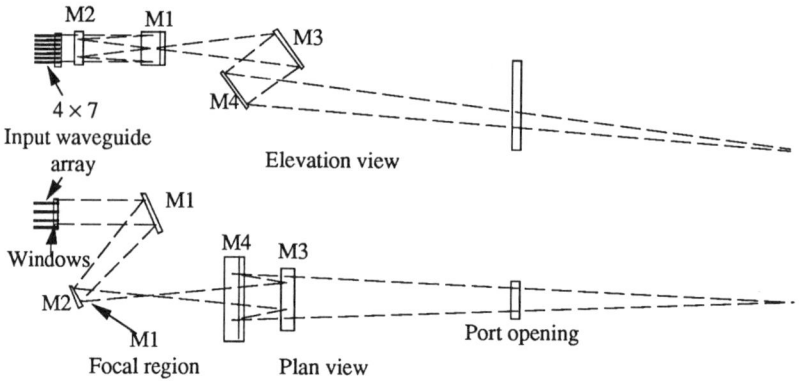

Fig. 1. ITER ECH launcher mirror optics.

uniform temperature across the mirror. A rather large thermal gradient is required to produce a significant change in the mirror focal length in the region of a single beam. A complicated variable temperature cooling system is proposed to maintain the surface temperature above 200°C for optimum vacuum conditions.[1] More detailed thermal stress and beam analyses are desired for the final mirror configuration.

Surface erosion due to sputtering of the last mirror will be caused by energetic neutral particles emitted from the plasma edge. The erosion rate was previously estimated[1] to be relatively small and is likely to be uniform, since the direct particle flux is uniform. Reflector distortion due to neutron damage is also very small due to the large separation from the plasma.

It appears that the reflectors are not likely to be a significant problem in the CDA launcher design, partly because of the large plasma-to-mirror separation used in the design and partly because of the large mirrors with low power density. The CDA team considered the launcher system (except for the windows) to be a very low maintenance item. Newer launcher configurations with reflecting mirrors located much closer to the plasma will require more work and careful design to avoid difficulty in these areas.

VACUUM WINDOWS

A vacuum barrier window is required to isolate the ITER vacuum chamber from the waveguide transmission and power generation portions of the ECH system. Since the window also serves as a tritium and radioactive particle containment barrier, a reliable, highly robust design is required. In the CDA system, a window and a second back-up window are included in each corrugated waveguide. The design of a reliable window that functions at 120 GHz, 1 MW cw for a HE_{11} Gaussian-like mode is a very difficult problem due to the high dielectric loss and centrally peaked power density. Sapphire, the window material of choice, has a loss tangent of ~0.0002 at 120 GHz and 27°C; a single 3.2-mm-thick disk of sapphire will absorb <0.1% of the power at 120 GHz in a large waveguide size. Three categories of window design options are available: (1) face-cooled double disk, (2) edge-cooled cryogenic, and (3) distributed. The additional ITER environmental factors described in Table 1 (such as neutron flux) complicate the design by degrading the window material properties. Although the window design for ITER is not yet finalized, it is possible to analyze the effect of the ITER environmental factors on sapphire material to determine the safety factor required in the design.

Neutron flux onto sapphire will cause atomic displacement and ionization of the crystal structure and lead to increased dielectric loss. Gamma flux causes only ionization. Measurements of the alumina loss tangent after irradiation to 10^{20} n/cm^2 have shown a factor of 2–4 increase after irradiation.[1] Measurements during radiation have not been widely reported at 100 GHz, although recent tests have been performed at 100 MHz on several materials with both neutron and gamma irradiation.[2] Displacive and ionizing radiation were both found to effect on loss tangent at 100 MHz, and sapphire (which had the lowest loss tangent) actually showed the greatest increase of the materials tested. These tests clearly indicate the need for loss measurements at >100 GHz under both types of irradiation.

Exact analysis of the neutron flux level that will exist at the window location is extremely difficult; however, an approximate numerical calculation of the flux and energy spectrum at the waveguide apertures was performed.[3] The model indicates a reduction in flux from 10^{14} to 10^{10} n/cm^2 s between the ITER first wall to the window location. The neutron energy spectrum becomes thermalized due to the multiple reflections between the plasma and mirror M1. Additional neutron attenuation is provided by the corrugated waveguide length and bends. Although the waveguide could provide some ducting of lower-energy neutrons, there is attenuation at each reflection. Therefore, there is a flux reduction factor of approximately 10 due to the small window solid angle inside the waveguide and an additional factor of 10 for each miter bend (there are at least two bends per waveguide before the window). These conservative estimates indicate that neutron flux is reduced to ~10^7 n/cm^2 s at the window location, which gives a total fluence of ~5×10^{14} n/cm^2 after a full power year of ITER operation. Neutron irradiation does not appear to be a particularly bad problem for windows in the ITER launcher design. Other material properties such as thermal conductivity are also only slightly degraded by the predicted neutron flux levels. Additional measurements of dielectric loss data under irradiation at realistic levels are still necessary. There does appear to be a need to produce a window design with a conservative power safety margin, which implies the need to test to failure. A technique has been proposed that will allow window testing at power levels up to 10 times higher than available power sources. This technique uses a resonant waveguide ring with a quasi-optical directional coupler, as shown in Fig. 2. This arrangement would allow window designs under development to be tested to

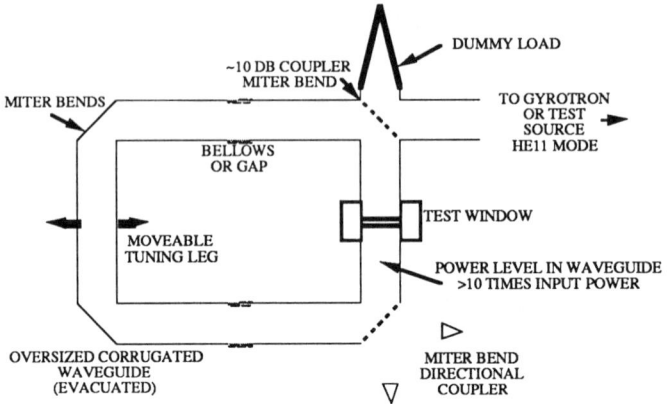

Fig. 2 HIGH POWER RESONANT RING TEST DEVICE

levels of nearly 2 MW with existing 200-kW cw gyrotrons.

ALTERNATE ECH APPLICATIONS ON ITER

There are several uses for ECH on ITER, shown in Table 2. For most applications, the launcher requirements are simpler than in the CDA design. A single system could be designed to provide many of these functions. The use of multiple launchers which are selectable with waveguide switches will provide optimum launch conditions for different applications.

Table 2. Possible applications of ECH on ITER

ECH application	Power level	Launcher requirements
Bulk plasma heating	>20 MW	Perpendicular low field launch
Plasma startup	2–4 MW	Perpendicular low field launch
Transport diagnostic	1 MW	Steerable low field launch
Alpha particle diagnostic	1 MW	Steerable top launch
Vacuum vessel conditioning	5 MW	Arbitrary launch
Edge heating	5 MW	Perpendicular low field launch
Disruption suppression with fixed launch angle	20 MW?	Fixed oblique launch

CONCLUSION

Several potential problems with the ITER CDA ECH launcher caused by the reactor environment have been investigated and found not to be severe because of the large separation between components and plasma. The window design appears to be the most significant problem at this time, and a concept for very high power testing was presented. This study has been useful in pointing out beneficial design features that could be incorporated into new designs.

[1] L. Rebuffi et al., "ITER Electron Cyclotron Wave System," ITER-IL-HD-6-0-31.
[2] R. E. Stoller, R. H. Goulding, and S. J. Zinkle, J. Nuc. Matl. **191–94**, 602–6,(1992).
[3] M. Gervaise, ITER-IL-HD-6-0-11.

MEGAWATT GYROTRONS FOR ECRH

M. Blank, T.L. Grimm, W.C. Guss, K.E. Kreischer, R.J. Temkin
MIT Plasma Fusion Center, Cambridge, MA 02139

ABSTRACT

Results from high frequency, megawatt gyrotron experiments at MIT will be presented. A 140 GHz gyrotron has produced output powers of 1.3 MW, corresponding to efficiencies of 38%, in the $TE_{15,2,1}$ mode. There is good agreement between these measurements and nonlinear simulations. In another set of experiments, a gyrotron was step tuned from 200 - 300 GHz, and produced output powers from 0.70- 0.97 MW with typical efficiencies of 17 - 20%.

A mode converter for the $TE_{16,2}$ mode was designed and tested. The converter consists of a helically cut waveguide launcher and a doubly curved reflector. The reflector focuses the launched radiation into a linearly polarized, Gaussian-like beam suitable for long path transmission in a corrugated waveguide or in an open mirror relay line. Experiments show that the converter focuses at least 95% of the incident radiation into a Gaussian-like focal spot in the far field.

140 GHz GYROTRON

Experiments have been performed on a 1 MW, 3μsec pulsed gyrotron operating near 140 GHz. This gyrotron is designed to model cw operation. Our recent goal has been to increase the efficiency to levels predicted by theory (40-50%). The best results have been achieved with a two-section cavity, which consists of two cylindrical sections of slightly different radii. This cavity provides a long rf tail in the first section, which pre-bunches the electron beam and increases the efficiency. The highest power emission was achieved in the $TE_{15,2,1}$ mode. Efficiencies reached 42% near 30 A, with power levels above 1 MW. Figure 1 shows power and efficiency with varying beam current. Output power was measured with a laser calorimeter modified to absorb millimeter wave radiation. The threshold for oscillation is 4 A, and efficiencies in excess of 30% can be achieved over a wide range of currents. Frequency measurements indicate that single mode emission is obtained. Power and efficiency agree well with the self-consistent non-linear theory predictions using the beam velocity ratios as measured by a capacitive probe. A mapping of the emission as a function of cathode and cavity magnetic fields is shown in Fig. 2. Single mode emission in the $TE_{15,2,1}$ mode is observed over a wide range of parameter space. In general, the highest powers were obtained at the highest velocity ratios possible before arcing occurs in the tube.

200-300 GHz GYROTRON

A 200-300 GHz, 1 MW gyrotron was run at 3 μsec pulses. The gyrotron uses a highly overmoded cavity in order to keep ohmic losses less than 2 kW/cm^2. Therefore the design is consistent with continuous operation. Two different electron guns were used in the experiments. The first gun produced a large annular electron beam and was used to excite whispering gallery modes, while the second gun produced a smaller beam and was used to couple to volume modes of the highly overmoded cavity. This allowed us to compare operation in a wide range of modes. The cavity, which consists of a straight section between two linear tapers, was designed for the TE$_{34,6,1}$ mode at 230 GHz and the TE$_{42,7}$ mode at 280 GHz. The cavity is $\sim$ 23 wavelengths in diameter and has an interaction length of $\sim$ 10 wavelengths at 280 GHz. The gyrotron was step tunable over a wide range in frequency. Tuning was accomplished by varying the cavity magnetic field. Output power measurements accounted for 13% losses in the 0.25 inch thick quartz window. Table I shows some typical results for the two guns. In general, both guns produced similar results, with maximum efficiencies of about 20%.

Table I Typical Results for Large and Small Guns

Large Gun		
Mode	TE$_{34,6,1}$	TE$_{41,8,1}$
Frequency (GHz)	229	290
Ohmic losses (kW/cm^2)	1.5	2
P (MW)	0.97	0.89
Efficiency (%)	18	18
$\Delta\omega / \omega$ (%)	2.0	1.6
Small Gun		
Mode	TE$_{25,13,1}$	TE$_{25,14,1}$
Frequency (GHz)	280	290
Ohmic losses (kW/cm^2)	1.5	1.5
P (MW)	0.78	0.72
Efficiency (%)	17	18
$\Delta\omega / \omega$ (%)	1.8	1.8

A detailed study of the TE$_{34,6,1}$ mode at 229 GHz was made. Figure 3 shows the optimum output power and efficiency. The experimental efficiency increases with increasing current to 18-19% at 35 A and remains roughly constant as current continues to increase. The nonlinear self-consistent theory, however, predicts that efficiency continues to rise with current to values above 40% at 58 A. In general, the experimental efficiencies observed were one half the value predicted by theory based on a single mode. Other modes gave similar results. The peak power measured was 1.2 MW in the TE$_{38,5,1}$ mode at 231 GHz with an efficiency of 20%. It is likely that mode competition with azimuthal modes is preventing operation at the optimum cavity magnetic field, thus reducing the efficiency. Multimode simulations by the University of Maryland indicate that mode competition would reduce the maximum efficiency of our cavity to about 30%.

QUASI-OPTICAL MODE CONVERTER

A quasi-optical mode converter for the TE$_{16,2}$ mode was designed and tested. The converter consists of a helically cut waveguide launcher and doubly curved focusing reflector. The output port of the gyrotron is comprised of a waveguide internal to the vacuum, which extends to within a few millimeters of the 0.025 cm quartz vacuum window, and an external waveguide segment, or sleeve, on which the launcher is mounted. The vacuum window produces a discontinuity in the waveguide, although the external waveguide is carefully aligned to minimize mode conversion caused by the disruption.

Several experiments were performed to assess the mode purity of the radiation at the launcher and the operation of the converter. First, polar radiation pattern diode scans were made to determine the mode content of the radiation exiting the external waveguide. Next, near field launcher pattern scans at the reflector position were made by rotating the launcher while sliding the detector along a platform mounted parallel to the launcher straight edge. Third, diode scans were made in the focal plane. The diode was mounted on a sliding platform which was aligned perpendicular to the line connecting the center of the reflector to the theoretical focus. Last, the calorimetric efficiency of the converter was measured by placing a 10 cm diameter laser calorimeter in the focal plane, perpendicular to the incoming radiation and comparing this power to the power measured at the output of the gyrotron.

Polar scans made with the sleeve in place showed that the radiation reaching the launcher is composed of 97.2% $TE_{16,2}$, 0.3% $TE_{16,1}$, 0.3% $TE_{16,3}$, 1.6% $TE_{15,2}$, and 0.6 % $TE_{17,2}$. The nonlinear uptaper of the gyrotron cavity virtually eliminates mode conversion from the $TE_{16,2}$ to adjacent radial modes, the $TE_{16,2}$ and $TE_{16,3}$. Mode conversion from the $TE_{16,2}$ mode to adjacent azimuthal modes, the $TE_{15,2}$ and $TE_{17,2}$, is a result of the break in the waveguide caused by the vacuum window.

Next, near field launcher pattern scans, taken at the reflector position, were compared with theoretical near field launcher patterns for the mode mix detailed above. The size and shape of the experimentally determined and theoretically predicted patterns were in good agreement. The experimental pattern also showed that > 97% of the radiation from the launcher is intercepted by the reflector.

A diode scan in the focal plane showed a small Gaussian-like focus for the $TE_{16,2}$ mode with two other foci in the theoretical positions for the $TE_{15,2}$ and $TE_{17,2}$ foci. Integrating the power in the focal plane shows 97.8% in the $TE_{16,2}$ focus, 1.6% in the $TE_{15,2}$ focus, and 0.6% in the $TE_{17,2}$ focus. This mode content agrees well with the mode mix determined by both the far field polar scans from the external waveguide and the near field launcher pattern scans. The quasi-optical converter is therefore a useful diagnostic in determining the mode content of the gyrotron output. Calorimetric efficiency measurements showed that at least 95% of the power leaving the gyrotron appears in the focal plane, within the 10 cm diameter of the laser calorimeter. The total powers and powers in each mode at several points within the quasi-optical converter system are shown in Table II.

The focused radiation was then launched into a 1.25 inch diameter corrugated waveguide. The waist of the Gaussian-like beam was incident on the waveguide entrance. The waveguide diameter was slightly smaller than the beam diameter, which resulted in a waveguide coupling efficiency of only 88%. The overall efficiency of the converter and corrugated guide, which is defined as the power in the HE_{11} mode at the exit of the guide divided by the total gyrotron output power, was measured to be 83%.

Table II Power and mode content in the quasi-optical converter.

	%TE$_{16,1,1}$	%TE$_{16,2,1}$	%TE$_{16,3,1}$	%TE$_{15,2,1}$	%TE$_{17,2,1}$	Total
Window	0.3	99.0	0.3	0.0	0.4	100
Sleeve	0.3	97.2	0.3	1.6	0.6	100
Reflector	0.3	94.3	0.3	1.5	0.6	97
Focus	0.0	92.9	0	1.5	0.6	95

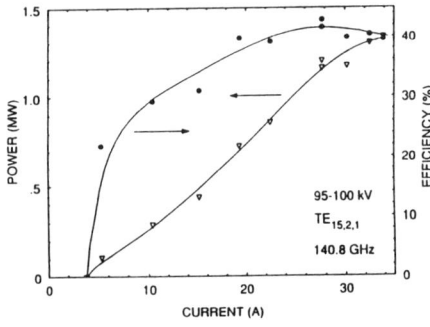

Fig. 1 Experimentally observed power and efficiency for the TE$_{15,2,1}$ mode.

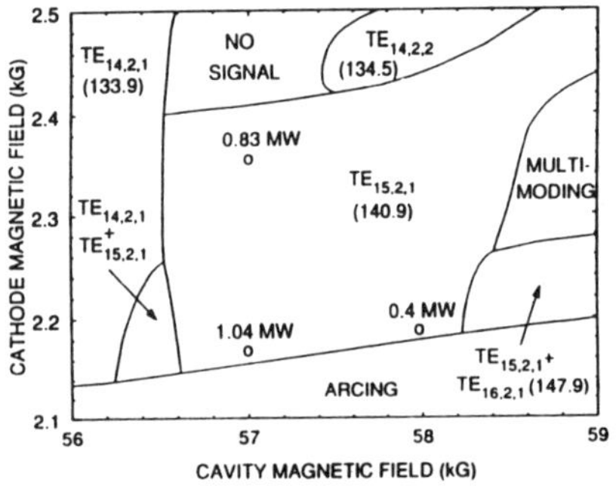

Fig. 2 Mapping of emission.

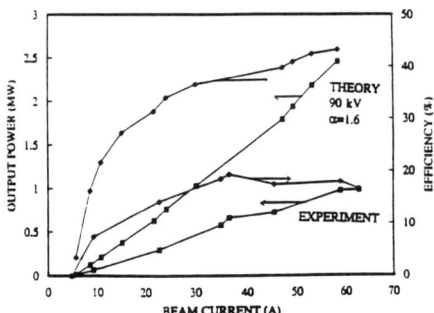

Fig. 3 Optimum power and efficiency for the TE$_{34,6,1}$ mode at 229 GHz.

GENERAL RF
PLASMA INTERACTIONS

Regimes for electron heating via mode conversion in TFTR.

R. Majeski, J.C. Hosea, C.K. Phillips, J.H. Rogers, G. Schilling, J.E. Stevens, J.R. Wilson,

Princeton Plasma Physics Laboratory, Princeton, NJ 08543.

ABSTRACT

Electron heating through mode conversion at the perpendicular ion cyclotron resonance in TFTR is explored. For a certain range of $k_\parallel$, in a two ion species plasma with a light ion majority species which is resonant near the axis and a large heavy ion minority fraction, the $n_\parallel^2 = L$, $n_\parallel^2 = S$, and $n_\parallel^2 = R$ surfaces lie in a closely spaced triplet. This situation is similiar to the cutoff-resonance-cutoff geometry at the Alfven resonance in a plasma with a single ion species, and may produce strong, localized electron heating through efficient mode conversion.

INTRODUCTION

Mode conversion at the perpendicular ion cyclotron resonance in plasmas with two ion species of roughly equal density fractions has long been considered an important heating process for electrons[1]. In hot reactor-like D-T plasmas where the fundamental deuterium cyclotron resonance is located on axis single-pass absorption at the fundamental is expected to be strong[2] and so the power incident on the mode conversion layer on the high field side of the torus is reduced. However for ion temperatures more typical of present-day large tokamaks mode conversion can be competitive with absorption at the fundamental ion cyclotron resonance. We consider here two heating scenarios for TFTR: a D-T plasma with the deuterium cyclotron resonance located on axis, and a model ^{3}He-D plasma (^{3}He resonant near the axis). Plasma density and species mix are chosen to obtain good mode conversion efficiency.

COLD PLASMA CASE

We first consider the geometry of the cutoff and resonant surfaces appropriate to TFTR in the cold plasma limit. Here the density profile is assumed to be parabolic and the toroidal field (and $k_\parallel$) vary as 1/R. The excitation frequency is 43 MHz. Calculations of the Bay L ICRF antenna $k_\parallel$ spectrum indicate that the peak in the vacuum spectrum occurs at 10 m^{-1} for 180° antenna phasing, and at 5 m^{-1} for 90° phasing, so that we model launched wavenumbers in this range.

Although H-^{3}He is the best model species mix for D-T, operational restrictions preclude majority H discharges in TFTR. Hence we consider experiments in ^{3}He-D plasmas. We first discuss the cutoff and resonant surfaces for a fast wave in a pure ^{3}He plasma. For a field of 4.0 T the ^{3}He fundamental resonance is near the axis. In this case the $n_\parallel^2=R$, S and L surfaces occur in a triplet on the high field side of the the torus where $\omega<\Omega_i$. The $n_\parallel^2=R$ surface lies to the low field side of the Alfven resonance, $n_\parallel^2=S$, and the $n_\parallel^2=L$ surface is on the high field side of the resonance. Variants of this case have been discussed by Heikkenen[3]. In the case of a two ion species plasma the $n_\parallel^2=S$ resonance is referred to as the ion-ion hybrid resonance, if it occurs between the cyclotron frequencies of the two species. For low minority concentrations and sufficiently low $k_\parallel$ the $n_\parallel^2=L$ cutoff now lies on the low field side of the resonance, and the $n_\parallel^2=R$ cutoff occurs on the high field side in the edge plasma. The position of the two cutoffs with respect to the resonance are therefore interchanged in comparison to the Alfven resonant case. Intermediate wavenumbers and concentrations of ^{3}He result in a closely spaced cutoff-resonance-cutoff triplet, as shown in Fig. 1 for $n_e(0)=4\times 10^{19}$ m^{-3}, a wavenumber of 8 m^{-1}, $n_{He}=0.3n_e$, and $B_0=4.0T$. Here the geometry resembles that of the Alfven resonance below the majority cyclotron frequency in that the $n_\parallel^2=R$ cutoff lies on the low field side of the ion-ion hybrid resonance, with the $n_\parallel^2=L$ cutoff on the high field side. In Fig. 2 the resonant and cutoff surfaces for a 50-50 D-T supershot plasma with a central density of 5×10^{19} m^{-3}, $k_\parallel=8$ m^{-1}, and $B_0=5.6T$ are shown. The resonance-cutoff landscape is seen to be similar to the intermediate concentration D-^{3}He case.

1-D MODELLING

Both the D-^{3}He and D-T cases have been modelled with a 1-D full-wave code developed by M. Brambilla[4]. It should be noted that caution is required in interpreting the results since these scenarios involve strong conversion to modes with high ($k_\perp \rho_i$) which are difficult to resolve numerically. The intent here is to identify experimental regimes of interest, where mode conversion and electron heating may be dominant and which are relevant to D-T operation in large tokamaks, i.e. TFTR.

Figure 3 is a plot of the computed coupled power as a function of $n_\parallel$ for a D-^{3}He plasma with $n_{He}=0.3\ n_e$, for $B_0=4.0$ T, $n_e(0)=4.0\times 10^{19}$ m^{-3} (parabolic profile), $T_e(0)=3$ keV, $T_i(0)=2$ keV in the TFTR geometry (R=2.62 m, a=0.98 m). The peaks in the antenna loading spectrum correspond to wavenumbers of 5, 8, and 10 m^{-1}. Figure 4 is a plot of the associated power deposition profile. Two curves are shown in Fig. 4; the solid line denotes electron power deposition while the dotted line denotes power in high ($k_\perp \rho_i$) modes which is artificially dissipated by a "stochastic damping" term. In the absence of a nearby cyclotron resonance this mode converted power should damp on electrons. Only 3% of the incident power is predicted to be coupled to ions; the remainder is damped on electrons or mode converted to ion Bernstein waves.

The computed power deposition profile is localized to the region near the ion-ion hybrid resonance. If plots of the cold plasma resonant and cutoff surfaces are made for the wavenumbers corresponding to the peaks in the antenna loading spectrum we find that for 5 m^{-1} the $n_\parallel{}^2$=L cutoff lies on the low field side of the $n_\parallel{}^2$=S resonance, while for 8 and 10 m^{-1} the $n_\parallel{}^2$=R cutoff lies on the low field side of the resonance. The ICRF antenna therefore excites parallel wavenumber components which span the cold plasma cutoff-resonance geometries discussed earlier.

Figure 5 is a plot of the coupled power as a function of $n_\parallel$ for a D-T plasma with $n_D = n_T$, for B_0=5.6 T, $n_e(0)$=5.0 × 10^{19} m^{-3} (peaked profile appropriate to a supershot), $T_e(0)$=10 keV, $T_i(0)$=20 keV, R=2.62 m, and a=0.98 m. For 5.6 T the D fundamental cyclotron resonance is on axis, which will maximize absorption at the fundamental. However, from the power deposition profile shown in Fig. 6 it is clear that direct electron heating and mode conversion to the ion Bernstein wave are competitive processes, even in this temperature range. 75% of the incident power is predicted to be absorbed on electrons or mode converted.

CONCLUSIONS

Mode conversion is shown to be important in ^{3}He-D and D-T plasmas with parameters appropriate to TFTR ohmic or supershot discharges, for wavenumber ranges excited by the TFTR ICRF antennas. In ^{3}He-D ohmic discharges mode conversion and electron heating should dominate over ion heating. In D-T supershot plasmas electron heating and mode conversion should be comparable to ion heating at the fundamental cyclotron frequency, for ion temperatures of order 20 keV. The ^{3}He-D plasma may also provide an opportunity to test mode conversion current drive without the need for high field side ICRF antennas.

We gratefully acknowledge the assistance of P. Bonoli and Y. Takase in providing the numerical modelling code and instructions in its use. This work was supported by U.S. DoE Contract DE-AC02-76-CHO-3073.

REFERENCES

1. F. W. Perkins, Nucl. Fusion 17, 1197 (1977).

2. V. P. Bhatnagar, J. Jacquinot, D. F. H. Start, B. J. D. Tubbing, Nucl. Fusion 33, 83 (1993).

3. J. A. Heikkinen, T. Hellsten, M. J. Alava, Nucl. Fusion 31, 417 (1991).

4. M. Brambilla, Nucl. Fusion 28, 549 (1988).

404 Electron Heating via Mode Conversion in TFTR

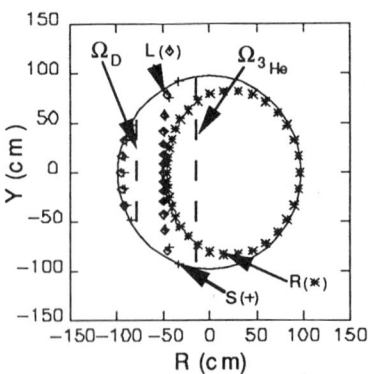

Figure 1. Resonant and cutoff surfaces for ^{3}He-D in the TFTR geometry.

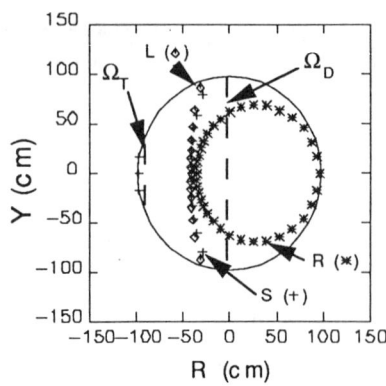

Figure 2. Resonant and cutoff surfaces for D-T in the TFTR geometry.

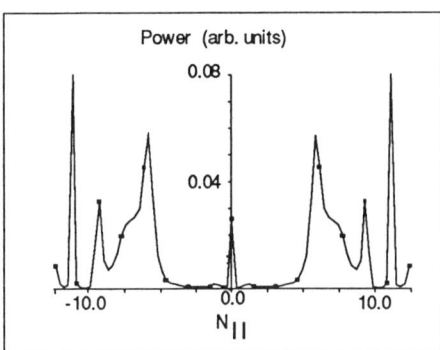

Figure 3. Coupled power as a function of parallel wavenumber for the ^{3}He-D case.

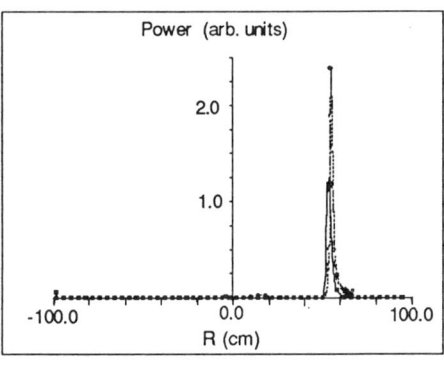

Figure 4. Power deposition profile for the ^{3}He-D case.

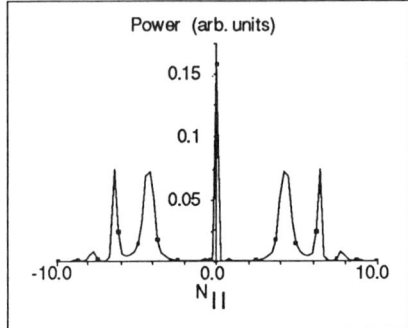

Figure 5. Coupled power as a function of parallel wavenumber for the D-T case.

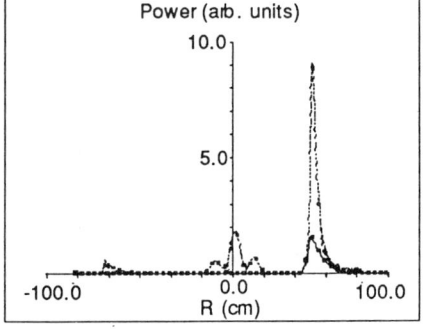

Figure 6. Power deposition profile for the D-T case. The central peak denotes fundamental D absorption.

DT Simulation of ICRF Heated Supershots in TFTR Using TRANSP[*]

R.C. Goldfinger, D.B. Batchelor
Oak Ridge National Laboratory, P. O. Box 2009, Oak Ridge, TN 37831-8058

C.K. Phillips, R. Budny, G. W. Hammett, J. C. Hosea, D. M. McCune, J. E. Stevens, J. R. Wilson, and the TFTR Team, Princeton Plasma Physics Laboratory, Princeton, NJ 08543

ABSTRACT

The principal goal of ion cyclotron range of frequency (ICRF) heating on the Tokamak Fusion Test Reactor (TFTR) is to enhance plasma performance during the deuterium-tritium (DT) physics phase of operations. Strongly centralized ICRF heating may play a critical role in obtaining high Q_{DT} and high β_α operation in TFTR, as well as in future fusion reactors. ICRF heating of a dilute minority species leads to the formation of an energetic ion population that, in turn, provides strong central electron heating. The corresponding rise in the central electron temperature translates into an increase in the slowing-down time of either neutral beam or alpha particles in the discharge. Preliminary DT simulations of the experimental results in deuterium-deuterium (DD) plasmas performed with the TRANSP code are presented in this paper.

INTRODUCTION

The main emphasis of the ICRF program on the TFTR is to enhance parameters in future DT experiments[1]. Strong central ICRF heating of supershots may play an important part in obtaining high Q and high β_α during the DT program in TFTR, as well as in future fusion devices, for the following reasons:

(1) The strong central heating that occurs with ICRF heating of supershot plasmas increases the central electron temperature. This causes an increase in the neutral beam and alpha slowing-down times and, hence, the density and beta of the alphas in the discharge.
(2) As neutral beam injection (NBI) power and the stored energy in the plasma increase, the plasma current must be increased to avoid MHD instability. At the highest currents, sawteeth appear. They may be able to be suppressed with the application of ICRF heating.

To investigate these issues, the TRANSP 1-1/2D time-dependent transport code[2,3] has been used to simulate the time evolution of ICRF and NBI heated supershot like plasmas in TFTR.

DT SIMULATIONS

The future DT experiments planned for TFTR are simulated with TRANSP using existing DD supershot data. In this simulation an equal mix of deuterium and tritium is substituted in the target deuterium plasma, as well as in the D^0 neutral beams, while the other measured beam and plasma parameters are unchanged. Thus, this conservative simulation neglects the effect that alpha heating enhances the plasma parameters or that confinement will increase with tritium. Also, TRANSP is not equipped at this time to predict second harmonic heating of the tritium ions; the heating of these ions would produce a hot tail in the distribution

---

[*]Research sponsored by Office of Fusion Energy, U.S. Department of Energy, under Contract No. DE-AC05-84OR21400 with Martin Marietta Energy Systems, Inc.

[**]The submitted manuscript has been authorized by a contractor of the U.S. Government under contract DE-AC05-84OR21400. Accordingly, the U.S. Government retains a nonexclusive, royalty-free license to publish or reproduce the published form of this contribution, or allow others to do so, for U.S. Government purposes.

enhancing the fusion rate. In this paper we discuss the analysis of a pair of NBI supershot like plasmas that differ by the presence or absence of ICRF He^3 minority heating. DT equivalent simulations, as just described, are performed for each shot. The pair of shots, 66785/66680, differed primarily in that 66785 had 22-MW NBI with no ICRF and 66680 had 18-MW NBI with 4.5 MW of ICRF (Fig. 1).

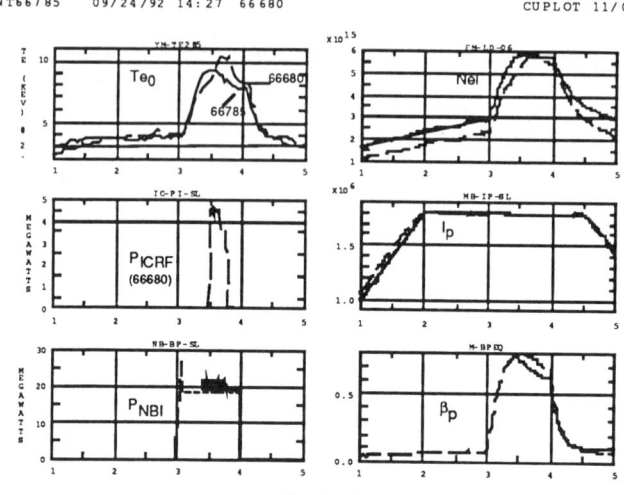

Fig. 1 Waveforms for TFTR shot 66785 (22 MW NBI) overlaid with 66680 (18-MW NBI + 4.5-MW ICRF). The DT equivalents of these two shots are discussed below.

Figure 2 shows heating from various sources in the DT equivalent shot, 66680. The heating rate by the 3.5-MeV alphas, as modeled by TRANSP, is ~5 x 10^5 W; thus the predicted Q_{DT} = (total fusion power)/(P_{NBI} + P_{ICRF}) = 0.12 for this shot.

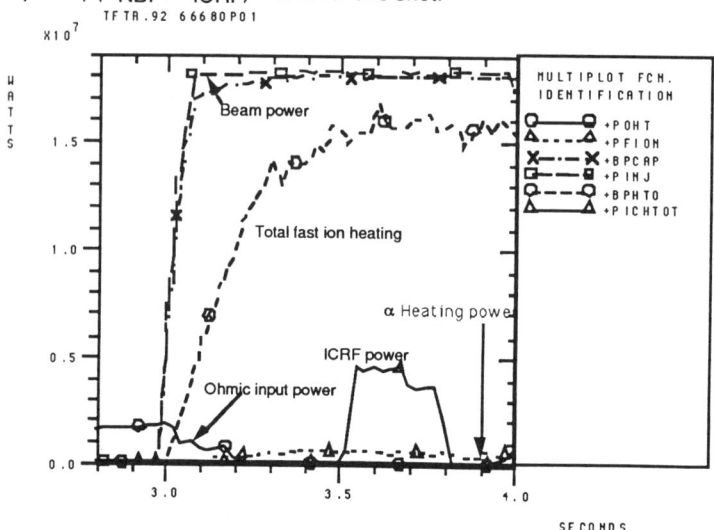

Fig. 2 Various heating terms in the DT equivalent shot, 66680p01 (NBI + ICRF).

The volume integrated heating powers are shown in Fig. 3. The beam fast ions heat mainly thermal ions, whereas the 3.5-MeV alpha particles mainly heat electrons, because the alpha energy is >> E_{crit}~14.8T. This alpha heating is overshadowed by the NBI during the NBI phase.

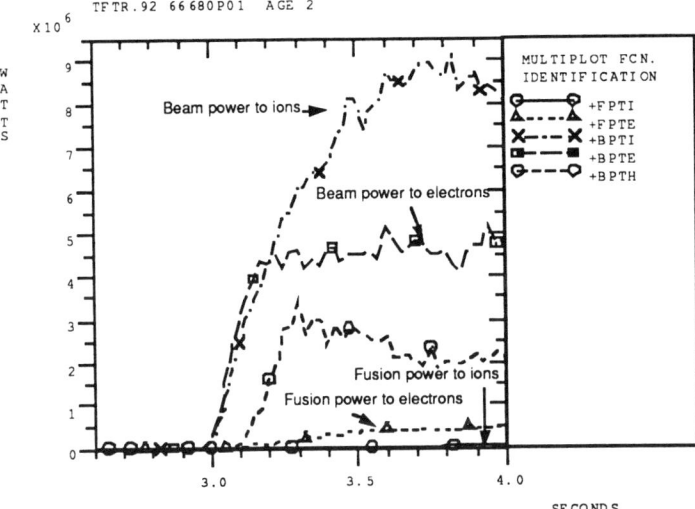

Fig. 3 66680p01(NBI + ICRF) Volume Integrated heating powers.

Figure 4 shows the alpha slowing-down time, volume averaged from r = 0 to r/a = 0.25. A rise of ~7% is seen in the ICRF shot, 66680, mainly due to the rise in the central electron temperature. This increase in the slowing-down time will lead to an increase in the density and β of the alpha particles, as seen in Fig. 5.

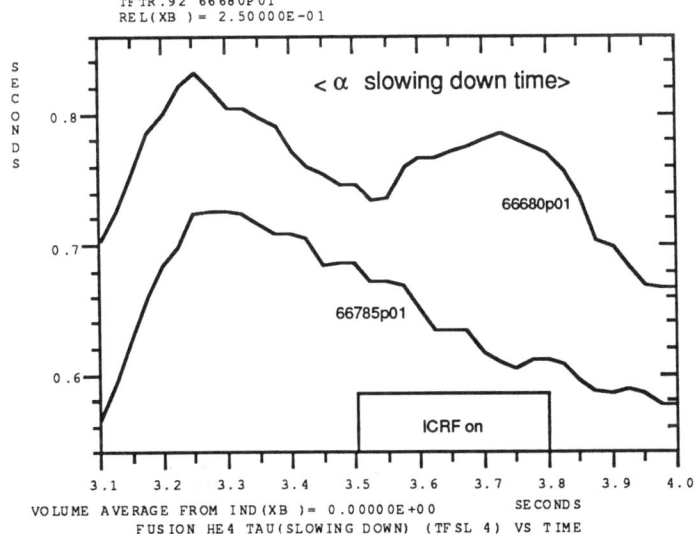

Fig. 4 Volume average, from r = 0 to r/a = 0.25, of the alpha slowing-down time; note that it increases significantly in 66680p01 as T_e increases with the onset of ICRF (see Fig. 1).

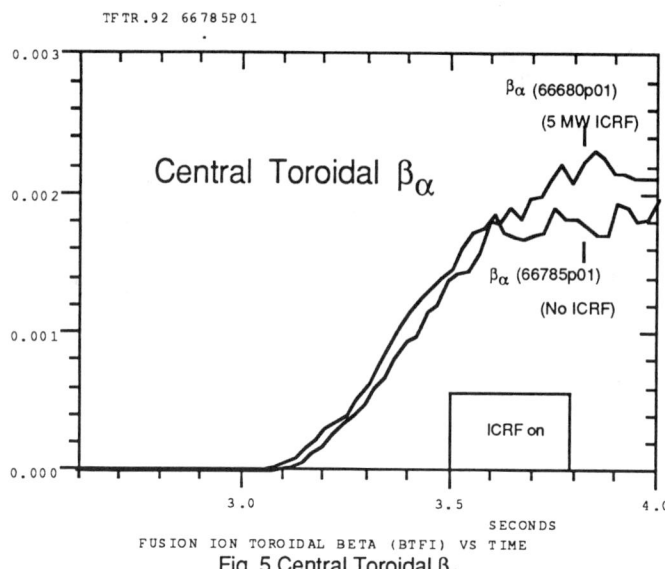

Fig. 5 Central Toroidal β_α

SUMMARY

Strong central electron heating is observed with ICRF in TFTR supershot 66680 [Te(0) increases ~2 KeV over the non-ICRF companion shot]. The higher Te is predicted to lead to higher β_α during DT operation in TFTR, due to the increase in the alpha slowing-down time. Because the alphas are born at energies far above the critical energy, the alphas will slow down primarily on the electrons. This increase in the slowing-down time will be approximately in the ratio of $Te^{3/2}/Ne$ (Spitzer slowing-down time).

DT equivalent simulations were performed with the TRANSP code for these discharges in which half the deuterium was replaced with tritium in the neutral beams and target plasmas. In the 66680 DT simulation, Q_{DT} is 0.12, and the central β_α increases over the non-RF companion by ~35% to β_α = 0.23%. The higher electron temperature due to the ICRF heating causes the alpha slowing down time to increase from 0.73 sec to 0.78 sec.

These DT simulations are conservative: alpha heating effects and isotopic improvement in confinement are neglected. Also, TRANSP is not equipped to handle second harmonic tritium heating yet. Heating of the tritium fast beam ions would also enhance the fusion rate. Therefore, actual DT plasma performance in TFTR for the same parameters might be better than indicated here.

REFERENCES

1. Wilson, J. R., Bell, M. G., et al.,"ICRF Heating on TFTR - Effect on Stability and Performance," Fourteenth International Conference on Plasma Physics and Controlled Nuclear Fusion Research 1992 (IAEA-CN-56/E-2-2), Wurzberg, Germany, 1992.

2. R. J. Goldston in, "Basic Physical Processes of Toroidal Fusion Plasmas" (Proc. Course and Workshop Varenna, 11085) EUR-10418-EN, (CEC, Brussells) **1**, p. 165, (1986).

3. R. J. Hawryluk, "An Empirical Approach to Tokamak Transport," Physics of Plasmas Close to Thermonuclear Conditions, ed. by B. Coppi, et al., (CEC Brussells), **1**, p. 100, (1980).

Behavior of Toroidal Alfven Eigenmodes during ICRF Heating[*]

K. L. Wong, J. R. Wilson, Z. Y. Chang, D. Darrow, E. Fredrickson,
and C. K. Phillips
Princeton Plasma Physics Laboratory, Princeton University,
Princeton, New Jersey 08543

Toroidal Alfven eigenmodes (TAE) may become unstable in ignited tokamaks, and the instability can eject energetic alpha particles and degrade the alpha heating efficiency. Simulation experiments with neutral beams show that the instability can indeed be excited. Its amplitude exhibits a bursting behavior. Each TAE burst is accompanied by a burst of energetic ions ejected from the plasma. When the number of fast particles is above the instability threshold, the mode grows rapidly; a few percent of the fast particles are ejected and the mode becomes stable again. The population of the energetic particles is clamped near the instability threshold. This phenomenon is very similar to the fishbone activities driven by perpendicular neutral beam injection. The pulsation state can be modelled by two coupled first order differential equations. The ejection of fast particles occurs in a very short time interval (comparable to the bounce period), and the damping time of TAE mode is short compared with the time required to raise the energetic particle population back to the instability threshold. Therefore, a pulsation scenario is expected from the theory of Berk, Breizman and Ye.

TAE modes can also be excited by energetic trapped ions. This was recently observed in TFTR during ICRF heating. Hydrogen minority ions were heated by fast magnetosonic waves to several hundred keV temperature. These energetic trapped ions have large drift velocities that can resonate with TAE modes. However, the mode amplitude behavior is distinctly different from those driven by tangential neutral beams. Instead of short bursts of TAE activities, we have very slow variation of the mode amplitude with a typical time scale of several hundred milliseconds. The mode amplitude can be modulated by sawtooth activities as shown in figure 1a, or it can be slowly varying as in figure 1b. We believe that this is due to the difference in the orbit characteristics of the fast ions which drive the instability.

In the hydrogen minority ICRF heating experiment, the fundamental cyclotron resonance zone for hydrogen ions is on a vertical plane passing through the plasma core. The hydrogen ions are heated to a perpendicular temperature of about 500 keV. We expect most of the energetic hydrogen ions to be magnetically trapped with reflection points near the cyclotron resonance zone. Following G. W. Hammett, we can represent their velocity distribution function in the following approximate form:

$$f(W, \xi) = f_0(W) [\exp(-|\xi - \xi_*|/\sigma_\xi) + \exp(-|\xi + \xi_*|/\sigma_\xi)] \quad (1)$$

where $\xi = v_\parallel / v$

ξ_* = pitch angle of particles with reflection points at the cyclotron resonance zone

$$\sigma_\xi = \{ Z_{eff} / [4A<Z_i^2/A_i>(1 + W^{3/2} / W_c^{3/2})] \}^{1/2}$$

$$W_c = 14.8 \, A \, T_e <Z_i^2/A_i>^{2/3}$$

= critical energy at which electron drag equals ion drag.

For our experimental parameter, $T_e=4$ keV, $Z_{eff}=3$, $W=(3/2)T_\perp=750$ keV, $A=1$ for hydrogen ions, $<Z_i^2/A_i> = 0.5$, $W_c=30$ keV. With $R=2.62$m and $r=0.4$m, we get $\sigma_\xi = 0.11$. Let ξ_1 be the pitch angle (evaluated at the resonance zone) for barely trapped particles. From equation (1), we obtain

$$f(W, \xi_1)/f(W, \xi_*) \sim 3 \times 10^{-2} \ll 1 \quad (2)$$

This means that very few of the energetic ions are barely trapped; most of them have their reflection points near the resonance zone, and it is not easy to change these particles into circulating particles in one bounce period. It requires a very large amplitude TAE mode with $\tilde{B}/B > 10^{-2}$ for this to be possible. In this experiment, preliminary estimates based on reflectometer measurements indicate that $\tilde{B}/B$ is of the order of 10^{-4}. Therefore, the effect of the low amplitude TAE modes on the energetic ions is quite similar to small pitch angle scattering due to Coulomb collisions. One only expects a sudden change of the energetic ion orbits in the transition from barely circulating to barely trapped orbits and vice versa. Particle loss mechanisms due to TAE modes driven by energetic circulating ions have been investigated by Hsu and Sigmar. It was found that shortly after the TAE modes were turned on, most of the energetic ion loss were due to transitions from barely circulating to barely trapped orbits. This loss mechanism which happens on a very fast time scale (the bounce period of the ion) is not applicable to the energetic trapped ions beause of the following reasons. Firstly, there are very few energetic ions near the trapped-circulating boundary. Secondly, a barely trapped particle has the largest guiding center orbit excursion from the flux surface on which the reflection points lie. Figure 2 depicts the particle orbits projected on a poloidal plane. When a barely trapped particle is changed into a circulating particle via pitch angle scattering, there are two possibilities. It can either go into orbit C1 which is almost the same as the outer half of the barely trapped orbit, or it can go into orbit C2 which is almost the same as the inner half of the barely trapped orbit. If the original barely trapped particle is confined, the barely circulating particle is also confined. Therefore, this process does not cause the prompt loss of energetic ions. Finally, let us consider the transition from the barely circulating orbit C2 to the barely trapped orbit which is the dominant prompt loss mechanism for the energetic beam ions in the previous TAE experiment. Unlike neutral beam injection, the barely circulating energetic ions in the ICRF experiment are produced from barely trapped energetic ions which are confined. In the time scale of the particle transit time, these barely circulating ions can only go back to a trapped orbit very similar to the original confined trapped orbit, i.e., small pitch angle scattering does not provide a fast escape channel for the energetic particles produced by ICRF. Because of the above reasons, the major effect of the TAE mode on the energetic ions is to move them from one trapped orbit to another. This would enhance the transport of the energetic ions across the magnetic field and reduce their density gradient. It is a relatively slow diffusion process which tends to reduce the driving force for the instability. Suppose the temporal evolution of the TAE amplitude A is governed by the equation

$$dA/dt = (\gamma_{drive} - \gamma_{damp})A = \gamma A \qquad (3)$$

If γ can be expressed in terms of A with no explicit time dependence, then Eq.(3) is integrable and the evolution of A can be expressed in terms of the initial condition at t=0. From the data shown in figure 1, it is apparent that the TAE modes do not grow exponentially. It appears that the driving term and the damping term almost cancel each other, i.e., $\gamma << \gamma_{drive} \sim \gamma_{damp}$. The amplitude changes on such a slow time scale that the plasma equilibrium can change. Although we cannot measure the evolution of the fast ion density distribution, the data suggest that the plasma parameters evolve in such a way that the TAE modes are marginally unstable and one can pick a slowly varying γ to model the evolution of the mode amplitude. We attribute this to the lack of a rapid loss mechanism for the energetic trapped ions. A fast particle probe outside the plasma edge detects a continuous flux of fast particles that correlates with the TAE mode amplitude. This observation is consistent with our interpretation.

*This project is supported by the U. S. Department of Energy under contract No. DE-AC02-76-CHO-3073.

Fig.1. Evolution of the Mirnov coil signal spectrum. The largest peak is the TAE mode signal.

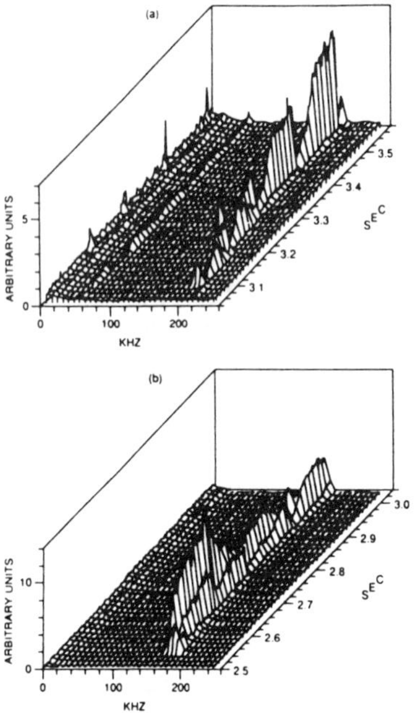

Fig.2. Projection of orbits on a poloidal plane.

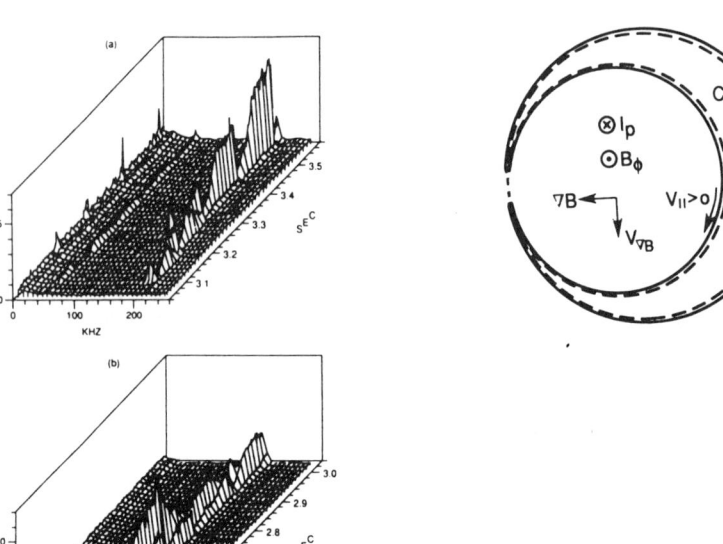

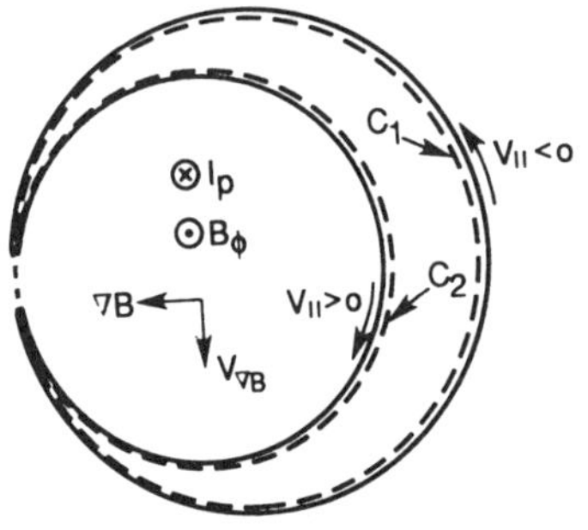

COMPUTATIONAL PATH TO SECOND STABILITY IN PBX-M

M. Okabayashi, D. W. Ignat, S. C. Jardin,
and Y.-C. Sun
Princeton Plasma Physics Laboratory
Princeton, NJ   08543-0451

ABSTRACT

MHD stability analysis[1-3] has shown that a bean-shaped configuration with indentation of 20% can attain a second stability regime by additional non-ohmic current and by increasing $q(0)$ to 1.3 or more. Recently, the increase of $q(0)$ with off-axis current drive by lower hybrid waves was demonstrated by PBX-M. We have studied possible paths for achieving $1.3 < q(0) < 2$ with current drive and heating with a simulation. We have identified several potential problems associated with LHCD off-axis current drive and increasing $q(0)$ in the high β and high $\beta_{poloidal}$ regime. We discuss the approaches used and the problems encountered in attaining parameters that are clearly in the second stable region.

INTRODUCTION

Plasmas for fusion power production need to operate at high β to achieve high power density at moderate capital cost. Computational studies by Troyon and Gruber[4] suggest that it is difficult to exceed $\beta \approx 0.035 I[MA]/(a[m]B[T])$, and this is generally consistent with the present experiments. A second stable region characterized by a multiplier of 0.050 (as opposed to 0.035) is suggested by other studies with specific profiles,[1,2] and can be reached experimentally if special care is taken to control profiles. Attaining proper profiles and high β in a dynamic tokamak plasma is sought in the PBX-M experiment with the help of lower hybrid current drive, and other tools. The present work reports an early computational attempt to evolve proper profiles using the free-boundary MHD equilibrium and transport code TSC[5] as coupled to various models of lower hybrid current deposition, including multiple ray tracing and evolution of a 1-dimensional velocity distribution[6] as well as simpler models.

APPROACH

We followed two basic approaches: (1) extreme edge current drive and high indentation; (2) current drive as predicted for our high-efficiency coupler plus some rapid current ramp to maintain edge current and bean shape. Each approach requires trial and error on timing of heating and current drive, power levels, and current in the plasma.

The first approach included the following steps. Establish a 100 kA current in a circular plasma; raise $q(0)$ by off-axis LHCD and bootstrap current; heat

© 1994 American Institute of Physics

to capture the q-profile at the plasma center; broaden the current profile with rapid current ramp; adjust shaping field to form the high-indentation bean shape; heat with even more power to raise β. These steps achieved second stability, as determined from $S - \alpha$ diagrams,[7] over 95% of the plasma profile.

The second approach included the following steps. Establish a 200 kA plasma at low indentation; raise central q by sweeping the phase angle at the LHCD coupler; heat with neutral beams to increase β without large current ramp.

EXTREME EDGE CURRENT DRIVE

In this set of simulations, the TSC code was run with lower hybrid power deposition specified by parameters. The amount of current drive was computed at each position according to a simple analytic formula containing the lowest and highest parallel phase velocities. Setting power and current deposition far out on the edge of the plasma was easily done with the analytic formula, The ramp of the plasma current, and the adjustment of the various currents in the poloidal coil system, were prepared by experimentation to achieve strong a strong bean shape — having high indentation. It was found that $q(0)$ rose to the range of 2, and that stability to ballooning modes was achieved over the entire plasma volume.

ACTUAL-SPECTRUM CURRENT DRIVE

We have started to simulate the LHCD behavior with a lower hybrid model which includes ray tracing and current deposition with the quasilinear theory.[6] The cases studied have $I_p = 200$kA, $B_t = 15$kG, neutral beam power of 3.5 MW, LHCD power of 0.8 MW and gradual I_p ramp. These power levels are within the present PBX-M hardware capability. The ferrite-fast-time-sweeping coupler was simulated with three couplers (135°, 150°, and 165°) in the calculation. The physical dimensions of the coupler now being installed on PBX-M were used to calculate the launched spectrum as in Fig. 2. Multiple rays were launched into the plasma and traced to find the location of energy absorption and current drive. Fast electron inertia and fast electron diffusion effects are not included for the current drive. A qualitative difference with the previous case is that the current drive tends to be narrowly deposited part way into the plasma, as opposed to broadly deposited on the outer region, as can be seen by comparing Fig. 3 with Fig. 1. Nonetheless, as shown in Fig. 4, the central $q(0)$ was increased to nearly 2 within 500 ms. The slow increase of $q(0)$ seems to be related to the gradual change of the deposition profile during the process of increasing n_e and β. According to the 'BALLOON' stability code[3] we find the central area has reached the deep second stability regime due to high $q(0)$ and small q' near the axis. Without large current ramp, the edge area is in the first stability regime. The improvement of edge area is in the progress with large I_p ramp and higher edge electron temperature.

DISCUSSION AND CONCLUSION

In this paper, we have studied LHCD issues related to the $q(0)$ increase required for achieving the second stability regime. We have demonstrated that the $q(0)$ can be increased to nearly 2 with the PBX-M hardware. Our work indicates that it is important to have the current density near the edge more than obtained with the present LHCD model. We have not optimized the path to the second stability regime. Our next steps include the optimization of neutral beam current drive and bootstrap current, and the synergistic effect with additional parallel E-field such as available from Ion Bernstein Wave heating..

ACKNOWLEDGMENTS

This work was supported by Department of Energy contract DE-AC02-76-CHO-3073. We thank N. R. Sauthoff, M. Chance, and N. Pomphrey for valuable advice and assistance.

REFERENCES

[1] M. S. Chance, S. C. Jardin, and T. H. Stix, "Ballooning Mode Stability of Bean-Shaped Cross Sections for High-β Tokamak Plasmas," Phys. Rev. Lett. **51** 1963 (1983).

[2] R. C. Grimm, M. S. Chance, A. M. M. Todd, J. Manickam, M. Okabayashi, W. M. Tang, R. L. Dewar, H. Fishman, S. L. Mendelsohn, D. A. Monticello, M. W. Phillips, and M. Reusch, "MHD Stability Properties of Bean-Shaped Tokamaks," Nucl. Fusion **25** 805 (1985).

[3] M. S. Chance, Y.-C. Sun, S. C. Jardin, C. E. Kessel, and M. Okabayashi, "MHD Stability of Tokamak Plasmas," in *Proceedings of the Second Symposium on Plasma Dynamics: Theory and Applications*, Trieste, Italy (July 8–10, 1992).

[4] F. Troyon and R. Gruber, "A Semi-Empirical Scaling Law for the β Limit in Tokamaks," Phys. Lett. **110A** 29 (1985).

[5] S. C. Jardin, N. Pomphrey, and J. DeLucia, "Dynamic modeling of transport and positional control of tokamaks," J. Comput. Phys. **66** 481 (1986).

[6] D. W. Ignat, E. J. Valeo, and S. C. Jardin, "Dynamic Modeling of Lower Hybrid Current Drive," in preparation.

[7] An $S - \alpha$ diagram plots a stability boundary in a plane with shear S on the ordinate and the pressure gradient α on the abscissa. These are related to β_p - ι graphs in the literature quoted above.

FIGURES

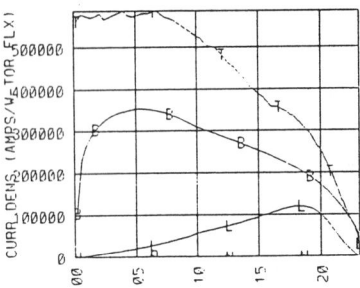

Fig. 1 Current densities driven in the plasma in the first method as a function of poloidal flux. The lower hybrid current 'L' is located according to an analytic form.

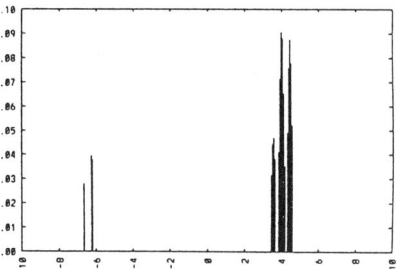

Fig. 2 Lower hybrid spectrum assumed launched. The three distinct peaks at positive $n_\parallel$ represent three different phasings of the waveguides, which can be approximated by sweeping phase.

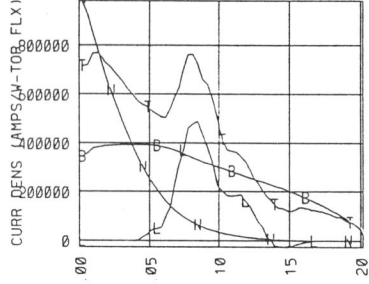

Fig. 3 Current densities driven in the plasma according to ray tracing, quasilinear absorption, and the current drive model. Note that the rf current is more toward the interior of the plasma relative to Fig. 1.

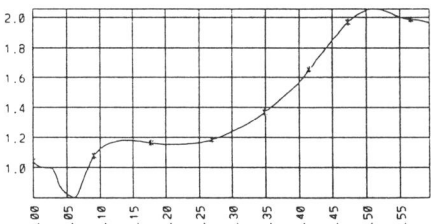

Fig. 4 Time development of $q(0)$ in the calculation with the wave physics. This shows that $q(0)$ can be strongly influenced by the current drive even in the case of the relatively central deposition computed with the more detailed wave-physics model.

QUANTITATIVE RADIATION LIMITS ON ICRF-SPECIFIC IMPURITIES

D. A. D'Ippolito and J. R. Myra
Lodestar Research Corporation, Boulder, Colorado, U.S.A.

ABSTRACT

ICRF-specific impurity influxes can be computed from rf sheath theory as one aspect of evaluating the design of an ICRF antenna system. Although these impurities can be reduced to negligible levels in heating experiments, this is not necessarily the case for other applications of ICRF antennas, such as fast wave current drive. Criteria are needed to determine tolerable levels of impurities for future tokamaks, including current drive and ignition experiments. Here, a model of the confined plasma thermal equilibrium is formulated, including a simple 1D impurity transport and radiation model, which can be used to establish quantitative radiation limits on ICRF-specific and other impurities. Applications to JET are discussed.

INTRODUCTION

It is now well-understood that ICRF-specific impurity generation is caused by sputtering of ions accelerated in rf sheaths,[1-7] which are in turn driven by the near and far fields of the powered antennas. In tokamaks with good single pass absorption the near field sheaths are the most important source of impurities; recent experiments in this regime on JET have shown rf-specific impurities can be virtually eliminated by proper design of the antennas, Faraday screens (FS), and limiters.[6,7] However, there are a number of circumstances in which one might wish to violate the "baseline" prescription of Refs. 5-7, including experiments on fast wave current drive (FWCD) and on scrape-off-layer (SOL) profile control by rf-sheath driven convective cells,[8] in which the use of toroidal dipole phasing is not appropriate. Thus, it would be useful to develop criteria for tolerable levels of impurities, and to use these criteria in the design of antennas and the planning of operational scenarios for future machines such as TPX, ITER and Ignitor. In this paper, we describe a thermal equilibrium model to calculate the impurity radiation limit for known impurity influxes from the limiters/divertors and the ICRF-specific impurity influx from the antenna screens.[5]

The antenna design should have the property that incremental impurity radiation δP_R caused by ICRF-specific impurities not substantially perturb the thermal equilibrium of the plasma, either globally or at the edge. It will be shown that this situation is approached in present experiments with low-Z first walls and Faraday screens. Ignition experiments require[11] that $\delta P_R \ll P_\alpha$ to allow a steady-state burning plasma. In principle, δP_R contains contributions from the direct FS impurity influx and the incremental wall impurity influx due to i) increased power flow to the edge and ii) far field sheath effects.[9] In the present paper, only the direct FS influx is explicitly modeled.

Calculations for JET are presented here to illustrate the advantages of low-Z Faraday screens in reducing the edge radiation and facilitating coupling to the H-mode. In future work, the model will be used to determine whether monopole phasing is consistent with ignition for the recently proposed[10] Ignitor antenna design.

© 1994 American Institute of Physics

THE MODEL

The model allows two species of impurities: the ICRF-specific impurity from the FS and the dominant "first wall" impurity species coming from the rest of the machine (limiters, X-point regions, or divertors). The FS impurity fluence Φ_{FS} is calculated for a given antenna design using an rf-sheath based model;[5] the wall impurity fluence Φ_w is an input parameter in our model which can also be calculated,[11-12] but here is chosen to give a reasonable value of $<Z_{eff}>$, the volume-averaged Z_{eff}. For each impurity species, the flux of neutral atoms across the last closed flux surface (LCFS) is calculated taking into account ionization in the SOL (screening). By balancing this neutral influx with the diffusion of impurity ions across the LCFS, a one-dimensional density profile $n_Z(r) = n_{Z0} \exp[-(S/2)(r^2/a^2)]$ is calculated for each impurity species following Engelhardt and Feneberg,[13] where S determines the ratio of the inward pinch velocity $v_\perp$ to the outward diffusion, i.e. $v_\perp = -S\, D_\perp r/a^2$.

The thermal equilibrium in our 1D model is given by

$$\frac{1}{r}\frac{\partial}{\partial r} r(Q_{\perp e} + Q_{\perp i}) - p_H(r) - p_\alpha(r) + p_R(r) = 0, \qquad (1a)$$

$$Q_{\perp j} = -\left(\kappa_j \frac{\partial T_j}{\partial r} + \frac{3}{2} T_j D_\perp \frac{\partial n_j}{\partial r}\right), \qquad (1b)$$

where $Q_{\perp j}$ is the heat flux for species j with particle and heat diffusion coefficients $D_\perp$ and κ_j [$D_\perp$ is assumed to be constant and independent of species, $\kappa_j = n_j \chi_j$ with thermal conductivity $\chi_j(r) = \chi_{jo}(1 + \alpha\, r^2/a^2)$], p_H is the heating power density including Ohmic and auxiliary heating, p_α is the alpha-particle heating power density, and p_R is the radiation power density including deuterium and impurity bremsstrahlung and impurity radiation due to line transitions, radiative and dielectronic recombination.

For simplicity, the radiation term p_R is computed in the "average ion" approximation of Post et al.,[14] where detailed solutions of the coupled rate equations for the various ionization stages are replaced for each element by a single, fictitious "average ion" which represents a statistical average of all possible charge states of the element. The steady-state, coronal equilibrium values for radiation function L_Z, charge state $<Z>$ and mean-square charge state $<Z^2>$ for the "average ion" are expressed as functions of the electron temperature for many low- and moderate-Z elements in Ref. 14. These simple polynomial fits are employed in our model to compute the radiation, dilution and Z_{eff} corresponding to the impurity fluences Φ_{FS} and Φ_w. It should be noted that the "average ion" formulas were computed for coronal equilibrium, neglecting transport of neutral atoms and ions, whereas the relevant transport times across the radiating edge layer are shorter than typical recombination times. Thus, for purposes of computing tokamak edge (e.g. impurity line) radiation, this model overestimates the ionic charge state and underestimates the radiation. Comparison of our numerical results with experimental data suggest p_R is underestimated by a factor of order unity.

The following procedure is used to obtain a thermal equilibrium. The function $p_H(r)$ is computed from Eq. (1) for given $D_\perp$ and $\kappa_j(r)$, assumed profiles $T_j(r)$ and $n(r) = n_e(r) = n_i(r)$, and computed impurity profiles $n_{FS}(r)$ and $n_w(r)$. The profiles are integrated numerically to obtain the stored energy $W = (3/2) \int dV \, (n_e T_e + n_i T_i)$, net input power $P_{net} = P_H + P_\alpha - P_R = \int dV \, (p_H + p_\alpha - p_R)$, and energy confinement time $\tau_E = W/(P_H + P_\alpha)$, where $\dot{W} \equiv \partial W/\partial t = 0$ in equilibrium. The transport coefficients and χ profile parameter are adjusted (within experimental constraints) to give reasonable values of W, τ_E and required P_H, consistent with the chosen central plasma parameters n_{jo}, T_{jo} and the other characteristics of the shot to be modeled.

We define two figures of merit for evaluating the effect of the impurity radiation on the thermal equilibrium: $\Lambda_G = \delta P_R / P_{tot}$ and $\Lambda_E = (\int dV_E \, \delta p_R) / |\int dV_E \, (rQ_\perp)'/r|$, where $\delta P_R = \int dV \, \delta p_R$ is the incremental radiation due to the impurities and $\int dV_E$ denotes the volume integral across the radiating edge layer. When $\Lambda_G \ll 1$ the impurity radiation does not perturb the global equilibrium, and $\Lambda_E \ll 1$ implies no perturbation of the edge temperature, i.e. the impurity line radiation at the edge is much smaller than the heat flow from the central plasma. The latter condition is postulated to be necessary for development of the H-mode.

RESULTS AND DISCUSSION

A number of simulations have been carried out with parameters modeling JET limiter and X-point discharges in both L- and H-modes. For reasonable choices of the input profiles the model qualitatively reproduces key experimental parameters measured on JET (stored energy, global energy confinement, required input power, radiated power, etc.) and should be useful for the quantitative interpretation of experimental shots for which measured density and temperature profiles are available.

The impurity line radiation in the model is very sensitive to the edge temperature just inside the LCFS. For a wide range of temperature profiles, our model shows that δP_R for typical influxes of low-Z impurities (Be or C) is a small contribution to the power balance in JET (except locally near the X-point, which we are not modelling here). The reason is that the threshold edge temperature above which the ion is fully stripped, and the line radiation vanishes, is quite small for low-Z atoms. For example, in the "average ion" model[14] the peak in the C line radiation function L_Z occurs at $T_e = 100$ eV and L_Z is an order of magnitude smaller for $T_e > 300$ eV. Thus, the radiating edge layer is a small fraction of the plasma radius for low-Z impurities, and typically $\Lambda_G \ll 1$ in L-mode plasmas. After the H-mode transition, the line radiation inside the LCFS from C and Be becomes negligible as the edge temperature pedestal develops. This pedestal develops easily in the absence of high-Z impurities because $\Lambda_E \ll 1$.

The edge power balance is quite different when high-Z impurities are involved. For example, the radiation function L_Z for Ni has a maximum at $T_e = 100$ eV which is 400 times larger than C, and L_Z for Ni drops more slowly with temperature, down by an order of magnitude at $T_e > 2$ keV. Thus, the condition for local cooling, $\Lambda_E \sim 1$, is much easier to obtain.

This observation is relevant to the understanding the dependence of the JET H-mode quality on the FS material. In the experiments with Be-gettered Ni screens, it was found[7] that a Ni influx of approximately $\Phi_{FS} \lesssim 4 \times 10^{18}$ Ni atoms MW^{-1} s^{-1} came from the uncoated gaps of the FS. ICRF H-modes with dipole phasing were obtained, but no monopole H-modes were obtained until the Be screens were installed. It was thought[15] that the Ni influx, combined with the increased edge convection[8] in monopole phasing, prevented the monopole H-mode. In a later experiment with the Be screens, a similar influx of Ni atoms was introduced by laser ablation of a Ni pellet in the plasma edge during an H-mode, resulting in a transition to the low particle confinement H-mode.[16] To understand the role of Ni radiation in these experiments, we have simulated a JET discharge with $P_H = 10$ MW and $\Phi_{FS} = 4 \times 10^{19}$ Ni atoms s^{-1}, for which $<\delta Z_{eff}>_{Ni} = 0.2$. The radiated power due to Ni was 1.6 MW yielding $\Lambda_G \sim 0.2$. More significant for the H-mode was the fact that the radiated power was highly peaked at the plasma edge and $\Lambda_E \sim 1$, which is consistent with the experimentally observed perturbation of the H-mode by the Ni influx.

In summary, the simple thermal equilibrium model described in this paper can be a useful tool in evaluating ICRF antenna designs and operating scenarios from the standpoint of impurity radiation. The effect of ICRF on limiter or divertor impurities could also be included by extending standard SOL models.[11,12]

ACKNOWLEDGEMENTS

We would like to thank J. Jacquinot and M. Bures for their collaboration on the impurity modeling which motivated this work, and for useful discussions on the subject of this paper. This work was supported by U.S. Department of Energy Grant No. DE-FG02-88ER53263.

REFERENCES

1. F. W. Perkins, Nucl. Fusion **29**, 583 (1989).
2. R. Chodura and J. Neuhauser, in *Controlled Fusion and Plasma Heating, Proc. 16th European Conf.*, Venice, Vol. 13B, Part III, p. 1089.
3. R. Van Nieuwenhove and G. Van Oost, J. Nucl. Mater, **162-164**, 288 (1989).
4. J. R. Myra, D. A. D'Ippolito and M. J. Gerver, Nucl. Fusion **30**, 845 (1990).
5. D. A. D'Ippolito, J. R. Myra, M. Bures, and J. Jacquinot, Plasma Phys. Contr. Fusion **33**, 607 (1991).
6. M. Bures, J. Jacquinot, K. Lawson, M. Stamp, H. P. Summers, D. A. D'Ippolito, and J. R. Myra , Plasma Phys. Contr. Fusion **33**, 937 (1991).
7. M. Bures, J. Jacquinot, M. Stamp, D. Summers, D.F.H. Start, T. Wade, D. A. D'Ippolito and J. R. Myra , Nucl. Fusion **32**, 1139 (1992).
8. D. A. D'Ippolito, J. R. Myra, J. Jacquinot, and M. Bures, in *Plasma Physics and Controlled Nuclear Fusion Research, 1992* (IAEA, Vienna, 1993) paper IAEA-CN-56/E-3-9; and D. A. D'Ippolito, J. R. Myra, J. Jacquinot, and M. Bures, submitted to Phys. Fluids.
9. J. R. Myra and D. A. D'Ippolito, presented at this conference.
10. J. Jacquinot, private communication.
11. M. Hugon, P. P. Lallia, P. H. Rebut, JET-R(89) 14.
12. P. C. Stangeby and G. M. McCracken, Nucl. Fusion **30**, 1225 (1990).
13. W. Engelhardt and W. Feneberg, J. Nucl. Mater. **76 & 77**, 518 (1978).
14. D. E. Post, R. V. Jensen, C. B. Tarter, W. H. Grasberger, and W. A. Lokke, Atomic Data and Nucl. Data Tables **20**, 397 (1977).
15. M. Bures, private communication.
16. M. Bures, D. J. Campbell, N. A. C. Gottardi, J. J. Jacquinot, M. Mattioli, P. D. Morgan, D. Pasini, and D. F. H. Start, Nucl. Fusion **32**, 539 (1992).

FAR FIELD ICRF SHEATH FORMATION ON WALLS AND LIMITERS

J.R. Myra and D.A. D'Ippolito
Lodestar Research Corp., Boulder, Colorado 80301

ABSTRACT

A body of theoretical and experimental evidence suggests that unabsorbed wave energy in ICRF fast wave heating or current drive experiments can result in deleterious edge interactions. In this paper, a model describing the formation of far field sheaths due to FW interaction with material surfaces is presented. Near conductors which do not conform to flux surfaces, an incoming fast wave (FW) causes the generation of a slow wave (SW) component. The $E_\parallel$ of the SW drives an rf sheath, in a manner similar to what has been previously discussed for antenna (near field) sheaths. To assess the importance of the proposed mechanism, a heuristic scaling model of the resultant sheath voltage V is developed. The model illustrates the important dependences of V on the single pass absorption, edge density, rf frequency, FW cutoff location and limiter/wall geometries.

INTRODUCTION

Experimental evidence on many tokamaks indicates that in regimes of poor single pass absorption, fast wave (FW) interaction with walls and limiters can lead to increased impurity influxes. On JET, before the introduction of a beryllium first wall, poor single pass absorption of the ICRF (ion cyclotron range of frequencies) waves was found to lead to disruption in severe cases.[1] Direct measurements in the edge plasma also showed that both magnetic and electric fields in the edge were larger in regimes of poor single pass absorption.[1,2] In ASDEX[3] rf induced impurities were found in the plasma that could not have come from near field sheath[4-10] processes. Theoretical work has been motivated by these and related observations[11-13]. In this paper, the interaction of ICRF with a conducting wall or limiter of a tokamak is considered, with attention focused on the generation of slow waves (SW) and subsequent formation of far field rf sheaths.

The present hypothesis may be summarized as follows. i) A launched FW from the antenna is not completely absorbed on one pass and therefore encounters the inner wall or limiter on the high field side (HFS), perhaps after tunneling through the HFS FW cutoff. More generally, a FW encounter with a HFS or LFS surface could occur after multiple reflections from either FW or minority cutoffs. ii) Because the flux surfaces at the edge do not match the conducting surface exactly, a SW component is generated at the wall,[11] with concomitant $E_\parallel$. iii) The $E_\parallel$ is screened along the field line by the plasma in a characteristic penetration length resulting in a net rf voltage drop along the field line $V = \int ds\, E_\parallel$. iv) This rf voltage V is the drive for rf sheaths which act to increase impurity sputtering[7,8] and generate convective cells[14] by the same mechanisms that are operative in the near field case[4-10]. The remainder of this paper expands upon the physics of items i) - iii).

© 1994 American Institute of Physics

THEORY

The radial (x) Poynting flux S launched from the antenna is related to the launched FW electric field by $S = k_x c^2 |E_{\perp 0}|^2 / 16\pi\omega$ where $E_{\perp 0}$ is the FW electric field amplitude, $\perp$ is perpendicular to the equilibrium magnetic field $\mathbf{B} = B\mathbf{b}$, and ω is the rf frequency. Note that in high B field tokamaks, the large ω implies larger $E_{\perp 0}$ to deliver the same S. The single pass transmission coefficient T_{sp} determines how much wave energy gets through to the other side of the torus. The wave may encounter a FW cutoff on the HFS, depending on n_e and $k_\parallel$. Just before the FW cutoff (if there is one) the FW electric field is $E_\perp = T_{sp}^{1/2} E_{\perp 0}$. A distance x from a flat conducting wall where $E_\perp$ vanishes, the FW electric field is approximately $E_\perp = \sin(k_x x) E_{\perp 0} \sim k_x x E_{\perp 0}$, or

$$\mathbf{E} = \mathbf{e}_{fw} (32\pi\omega k_x T_{sp} S)^{1/2} x a/c \qquad (1)$$

where $a \sim \exp(-\int dx |\text{Im } k_x|)$ is the tunneling factor from the FW cutoff to the wall, and $\mathbf{e}_{fw}$ is the FW polarization unit vector with $\mathbf{b} \cdot \mathbf{e}_{fw} \sim 0$. Here $\mathbf{B} \cdot \mathbf{n} = 0$ where $\mathbf{n}$ is the unit normal to the wall which lies in the plane $x = 0$.

The flux surfaces at the edge generally do not match the conducting surface due to the presence of bumper limiters or simply the effect of plasma shaping. A simple model describing this feature is a wall given by the function $x = \xi(y)$, where $\xi \sim 0$ except near a bump of height $\xi \sim h$. Thus h represents the characteristic mismatch of the flux surface and the conducting wall. The significance of the bump is that the tangential component of E given by Eq. (1) is no longer zero on the conducting surface $x = \xi$. Consequently, as noted by Perkins[11], a SW component must be introduced so that

$$(\mathbf{E}_{fw} + \mathbf{E}_{sw}) \cdot \mathbf{e}_t = 0 = (\mathbf{E}_{fw} + \mathbf{E}_{sw}) \cdot \mathbf{e}_z. \qquad (2)$$

Here $\mathbf{e}_t$ and $\mathbf{e}_z$ are the two unit vectors tangent to the wall. SW generation in the edge plasma has also been discussed by Brambilla,[13] invoking a different mechanism. In the present model, we can estimate $E_{\perp sw} \sim E_{\perp fw}(x=h)$, where the latter is given by Eq. (1). Once there is a SW, then the SW polarization $\mathbf{e}_{sw}$ introduces a nonzero component $E_\parallel \equiv \mathbf{E} \cdot \mathbf{B}$. Equation (2), when applied to the model geometry at a characteristic point $x \sim h$, leads to an explicit, albeit complicated, algebraic result for $E_\parallel$ that is easily coded and is roughly of order

$$E_{\parallel 0} \sim E_{\perp fw} (\epsilon_\perp/\epsilon_\parallel)^{1/2} (n_\parallel^2 - \epsilon_\perp)^{1/2}/n_\parallel. \qquad (3)$$

Here $n_\parallel \equiv k_\parallel c/\omega$, $\epsilon_\perp \equiv 1 - \omega_{pi}^2/(\omega^2 - \Omega_i^2)$, and $\epsilon_\parallel \equiv 1 - \omega_{pe}^2/\omega^2$.

Finally, the SW dispersion relation determines the penetration of $E_\parallel$ along the field line, accounting for its screening by the plasma. This calculation is done for specified $k_\perp \sim \pi/w$ where w is the typical bump width (characteristic poloidal dimension of the flux surface mismatch) and results in an evanescent $|k_\parallel| \equiv 1/L_\parallel$. Once the size and parallel penetration of $E_\parallel$ are known, the driving voltage along the field line for rf sheaths is computed as

$$V = 2\int ds\, E_\parallel = 2\, E_{\parallel 0}\, L_\parallel, \qquad (4)$$

where the factor of 2 arises because we assume that a given field line intersects two similar limiting surfaces.

The voltage V has been computed from Eqs. (1), (2), (4) and the full expression in place of Eq. (3). The results have been compared to a 2D code with surprisingly good agreement. The 2D code work will be presented elsewhere.

QUANTITATIVE EXAMPLE

The normalized V is shown in Fig. 1 for parameters characteristic of JET. Note the exponential decay of V below the cutoff density for the dipole case (the monopole case is not cut off), and the gradual reduction of V at very high edge densities due to increased plasma screening (reduced $L_{\|}$). Anticipating that V is the driving parameter for rf sheath edge interactions,[4-10] these results correspond qualitatively to the JET experiments as follows. i) Monopole operation produces more edge interaction than dipole. ii) Edge interaction is stronger in cases of poor single pass absorption ($T_{sp} \sim 1$) than in cases of good absorption ($T_{sp} \ll 1$). Quantitatively, employing $S \sim 0.3$ kW/cm^2 and $T_{sp} \sim 0.9$ in poor single pass regimes, $V \lesssim 40$ volts, corresponding to a sputtering energy $\sim Z(eV+3T_e)$ including the Bohm sheath, which exceeds typical sputtering thresholds of order 25 - 100 eV.

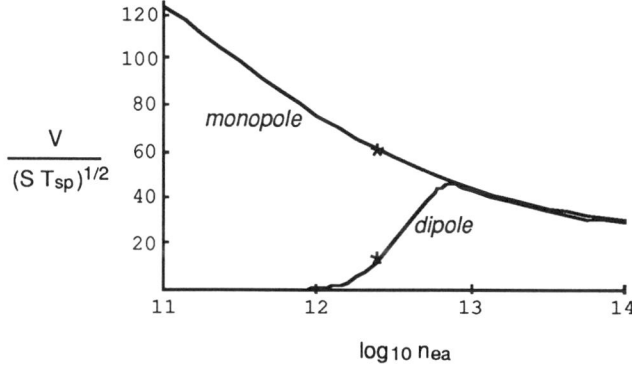

Fig.1 Sheath driving voltage V(volts) vs. edge density n_{ea}(cm^{-3}). V is normalized to the launched Poynting flux S(kW/cm^2) and the single pass transmission coefficient T_{sp}. Parameters are chosen to be characteristic of JET, except for n_{ea}, where a large range is shown to better illustrate the underlying physics. The asterisks show characteristic points for JET.

DISCUSSION

In the preceding sections, it has been shown that linear wave physics, in the presence of plasma and a general conducting boundary, results in an induced rf voltage drop along the equilibrium magnetic field lines in the vicinity of the bounding surface. This voltage V is of interest, based on previous studies of near field sheaths[5], because it is expected to drive rf sheaths which cause rectification and the generation of a dc sheath voltage of a comparable magnitude. However, the present situation is more complicated than the near field case, because the rf sheaths so produced modify the boundary conditions seen by the waves. A model of wave interaction with edge sheaths has been proposed by Berro et al.[12]. In the far field case of interest here, it may be necessary to further treat SW generation and sheath formation self-consistently.

Self-consistency of plasma-wall interactions may also be of importance in describing the far field sheaths. Absorption and desorption processes determine the plasma density near the surface. Thus conditioning could be important because if n_e at the surface were very low, rf sheaths would be mitigated by space charge depletion.

We have not discussed here mechanisms by which the driving fields $E_{\perp fw}$ could be enhanced, e.g. locally by nonuniform focusing and globally by eigenmode formation in the multiple pass regime, effects which work to increase V.

In spite of the simplicity of the present model, it seems clear that far field sheath processes may play an important role in low single pass absorption experiments. The typical driving voltage V is seen to be comparable to both the usual $3T_e$ Bohm sheath voltage as well as to sputtering thresholds. It is therefore expected that far field sheath processes will contribute to impurity sputtering and modifications of edge potentials, transport and $E \times B$ convection.

ACKNOWLEDGEMENTS

The authors are indebted to M. Bures and J. Jacquinot for many stimulating conversations on ICRF interaction with the plasma edge, and for sharing JET data and interpretations thereof which were instrumental in motivating the present study. This research was supported by U. S. Department of Energy Grant No. DE-FG02-88ER53263.

REFERENCES

1. M. Bures, K. Avinash, H. Brinkschulte, G. Devillers, J. Jacquinot, S. Knowlton, A. Pochelon, D. Start, J. Tagle, Bul. APS **33**, 2032 (1988).
2. J.A. Tagle, M. Laux, S. Clement, S.K. Erents, H. Brinkschulte, M. Bures, and L. DeKock, Fusion Eng. Des. **12**, 217 (1990).
3. J.V. Hofmann, et al., Fusion Eng. Des. **12**, 185 (1990).
4. F.W. Perkins, Nucl. Fusion **29**, 583 (1989).
5. J.R. Myra, D. A. D'Ippolito and M. J. Gerver, Nucl. Fusion **30**, 845 (1990).
6. R. Chodura, Fusion Eng. Des. **12**, 111 (1990); R. Chodura, Phys. Fluids **25**, 1628 (1982).
7. D.A. D'Ippolito, J.R. Myra, M. Bures, and J. Jacquinot, Plasma Phys. Cont. Fusion **33**, 607 (1991); and references therein.
8. M. Bures, J. Jacquinot, K. Lawson, M. Stamp, H.P. Summers, D.A. D'Ippolito, J.R. Myra, Plasma Phys. Cont. Fusion **33**, 937 (1991).
9. R. Majeski, P. Probert, T. Tanaka, D. Diebold, R. Breun, M. Doczy, R. Fonck, N. Hershkowitz, T. Intrator, G. McKee, J. Sorenson, in *AIP Conference Proceedings 244 - Radio Frequency Power in Plasmas*, Charleston SC, USA (AIP, New York, 1992), p. 322.
10. R. Van Nieuwenhove and G. Van Oost, J. Nucl. Mater. **162-164**, 288 (1989).
11. F.W. Perkins, Bull. APS **34**, 2093 (1989), paper 6S6.
12. E. Berro, B. Fried, D. Holland, G. Morales, presented at the Boulder Workshop on ICRF/Edge Physics, March 30 - April 1, 1988 (unpublished).
13. M. Brambilla et al., in *Plasma Physics and Controlled Nuclear Fusion Research 1990* (IAEA, Vienna, 1991), Vol. I, p. 723.
14. D.A. D'Ippolito, J.R. Myra, J. Jacquinot and M. Bures, in *Plasma Physics and Controlled Nuclear Fusion Research 1992* (IAEA, Vienna, 1993), paper IAEA-CN-56/E-3-9; and D.A. D'Ippolito, J.R. Myra, J. Jacquinot and M. Bures, submitted to Phys. Fluids.

ICRF SYSTEM AND PLASMA PERFORMANCE OF THE IGNITOR EXPERIMENT*

F. Carpignano, B. Coppi, P. Detragiache, M. Nassi and L. Sugiyama

Massachusetts Institute of Technology, Cambridge, Ma 02139

ABSTRACT

An ICRF System, with a wide range of frequencies ($\nu \lesssim 210$ MHz) and a power delivered to the plasma $P_J \lesssim 18$ MW, has been adopted for the high field, high density, Ignitor experiment. The resulting injected heating can be employed to optimize the current density distribution for the attainment of D-T ignition, as well as to increase the range of plasma regimes that can be investigated by the machine. Thus, for example, a significant level of fusion power from the D-He^3 reaction can be produced and the access to high-β second stability region can be investigated.

MOTIVATION FOR USING ICRH

The Ignitor experiment[1] involves an advanced, compact, high magnetic field machine (see Fig. 1 and Table I) with the purpose of investigating D-T fusion burning and ignition conditions. The main goal of this experiment is to attain fusion burning ignition at relatively low peak temperature ($T_{io} \simeq T_{eo} < 15$ keV) avoiding the need for reliance on an injected heating system.

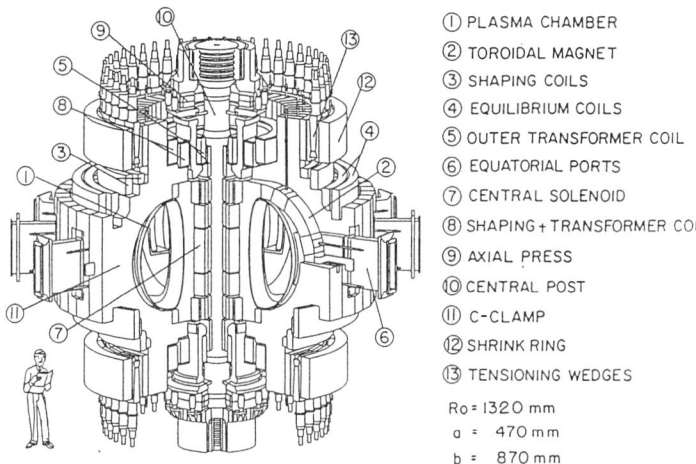

① PLASMA CHAMBER
② TOROIDAL MAGNET
③ SHAPING COILS
④ EQUILIBRIUM COILS
⑤ OUTER TRANSFORMER COIL
⑥ EQUATORIAL PORTS
⑦ CENTRAL SOLENOID
⑧ SHAPING + TRANSFORMER COIL
⑨ AXIAL PRESS
⑩ CENTRAL POST
⑪ C-CLAMP
⑫ SHRINK RING
⑬ TENSIONING WEDGES

Ro = 1320 mm
a = 470 mm
b = 870 mm

Figure 1. The Ignitor Machine

Since ignition can be attained by ohmic heating alone, the adopted ICRH system has the role of a backup[1]; in fact this is the only kind of injected heating that has been shown to be effective in high density regimes. This system should be available[1-2] to control the temperature evolution and the current density profiles by local heating of the edge region or global heating of the entire plasma column, in order to keep the volume of the region where $q < 1$ ($Vol_{q=1}$) small by reducing current penetration (inducing higher temperatures in the outer region of the plasma) and "freezing in" the central current density, to accelerate the attainment of ignition, and to investigate the suppression of possible sawtooth oscillations by fast particle stabilization in relatively low density discharges. In addition this system can be employed to extend the conditions under which ignition is possible, to investigate regimes where the second stability

---

* Sponsored in part by the U.S.Department of Energy, by E.N.E.A. of Italy and A.S.P. of Piedmont.

region of finite-β plasmas can be achieved, to produce a significant power level from D-He^3 fusion reactions, to reduce the Volt-sec required to reach high currents I_p in dense plasmas, to achieve significant level of α particles in low plasma current and field discharges and to perform current drive experiments in low density discharges ($n_e < 2 \times 10^{20}$ m^{-3}).

Table I. Reference Design of the Ignitor Ult Configuration

$R_o \simeq 1.32$ m	Major radius of the plasma column
$a \times b \simeq 0.47 \times 0.87 \text{m}^2$	Minor radii of the plasma cross section
$\delta_G \simeq 0.4$	Triangularity of the plasma cross section
$I_p \lesssim 12$ MA	Plasma current in the toroidal direction
$I_\theta \lesssim 9$ MA	Plasma current in the poloidal direction
$B_T \lesssim 13$ T	Vacuum toroidal field at R_o
$\Delta B_T \lesssim 1.5$ T	Paramagnetic (additional) field produced by I_θ
$\langle J_\phi \rangle \lesssim 9.3$ MA/m^2	Volume-average toroidal current density
$\bar{B}_p \lesssim 3.75$ T	Mean poloidal field
$I_p \bar{B}_p \lesssim 45$ MN/m	Confinement strength parameter
$q_\psi \simeq 3.5$	Edge magnetic safety factor at $I_p = 12$MA
$V_o \simeq 10$ m^3	Plasma volume
$S_o \simeq 36$ m^2	Plasma surface area
$P_J \lesssim 18$ MW	Injected heating power (ICRF at $100 \lesssim \nu \lesssim 210$ MHz)

ICRF SYSTEM DESCRIPTION

An ICRF system with a relatively wide range of frequency (100 MHz $\lesssim \nu \lesssim$ 210 MHz) and maximum power of 18 MW has been proposed[2] for Ignitor. The wide band of frequencies is consistent with different values of B_T and allows operation with different scenarios ($\omega = \omega_{CD}, \omega_{CHe^3}, 2\omega_{CT}$ or ω_{CH} at full field). The absorbed power in the different scenarios is showed in Table II.

Table II. Comparison of the Heating Scenario[4] ($n_{eo} = 10^{21}$ m^{-3}, $T_{eo} = T_{io}$, $k_\parallel = 10$ m^{-1})

Plasma Composition	Heating Scenario	Frequency (MHz)	Damping per pass	
			$T_o \simeq 4$ keV	$T_o \simeq 10$ keV
49% D 49% T 1% α	$\omega = 2\omega_{CT}$	140	16%	51%
47% D 47% T 1% α 5% H	$\omega = \omega_{CH}$	210	> 37%	> 70%
49%D 49% T 1% α	$\omega = \omega_{CD}$	105	> 8%	> 46%

The ICRF system is composed of 6 antennae completely inserted into recesses in the plasma chamber with RF return currents in the vessel and the port walls. Recess sides are fanned to decrease the importance of return current, deleterious for coupling. Four amplifiers (1.5 MW each), phased to achieve the required dipole radiation, drive the antenna module. In this way the system is intrinsically wide-band and there is automatic matching by frequency and tuning stub variation. A power up to 4 MW can be coupled by each antenna module, provided the antenna can withstand 40 kV and the last closed flux surface is less than 6 cm away from the outward face of the screen. The antenna is composed of straps grouped in poloidal pairs; each strap is fed by a pair of coaxial lines of 50 Ω. The strap length is less or comparable with $\lambda/4$ at an appropriate intermediate frequency and its width is chosen to give 50 Ω impedance. The straps are composed of Inconel tube and plates plated with 50 μm of Ag. The strap currents are in phase poloidally and out of phase toroidally.

NUMERICAL SIMULATIONS

To study the effects of ICRF heating we consider a free boundary plasma with an axisymmetric plasma transport and MHD model (TSC code[3]). We analyze different plasma mixtures, ranging from all D to 50:50 D:T. ICRH is applied during the initial current ramp and the flat top phase of the discharge over the following range of plasma parameters: density $n_{eo} \simeq 2.5$ to 11×10^{20} m^{-3}, plasma current $I_p \simeq 5$ to 12 MA, toroidal magnetic field $B_T(R_o) \simeq 7$ to 13 T, peak plasma temperature $T_o \simeq 2$ to 20 keV and edge plasma temperature $T_a \gtrsim 50$ eV. We consider current ramps where the plasma is started from the inside or the outside limiter. In each case ICRF heating is applied when the plasma shape and dimensions allow good coupling with the antenna. This happens about 1 s after the breakdown.

For D-T discharges with a 12 MA maximum plasma current the limiting conditions for ohmic ignition are represented by a value of the plasma effective charge of $Z_{eff} \simeq 1.6$ or by a total nonohmic thermal transport $\chi^{noh} \simeq 1.7$ times the reference value ($\chi^{noh} \simeq \chi_i^{noh} + \chi_e^{noh} = \chi_e^{noh}(1+\gamma_i); \gamma_i^{ref} = 0.5; \gamma_i^{max} \simeq 1.5$) see Table III, cases 1 and 5. The use of ICRF heating clearly improves the plasma performances, as shown cases 2-3 and 6, and allows ignition under significantly more degraded conditions ($\tau_E \simeq 0.4$ s) as reported in cases 4 and 7.

Table III. Effect of ICRH on ignition for different cases.

Case	1	2	3	4	5	6	7
		Large Z_{eff}			Large Transport		
	Ohmic	Low P_J	Large P_J		Ohmic	Large P_J	
P_J (MW)	/	5.0†	15.0‡	15.0‡	/	15.0‡	15.0‡
Z_{eff}	1.6	1.6	1.6	1.9	1.2	1.2	1.2
γ_i	0.5	0.5	0.5	0.5	1.5	1.5	1.75
I_p (MA)	12.0	12.0	11.2*	12.0	12.0	12.0	12.0
n_{eo} (10^{20}m^{-3})	11.0	11.0	10.0*	11.0	14.5	11.0	11.0
t_{ign} (s)	5.3	3.3	2.7	3.0	5.0	3.0	3.0
β_p	0.14	0.15	0.18	0.20	0.16	0.19	0.21
W_{int} (MJ)	12.8	14.0	13.6	18.0	14.1	16.6	19.5
T_{eo} (keV)	13.0	13.2	13.4	15.5	10.7	15.1	16.7
τ_E (s)	0.570	0.555	0.525	0.485	0.545	0.425	0.390
P_{OH} (MW)	8.7	8.8	7.2	8.1	9.6	7.0	7.0
P_α (MW)	22.5	25.2	25.9	37.0	25.8	39.2	50.0
$Vol_{q=1}$ (%)	> 10	2.3	0.5	1.5	> 10	1.4	1.5

† $P_J \simeq 5$ MW for $t > 1.2$ s.
‡ $P_J \simeq 5$ MW for $1.2 < t < 1.8$ s, 10 MW for $1.8 < t < 2.4$ s and 15 MW for $t > 2.4$ s.
* Lower values due to ignition during the ramp.

The use of ICRH allows the achievement of significant level of α-heating power ($P_\alpha \simeq P_J \simeq 10$ MW) even in the lower current (8 MA) plasma scenario, the attainment of burning plasma conditions ($P_\alpha \simeq 38$ MW) or ignition in the 10 MA scenario, and to obtain high levels of performance at lower fields and currents, and thus operation at lower mechanical and thermal stresses and longer discharges (see Table IV where the data are reported at the time t_I when ignition or the maximum P_α is achieved).

Pure Deuterium plasmas will be used to test and adjust the machine operation as well as to carry out studies on high temperature plasma physics. In particular, they will be used to explore the access to the high-β second stability region, which is important to identify the parameters of advanced fusion reactors. In this case, Ignitor will operate with high q_o in relatively low B_T and I_p regimes. The low aspect ratio ($\simeq 2.8$) and the elongated cross section for which Ignitor is designed makes it well suited to reach finite-β conditions, in the central part of the plasma column, with relatively high plasma pressure. In this case ICRH must be relied upon as the primary heating system and the plasma can be maintained over relatively long times (10 to 40 s), in view of the fact that the Ignitor magnets are initially cooled down to 30 K. In this respect Ignitor is as suitable as a superconducting facility to explore long pulse operation, while having a set of desirable, such as the low aspect ratio and the ability to operate over a wide range of B_T values.

D-He^3 plasmas will be used to test the capability of a compact, high magnetic field experiment to approach burning conditions with advanced fuels. The machine will be operated at high B_T and I_p, with moderate D density $(n_o \simeq 3 \times 10^{20}$ m$^{-3})$ and He^3 minority up to 5%. The strong ohmic heating system that can bring a D-T plasma to ignition is used in conjunction with the ICRF system to raise the peak temperature of the thermal plasma up to 10 keV. An high ICRF power density will be used to create an energetic tail of He^3 ions, in order to enhance the D-He^3 fusion reactivity. A preliminary analysis, based on current experimental results[4], shows that a few MW of D-He^3 fusion power should be produced in these regimes.

Table IV. 8 and 10 MA Plasma Current Scenarios.

I_p (MA)	8	8	10	10	10^a	10.6
P_J (MW)	0	10	0	10	0	15
B_T (T)	10	10	11	11	11	11
t_I (s)	6.0	4.0	5.6	5.0	5.5^b	2.5^b
n_{eo} (10^{20})	8.0	8.0	9.0	9.0	9.0^b	9.5^b
T_{eo} (keV)	6.4	11.4	9.7	17.2	13.2^b	12.5^b
τ_E (s)	0.71	0.35	0.53	0.28	0.57^b	0.53^b
P_α (MW)	2.3	10.3	10.8	38.0	20.0^b	23.3^b
P_{oh} (MW)	6.1	4.0	7.1	4.0	5.8^b	6.7^b
Z_{eff}	1.2	1.2	1.2	1.2	1.2	1.2
Q^c	2.0	4.0	8.0	14.0	∞^b	∞^b

^{a} $\gamma_i = 0$
^{b} at ignition.
^{c} $Q \equiv (5 \times P_\alpha/(P_{OH} + P_J - dW/dt)$ under transient conditions.

ICRF heating may save up to $\simeq$ 6 Vs of the ignition Volt-sec requirement (that can be as high as 35 Vs) because it allows earlier ignition by 2 to 3 s, sometimes during the current ramp, with a lower value of I_p. Therefore the internal inductive consumption (lower l_i, less penetrated current density profile) and the resistive consumption (due to earlier ignition, larger temperature and therefore lower plasma resistivity, and also a larger fraction of bootstrap current) are reduced. However, if the plasma current follows that of the ohmic discharge, then the saving in Volt-sec at the end of the pulse is of the order of 2 Vs, since it is due only to the reduction in the resistive part (lower plasma resistivity and larger bootstrap current).

SUMMARY AND CONCLUSION

The ICRH system adopted for Ignitor allows it to extend the conditions under which ignition is possible, it can help to avoid deleterious effects of possible sawtooth oscillations by reducing the size of the region where $q < 1$, and allows the achievement of burning or even ignition conditions in low plasma current scenarios. Furthermore it significantly reduces the Volt - sec required to reach a certain level of plasma performance, it allows the investigation of the second stability region of finite-β plasmas in long-pulse discharges, and it can lead to the production of significant level of power from D-He^3 fusion reactions.

ACKNOWLEDGEMENT

It is a pleasure to thank C. Bolton and W. Sadowski for promoting the 6th Boulder Workshop on the Ignitor High Power ICRF Antenna Design and Physics, A. Aamodt and the Lodestar Research Corporation for organizing it, and all the participants of the Workshop for their valuable contributions, many of which have been incorporated into this work.

REFERENCES

1. B. Coppi, M. Nassi, and L.E. Sugiyama, *Physica Scripta*, **45**, 112 (1992).
2. J. Jaquinot, "ICRH Heating in Ignitor," 6th Boulder Workshop, Boulder (CO), (1993).
3. S.C. Jardin, N. Pomphrey, and J. Delucia, *J. Comp. Phys.*, **66**, 481 (1986).
4. J. Jacquinot, G.J. Sadler, and The Jet Team, *Fusion Technology*, **21**, 2254 (1992).

PHAEDRUS-T TOKAMAK PROBE MEASUREMENTS

J. Sorensen, D. Diebold, N. Hershkowitz, T. Intrator, R. Majeski*,
J. Meyer, P. Probert, G. Winz
Dept. Nuc. Eng. and Eng. Physics, University of Wisconsin-Madison

ABSTRACT

Phaedrus-T is an ISX class tokamak ($B \leq 1T$, $R \simeq 0.9m$, $a \simeq 0.25m$, $I_p \simeq 100$ kA, $n_e(0) \simeq 2 \times 10^{19}/m^3$ and $T_e(0) \sim 500$ eV), which began operation in 1989. Measurements of the Phaedrus-T edge plasma parameters have been made with triple probes [1-6], swept Langmuir probes[2], swept emissive probes, self- emissive probes, reciprocating probes [7,8], capacitive probes [9] and shadowed probes of various sizes and materials. ICRF current drive studies are the focus of the Phaedrus-T program; and, consequently, most Phaedrus-T probe measurements have been of ICRF/edge plasma interactions. Data showed strong ICRF modifications of the edge plasma when an antenna with a conventional stainless steel Faraday Shield was used; greatly reduced ICRF/edge coupling when 1/4 inch boron nitride side limiters were placed on the Faraday Shield; and no significant change in the ICRF/edge coupling when the Faraday Shield was later removed (even though there was a significant reduction in plasma impurities).

BEGINNING OF OPERATION

Assembly of the Phaedrus-T tokamak began in the summer of 1988. The first plasma was achieved in the fall of 1989. During 1990, the rf system (low field side, two current strap fast wave antenna encased in a stainless steel Faraday Shield (FS)) and probes became operational. By the end of 1990, 100 kW of rf power had been coupled to the plasma and plasma edge parameters had been measured with triple probes[1] and a capacitive probe[9]. The triple probes are able to measure ion saturation current (I_{is}), floating potential (ϕ_f) and electron temperature (T_e) with a frequency response $\leq$ 1 MHz. The capacitive probe measures ϕ_f with a frequency response up to 40 MHz.

At the beginning of 1991, systematic measurements of the Scrape Off Layer plasma parameters during rf were made by three poloidally spaced triple probes all located 80 cm away toroidally from the antenna[2,3,4]. The "Top Probe" scanned vertically along field lines that did not map to the antenna. The "Out Probe" came in horizontally on the low field side and mapped to the antenna 10° above midplane. The "Bottom Probe" was vertically positioned and mapped to the bottom of the antenna. The probes that mapped to the antenna typically saw a ϕ_f increase of 10-30 V and a T_e increase of 5-15 eV. The top probe saw few effects due to rf and the intensity of H_α from the carbon limiter

* Present address: Princeton Plasma Physics Laboratory, Princeton, NJ

was also only slightly affected by the rf.

In addition, the measured I-V characteristics of a swept single tip Langmuir probe showed an increase in ϕ_f during rf. In ohmic plasmas, direct measurements of plasma potential (ϕ_p), made by sweeping an emissive probe, were found to be in rough agreement with both ϕ_p as inferred from the swept Langmuir probe data and ϕ_p as calculated using triple probe data and the formula $\phi_p=3.5T_e/e+\phi_f$. The ϕ_p calculated by this formula increased during rf and such an increase lead to increased sputtering and impurities[3].

Carbonization was tried in February 1991 and boronization with Trimethylboron in March. Boronization was more effective in reducing impurities and the recycling coefficient than carbonization. Since then, the machine has been periodically boronized, approximately once a week.

ADDITION OF BN SIDE LIMITERS

During the summer of 1991, boron nitride (BN) side limiters were installed on each side of the antenna and rf edge effects were greatly reduced[3,4,6]. In particular, the measured ϕ_f during rf decreased by as much as 20 V and the increase in T_e was usually less than 10 eV. The calculated ϕ_p was not significantly changed by rf. The increase in the intensity of iron light with rf decreased by a factor of 5. Again, only probes that mapped to the antenna saw rf effects.

Also at this time, in collaboration with the University of Texas, measurements of ohmic plasma parameters were made with reciprocating triple probes[7,8] of various tip sizes and materials. Measurements made by the stand alone University of Texas probe system and Phaedrus-T probe system were compared and found to be in approximate agreement. Earlier that year, fluctuation measurements made by probes and Beam Emission Spectroscopy were found to be in good agreement[10].

REMOVAL OF FS

In the fall of 1991, the stainless steel FS was removed, BN tiles were placed on the current straps and a BN septum was installed between the straps. This resulted in essentially no change in the plasma edge parameters[5]. The main benefit was a reduction in the intensity of iron impurity light by a factor of 10. Because the relative reduction in iron was the same in ohmic as it was during rf, it is thought that this reduction was mainly due to the removal of the nearest source of iron to the plasma rather than rf sheath effects between the FS blades, which were aligned toroidally and not parallel to the total magnetic field.

ALFVÉN WAVE HEATING

Near the end of 1991, it was shown that by creating a 5 minute helium glow discharge between discharges resulted in a lower recycling coefficient, and as a result, greater density control and a cleaner plasma. A 5 minute helium glow discharge between main plasma discharges became standard practice after this.

At the start of 1992, heaters were installed on the BN antenna side limiters so that they could be heated to remove trapped impurities. Also, the antenna current straps were moved closer to the plasma. Soon after the side limiters were baked, apparent Alfvén wave heating (a 30% drop in loop voltage and 15% increase in $T_e(0)$) was observed. Edge parameters during Alfvén wave heating are shown in Fig. 1. For the top probe, I_{is} and T_e increased, which suggests and is consistent with global heating. ϕ_f also increased. For a probe that maps to the antenna, I_{is} and T_e go up more and ϕ_f drops. This drop in ϕ_f is probably due to the slow time response of the probe and the presence of a ϕ_f oscillation at the rf frequency. The intensity of H_α light from the carbon limiter doubles, which is further indication of a global increase of particle flux to the limiter. Uniquely, for Phaedrus-T, during Alfvén wave heating, the line averaged density was seen to decrease slightly during rf, even though the fueling rate was held constant. These Alfvén wave heating results were obtained at 7.2 MHz in H_2.

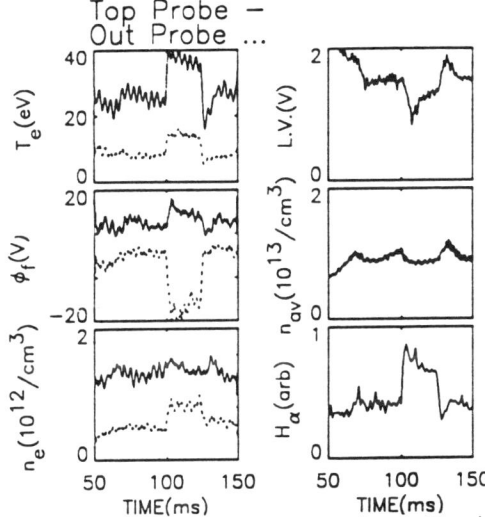

Fig. 1 The left hand column shows T_e, ϕ_f and n_e = electron density for the top probe (solid line) and out probe (dotted line). In the right hand column is L.V.=loop voltage, n_{av} = line averaged density and the intensity of H_α light from the carbon limiter.

FOUR STRAP ANTENNA

In the summer of 1992, a 4 strap antenna enclosed with BN tiles, (many of them unheated) was installed. As a result, even in ohmic discharges more impurities were present. During this time different techniques of boronizing were tried (different flow rates) in order to obtain better conditioning. As the machine surface conditions improved, edge characteristics similar to those seen for the Alfvén wave heating with the two strap antenna were observed. With deuterium at 4.2 MHz, H_α and both the top and out probes' I_{is} and T_e all increased which was evidence of a global rf-plasma interaction. However, no drop in loop voltage was observed. The density increase during rf was greater than with the old two strap antenna and this may have been the reason no loop voltage drop was seen. The large amount of BN covering the antenna was suspected of causing this larger density increase.

Also during this time, it was found that when a triple probe was inserted far enough into the plasma that it went into thermionic electron emission, the plasma did not disrupt as it had previously. This presents the possibility of using such self-emissive probes on Phaedrus-T for direct plasma potential relatively deep (for probes) in the plasma.

At the end of 1992, some of the BN was removed to reduce the density rise during rf and the current straps were moved farther away from the plasma. With this configuration, measurements of the edge parameters showed varied results. Also, shadowed probe tips were used for the first time. Differences between ϕ_f and T_e as measured from the two tips were found.

In early February, 1993, pure He discharges were obtained in Phaedrus-T. The uncontrolled density rise during rf disappeared even though the He plasmas had more impurities present than hydrogen and deuterium plasmas. This strongly suggests that for Phaedrus-T the uncontrolled rf density rise is due to fueling by hydrogen/deuterium rather than impurities.

It is possible that the large amount of BN is outgassing throughout the discharge and during rf produces a low density plasma near the antenna and possibly between the current straps. To investigate this, single tip probes were positioned poloidally above and below and at the same toroidal position as the center of the antenna. I_{is} for the top antenna probe saw no change with rf, while the probe directly below the antenna saw an increase in I_{is} of up to 1 mA, which corresponds to n_e of $\approx 5 \times 10^{11}/\text{cm}^3$. At some locations below the antenna where there was no measurable I_{is} current before rf, during rf I_{is} of $\leq .5$ mA was measured. From swept probe data, T_e was ≈ 15 eV and did not change during rf. When a resistive tip was used, an oscillating ϕ_f of less than 15 V in amplitude was measured near the antenna. n_e of $5 \times 10^{11}/\text{cm}^3$, measured below the antenna, if present between the septa, would be sufficient to account for ~ 15 kW of rf power.

Supported by U.S. DOE Grant No. DE-FGO2-88ER53264

1 D. Diebold et al., Rev. Sci. Instrum. **61**, 2870 (1990).
2 D. Diebold et al., Nucl. Fus. **32**, 2040 (1992).
3 R. Majeski et al., accepted by Fusion Eng. and Design (1993)
4 R. Majeski et al., AIP Conf. Proc. **244**, 322 (1991).
5 J. Sorensen et al., accepted by Nucl. Fus. (1993).
6 T.Tanaka, Ph.D. Thesis, University of Wisconsin, Madison (1993).
7 D. Diebold et al., J. Nucl. Mater. **196-198**, 789 (1992).
8 J. Pew et al., Bull. Am. Phys. Soc., **36**(9),(1991).
9 E-Y. Wang et al., Rev. Sci. Instrum. **62**, 1494 (1991).
10 H. Evensen et al., Rev. Sci. Instrum. **63**, 4928 (1992).

ANOMALOUS ELECTRON STREAMING DUE TO WAVES IN TOKAMAK PLASMAS

S.D. Schultz, A. Bers, and A.K. Ram
Plasma Fusion Center, MIT, Cambridge, MA 02139

1. Abstract

The motion of circulating electrons in a tokamak interacting with electrostatic (lower hybrid) waves is given by a guiding center Hamiltonian and studied by numerical integration. On surfaces with rational safety factor q, superposition of modes with degenerate values of the parallel mode number $n + (m/q)$ is shown to result in electron streaming perpendicular to the magnetic field.

2. Introduction

Recent work by Kupfer [1,2] on the chaotic electron dynamics induced by waves demonstrated that in the presence of four waves wherein two are of identical parallel phase velocity, an electron in resonance with the waves will have a drift in the radial direction. The work we present expands upon these results, with the intent of formulating a theory explaining the existence and properties of this "streaming". The problem is complicated by the necessity of visualizing dynamics in a four-dimensional phase space including both parallel and radial motion.

3. Hamiltonian Guiding Center Theory

We use a Hamiltonian description of the electron guiding center motion that was derived in detail in [2], which makes use of several previous guiding center theories [3,4,5]. The model is based on a tokamak with an MHD equilibrium in the low-beta limit, and a low inverse-aspect ratio $\epsilon = r/R$. The electron is circulating, with an orbit which remains near its initial flux surface. Although magnetic field shear can be included in the derivation, it is neglected for purposes of this work. The phase space for the guiding center motion is reduced to four dimensions by gyroaveraging and assuming the magnetic moment μ to be a constant: this requires that our study be restricted to waves in the low-frequency, long-wavelength limit.

The derivation gives two sets of canonical coordinates for the guiding center, (z_1, p_1) and (z_2, p_2). In terms of the familiar guiding center position (ψ, θ, ϕ) and parallel velocity $v_\parallel$, and omitting correction terms of order ϵ, one finds

$$z_1 \approx \phi \qquad p_1 \approx mR_o v_\parallel$$
$$z_2 \approx \phi - q_o \theta \qquad p_2 \approx e\psi \qquad (1)$$

Here q_o is the safety factor. The Hamiltonian in the absence of wave perturbations is

$$H_o = \frac{1}{2m_e R_o^2} p_1^2 + \mu B_o. \qquad (2)$$

This is simply the form of a free particle Hamiltonian in one dimension; drifts are accounted for in the order-ϵ corrections of (z_1, p_1, z_2, p_2), which are too complicated to give in the limited space here.

To this Hamiltonian we add a small electrostatic field perturbation, $-e\Phi$, in the form of a few discrete plane waves with identical frequency.

$$\Phi = \sum_{n,m} \Phi_{nm} \cos(n\phi + m\theta + k_\psi \psi - \omega t) \tag{3}$$

n and m are mode numbers, and k_ψ is found using the dispersion relation of lower-hybrid waves. Using the approximate relations given in (1), we obtain

$$\Phi = \sum_{n,m} \Phi_{nm} \cos(k_1 z_1 + k_2 z_2 + k_\psi p_2 - \omega t) \tag{4}$$

with $k_1 = n + (m/q_o)$ and $k_2 = -(m/q_o)$. The dynamics of interest occur when the safety factor q_o is a rational number, which makes the perturbation periodic in z_1 and z_2, and allows us to choose different integers n, m giving the same value of the parallel mode number k_1 but different values of k_2 and k_ψ. The electron is in resonance with a wave when $d(k_1 z_1 + k_2 z_2 + k_\psi p_2 - \omega t)/dt = 0$. The unperturbed Hamiltonian gives $\dot{z}_1 = (m_e R_o^2)^{-1} p_1$ and the other three coordinates constant, so resonance occurs for $p_1 = (m_e R_o^2)(\omega/k_1)$.

4. Numerical Integration

The equations of motion derived from the perturbed Hamiltonian were used to evolve the coordinates $z_1(t), p_1(t), z_2(t), p_2(t)$ in time. We select units so that m_e, ϵ, R_o, and ω are all unity. The safety factor is chosen to be $q_o = 2$. To reproduce the radial streaming observed by Kupfer [2], we choose four electrostatic modes with mode numbers $(n, m) = (1,1)(1,2)(2,0)(2,2)$; in the new coordinates $(k_1, k_2) = (3/2, -1/2)(2, -1)(2, 0)(3, -1)$. This case is degenerate: two modes have the same parallel mode number $k_1 = 2$ with different values of k_2 and k_ψ. The resonance condition in simplified units is $p_1 = 1/k_1$, so that the resonance surfaces in phase space are at $p_1 = 2/3, 1/2,$ and $1/3$. The amplitude of each wave is the same and was chosen so that the separatrix layers for these three waves barely overlap, which creates a stochastic layer around all three of the resonance surfaces in (z_1, p_1) phase space. The initial condition of the coordinates is chosen to lie in this resonance region.

The figures describe the motion observed for two cases with different initial conditions in (z_2, p_2). The two points are started on the same flux surface ($p_{2o} = 0$) but at different poloidal angles ($z_{2o} = 0, \pi$) separated by $\Delta\theta = \pi/2$.

Figures 1(a) and 1(b) show the time series $p_1(t)$ (which corresponds to parallel velocity) for these two cases. In the $z_{2o} = 0$ case, p_1 is observed to fluctuate rapidly throughout the resonance region, spending a roughly equal amount of time near each of the three resonances. However, in the $z_{2o} = \pi$ case, the electron quickly moves into an orbit close to the $p_1 = 1/2$ resonance, which is degenerate, and stays there.

Figures 2(a) and 2(b) show the time series $p_2(t)$ (corresponding to the flux coordinate) in the two cases. In the first case, motion in the p_2 direction is wildly fluctuating, but with occasional periods of directed, non-chaotic motion. Over long times, these periods of streaming add up to a slow drift in the radial direction. In the second case, this streaming motion is nearly continuous, and the rapid fluctuations are no longer visible on this scale, which is an order of magnitude larger than on the previous figure.

5. Interpretation of Results

The electron streaming appears to be related to the patterns of constructive and destructive interference of the two degenerate waves. Let us explicitly add two such waves:

$$\Phi_o \cos(k_1 z_1 + k_{2a} z_2 + k_{\psi a} p_2 - t) + \Phi_o \cos(k_1 z_1 + k_{2b} z_2 + k_{\psi b} p_2 - t)$$
$$= 2\Phi_o \cos(\Delta k_2 z_2 + \Delta k_\psi p_2) \cos(k_1 z_1 + \bar{k}_2 z_2 + \bar{k}_\psi p_2 - t), \qquad (5)$$

where $\bar{k}_2 = (k_{2a} + k_{2b})/2$, $\Delta k_2 = (k_{2a} - k_{2b})/2$, $\bar{k}_\psi = (k_{\psi a} + k_{\psi b})/2$, and $\Delta k_\psi = (k_{\psi a} - k_{\psi b})/2$.

We would like to see how this interference pattern affects the phase space of the guiding center motion. Unfortunately, a surface of section in this phase space is given by a four-dimensional mapping, which is impossible to visualize. But if a canonical transformation can be found so that two of the four phase space variables are nearly constants of the motion, a plot of this mapping in the phase plane of the other two coordinates is an approximate surface of section. From these numerical integration experiments, it was discovered that a plot of p_1 versus $(k_1 z_1 + \bar{k}_2 z_2 + \bar{k}_\psi p_2)$ taken as a surface of section contains what appear to be KAM surfaces and islands on the three resonance lines. Figures 3(a) and 3(b) show these surfaces for the two cases described above.

We observe that, in the $z_{2o} = \pi$ case, a set of invariant tori appear near the degenerate resonance on this "surface of section". It is easy to show with (5) that $z_{2o} = \pi, p_{2o} = 0$ is a point where the degenerate waves interfere destructively. Thus the extra KAM surfaces appear because the electron perceives no waves to interact with at this resonance.

The details of the canonical transformation giving this approximate surface of section are under investigation. This is expected to reveal that there are quantities which are very nearly conserved by the streaming motion.

Work supported by DoE Grant No. DE-FG02-91ER-54109, NSF Grant No. ECS-88-22475, and in part by the Magnetic Fusion Science Fellowship Program.

References:
[1] K. Kupfer, A. Bers, and A.K. Ram, in *Research Trends in Physics: Nonlinear and Relativistic Effects in Plasmas*, V. Stefan (ed.), (New York: Amer. Inst. of Phys.) 1992, pp. 670-715.
[2] K. Kupfer, Ph. D. thesis, Massachusetts Institute of Technology, May 1991.
[3] R.G. Littlejohn, J. Plasma Phys. **29**, 111 (1983).
[4] J.R. Cary and R.G. Littlejohn, Annals of Physics **151**, No. 1, 1 (1983).
[5] J.D. Meiss and R.D. Hazeltine, Phys. Fluids B **2**, 2563, (1990).

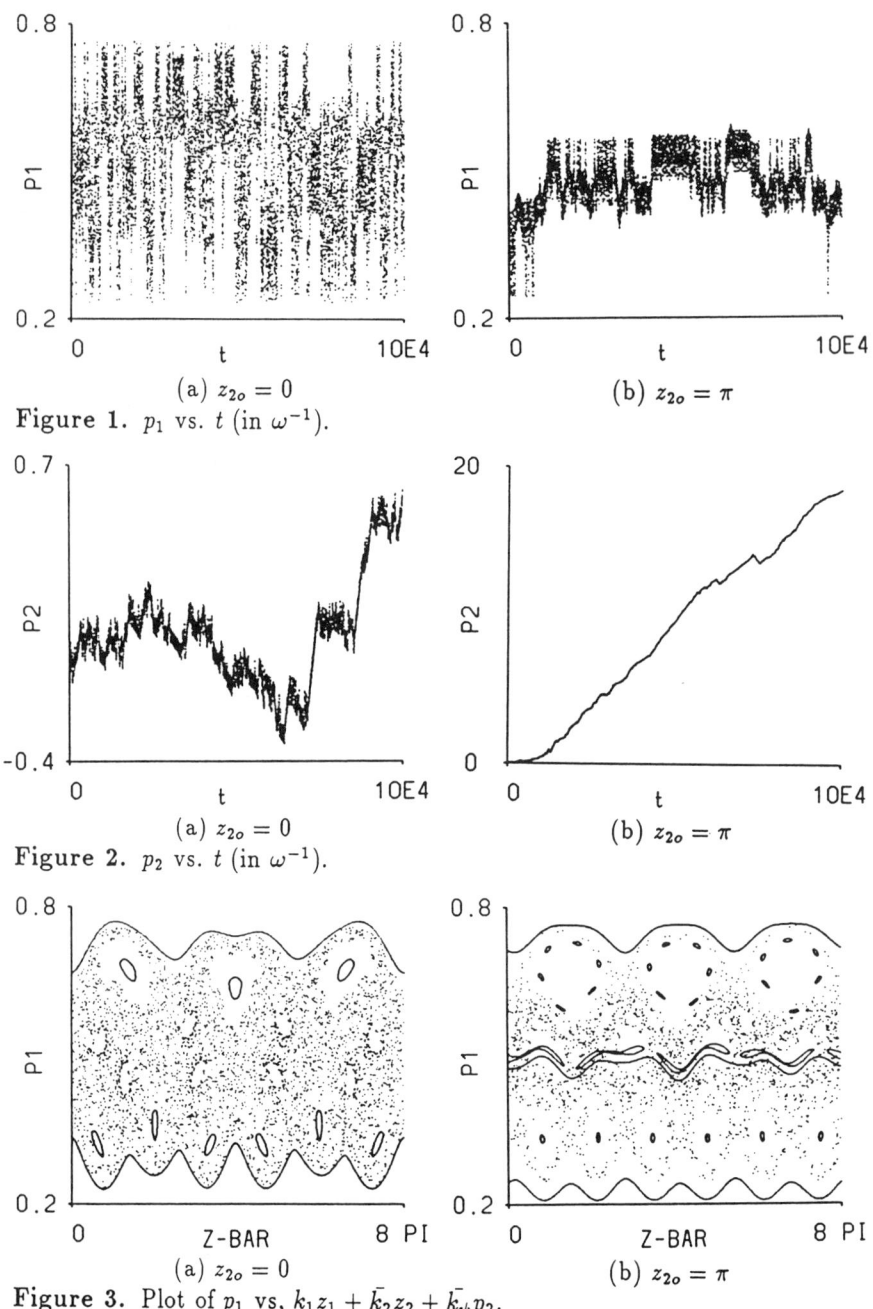

Figure 1. p_1 vs. t (in ω^{-1}).

Figure 2. p_2 vs. t (in ω^{-1}).

Figure 3. Plot of p_1 vs, $k_1 z_1 + \bar{k}_2 z_2 + \bar{k}_\psi p_2$.

MICROWAVE REFLECTOMETRY FOR ICRF COUPLING STUDIES ON TFTR

J. B. Wilgen, G. R. Hanson[†], T. S. Bigelow, D. B. Batchelor, I. Collazo[‡],
D. J. Hoffman, M. Murakami, D. A. Rasmussen, and D. C. Stallings
Oak Ridge National Laboratory, Oak Ridge, TN 37831-8072

S. Raftopoulos and J. R. Wilson
Plasma Physics Laboratory, Princeton University, P. O. Box 451, Princeton, NJ 08543

ABSTRACT

A dual-frequency differential-phase reflectometer has been developed for use in ICRF power coupling studies on TFTR. This system has been optimized for measurements of the electron density profile in the edge-gradient region, where density fluctuations are large. Initial proof-of-principle measurements demonstrate that this is an effective way to measure the electron density profile in the plasma-edge region. A new reflectometer launcher is presently being installed on the center axis of the bay-K ICRF antenna on TFTR, along with the associated waveguide transmission line. This will allow direct measurement of the edge-density profile within the high-power-density environment of the ICRF antenna where density profile modification might be expected.

INTRODUCTION

The coupling of ICRF power to the plasma is sensitive to details of the density profile in the plasma edge region, and the high power density ICRF environment can potentially alter the edge-density profile, at least locally in the immediate vicinity of the ICRF antenna. Theoretical ICRF antenna coupling calculations show that density profile changes immediately in front of the ICRF antenna can result in changes in the antenna loading by a factor of 2 to 3. Similar loading changes have been observed experimentally during ICRF heating. Reflectometer measurements of the edge-density profile can be used to correlate changes in antenna loading with shifts of the fast wave cutoff density. Consequently, there is significant interest in obtaining detailed measurements of the shape of the edge-density profile, and it is particularly important that the measurement be performed in the ICRF antenna environment.

DUAL-FREQUENCY DIFFERENTIAL-PHASE REFLECTOMETER

A dual-frequency differential-phase reflectometer has been developed for use in ICRF power coupling studies on TFTR[1]. This system has been optimized for measurements of the electron density profile in the edge-gradient region where density fluctuations are large. A differential-phase measurement was chosen because the multiplicity of fringes is thereby greatly reduced, and phase fluctuations arising from density fluctuations in the plasma are also significantly reduced. Both of these attributes are essential for reliable phase-tracking of multiple-fringe phase data.

A block diagram of the reflectometer as configured for a proof-of-principle measurement on TFTR is shown in Fig. 1. To provide the capability to measure the edge-density profile in the range between 1.0×10^{12} and 3.0×10^{13} cm^{-3} in high field (4.5–4.9 T) IRCF-heated TFTR plasmas, the frequency range of 91–117 GHz was chosen, corresponding to extraordinary mode polarization. Starting with a swept frequency source at low frequency, (8.0–12.4 GHz), upconversion and frequency multiplication

[†]Oak Ridge Associated Universities
[‡]Georgia Institute of Technology

(doubler and tripler) are used to provide the frequencies of interest. In this way, the frequencies of two probing signals are simultaneously swept from 91 to 117 GHz while maintaining a fixed frequency separation of 250 MHz. This frequency spacing determines the radial separation of the dual cutoff layers in the plasma, which should be small in comparison with the radial correlation length of the plasma density fluctuations if a reduction in the differential-phase fluctuation level is to be effected. Amplitude fading in the reflected signal amplitude is removed through the use of constant-phase limiting amplifiers. Heterodyne detection is used to measure the differential phase delay between the two signals, which can then be used to reconstruct the shape of the density profile.

INITIAL RESULTS ON TFTR

To facilitate testing of this reflectometer in a realistic environment, it was attached via waveguide switches to share diagnostic access with the existing TFTR fluctuation reflectometer[2], an instrument that was specifically designed to investigate density fluctuations in the interior of the plasma. This system utilizes corrugated cylindrical waveguide to launch highly directional gaussian beams that are focused and directed into the plasma with scannable mirrors.

Differential-phase data obtained with this quasi-optical viewing system are shown in Figs. 2 and 3. For these measurements, the gaussian beams (transmitting and receiving) are aimed to intersect at $R = 3.3$ m, resulting in a saturated amplitude for the received signal whenever the reflection surface is in the range of 3.1 to 3.5 m, the region characterized by good overlap of the two beams. When the location of the plasma edge region, $R_0 + a$, is systematically scanned from 3.25 to 3.53 m, the differential-phase data changes in the expected fashion, as illustrated by the phase data in Fig. 2a. Note in particular that the differential phase typically shows a variation of only 2 to 4 fringes as the frequency is swept from 91 to 117 GHz. For most of this data, even the differential phase exhibits substantial phase fluctuations, with a typical magnitude of 1 radian rms or larger, but this does not present a serious problem in tracking the average trend in the phase. Edge-density profiles reconstructed from this data using an algorithm based on an extension of Doyle's method[3] are shown in Fig. 2b. Similar data for a selection of shots with the same plasma size but different density (resulting from variation in NBI power from 0 to 27.5 MW) are shown in Fig. 3, demonstrating the expected variation in the differential phase as the density is varied.

The differential phase can be considered as consisting of two contributions, one associated with the shape of the profile through the local density gradient-dependent plasma dispersion, and another arising from the location of the plasma edge region. For the data shown in Fig. 2, shifting the location of the plasma edge by ≤ 30 cm while maintaining nearly the same profile shape is expected to contribute $\leq \frac{1}{2}$ fringe to the total phase shift (i.e., the beat wavelength for 250 MHz frequency spacing is 1.2 m). This indicates that dispersive effects related to the profile shape represent the largest contribution to the differential phase. At the present time it is not clear to what extent these two contributions to the total phase can be separated. Although the reflectometer provides good information on the shape of the density profile in the edge region, it appears that the location of the plasma edge can be resolved only through detailed comparisons with other TFTR diagnostics.

Data obtained from full-sized ($R_0 + a = 3.6$ m) ICRF-heated plasmas show a similar shape for the edge-density profile. Attempts to observe modification of the edge-density profile during ICRF heating have not revealed any measurable changes, suggesting that if changes are occurring they are local to the antenna environment.

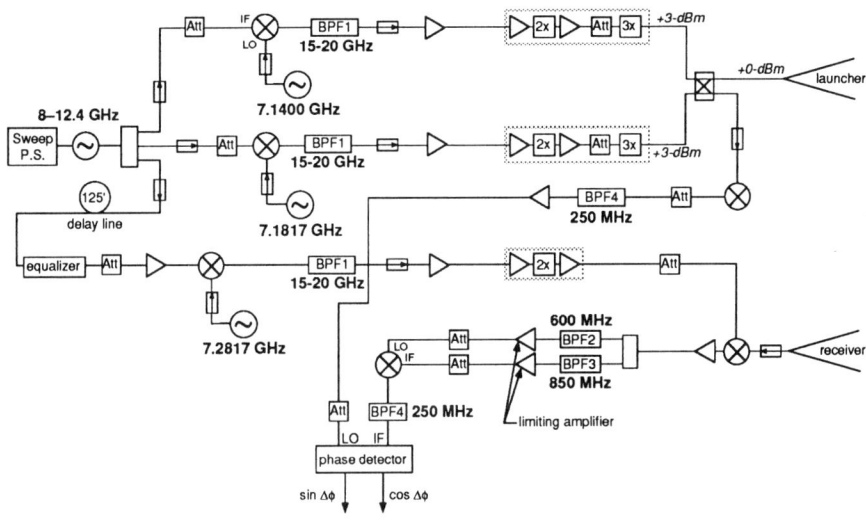

Fig. 1. A block diagram of the dual-frequency differential-phase reflectometer as configured for the proof-of-principle demonstration measurement on TFTR.

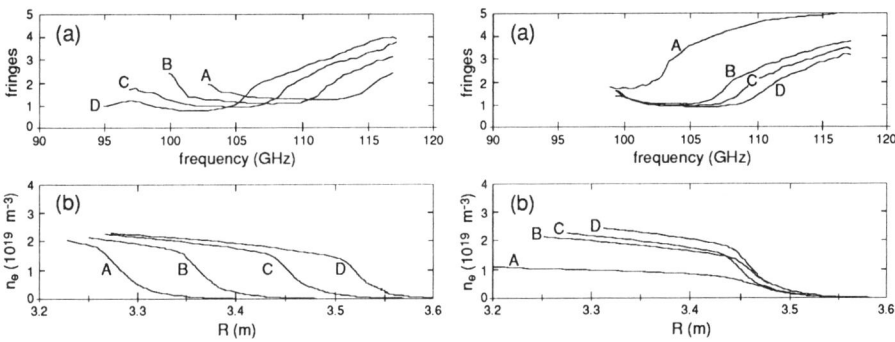

Fig. 2. (a) Measured differential-phase shift as a function of frequency and (b) reconstructed edge-density profiles for a plasma size scan where, for cases A, B, C, and D, the outer edge of the plasma ($R_0 + a$) is located at 3.25, 3.35, 3.44, and 3.53 m, respectively. Plasma conditions include $B = 4.5$ T ($I_{tf} = 67$ kA), $I_p = 1.6$ MA, $\bar{n}_e = 3.2$–3.6×10^{19} m^{-3}, $P_{NBI} = 22.5$ MW, and $P_{ICRF} = 0$.

Fig. 3. (a) Differential-phase shift and (b) reconstructed edge-density profiles for a collection of shots where the density varies with the neutral beam power. Cases A, B, C, and D correspond to $P_{NBI} = 0$, 17.5, 22.5, and 27.5 MW and $\bar{n}_e = 1.2$, 2.8, 3.2, and 3.7×10^{19} m^{-3}, respectively. Other plasma conditions include $R_0 + a = 3.44$ m, $B = 4.5$ T ($I_{tf} = 67$ kA), $I_p = 2.0$ MA, and $P_{ICRF} = 0$.

NEW LAUNCHER AND WAVEGUIDE TRANSMISSION LINE

A new launcher and waveguide transmission line are presently being installed on TFTR. The bay-K ICRF antenna was designed with a central diagnostic port that provides access for the reflectometer launchers on the center axis of this two-strap antenna. To eliminate the effects of spurious reflection on the reflectometer phase measurements, a pair of oversized WR-90 rectangular waveguides are used for the transmitting and receiving antennas. The launcher apertures are recessed 3 mm behind the front surface of the Faraday shield for the ICRF antenna. Stainless steel waveguides are necessary to limit disruption forces; a single quartz window is used for the vacuum feedthrough. "Tall-guide" polarization is used to achieve acceptable waveguide cross-coupling (<- 40 db measured) between the closely spaced waveguides at the vacuum window.

With the exception of a downtapered section of WR-10 waveguide immediately outside the vacuum window, which serves to filter higher-order modes, the remainder of the waveguide run consists of 26 m of WR-90 waveguide leading to the reflectometer electronics located in the test cell basement. Transmission losses are reduced more than 50% by using the tall-guide polarization, resulting in an estimated round-trip transmission loss of 12 db, excluding the window/launcher assembly. For the tall-guide polarization, miter bend losses are measured to be approximately 0.4 and 0.8 db for the H-plane and E-plane bends, respectively.

CONCLUSION

Initial proof-of-principle measurements demonstrate that the dual-frequency differential-phase reflectometer is an effective way to measure the electron density profile in the plasma edge region where density fluctuations are large. An ICRF-antenna-mounted launcher and the associated waveguide transmission line are presently being installed in the bay-K ICRF antenna on TFTR. This will allow direct measurement of the edge-density profile within the high power density environment of the ICRF antenna, where density profile modification might be expected.

ACKNOWLEDGMENTS

We are indebted to E. Mazzucato, R. Nazikian, and M. McCarthy of PPPL, who generously shared their diagnostic access, without which the proof-of-principle measurement would not have been possible. We thank K. Young and N. Bretz of PPPL and S. L. Milora of ORNL for their support and encouragement. This research was sponsored by the Office of Fusion Energy, U.S. Department of Energy, under contract DE-AC05-84OR21400 with Martin Marietta Energy System, Inc. This research was supported in part by an appointment to the U.S. Department of Energy Fusion Energy Postdoctoral Research Program administered by the Oak Ridge Institute for Science and Education.

REFERENCES

1. G. R. Hanson, J. B. Wilgen, T. S. Bigelow, I. Collazo, and C. E. Thomas, Rev. Sci. Instrum. **63**, 4658 (1992).
2. E. Mazzucato, R. Nazikian and the TFTR group, Proc. 19th EPS Conference on Controlled Fusion and Plasma Physics, Innsbruck, June 1992, V2, p. 1055. [E. Mazzucato, R. Nazikian, N. Bretz, M. McCarthy, and A. Nagy, Rev. Sci. Instrum. **63**, 4657 (1992).]
3. E. J. Doyle, T. Lehecka, N. C. Luhmann, Jr., W. A. Peebles, and the DIII-D Group, Rev. Sci. Instrum. **61**, 2896 (1990).

ION CYCLOTRON RESONANCE HEATING
FOR THE PLASMA SEPARATION PROCESS

A. Compant La Fontaine
DCC/DPE/SPEA, CEA-C.E./Saclay
91191 Gif Sur Yvette Cedex - France

ABSTRACT

The selective heating near the ion cyclotron resonance (ICR) in order to separate isotopes by the plasma separation process is investigated both experimentally and theoretically on the ERIC device[1,2]. The electromagnetic field induced by an half-turn helicoidal antenna is calculated for an uniform hot and collisional plasma. The axial wave number k_z spectrum of the rotating electric field E_+ exhibit the presence of two waves which characteristics determine the heating. The motion equation of the ions is resolved in the approximation of the guiding center hypothesis, so that the final heating can be directly obtained ; also the comparison with the experiment is satisfactory.

1. INTRODUCTION

On the ERIC device the separation of isotopes of various stable elements (11) has been carried out with high separation factors for some of them and the selectivity of the heating has been measured[3]. In order to study the limitation effects on the selectivity of the heating leading to smaller separation factors observed for some isotopes, we intend here to calculate the heating of the resonance species and compare it to experimental results.

2. THEORY

2.1. Equations of the field

The electromagnetic field excited by an inductive antenna is calculated following the method developed by Mc Vey[4]. The plasma is supposed cylindrical, homogeneous, hot, collisional and the antenna is disposed out of the plasma, and the axial magnetic field $\vec{B_0}$ is assumed constant. We use the Fourier transform decomposition for the fields to solve Maxwell equations in the zones I (plasma) and II and III (vacuum) defined on Fig.1. The Fourier components of the field are expressed in the plasma by the complex Bessel funcions J_n, and in the vacuum by the second species modified Bessel functions I_n and K_n. For example, in the plasma the Fourier components E_z of the axial electric field write as function of the radius r :

$$E_z(r,n,k) = c_1 \cdot J_n(k_+ r) + c_2 \cdot J_n(k_- r) \quad (1)$$

, where n and k are the azimuthal and axial mode and k_+ and k_- the transverse wave numbers of the fast and slow wave. We get the 10 constants c_j by taking the tangential fields continuous at the interfaces.

2.2. Movement equations

The ion trajectories are calculated according to the guiding center approximation[5] (i.e. $\partial E/\partial r \ll E/r_L$). The motion equations can be solved by setting $X = x+iy$:

$$\ddot{X}+i\dot{X} = \frac{\omega}{B_0}[\Re e(E_x)+i\Re e(E_y)] = \frac{\omega}{B_0}\int_{-\infty}^{+\infty}[f(k)\cos(\Omega-kv)t + g(k)\sin(\Omega-kv)t]\cdot dk \quad (2)$$

The solution of (2) gives the Larmor radius of the ion ($r_L(t=0) = r_{L_0}$)

$$r_L = |-i\cdot r_{L_0} e^{-i(\omega t+\theta)} + \frac{1}{2\pi B_0}\int_{-\infty}^{+\infty}\frac{L(\omega,\Omega,t,k)}{(\omega+\Omega-kv)\cdot(\omega-\Omega+kv)}\cdot dk| \quad (3)$$

At the resonance, i.e. for $\Omega = \omega+kv$, we have :

$$r_L = |-i\cdot r_{L_0} e^{-i(\omega t+\theta)} + \frac{1}{\omega B_0}[E_+\omega t\cdot e^{i[(\Omega-\omega)t-(\theta+\varphi)]}+E_-^*\sin\omega t\cdot e^{-i(\Omega t+\theta-\varphi)}]|$$
$$(4)$$

, where the electric field can be decomposed into two components : one, E_+ rotating in the same direction of the ions and the other, E_- in the opposite direction, with $E_\pm = (E_x \pm iE_y)/2$.

3. RESULTS

The experimental measurement of the heating of ^{44}Ca obtained with an energy analyser versus the antenna frequency is shown on Fig.2. The calculation of $|E_+|$ is represented on Fig.3 versus the axial (a) and radial (b) positions. The penetration of the field is good in the main plasma region inside the antenna zone. At higher densities i.e. for $n_e \gtrsim 5\cdot 10^{12}$ cm^{-3} (broken lines), the screening of the plasma begins to be strong as expected. It is found that the strongest mode is n=-1 (rotates with the ions) and the slow wave (smaller values of $k_\pm$) is in general dominant. In addition the dissymmetry of E_+ at the center of the antenna is due to that of the spectrum in k_z (in the cold plasma hypotesis, E_+ is symmetrical along z). The spectrum of $|E_+(k_z)|$ in k_z, where $E_+ = \int E_+(k_z)\cdot e^{ik_z z} dk_z$, is shown on Fig. 4, where two waves propagating in the opposite direction appear for $k_z = -2.3$ and $+1.1$ m^{-1}.

The final transverse ionic temperature at the collector calculated according (3) by integration over phases φ and velocity distribution functions is represented on Fig.2, and the agreement with the

experimental results is satisfactory. In addition it is found that the heating versus f_a and around the maximum value is due to two effects : the resonant (or in phase) and non-resonant (or dephased) heating. The resonant heating occurs for $f_a \simeq f_{c_{44}}$, which is more efficient for the maximum of E_+ with the smallest values of k_z (here $+1.1$ m^{-1}) so that the major part of the axial velocity distribution function corresponds to the resonant case ($\Omega \rightarrow \omega + kv$). The non-resonant heating takes place for f_a values not close to $f_{c_{44}}$, but for higher values of E_+ ; indeed, E_+ shows important variations with f_a.
This effect is illustrated on Fig.5 where the mean Larmor radius $\langle r_L \rangle$ of ^{44}Ca is represented versus the transit length L_t for the two above cases, i.e. the resonant and non-resonant cases for respectively $f_a =$ 374 and 371 KHz. The saturation of $\langle r_L \rangle$ is found to be quite different in the two cases. Indeed according to (4) we have at the resonance : $r_L \simeq E_+ t/B_0$, but the width of the resonance varies with t^{-1}, see (3), so that the total heating saturates for the transit length : $L_t \gg L_R = k_{res}^{-1}$; for $f_a =$ 374 KHz, we have $L_R \simeq 1$ m, with $E_+ =$ 20 V/m. For the non-resonant case, the saturation length writes according to (3) : $L_t \simeq L_{N.R.} = \langle k \rangle^{-1}$; for $f_a =$ 371 KHz, we find $E_+ =$ 70 V/m and $L_{N.R.} \simeq 1$ m.
In summary the width of the resonance is determined by the $E_+(k_z)$ spectrum and by the E_+ dependance with f_a.

4. CONCLUSION

The model gives a satisfactory estimation of the heating of the resonant species in spite of a high sensitivity of the dielectric tensor $\vec{\epsilon}$ with the plasma parameters. Also the heating of the minor species can be calculated with this method provided that the non-resonant major species be only weakly heated (that is also the case here), so that the change of $\vec{\epsilon}$ along the z axis be negligible.

REFERENCES

1. P. Louvet, in CEA-Saclay, ed. 1989 (Proc. 2nd Workshop on Separation Phenomena in Liquids and Gases, Versailles-France, 1989) Vol.1,p.5.
2. A. Compant La Fontaine and P. Louvet, ibid., Vol.1, p.139.
3. P. Louvet and A. Compant La Fontaine, in Bull. Res. Lab. for Nucl. Reactors, 1992 (Proc. International Symposium on Isotope Separation and Chemical Exchange Uranium Enrichment, Tokyo-Japan, 1990).
4. B. Mc Vey, ICRF Antenna Coupling Theory for a Cylindrically Stratified Plasma, DOE/ET/51013--129, 1984.
5. C. Gil, private communication, 1987.

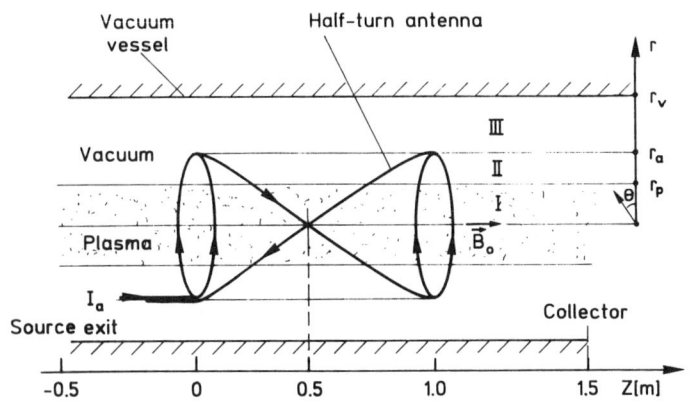

Figure 1 - EXCITATION BY AN EXTERNAL
HALF-TURN INDUCTIVE ANTENNA

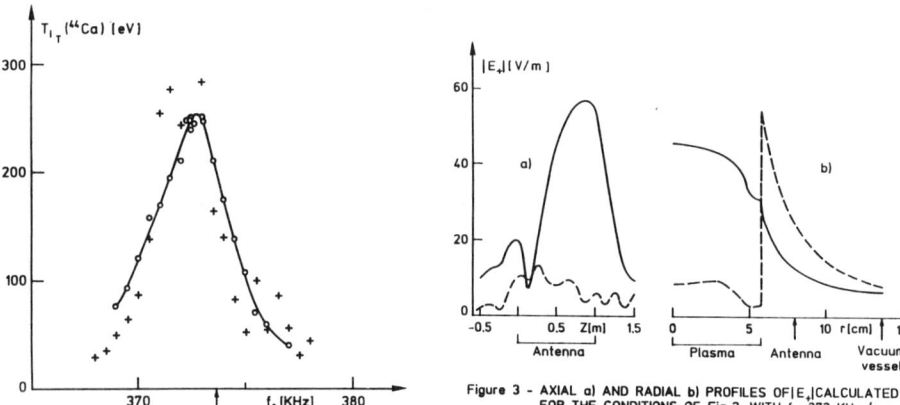

Figure 2 - TRANSVERSE IONIC TEMPERATURE OF ^{44}Ca AT
THE CENTER OF THE COLLECTOR (Z=1.5 m, r=0)
OPERATING CONDITIONS: B_0=1.071T, I_a=50 A; T_{i_0}=3eV
T_e=2.5 eV, n_e=5 × 10^{11} cm^{-3}
EXPERIMENT (o) AND NUMERICAL CALCULATION (+)

Figure 3 - AXIAL a) AND RADIAL b) PROFILES OF $|E_+|$ CALCULATED
FOR THE CONDITIONS OF Fig. 2, WITH f_a=373 KHz (———)
AND FOR n_e=5 × 10^{12} cm^{-3} (– – – –)
a) r =0 , b) Z = +0.5 m

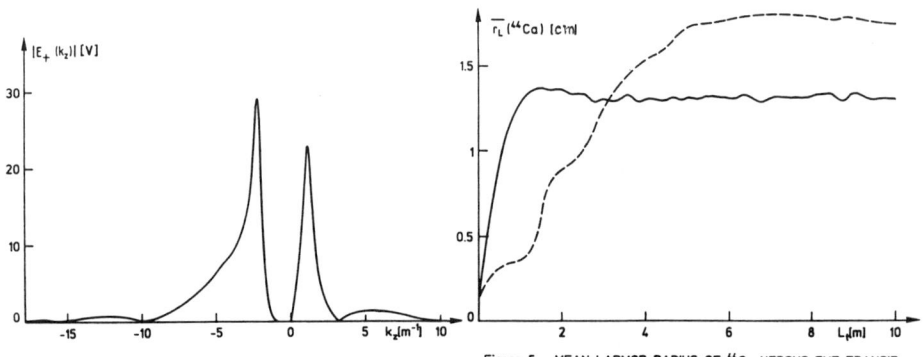

Figure 4 - SPECTRUM OF $|E_+|$ VERSUS k_z FOR THE CONDITIONS
OF Fig. 2, WITH f_a=373 KHz , r=0 AND Z=+0.5 m

Figure 5 - MEAN LARMOR RADIUS OF ^{44}Ca VERSUS THE TRANSIT
LENGTH L_t FOR THE CONDITIONS OF Fig. 2 FOR THE
RESONANT CASE f_a=374 KHz (– – – –) AND THE NON
RESONANT CASE f_a=371 KHz (———)

POLOIDAL ELECTRIC FIELD DUE TO ICRH AND ITS EFFECT ON NEOCLASSICAL TRANSPORT.

Luigi Vacca, PLASMA FUSION CENTER, MIT CAMBRIDGE 01239 MA.

Abstract

We study the transport of a plasma in which a minority ion species is heated by fast Alfven waves. The strong anisotropy of the minority distribution function gives origin to a poloidal electric field [1]. We calculate the poloidal dependence of the electric potential by numerically integrating the leading order minority distribution function. When the amplitude of this field is such that electrostatic trapping is not negligible in comparison to the magnetic trapping then neoclassical transport can be enhanced as found in previous work [2]. The linearized kinetic equations are solved using a variational method in the banana regime. Approximate analytic expressions for the transport coefficients are given.

Kinetic equations and electrostatic field

We study the problem of neoclassical transport of plasma made of two ion species:majority and minority species. The assumptions are:1) only the minority ions are heated by RF 2)the minority density is less than the electron density: $n_e/n_m = 0(\lambda)$, where λ is a small expansion parameter. The following expansion scheme applies: $n_e = n_0 + n_{e1}$; $n_i = n_0 + n_{i1}$; $n_m = 0 + n_{m1}$, where the first order quantities are proportional to λ.
The kinetic equations for ions and electrons are the usual drift drift-kinetic equations:

$$v_\| \mathbf{b} \cdot \nabla f - C(f) = -\mathbf{v_d} \cdot \nabla f - q v_\| E_\| \frac{\partial f}{\partial \epsilon} \quad (1)$$

where the energy is $\epsilon = v^2/m + q\phi/m$. At leading order the minority distribution function is given by the solution of $<C> + <Q> = 0$, where $<C>$ and $<Q>$ are the bounce averaged collision operator and the bounce averaged quasilinear operator. We consider a tokamak with circular flux surfaces in the large aspect-ratio limit: $r/R_0 \ll 1$. The toroidal coordinates are the radial coordinate r, the poloidal angle θ and the toroidal angle ζ.
The zero-th order equation in the poloidal gyroradius for electrons and ions is: $v_\| \mathbf{b} \cdot \nabla f_{a0} - C(f_{a0}) = 0$. This equation does not contain the minority distribution function since we assumed the minority density to be a perturbation with respect to the background. The solution to the zero-th order equation is a Maxwellian of the form

$$f_{a0} = \frac{N_{a0}(r)}{(2\pi T_{a0}(r)/m_a)^{3/2}} exp(-\frac{\epsilon}{T_{a0}}) \quad (2)$$

446 Poloidal Electric Field Due to ICRH

If we integrate the zero order solution over velocity space we obtain the density:

$$n_{a0}(r,\theta) = N_{a0}(r)exp(-\frac{q\phi(r,\theta)}{T_{a0}(r)}) \quad (3)$$

We expand the electrostatic potential in a small parameter Δ: $\phi(r,\theta) = \phi_0(r) + \Delta\phi_1(r,\theta)$. The correction to the leading order Maxwellian is given by the adiabatic response to the poloidally dependent electrostatic field: $n_{a1} \simeq -n_{a0}\frac{q\phi_1}{T_0}$. From now on we assume a strong poloidal electric potential: $q\phi_1/T \simeq O(r/R)$ This is equivalent to set $\Delta = r/R$. The solution of the kinetic equation to first order will yield the perturbed function f_{a1}, the perturbed density of which will be $N_{a1}(r,\theta) = \int f_{a1} d^3v$. Assuming quasineutrality we obtain

$$-\sum \frac{n_{a0}q_a^2}{T_{a0}}\phi_1 = -q_m n_{m1}(r,\theta) \quad (4)$$

Since most trapped particles belong to the minority species we can neglect the majority perturbed density contribution. We adopt the pitch-angle variables (v, ξ_m). The minority density is given by

$$n_m = \int_0^\infty 2\pi dv [\int_{-1}^{-\sqrt{1-\frac{B_m}{B}}} + \int_{\sqrt{1-\frac{B_m}{B}}}^1] \cdot \frac{B(\theta)}{B_m} \frac{v^2|\xi|d\xi f_m(v,\xi_m)}{\sqrt{1-(1-\xi_m^2)B/B_m}}$$

(5)

The reason why the density is dependent on the angle is due to the fact that the distribution function is a function of ξ_m.

Numerical evaluation of minority poloidally varying density.

Minority density is evaluated by integrating an analytic fit [3]

$$f(v,\xi_m) = f^{iso}\frac{2}{\sqrt{\pi}}(\frac{v}{\gamma v_c})^{3/2}\frac{\sum_\sigma exp[-(v/\gamma v_c)^3(\xi_m - \sigma\xi_0)^2]}{\sum_\sigma \Phi((1-\sigma\xi_0)(v/\gamma v_c)^{3/2})} \quad (6)$$

For further detail see reference [3]. List of JET parameters:

- $B_0 = 3.4T; T_e = T_{ions} = T_l = 5.Kev; r/R_o = 0.2; <P> = 0.28Wcm^{-3}$
- $T_m = 164.0Kev; T_h = 533.0Kev; n_e = 5. \times 10^{19}m^{-3}$
- relative minority densities: $n_l = 0.02; n_m = 0.51; n_h = 0.46$

Numerical results are given in fig.1,2, and 3. Fig.1 gives the poloidal potential as a function of the poloidal angle. Fig.2 shows increased trapping of the electrons. Fig.3 shows decreased trapping of the ions.

Neoclassical transport modification due to poloidal potential

In the banana regime a rough estimate of the diffusion coefficient is

$$D_b \approx f_{tr}\nu_t \delta r^2 \qquad (7)$$

Neglecting any change in collision frequency, for the ions we get:

$$D_b \approx exp(\frac{-q\delta\phi}{2r/RT})D_{neocl}. \qquad (8)$$

For the electrons:

$$D_b \approx erf(\frac{q\delta\phi}{T})\sqrt{R/r}D_{neocl}. \qquad (9)$$

In order to obtain more accurate results we need kinetic theory. The first-order version of the drift kinetic equation for electrons or ions is

$$\mathbf{v}_\| \cdot \nabla f_{a1} - (C_{ae}^l + C_{ai}^l)f_{ai} = -\mathbf{v_{da}} \cdot \nabla f_{ao} - (q/T_a)v_\| E_\| f_{a0} + C(f_{a0}, f_m) \quad (10)$$

Two different sources of flux:collisions with minority and poloidal field. Isolating the collision source, the linearized kinetic equation is

$$\mathbf{v}_\| \cdot \nabla f_{a1} - (C_{ae}^l + C_{ai}^l)f_{ai} = C(f_{a0}, f_m) \qquad (11)$$

For ions this contribution should be negligible since the most energetic ions damp their energy on the electrons. The first-order drift kinetic equation not including collisions with minority is:

$$\mathbf{v}_\| \cdot \nabla f_{a1} - (C_{ae}^l + C_{ai}^l)f_{ai} = -\mathbf{v_{de}} \cdot \nabla f_{ao} - (q/T_a)v_\| E_\| f_{a0} \qquad (12)$$

Results of a previous work [2] are given for $X_0 \approx \frac{q\phi}{T}R/r$:
Ware pinch coefficient $\propto (1 + 1.25X_0/)^{1/2}$; reduction in conductivity $\propto (1 + 0.75X_0/)^{1/2}$; ion energy conduction is not practically affected by X_0.

Conclusions and future objectives

- A broadening of the trapped region for electrons results in in an increased Ware pinch coefficient and reduced conductivity. The change in energy conduction is not so dramatic as to be compared to the anomalous part.

- The ion energy conduction might actually slightly improve due to detrapping effect provided by ICRF.

- Collisions effects due to minority for both ions and electrons should be examined by solving the Spitzer problem.

*Work supported by DOE Grant No. DE-FG02-91ER-54109
[1] J.Y.Hsu,V.S.Chan,R.W.Harvey,R.Prater,S.K.Wong,Phys.Rev.Lett. ,53 (1984) pg.564.
[2] C.S.Chang,Phys.Fluids 26(1983) pg.2140.
[3] C.H.Park,W.H Koh,C.S. Chang and D.I. Choi,Plasma Phys.Controlled Fusion 34 (1992) pg.77.

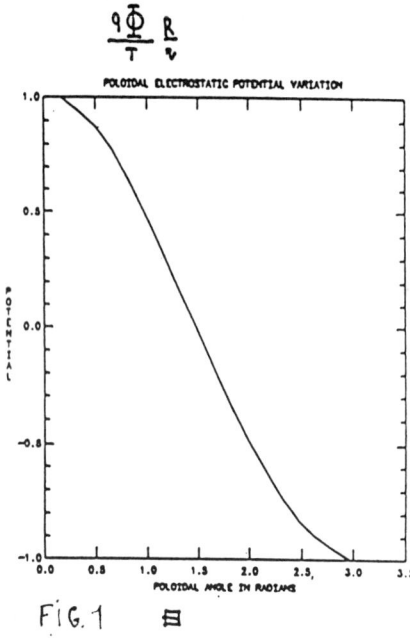

Fig. 1

Fig.1 shows the poloidal dependence of the normalized potential with respect to the aspect ratio.

Fig.2 shows the poloidal dependence of both electrostatic and magnetic potential for electrons.

Fig.3 shows the poloidal dependence of both electrostatic and magnetic potential for ions.

The ratio between the poloidal electrostatic part and the poloidal magnetic part of the total potential is equal to 1/2.

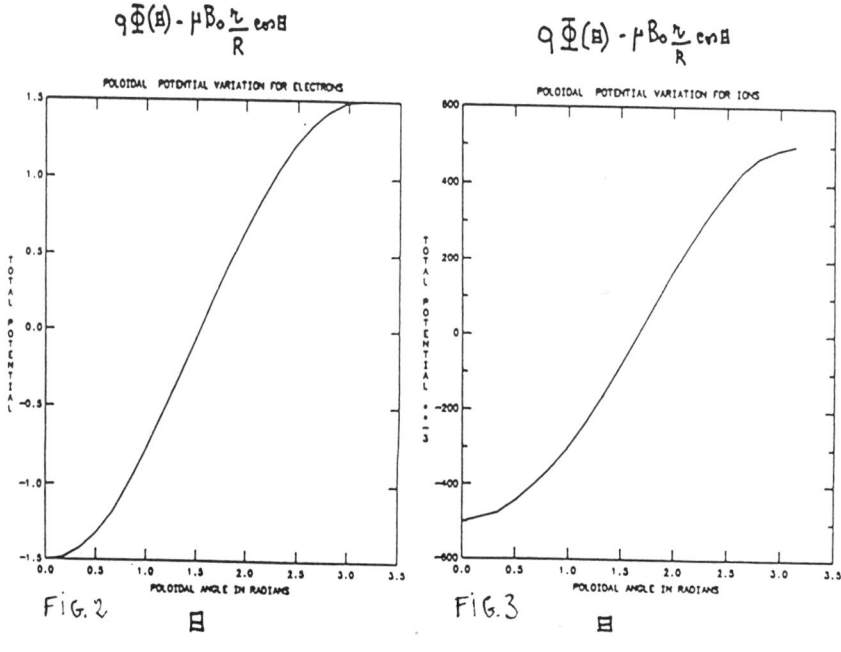

Fig. 2

Fig. 3

MEASUREMENT OF THE CORRELATION SPECTRUM OF ELECTROSTATIC POTENTIAL FLUCTUATIONS IN A TOROIDAL ECRH PLASMA

E. D. Zimmerman, S. C. Luckhardt, J. C. Rost
MIT Plasma Fusion Center

Abstract. Measurements of the length and time scales of turbulent potential fluctuations were carried out on the MIT Versatile Toroidal Facility (VTF). Plasmas were produced by ECRF heating in a HELIMAK configuration. The HELIMAK toroidal field B_ϕ was 800 gauss, vertical field B_z typically 10 gauss, and $T_e \approx 10$ eV. The density exhibited a peaked profile having $n_{max} \approx 2 \times 10^{10}$ cm^{-3}. The fluctuation experiments were conducted using a moveable vertical array of eight Langmuir probes. At major radii outside the density peak, the correlation lengths of fluctuations were found to be on the order of 5–10 cm for fluctuation frequencies below 5–8 kHz, and on the order of 1 cm at higher frequencies. At major radii on the inner slope of the density peak, a new feature appears in the spatial coherence function consisting of a second peak at a probe separation of 5 cm. This observation suggests that the fluctuations have a toroidal extent greater than $2\pi R_0$.

1 BACKGROUND

Extensive theoretical and experimental work has been published which predicts conditions for turbulence in tokamak edge plasmas. This work characterizes some aspects of potential fluctuations and turbulence in a novel toroidal configuration which we have termed the Helimak.

The Versatile Toroidal Facility (VTF) at the MIT Plasma Fusion Center is a toroidal plasma chamber designed for fundamental plasma physics experiments. The outer radius of the VTF chamber is 126 cm; its inner radius is 60 cm, and its height is 105 cm. The basic magnetic configuration consists of eighteen toroidal field magnets setting up a field B_ϕ which varies inversely with major radius R between 600 and 1300 gauss. In addition, a set of coils approximating a Helmholtz coil produces a vertical field B_z which is constant to within 5% throughout the chamber. For the experiments described here, the vertical field was set to values in the range between 8 and 12 gauss (10 gauss unless otherwise specified).

A hydrogen plasma is produced by 3 kW of 2.45 GHz radiation injected at the midplane of the chamber. The electron cyclotron resonance occurs at a major radius $R = 89$ cm, where $B_\phi = 875$ gauss. The resulting plasma has a peaked density profile which has been measured using a radially sweeping Langmuir probe. With $B_z = 10$ gauss, the measured maximum density is 2.1×10^{10} cm^{-3}, located at a major radius $R = 99$ cm. The density peak has a full width at half maximum of 18 cm[1].

Potential fluctuation data were gathered using a linear array of eight Langmuir probes oriented parallel to the vertical axis of the chamber. Each probe consists of a stainless steel wire 0.9 mm in diameter and 6 mm in length. The array permits simultaneous measurements of the floating potential at different vertical positions in the plasma. It can be moved to various radial positions as well. The probe electronics and digitizing equipment can record data up to frequencies of 40 kHz.

2 SINGLE-POINT MEASUREMENTS

An important aspect of potential fluctuations is their amplitude, defined in this case to be the standard deviation $\Delta\phi$ of the floating potential:

$$\Delta\phi = \sqrt{\langle\phi^2\rangle - \langle\phi\rangle^2} \qquad (1)$$

The fluctuation amplitudes of the measured floating potentials were obtained by calculating the standard deviation of time series data over 0.82-second digitizing periods during shots. The radial dependence of the fluctuation amplitude (with $B_z = 10$ gauss) is shown as Fig. 1. The fluctuation amplitude is clearly higher on the outer slope of the density peak, where the magnetic curvature is toward the higher-density region, i. e. the bad curvature region.

The power spectrum of potential fluctuations in the VTF plasma shows strikingly different behavior in different frequency regions. At low frequencies (below 2 kHz), the spectrum is dependent on the radial position of the probes: in the low-density regions, the spectrum is fairly flat below 1 kHz. Closer to the density peak, however, the low-frequency end of the spectrum exhibits a very broad peak around 500 Hz. Above a roll-off frequency (1 to 2 kHz, depending on major radius), the power spectrum drops off as $1/\omega^2$ (Fig. 2).

3 CORRELATION ANALYSIS

Plasma fluctuations are correlated on a finite spatial scale which has been measured using correlation analysis on simultaneous signals from spatially separated probes. Coherence functions have been calculated for sets of data obtained from the eight-probe array.

By measuring the coherence between signals from probes at various locations along the array, the correlation lengths of vertical fluctuations can be inferred. With the vertical field fixed at 10 gauss, measurements were taken at four different major radii. For fluctuation frequencies below a characteristic range of about 3–7 kHz, coherence remains significant even at a separation of 8.5 cm, the length of the probe array. Correlation lengths are therefore on the order of 5–10 cm. As frequency increases through this range, the coherence length drops rapidly to about 2 cm. Beyond 7 kHz, the correlation length remains constant or drops very slowly through the resolution limit of 20 kHz. Thus, fluctuations in the low-frequency range exhibit a large spatial scale length and fluctuations in the high-frequency range (> 7 kHz) exhibit a scale length smaller by nearly an order of magnitude. A contour plot showing the behavior of the coherence as a function of both frequency and probe separation, at a major radius of 117 cm, is presented as Fig. 3. The contours represent paths of constant coherence.

Under certain conditions, particularly at radii on the inner slope of the density peak, a new feature appears in the coherence spectrum. Rather than continuously decreasing as probe separation increases, the coherence of higher frequency fluctuations actually *increases* briefly at certain probe separations. A contour plot exhibiting this coherence peak is displayed as Fig. 4. The coherence peak is not dependent upon vertical position— it is equally apparent between probes on the lower and middle parts of the array as on the middle and upper sections.

Our interpretation of the coherence peak is that it is the result of "worm-like" fluctuations which have a very long extent along the torus, and wrap completely around the chamber. However, the coherence peak does not occur at the probe separation which would be expected if the fluctuations followed the helical magnetic field lines exactly. The expected separation,

$$\Delta z = 2\pi R \frac{B_z}{B_\phi} \quad (2)$$

is, for example, approximately 7.5 cm for $R = 94$ cm and $B_z = 10$ gauss. The actual coherence peak occurs at a probe separation of (4.8 ± 0.5) cm.

The dependence of the location of the coherence peak on vertical field and major radius is consistent with the idea of fluctuations following curves with a pitch proportional to that of the magnetic field lines. Varying the magnetic field slightly alters the probe separation at which the coherence peaks. This is shown as Fig. 5. While too few data points could be obtained to demonstrate conclusively that that the peak separation varies linearly with B_z, it is apparent that the data are not in conflict with that interpretation. In addition, a line drawn through the origin passes through all the error bars.

Fig. 6 shows the dependence, which is linear within the estimated errors, of the peak separation on R^2. This is expected because B_ϕ in Eq. 2 is proportional to $1/R$.

4 INTERPRETATON AND CONCLUSIONS

Low-frequency (≤ 3 kHz) fluctuations in the Helimak plasma have a large spatial scale in the vertical direction of 5-10 cm, and higher frequency fluctuations (≥ 7 kHz) have a smaller scale of 1-2 cm. Higher-frequency fluctuations also have a very long extent in the toroidal direction, although they do not exactly follow the magnetic field lines. Rather, they appear to follow curves having a pitch approximately 2/3 that of the actual field lines. Further study is needed to determine the cause of this deviation.

REFERENCES

[1] J. C. Rost, S. C. Luckhardt, R. R. Parker, E. D. Zimmerman, *Density and Potential Fluctuation Measurements in a Toroidal ECRH Sustained Plasma*, Poster, APS Plasma Physics Meeting, Seattle, November 1992.

[2] E. D. Zimmerman, *Measurement of the Correlation Spectrum of Electrostatic Potential Fluctuations in a Toroidal HELIMAK Plasma*, S. B. Thesis, Department of Physics, Massachusetts Institute of Technology, January 1993

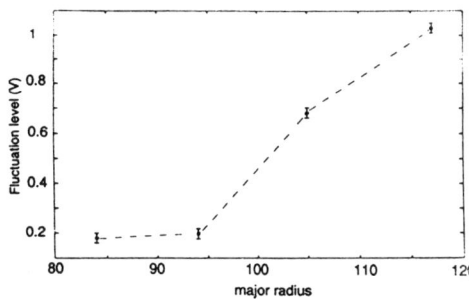

Figure 1: Measured fluctuation amplitudes

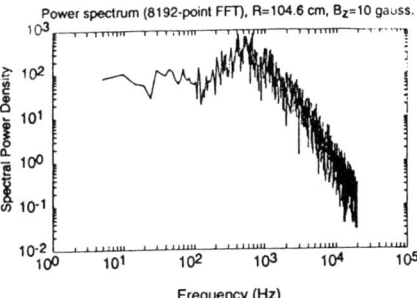

Figure 2: Full logarithmic plot of fluctuation power spectrum. Major radius 104.6 cm.

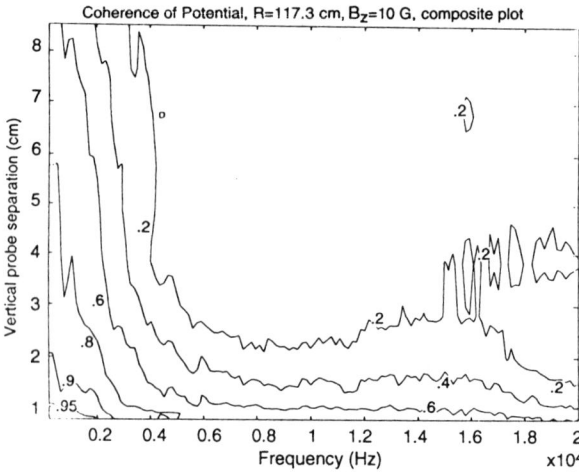

Figure 3: Contour plot of coherence of potential. Major radius = 117.3 cm, B_z = 10 gauss. Composite plot using data from 26 sets of two arrays each.

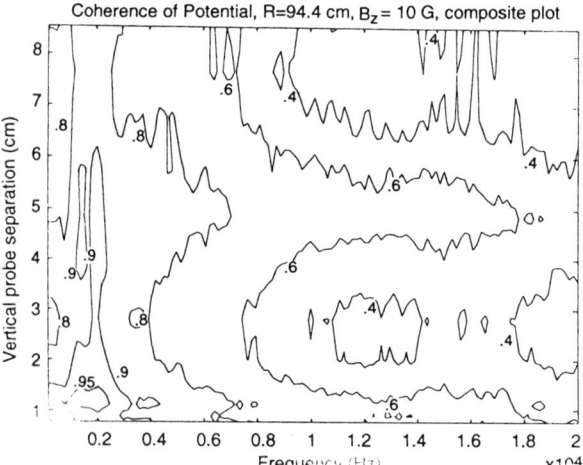

Figure 4: Composite contour plot of coherence of potential. Major radius = 94.4 cm.

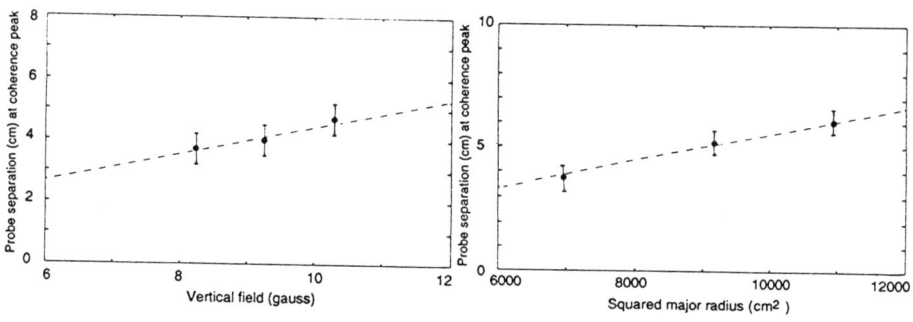

Figure 5: Coherence peak vs. vertical field. Dotted line passes through origin.

Figure 6: Coherence peak vs. R2. Dotted line passes through origin.

ANALYTICAL ESTIMATIONS ON THE AXIAL STRUCTURE OF PLASMA–WAVEGUIDE DISCHARGES*

Yu.M. Aliev

Lebedev Institute, Russian Academy of Sciences, 117924 Moscow, Russia

I. Ghanashev, A. Shivarova

Faculty of Physics, Sofia–University, 1126 Sofia, Bulgaria

H. Schlüter, M. Zethoff

Experimental Physics II, Ruhr–University, 4630 Bochum, Germany

ABSTRACT

Plasma–waveguide discharges produced in the diffusion controlled regime by surface wave propagation at weak collisions are studied. The widely discussed problem of the axial variation of electron density, which reflects the non–linear self–consistent behaviour of wave and discharge properties, reduced to analyzing the behaviour of the space damping rate of the wave is solved analytically in the whole region of wave existence. The proper normalizing parameters used for presenting the behaviour of the surface wave characteristics in the different regions of their existence are extrapolated to electrodynamical similarity relations of the discharges.

I. INTRODUCTION

During the last twenty years, a new branch in the field of high–frequency and microwave discharges — discharges sustained by surface wave (SW) propagation — was extensively developed[1-3]. They are particular cases of plasma–waveguide discharges with the plasma produced by the wave constituting the main component of the guiding structure. From the theoretical point of view, the production of gas discharges by travelling waves is a non–linear problem requiring self–consistent description of discharge and wave properties. It is based on simultaneous solution of the wave power balance equation and the electron energy balance equation which associate — through the Joule losses of the wave — the wave power flux (i.e. wave characteristics) with the energy losses of the electrons in the discharge sustained (i.e. discharge charateristics).

Here previous analytical results[4] for electrodynamical similarity laws giving — in a region close to the quasi-static one — the relationship between the density variation along the plasma column created and the space damping rate of SW propagation, are extended to the whole electromagnetic region of wave existence. The discharge regime considered is controlled by ambipolar diffusion losses and the wave propagation is under the conditions of weak collisions ($\nu/\omega_0 < 1$ where

---

*Support by the National Foundation for Scientific Research in Bulgaria (project F27), Alexander von Humboldt Foundation and Deutsche Forschungsgemeinschaft (exchange program, SFB 191) is gratefully acknowledged.

ω_0 is the wave frequency and ν is the electron–neutral collision frequency for momentum transfer).

II. GENERAL BEHAVIOUR OF PLASMA–WAVEGUIDE DISCHARGES

Discharges are considered created by propagating SWs with field variation of the form $\sim \exp(-i\omega + ik\zeta)$ where $\omega = \omega_0 + i\gamma$ and $k = k_0 + i\kappa$; ω_0 and k_0 are the frequency and the wavenumber, respectively; the time damping rate γ ($\gamma < 0$) and the space damping rate κ ($\kappa > 0$) are related to each other through $\kappa = -\gamma/v_{gr}$ where $v_{gr} = \frac{\partial \omega}{\partial k}$ is the wave group velocity. Under conditions of the diffusion controlled regime the wave power balance and electron energy balance reduce to

$$\frac{\partial}{\partial \zeta}\left(\frac{\bar{n}}{\kappa}\right) = -2\bar{n} \quad \text{or} \quad \frac{d|\varepsilon|}{d\zeta} = -\frac{2(|\varepsilon|+1)\kappa}{1 - \frac{|\varepsilon|+1}{\kappa}\frac{d\kappa}{d|\varepsilon|}} \qquad (1,2)$$

where $\varepsilon = 1 - \bar{\omega}_p^2/\omega_0^2$ is the plasma permittivity with $\bar{\omega}_p^2 \sim \bar{n}$ the electron plasma frequency squared (averaged over the discharge cross–section).

Thus the nonlinear problem of sustaining discharges by travelling electromagnetic waves guided along a structure (the main component of which is the plasma created by the wave itself) is reduced to analyzing the behaviour of the linear wave characteristics. In particular the information for the variation of the electron density of the created plasma along the direction of propagation (z) is contained in the local linear dispersion relation of the waves from which the time damping rate and the wave group velocity or, equivalently, the space damping rate κ used above should be obtained.

III. RESULTS FOR SURFACE WAVE PROPAGATION CHARACTERISTICS

Concerning azimuthally symmetric SW propagation at fixed frequency $\omega_0 \equiv \omega$ along a cylindrical plasma column (radius R) with axially varying electron density, the wave behaviour is apparent from the so called phase diagrams which have been presented up to now as a dependence of ω/ω_p vs. $z \equiv k_{(0)}R$ at $\omega = const.$ or, equivalently, at $\sigma = \omega R/c = const.$ (fig. 1). They obey the well–known dispersion relation

$$\frac{x}{|\varepsilon_p|}\frac{I_0(x)}{I_1(x)} = y\frac{K_0(y)}{K_1(y)} \qquad (3)$$

where $|\varepsilon_p| = |\varepsilon(\omega)| - i(|\varepsilon|+1)(\nu/\omega)$ is the plasma permittivity at weak collisions ($\nu/\omega < 1$), $I_{0,1}(x)$, $K_{0,1}(y)$ are the modified Bessel functions and $x \equiv k_{\perp p}R \equiv [k^2 + (\omega^2/c^2)|\varepsilon_p|]^{1/2}R$ and $y \equiv k_{\perp v}R \equiv [k^2 - (\omega^2/c^2)]^{1/2}R$ describe the field decay in radial direction into the plasma and the vacuum, respectively. The curves $x = 1$ and $y = 1$ separate the plane $((\omega/\omega_p), z)$ into three regions of different possible values of the variables x and y denoted by I ($x, y < 1$), II ($x > 1, y < 1$) and III ($x > 1, y > 1$) in fig. 1. Regions I and III are limited by the quasi-static approximation ($\sigma = 0$). With increasing σ–values, region II appears at the beginning of the phase diagrams. However, at small σ–values ($\sigma < 0.3$), although the phase diagram starts from region II, region I still constitutes its main part; the end of the phase diagram is in region III. With further increasing of $\sigma(\sigma > 0.3)$, region I disappears from the phase diagrams and region II enlarges. In this case the phase diagrams are positioned in regions II and III.

The well known exact analytical solutions for the SW characteristics in this case are as follows: (a) quasi-static SWs along a plasma column; (b) electromagnetic surface waves in semi–bounded plasmas. However, it appears that analytical solutions (in explicit form) for the SW behaviour can be obtained not only in the limiting cases (a) and (b) but for the phase diagrams at arbitrary σ as well. In such a way the SW properties (group velocity, space and time damping rates) and, respectively, the axial density profiles of SW produced plasmas [eq. (1)] can be obtained in an analytical form. The result for κ at arbitrary σ–values are:

$$\kappa = \frac{\nu}{\omega}\frac{\omega}{c}\frac{1}{2\sigma\sqrt{y^2+\sigma^2}}\frac{|\varepsilon|+1}{|\varepsilon|}\frac{1-\sigma^2|\varepsilon|\frac{M}{x^2}}{T} \qquad (4)$$

where $$T = \frac{L}{y^2} - \frac{M}{x^2}, \quad M = 1 - \frac{x}{2}\left[\frac{I_0(x)}{I_1(x)} - \frac{I_1(x)}{I_0(x)}\right],$$

$$L = 1 + \frac{xI_0(x)}{2|\varepsilon|I_1(x)} - \frac{y^2|\varepsilon|}{2x}\frac{I_1(x)}{I_0(x)}.$$

IV. AXIAL PROFILE OF THE ELECTRON DENSITY IN SW PRODUCED PLASMAS

The axial gradient of the electron density can be obtained straightforward from expression (2) by using the result for the space damping rate κ [expression (4)]. In fig. 2 the results are shown with a normalization $(d|\varepsilon|/d\zeta)(\omega R/\nu) \equiv (d|\varepsilon|/d\zeta)(\omega/\nu)R$ appropiate for quasi–static approximation [a suitable normalization for electromagnetic waves would be $(d|\varepsilon|/d\zeta)(c/\nu) \equiv (d|\varepsilon|/d\zeta)(c/\omega)(\omega/\nu)$].

In connection with the application of the results to studies on the behaviour of SW produced plasmas, the regions on the phase diagrams where the waves are weakly damped are of interest. In the cases of practical interest the values of the parameter σ should not exceed 2 (or 3). Therefore, approximate expressions for the SW characteristics applicable to regions I and to the bottom of regions II and III (fig. 1) could be important for easy and quick estimation of the density gradient and, respectively, of the length of the produced plasmas.

The generalized approximated result for the normalized density gradient is:

$$\frac{d|\varepsilon|}{d\zeta} = -\frac{1}{2}\beta\frac{1}{F}, \quad \text{where} \quad \frac{1}{F} = \frac{y^2|\varepsilon|}{G\sqrt{y^2+\sigma^2}} \qquad (5)$$

456 Analytical Estimations on the Axial Structure

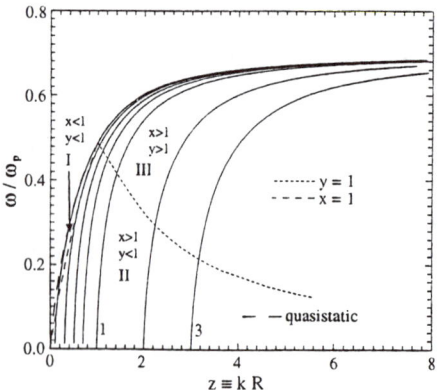
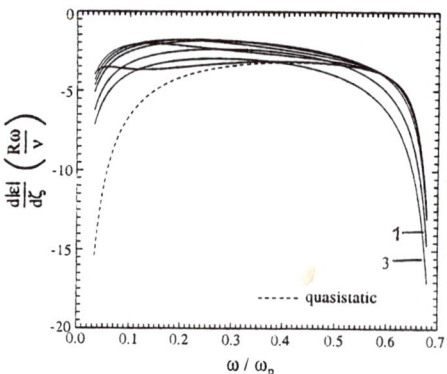

Figure 1: Phase diagrams of SW existence for $\sigma = 0.1, 0.3, 0.5, 0.7, 1, 2, 3$. The curves $x = 1$, $y = 1$ separate the plane into three regions.

Figure 2: Normalized axial density gradient vs. ω/ω_p. The curve parameters σ are the same as in fig. 1.

$$G = 1 - \frac{y^2}{4}|\varepsilon| + \frac{1}{|\varepsilon|} \quad \text{at} \quad x < 1 \quad \text{(case A)}$$

$$G = 2 + \frac{\sigma}{\sqrt{|\varepsilon|}} - \frac{y^2\sqrt{|\varepsilon|}}{\sigma} \quad \text{at} \quad x > 1,\ \sigma/y > 1 \quad \text{(case B)}$$

The coefficient β is $\beta = (\nu/\omega)(\omega/c\sigma)$, i.e. $\beta = (\nu/\omega)(h/\sigma)$ and the combination of the vacuum wave number h and the parameter σ results in the normalizing parameter $(1/R)$ for the quasi–static waves. It appears in such a form not only for case A which is close to the quasi–static dispersion curve, but also for case B since the latter is for $\sigma/y > 1$ and intermediate values of σ. With increasing σ–values or at $\sigma/y < 1$ the coefficient β transforms into $\beta = (\nu/\omega)(\sigma/R)$ and the combination of σ and the normalizing parameter for quasi–static waves $(1/R)$ results in the vacuum wavenumber, i.e. in the normalizing parameter for electromagnetic waves.

REFERENCES

[1] M. Moisan, R. Pantel, A. Ricard, V.M.M. Glaude, P. Leprince and W.P. Allis, Revue Physique App. **15**, 1383 (1980).
[2] M. Moisan and Z. Zakrzewski,in: Radiative processes in Discharge Plasmas, ed. by J.M. Proud and L.H. Luessen (Plenum Press, New York, 1986).
[3] M. Moisan, J. Hubert, J. Margot, G. Sauve and Z. Zakrzewski, in: Microwave Discharges: Fundamentals and Applications, ed. by C.M. Ferreira and M. Moisan (Plenum Press, London, NATO ARW series, 1993).
[4] Yu.M. Aliev, K.M. Ivanova, M. Moisan and A.P. Shivarova, Plasma Sources Sci. Technol. 1993 (in press).

OBSERVATION OF EDGE ELECTRON HEATING DURING 800 MHZ LOWER HYBRID FAST WAVE EXPERIMENTS ON THE VERSATOR II TOKAMAK

J. Villaseñor, M. Porkolab, G. Gibson, J. Colborn, J. Squire
MIT Plasma Fusion Center and Research Laboratory of Electronics,
Cambridge, MA 02139

ABSTRACT

High power injection of fast waves (P_{RF} < 25 kW) at the lower hybrid frequency of 800 MHz using a dielectrically loaded waveguide array has failed to produce any form of current drive or central heating, as shown by measurements using hard x-ray detectors[1]. Miniature retarding potential analyzer probes have detected a thin region at the plasma edge where electrons are heated from 5-10 eV to as high as 100 eV. This region has a spatial extent of ~1 cm in depth and 2 cm in height, and is located just behind the limiter edge along the midplane of the tokamak (coplanar with the antenna array). No heating was observed elsewhere. Parametric decay spectra was also measured at different toroidal and poloidal locations. The parametric decay activity has a measured threshold of $P_{RF} \cong 200$ W and corresponds with that of edge electron heating.

INTRODUCTION

Fast wave current drive in the lower hybrid range of frequencies $\omega_{ci} \ll \omega \ll \omega_{ce}$ and $\omega \cong \omega_{lh} = \omega_{pi}/(1+\omega_{pe}^2/\omega_{ce}^2)^{1/2}$ was attempted on the Versator II tokamak (R_0=40.5 cm, a=13.0 cm, $B_t \leq 12$ kG, 2×10^{12}cm^{-3}<n_e<3×10^{13}cm^{-3}, I_p~25-50 kA) using a dielectrically loaded waveguide array (ε=80). The experimental setup is depicted in Fig. 1. In the absence of tuning elements, theory predicts poor plasma coupling with this antenna compared with slow wave launching. This is verified by excellent agreement with low power experiments. The poor coupling is due to the high impedance mismatch (~85-90% reflectivity) at the antenna face. We have improved coupling using tuning stubs and phase shifters in a manner similar to ICRF impedance matching[2]. The low power data shows that 65-80% of the power is coupled to the plasma with this retuning.

High power experiments (P_{RF}<25kW) were performed at antenna phasings of 0°, 90°,-90° and 180° in both ohmic, and in S-band (2.45 GHz, P_{RF}<100 kW) current driven plasmas. In both cases, no evidence of bulk current generation was observed by most of the machine diagnostics (e.g. increased I_p, V_{loop} drops, increased $2\omega_{ce}$ emission) for a wide range of densities and magnetic fields. Furthermore, tangential and perpendicular hard x-ray arrays do not indicate any increase in suprathermal electron activity with the injection of 800 MHz power. Instead, we only note increased edge electron activity as evidenced by the electron saturation of Langmuir probes.

To study this phenomenon, miniature energy analyzer probes were inserted at different locations in the tokamak, including top and bottom locations.(Fig. 2) Each analyzer has a .020" aperture and is housed in a .125" SS tube. These probes could be inserted radially and rotated along their axes. One probe located on the midplane of the tokamak can tilt by 22.5°, which allows it to sample the electron distribution function over a small cross-sectional area.

Electrostatic and electromagnetic probes were also installed at different locations in the tokamak to measure parametric decay spectra. In particular, a poloidal probe array consisting of eight electrostatic probes was installed at a toroidal location 120° away from the 800 MHz fast wave antenna.

RESULTS

The energy analyzers typically measure an edge electron temperature of 5-10 eV in ohmic and 20-25 eV in LHCD driven plasmas. The floating potential is ~2-5 V. Upon injection of 800 MHz power, the electron temperature measured by a midplane analyzer at the port immediately adjacent to the 800 MHz antenna rises by a factor of 2-5 (Fig.3). A greater electron temperature rise is noted when the 800 MHz power is injected in conjunction with the S-band. No concomitant ion temperature increases were detected. Furthermore, the floating potential measured at the probe drops as the temperature rises.

The energy analyzers located at different locations did not show any sign of electron activity, except for those located at the midplane. Rotating the midplane analyzers 180° show reduced activity (ΔT_e =1/3). Other analyzer orientations did not yield any collecter current, since the electrons flow along the magnetic field lines. A radial scan of the analyzer probe reveals that the hot electrons are limited at the edge. Using the tilting analyzer, we find that the area of increased activity is limited to a region 1 cm in depth and 2 cm in height, or roughly the width of the antenna (Fig.4). Moreover, a limiter located 90° away from the antenna and composed of eight independent floating sections did not show any change in the floating potential.

The electron temperature rise also varied with the antenna phasing. In particular, an antenna phasing of 0° produced the largest temperature rise, followed by -90°, and then 180° phasings (54 eV:43 eV:33 eV).

By varying the input RF power, we note that the hot electron production begins when a threshold of approximately 200 W is passed (Fig. 5). The temperature rise does not increase with added RF power, and instead saturates at a certain level. The number of hot electrons, however, increases with the added RF power.

The RF probes show asymmetric broadening of the 800 MHz pump frequency consistent with the onset of parametric decay activity (Fig. 6). This condition is amplified when S-band power is applied simultaneously. Ion cyclotron peaks also appear when the S-band antenna is applied. The poloidal probe array shows no asymmetry in the decay activity. The ion cyclotron peaks detected are separated by a value which corresponds to a magnetic field at the outside edge of the tokamak, and does not vary as a function of the radial position. Furthermore, the threshold for pump broadening appears to coincide with the measured threshold for the generation of energetic electrons. The saturation of this signal also appears at a relatively low power level.

CONCLUSIONS

The generation of the hot electrons appears to be confined to the region just in front of the antenna. The electric fields excited at the antenna face are apparently confined to the plasma surface and cause the observed parametric decay. Although the parametric decay activity is detected at all poloidal locations, there is evidence to show that it originates at close to the antenna face. The decrease in the floating potential is most probably due to RF sheath rectification.

This work was supported by US DOE Contract No. DE-AC02-78ET51013

REFERENCES

1. J. Villasenor, *et al*, Bull. Am. Phys. Soc. **36** 2440 (1991)
2. J. Villasenor, *et al*, Bull. Am. Phys. Soc. **35** 1997 (1990)

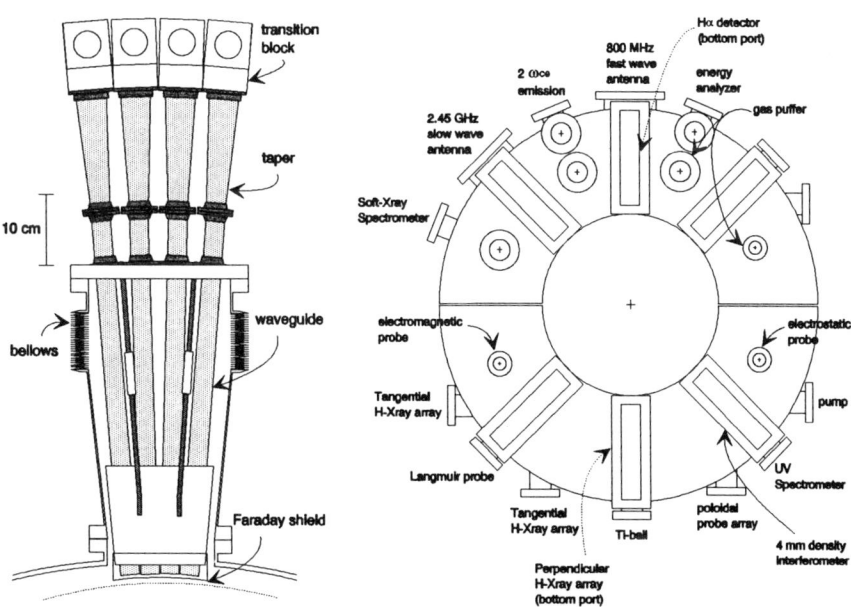

Figure 1. Top view of the 800 MHz Dielectric loaded fast wave launcher

Figure 2. Versator II and the location of the diagnostics

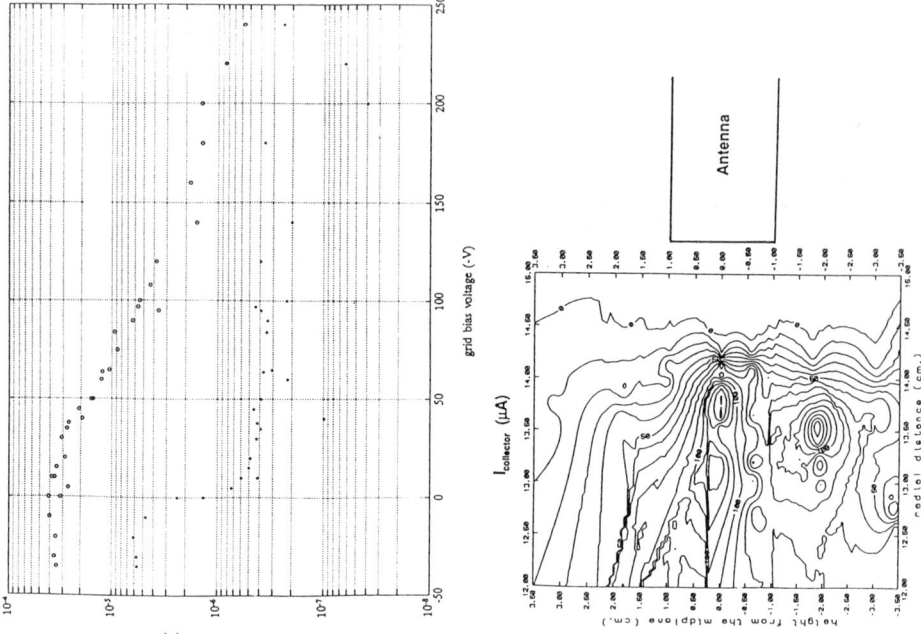

Figure 3. Comparison of the energy analyzer trace with (o) and without (*) 800 MHz power. The electron temperatures are ~40 and ~7 eV.

Figure 4. Spatial distribution of the analyzer collector current

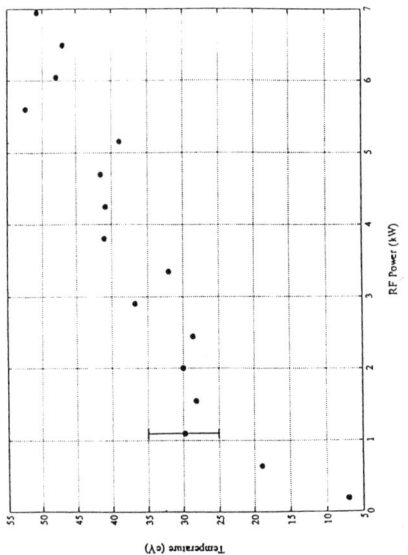

Figure 5. Threshold for the generation of hot electrons. The y-axis is the increase in electron temperature.

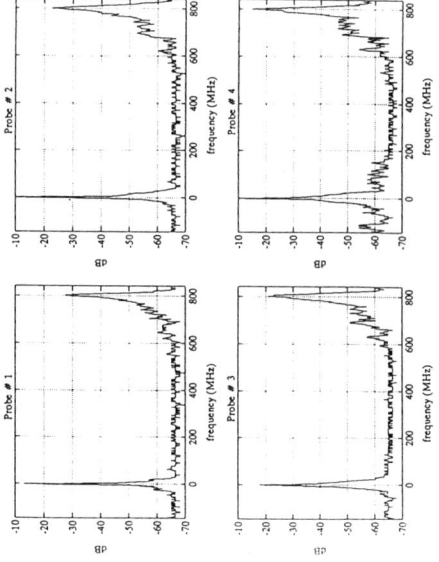

Figure 6. Parametric decay spectra at different poloidal locations. All probes are on the same poloidal plane.

Author Index

A

Aikawa, H., 149
Aliev, Yu. M., 309, 453
Alikaev, V. V., 234
Alladio, F., 115
Allen, S. L., 157
Apicella, M. L., 115
Apruzzese, G., 115
Argenti, L., 173
Asakura, N., 64, 72
ASDEX Upgrade Team, 12

B

Bagdasarov, A. A., 234
Baity, F. W., 343
Barbato, E., 115, 327
Bartiromo, R., 115
Batani, D., 80
Batchelor, D. B., 48, 60, 218, 301, 335, 343, 359, 371, 405, 437
Batha, S., 131
Beaumont, B., 24, 347
Bécoulet, A., 24, 347
Bell, M., 44, 48
Bell, R., 64, 68, 72, 76, 202
Berger-By, G., 315
Bernabei, S., 64, 119, 123, 131, 202, 375
Berndt, J., 309
Bers, A., 293, 433
Bettenhausen, M. H., 363
Bhatnager, V., 355
Bibet, P., 315
Bigelow, T. S., 327, 391, 437
Bills, K. C., 343
Bizarro, J. P., 24, 315
Blank, M., 395
Blush, L., 64, 76, 202
Bombarda, F., 115
Bonoli, P. T., 56, 192, 210, 222, 230
Bornatici, M., 269
Borshegovskij, A. A., 234

Bosia, G., 351
Bracco, G., 115
Brambilla, M., 12
Brizard, A. J., 253
Brouchous, D., 387
Bruschi, A., 173
Buceti, G., 115
Büchl, K., 12
Budny, R., 48, 107, 405
Buratti, P., 115
Bures, M., 351
Burhenn, R., 137
Byers, J. A., 157

C

Cairns, R. A., 265, 289
Campbell, D. J., 52, 99
Capitain, J. J., 315
Carasco, J., 315
Cardinali, A., 115
Caron, X., 111
Carpignano, F., 425
Carter, M. D., 327, 335, 347, 359
Casper, T. A., 157
Caughman, J. B. O., 327, 331, 379
Cesario, R., 64, 72, 115
Chang, Z. Y., 409
Chiu, S. C., 305
Chu, T. K., 64, 68, 72, 202
Cincotti, G., 285
Cio Hi, M., 115
Cirant, S., 173
Cocilovo, V., 115
Cohen, B. I., 157
Cohen, R. H., 157
Cohen, S. A., 383
Colborn, J. A., 226, 457
Coletti, A., 115
Collazo, I., 437
Compant La Fontaine, A., 441
Condrea, I., 115
Conn, R., 64, 76
Conrads, H., 32
Cook, D. R., 253

Coppi, B., 425
Crisanti, F., 115

D

Darrow, D., 409
Davis, W., 123
De Angelis, R., 115
Decyk, V. K., 277
De Marco, F., 115
Demers, Y., 315
Detragiache, P., 425
Diebold, D., 387, 429
Diesso, M., 383
D'Ippolito, D. A., 417, 421
Doczy, M., 387
Doerner, R., 64, 76, 202
Dremin, M. M., 234
Drobot, A., 359
Dumortier, P., 32
Dunlap, J., 64, 68, 202
Durodie, F., 32

E

Eberhagen, A., 12
Ehst, D. A., 214
Engelmann, F., 269
England, A., 64, 68, 72, 202
Erckmann, V., 137
Esipchuk, Yu. V., 234, 238
Esposito, B., 115
Euringer, H., 32

F

Fadnek, A., 347
Fenstermacher, M. E., 157, 210, 222
Field, A., 12
Fishman, H., 131
Flynn, W. G., 261
Fogelman, C. H., 327, 343
Fontanesi, M., 80
Foote, J. H., 157
Forest, C. B., 169, 234, 238

Fortunato, T., 115
Fraboulet, D., 24
Fredd, E., 371
Fredrickson, E., 44, 48, 409
Frezza, F., 285
Friedland, L., 257
Frigione, D., 115
Fuchs, C., 12
Fuchs, G., 32
Fuchs, V., 107, 289, 293
Fujii, T., 52, 99

G

Gabellieri, L., 115
Galassi, A., 80
Garbet, X., 24
Gardner, W. L., 327, 331
Gehre, O., 12
Geist, T., 137
Gernhardt, J., 12
Gettelfinger, G., 202
Ghanashev, I., 453
Gibson, G., 457
Giesen, B., 32
Giovannozzi, E., 115
Giruzzi, G., 111
Goldfinger, R. C., 48, 405
Golovato, S. N., 56
Goniche, M., 24, 315
Goranson, P., 375
Gorelov, Yu. A., 234
Gori, F., 285
Gormezano, C., 87
Gould, R. W., 323
Goulding, R. H., 60, 339, 351, 355, 371
Granucci, G., 115, 173
Greenough, N., 202, 375
Grimm, T. L., 395
Grolli, M., 115
Grossman, A., 64, 76
Grossman, W., 359
Gruber, O., 12
Guilhem, D., 315
Guiziou, L., 24
Guss, W. C., 395

H

Haas, G., 12
Hamamatsu, K., 52, 99
Hammett, G., 48, 405
Hanson, G. R., 437
Harris, J., 64, 202
Hartfuss, H. J., 137
Harvey, R. W., 169, 222, 234, 238, 293
Haste, G., 347
Hatcher, R., 64, 131, 202
Hermann, A., 12
Herrmann, H., 64, 72
Hershkowitz, N., 242, 429
Hillis, D., 32
Hirshman, S., 131, 202
Ho, Y. L., 359
Hoang, G. T., 24, 107, 315
Hoenen, F., 32
Hoffman, D. J., 48, 60, 327, 331, 339, 351, 379, 437
Hoffmann, C., 12
Hofmeister, F., 12
Hooper, E. B., 157
Hosea, J. C., 36, 40, 44, 48, 401, 405
Hoshino, K., 149, 157
Hubbard, A. E., 375
Hutter, T., 24
Hwang, Y. S., 246

I

ICRH Team, 12
Ide, S., 115
Ignat, D., 123, 127, 131, 202, 413
Ikeda, Y., 99, 319
Il'in, V. I., 137
Imai, T., 99, 319
Imparato, A., 115
Intrator, T., 387, 429
Isler, R., 64, 68, 72, 202
Ivanov, N. V., 234

J

Jaeger, E. F., 48, 60, 218, 301, 343
Janos, A., 48

Jardin, S., 202, 413
Joffrin, E., 24, 107
Jones, S. E., 119, 123, 131, 202

K

Kaita, R., 64, 72, 119, 123, 131, 202
Kallenbach, A., 12
Kamada, Yu., 99
Karney, C. F. F., 305
Kasai, S., 149
Kasparek, W., 137
Kaufman, A. N., 253, 257
Kawakami, T., 149
Kawashima, H., 149
Kaye, S., 64, 68, 107, 131, 202, 355
Kesner, J., 119, 131, 202
Kessel, C., 131
Khudaleev, A., 44
Kick, M., 137
Kimura, H., 52, 99
Kislov, A. Y., 234
Koch, R., 32
Kondoh, T., 52, 99, 319
Koslowski, H. R., 32
Kozub, T., 202
Krämer-Flecken, A., 32
Kreischer, K. E., 395
Kritz, A. H., 273
Kroegler, H., 115
Kugel, H., 64, 72, 76, 202
Kupfer, K., 281
Kurbatov, V. I., 137
Kuus, H., 24, 347
Kuznetsova, L. K., 234

L

Lam, N. T., 367
Lashmore-Davies, C. N., 265
Lasnier, C. J., 157
LeBlanc, B., 64, 68, 72, 131, 202
Lee, B. J., 214
Lee, D., 131, 202
Levinton, F., 131, 202
Lieder, G., 12

Lin-Liu, Y. R., 169, 238, 305
Litaudon, X., 24, 107, 281, 315
Littlejohn, R. G., 261
Litwin, C., 242
Liu, W. D., 111
Lochter, M., 32
Lohr, J., 169, 234, 238
Lovisetto, L., 115
Luce, T. C., 165, 234
Luckhardt, S. C., 119, 131, 202, 449

M

Maassberg, M., 137
Maddaluno, G., 115
Maeda, H., 149
Maeno, M., 149
Magne, R., 315
Majeski, R., 36, 40, 44, 48, 383, 401, 429
Makowski, M. A., 157
Malygin, S., 137
Malygin, V. I., 137
Mansfield, D., 48
Marinucci, M., 115
Matsuda, T., 149
Matsuoka, M., 52
Mau, T. K., 214, 218
Mazzitelli, G., 115
McCune, D., 68, 405
McDonald, D. C., 265
McNeill, D., 115
Mertens, V., 12
Messiaen, A. M., 32
Meyer, J., 429
Meyer, R. L., 111
Meyer, W. H., 157
Micozzi, P., 115
Mikkelsen, D. R., 218
Mirizzi, F., 115
Miura, Y., 149
Moeller, C. P., 323, 339
Moleti, A., 115
Mollard, P., 24
Moller, J. M., 157
Morales, G. J., 277
Moreau, D., 24, 107, 281, 315

Morehead, J. J., 257
Mori, M., 149
Moriyama, S., 52, 99
Moroz, P. E., 218
Motley, R. W., 383
Müller, G. A., 137
Mullier, B., 32
Murakami, M., 36, 44, 48, 301, 437
Murmann, H., 12
Myra, J. R., 417, 421

N

Nagashima, K., 149
Naito, O., 99, 319
Nassi, M., 425
NBI Team, 137
Nemoto, M., 52, 99
Nevins, W. M., 222
Nguyen, F., 24
Noterdaeme, J.-M., 12
Notkin, G. E., 234, 238
Nowak, S., 173

O

Oasa, K., 149, 157
Oda, T., 157
Odajima, K., 149, 157
Ogawa, H., 149
Ogawa, T., 149, 157
Ogo, T., 157
Okabayashi, M., 64, 68, 72, 76, 131, 202, 413
Oliver, H., 64
Ongena, J., 32
Ono, M., 64, 68, 72, 76, 246
Orsitto, F., 115
Oyevaar, T., 32

P

Panaccione, L., 115
Panella, M., 115
Paoletti, F., 64, 119, 131, 202

Paul, S., 64, 72, 202
Pavlov, Yu. D., 234
PBX-M Group, 76
Pecquet, A. L., 24
Pericoli, V., 115
Petravich, G., 123
Petrov, M. P., 44
Petty, C. C., 165, 339, 351
Peysson, Y., 24, 107, 315
Phelps, D. A., 323, 339
Phillips, C. K., 36, 44, 48, 401, 405, 409
Ping, J. L., 343
Pieroni, L., 115
Pinsker, R. I., 165, 179, 323, 339, 351
Podda, S., 115
Porkolab, M., 56, 210, 218, 222, 226, 375, 457
Poschenrieder, W., 12
Post-Zwicker, A., 68, 72, 123
Probert, P. H., 327, 331, 387, 429

R

Raftopoulos, S., 437
Ram, A. K., 293, 433
Ramos, J. J., 210
Ramsey, A., 44
Rasmussen, D. A., 36, 40, 48, 60, 437
Razumova, K. A., 234, 238
Reitzel, K. J., 277
Rey, G., 24, 315
Riccardi, C., 80
Rice, B. W., 157
Richter, Th., 12
Riemer, B. W., 343
Righetti, G. B., 115
Rimini, F. G., 48, 123, 202
Roccon, M., 115
Rogers, J. H., 36, 40, 44, 48, 401,
Rognlien, T. D., 157
Rome, J. A., 60
Roney, P., 123
Rost, J. C., 449
Roy, I. N., 234
Ryan, P. M., 339, 343, 351, 355, 359
Ryter, F., 12

S

Sabbagh, S. A., 107
Saigusa, M., 52, 99
Salmon, N., 12
Salzmann, H., 12
Santarsiero, M., 285
Santi, D., 115
Santini, F., 115, 285
Saoutic, B., 24, 347
Sassi, M., 115
Sato, M., 52, 99
Sauter, O., 169, 367
Sauthoff, N., 64, 202
Sawan, M., 391
Scharer, J. E., 363, 367
Schettini, G., 285
Schilling, G., 3, 36, 40, 44, 48, 401
Schlosser, J., 315
Schlüter, H., 309
Schmitz, L., 64, 76, 202
Schneider, W., 12
Schüller, P. G., 137
Schultz, S. D., 433
Ségui, J. L., 111
Seike, T., 149
Seki, M., 99, 315, 319
Seki, T., 64
Serrecchia, R., 285
Sesnic, S., 64, 68, 131, 202
Shiina, T., 149
Shipley, S., 371
Shivarova, A., 309, 453
Shoji, T., 149
Simonetto, A., 173
Sindoni, E., 80
Skinner, C. H., 44
Smith, G. R., 157, 273
Solari, G., 173
Soltwisch, H., 32
Sorenson, J., 429
Squire, J. P., 226, 457
Stallard, B. W., 157
Stallings, D. C., 335, 343, 371, 437
Start, D., 351
Sternini, E., 115
Stevens, J., 36, 40, 44, 48, 123, 401, 405
Sugiyama, L., 425

Sun, Y., 131, 202, 413
Sund, R. S., 367
Suzuki, N., 149
Swain, D. W., 371, 375, 391

T

T-10 Team, 234, 238
Takahashi, H., 64, 68, 72, 131, 202
Takase, Y., 56
Takeuchi, H., 52
Tamai, H., 149
Tammen, H. F., 32
Taylor, D. J., 343
Taylor, G., 36, 44, 48
Telesca, G., 32
Temkin, R. J., 395
TFTR Group (PPPL) 36, 44, 405
Thomas, C. E., 60, 347
Thomassen, K. I., 157
Tighe, W., 64, 68, 72, 202
Tonini, G., 115
Tonon, G., 315
Tore Supra Team, 107
Tracy, E. R., 253
Tuccillo, A. A., 115
Tudisco, O., 115
Tynan, G., 64, 76, 202

U

Uehara, K., 149
Uhlemann, R., 32
Ushigusa, K., 52, 99, 319

V

Vacca, L., 445
Vahala, G., 230
Vahala, L., 230
Valente, F., 115
Valeo, E. J., 127, 202
Vandenplas, P. E., 32
Van Eester, D., 32, 297
Van Houtte, D., 24

Van Nieuwenhove, R., 32
Van Oost, G., 32
Van Wassenhove, G., 32
Vasin, N. L., 234
Vershkov, V. A., 234
Vervier, M., 32
Vézard, D., 111
Villaseñor, J., 226, 457
Vitale, V., 115
von Goeler, S., 123, 127, 131, 202

W

W7-AS Team, 137
Wade, T., 351, 355
Waidmann, G., 32
Wesner, F., 12
Weynants, R. R., 32
Wilgen, J. B., 437
Wilson, J. R., 36, 40, 44, 48, 383, 401, 405, 409, 437
Winz, G., 429
Wolfe, S. W., 319
Wong, K.-L., 44, 409
Wood, R. D., 157

Y

Yamamoto, T., 149
Yamauchi, T., 149
Yoshino, R., 99
Yue, S., 273
Yugo, J. J., 327, 343, 371, 375

Z

Zabiégo, M., 24
Zanza, V., 115
Zarnstorff, M. C., 48
Zehrfeld, H.-P., 12
Zerbini, M., 115
Zethoff, M., 453
Zimmerman, E. D., 449
Zohm, H., 12

AIP Conference Proceedings

		L.C. Number	ISBN
No. 108	The Time Projection Chamber (TRIUMF, Vancouver, 1983)	83-83445	0-88318-307-2
No. 109	Random Walks and Their Applications in the Physical and Biological Sciences (NBS/La Jolla Institute, 1982)	84-70208	0-88318-308-0
No. 110	Hadron Substructure in Nuclear Physics (Indiana University, 1983)	84-70165	0-88318-309-9
No. 111	Production and Neutralization of Negative Ions and Beams (3rd Int'l Symposium) (Brookhaven, NY, 1983)	84-70379	0-88318-310-2
No. 112	Particles and Fields – 1983 (APS/DPF, Blacksburg, VA)	84-70378	0-88318-311-0
No. 113	Experimental Meson Spectroscopy – 1983 (7th International Conference, Brookhaven, NY)	84-70910	0-88318-312-9
No. 114	Low Energy Tests of Conservation Laws in Particle Physics (Blacksburg, VA, 1983)	84-71157	0-88318-313-7
No. 115	High Energy Transients in Astrophysics (Santa Cruz, CA, 1983)	84-71205	0-88318-314-5
No. 116	Problems in Unification and Supergravity (La Jolla Institute, 1983)	84-71246	0-88318-315-3
No. 117	Polarized Proton Ion Sources (TRIUMF, Vancouver, 1983)	84-71235	0-88318-316-1
No. 118	Free Electron Generation of Extreme Ultraviolet Coherent Radiation (Brookhaven/OSA, 1983)	84-71539	0-88318-317-X
No. 119	Laser Techniques in the Extreme Ultraviolet (OSA, Boulder, CO, 1984)	84-72128	0-88318-318-8
No. 120	Optical Effects in Amorphous Semiconductors (Snowbird, UT, 1984)	84-72419	0-88318-319-6
No. 121	High Energy e^+e^- Interactions (Vanderbilt, 1984)	84-72632	0-88318-320-X
No. 122	The Physics of VLSI (Xerox, Palo Alto, CA, 1984)	84-72729	0-88318-321-8
No. 123	Intersections Between Particle and Nuclear Physics (Steamboat Springs, CO, 1984)	84-72790	0-88318-322-6
No. 124	Neutron-Nucleus Collisions: A Probe of Nuclear Structure (Burr Oak State Park, 1984)	84-73216	0-88318-323-4

No. 125	Capture Gamma-Ray Spectroscopy and Related Topics – 1984 (Int'l Symposium, Knoxville, TN)	84-73303	0-88318-324-2
No. 126	Solar Neutrinos and Neutrino Astronomy (Homestake, 1984)	84-63143	0-88318-325-0
No. 127	Physics of High Energy Particle Accelerators (BNL/SUNY Summer School, 1983)	85-70057	0-88318-326-9
No. 128	Nuclear Physics with Stored, Cooled Beams (McCormick's Creek State Park, IN, 1984)	85-71167	0-88318-327-7
No. 129	Radiofrequency Plasma Heating (Sixth Topical Conference) (Callaway Gardens, GA, 1985)	85-48027	0-88318-328-5
No. 130	Laser Acceleration of Particles (Malibu, CA, 1985)	85-48028	0-88318-329-3
No. 131	Workshop on Polarized ^{3}He Beams and Targets (Princeton, NJ, 1984)	85-48026	0-88318-330-7
No. 132	Hadron Spectroscopy – 1985 (International Conference, Univ. of Maryland)	85-72537	0-88318-331-5
No. 133	Hadronic Probes and Nuclear Interactions (Arizona State University, 1985)	85-72638	0-88318-332-3
No. 134	The State of High Energy Physics (BNL/SUNY Summer School, 1983)	85-73170	0-88318-333-1
No. 135	Energy Sources: Conservation and Renewables (APS, Washington, DC, 1985)	85-73019	0-88318-334-X
No. 136	Atomic Theory Workshop on Relativistic and QED Effects in Heavy Atoms (Gaithersburg, MD, 1985)	85-73790	0-88318-335-8
No. 137	Polymer-Flow Interaction (La Jolla Institute, 1985)	85-73915	0-88318-336-6
No. 138	Frontiers in Electronic Materials and Processing (Houston, TX, 1985)	86-70108	0-88318-337-4
No. 139	High-Current, High-Brightness, and High-Duty Factor Ion Injectors (La Jolla Institute, 1985)	86-70245	0-88318-338-2
No. 140	Boron-Rich Solids (Albuquerque, NM, 1985)	86-70246	0-88318-339-0
No. 141	Gamma-Ray Bursts (Stanford, CA, 1984)	86-70761	0-88318-340-4
No. 142	Nuclear Structure at High Spin, Excitation, and Momentum Transfer (Indiana University, 1985)	86-70837	0-88318-341-2
No. 143	Mexican School of Particles and Fields (Oaxtepec, México, 1984)	86-81187	0-88318-342-0

No. 144	Magnetospheric Phenomena in Astrophysics (Los Alamos, NM, 1984)	86-71149	0-88318-343-9
No. 145	Polarized Beams at SSC & Polarized Antiprotons (Ann Arbor, MI & Bodega Bay, CA, 1985)	86-71343	0-88318-344-7
No. 146	Advances in Laser Science—I (Dallas, TX, 1985)	86-71536	0-88318-345-5
No. 147	Short Wavelength Coherent Radiation: Generation and Applications (Monterey, CA, 1986)	86-71674	0-88318-346-3
No. 148	Space Colonization: Technology and The Liberal Arts (Geneva, NY, 1985)	86-71675	0-88318-347-1
No. 149	Physics and Chemistry of Protective Coatings (Universal City, CA, 1985)	86-72019	0-88318-348-X
No. 150	Intersections Between Particle and Nuclear Physics (Lake Louise, Canada, 1986)	86-72018	0-88318-349-8
No. 151	Neural Networks for Computing (Snowbird, UT, 1986)	86-72481	0-88318-351-X
No. 152	Heavy Ion Inertial Fusion (Washington, DC, 1986)	86-73185	0-88318-352-8
No. 153	Physics of Particle Accelerators (SLAC Summer School, 1985) (Fermilab Summer School, 1984)	87-70103	0-88318-353-6
No. 154	Physics and Chemistry of Porous Media—II (Ridgefield, CT, 1986)	83-73640	0-88318-354-4
No. 155	The Galactic Center: Proceedings of the Symposium Honoring C. H. Townes (Berkeley, CA, 1986)	86-73186	0-88318-355-2
No. 156	Advanced Accelerator Concepts (Madison, WI, 1986)	87-70635	0-88318-358-0
No. 157	Stability of Amorphous Silicon Alloy Materials and Devices (Palo Alto, CA, 1987)	87-70990	0-88318-359-9
No. 158	Production and Neutralization of Negative Ions and Beams (Brookhaven, NY, 1986)	87-71695	0-88318-358-7
No. 159	Applications of Radio-Frequency Power to Plasma: Seventh Topical Conference (Kissimmee, FL, 1987)	87-71812	0-88318-359-5
No. 160	Advances in Laser Science—II (Seattle, WA, 1986)	87-71962	0-88318-360-9
No. 161	Electron Scattering in Nuclear and Particle Science: In Commemoration of the 35th Anniversary of the Lyman-Hanson-Scott Experiment (Urbana, IL, 1986)	87-72403	0-88318-361-7

No. 162	Few-Body Systems and Multiparticle Dynamics (Crystal City, VA, 1987)	87-72594	0-88318-362-5
No. 163	Pion–Nucleus Physics: Future Directions and New Facilities at LAMPF (Los Alamos, NM, 1987)	87-72961	0-88318-363-3
No. 164	Nuclei Far from Stability: Fifth International Conference (Rosseau Lake, ON, 1987)	87-73214	0-88318-364-1
No. 165	Thin Film Processing and Characterization of High-Temperature Superconductors (Anaheim, CA, 1987)	87-73420	0-88318-365-X
No. 166	Photovoltaic Safety (Denver, CO, 1988)	88-42854	0-88318-366-8
No. 167	Deposition and Growth: Limits for Microelectronics (Anaheim, CA, 1987)	88-71432	0-88318-367-6
No. 168	Atomic Processes in Plasmas (Santa Fe, NM, 1987)	88-71273	0-88318-368-4
No. 169	Modern Physics in America: A Michelson-Morley Centennial Symposium (Cleveland, OH, 1987)	88-71348	0-88318-369-2
No. 170	Nuclear Spectroscopy of Astrophysical Sources (Washington, DC, 1987)	88-71625	0-88318-370-6
No. 171	Vacuum Design of Advanced and Compact Synchrotron Light Sources (Upton, NY, 1988)	88-71824	0-88318-371-4
No. 172	Advances in Laser Science—III: Proceedings of the International Laser Science Conference (Atlantic City, NJ, 1987)	88-71879	0-88318-372-2
No. 173	Cooperative Networks in Physics Education (Oaxtepec, Mexico, 1987)	88-72091	0-88318-373-0
No. 174	Radio Wave Scattering in the Interstellar Medium (San Diego, CA, 1988)	88-72092	0-88318-374-9
No. 175	Non-neutral Plasma Physics (Washington, DC, 1988)	88-72275	0-88318-375-7
No. 176	Intersections Between Particle and Nuclear Physics (Third International Conference) (Rockport, ME, 1988)	88-62535	0-88318-376-5
No. 177	Linear Accelerator and Beam Optics Codes (La Jolla, CA, 1988)	88-46074	0-88318-377-3
No. 178	Nuclear Arms Technologies in the 1990s (Washington, DC, 1988)	88-83262	0-88318-378-1
No. 179	The Michelson Era in American Science: 1870–1930 (Cleveland, OH, 1987)	88-83369	0-88318-379-X
No. 180	Frontiers in Science: International Symposium (Urbana, IL, 1987)	88-83526	0-88318-380-3

No. 181	Muon-Catalyzed Fusion (Sanibel Island, FL, 1988)	88-83636	0-88318-381-1
No. 182	High T_c Superconducting Thin Films, Devices, and Applications (Atlanta, GA, 1988)	88-03947	0-88318-382-X
No. 183	Cosmic Abundances of Matter (Minneapolis, MN, 1988)	89-80147	0-88318-383-8
No. 184	Physics of Particle Accelerators (Ithaca, NY, 1988)	89-83575	0-88318-384-6
No. 185	Glueballs, Hybrids, and Exotic Hadrons (Upton, NY, 1988)	89-83513	0-88318-385-4
No. 186	High-Energy Radiation Background in Space (Sanibel Island, FL, 1987)	89-83833	0-88318-386-2
No. 187	High-Energy Spin Physics (Minneapolis, MN, 1988)	89-83948	0-88318-387-0
No. 188	International Symposium on Electron Beam Ion Sources and their Applications (Upton, NY, 1988)	89-84343	0-88318-388-9
No. 189	Relativistic, Quantum Electrodynamic, and Weak Interaction Effects in Atoms (Santa Barbara, CA, 1988)	89-84431	0-88318-389-7
No. 190	Radio-frequency Power in Plasmas (Irvine, CA, 1989)	89-45805	0-88318-397-8
No. 191	Advances in Laser Science—IV (Atlanta, GA, 1988)	89-85595	0-88318-391-9
No. 192	Vacuum Mechatronics (First International Workshop) (Santa Barbara, CA, 1989)	89-45905	0-88318-394-3
No. 193	Advanced Accelerator Concepts (Lake Arrowhead, CA, 1989)	89-45914	0-88318-393-5
No. 194	Quantum Fluids and Solids—1989 (Gainesville, FL, 1989)	89-81079	0-88318-395-1
No. 195	Dense Z-Pinches (Laguna Beach, CA, 1989)	89-46212	0-88318-396-X
No. 196	Heavy Quark Physics (Ithaca, NY, 1989)	89-81583	0-88318-644-6
No. 197	Drops and Bubbles (Monterey, CA, 1988)	89-46360	0-88318-392-7
No. 198	Astrophysics in Antarctica (Newark, DE, 1989)	89-46421	0-88318-398-6
No. 199	Surface Conditioning of Vacuum Systems (Los Angeles, CA, 1989)	89-82542	0-88318-756-6
No. 200	High T_c Superconducting Thin Films: Processing, Characterization, and Applications (Boston, MA, 1989)	90-80006	0-88318-759-0

No. 201	QED Structure Functions (Ann Arbor, MI, 1989)	90-80229	0-88318-671-3
No. 202	NASA Workshop on Physics From a Lunar Base (Stanford, CA, 1989)	90-55073	0-88318-646-2
No. 203	Particle Astrophysics: The NASA Cosmic Ray Program for the 1990s and Beyond (Greenbelt, MD, 1989)	90-55077	0-88318-763-9
No. 204	Aspects of Electron-Molecule Scattering and Photoionization (New Haven, CT, 1989)	90-55175	0-88318-764-7
No. 205	The Physics of Electronic and Atomic Collisions (XVI International Conference) (New York, NY, 1989)	90-53183	0-88318-390-0
No. 206	Atomic Processes in Plasmas (Gaithersburg, MD, 1989)	90-55265	0-88318-769-8
No. 207	Astrophysics from the Moon (Annapolis, MD, 1990)	90-55582	0-88318-770-1
No. 208	Current Topics in Shock Waves (Bethlehem, PA, 1989)	90-55617	0-88318-776-0
No. 209	Computing for High Luminosity and High Intensity Facilities (Santa Fe, NM, 1990)	90-55634	0-88318-786-8
No. 210	Production and Neutralization of Negative Ions and Beams (Brookhaven, NY, 1990)	90-55316	0-88318-786-8
No. 211	High-Energy Astrophysics in the 21st Century (Taos, NM, 1989)	90-55644	0-88318-803-1
No. 212	Accelerator Instrumentation (Brookhaven, NY, 1989)	90-55838	0-88318-645-4
No. 213	Frontiers in Condensed Matter Theory (New York, NY, 1989)	90-6421	0-88318-771-X 0-88318-772-8 (pbk.)
No. 214	Beam Dynamics Issues of High-Luminosity Asymmetric Collider Rings (Berkeley, CA, 1990)	90-55857	0-88318-767-1
No. 215	X-Ray and Inner-Shell Processes (Knoxville, TN, 1990)	90-84700	0-88318-790-6
No. 216	Spectral Line Shapes, Vol. 6 (Austin, TX, 1990)	90-06278	0-88318-791-4
No. 217	Space Nuclear Power Systems (Albuquerque, NM, 1991)	90-56220	0-88318-838-4
No. 218	Positron Beams for Solids and Surfaces (London, Canada, 1990)	90-56407	0-88318-842-2
No. 219	Superconductivity and Its Applications (Buffalo, NY, 1990)	91-55020	0-88318-835-X

No. 220	High Energy Gamma-Ray Astronomy (Ann Arbor, MI, 1990)	91-70876	0-88318-812-0
No. 221	Particle Production Near Threshold (Nashville, IN, 1990)	91-55134	0-88318-829-5
No. 222	After the First Three Minutes (College Park, MD, 1990)	91-55214	0-88318-828-7
No. 223	Polarized Collider Workshop (University Park, PA, 1990)	91-71303	0-88318-826-0
No. 224	LAMPF Workshop on (π, K) Physics (Los Alamos, NM, 1990)	91-71304	0-88318-825-2
No. 225	Half Collision Resonance Phenomena in Molecules (Caracas, Venezuela, 1990)	91-55210	0-88318-840-6
No. 226	The Living Cell in Four Dimensions (Gif sur Yvette, France, 1990)	91-55209	0-88318-794-9
No. 227	Advanced Processing and Characterization Technologies (Clearwater, FL, 1991)	91-55194	0-88318-910-0
No. 228	Anomalous Nuclear Effects in Deuterium/Solid Systems (Provo, UT, 1990)	91-55245	0-88318-833-3
No. 229	Accelerator Instrumentation (Batavia, IL, 1990)	91-55347	0-88318-832-1
No. 230	Nonlinear Dynamics and Particle Acceleration (Tsukuba, Japan, 1990)	91-55348	0-88318-824-4
No. 231	Boron-Rich Solids (Albuquerque, NM, 1990)	91-53024	0-88318-793-4
No. 232	Gamma-Ray Line Astrophysics (Paris-Saclay, France, 1990)	91-55492	0-88318-875-9
No. 233	Atomic Physics 12 (Ann Arbor, MI, 1990)	91-55595	088318-811-2
No. 234	Amorphous Silicon Materials and Solar Cells (Denver, CO, 1991)	91-55575	088318-831-7
No. 235	Physics and Chemistry of MCT and Novel IR Detector Materials (San Francisco, CA, 1990)	91-55493	0-88318-931-3
No. 236	Vacuum Design of Synchrotron Light Sources (Argonne, IL, 1990)	91-55527	0-88318-873-2
No. 237	Kent M. Terwilliger Memorial Symposium (Ann Arbor, MI, 1989)	91-55576	0-88318-788-4
No. 238	Capture Gamma-Ray Spectroscopy (Pacific Grove, CA, 1990)	91-57923	0-88318-830-9

No. 239	Advances in Biomolecular Simulations (Obernai, France, 1991)	91-58106	0-88318-940-2
No. 240	Joint Soviet-American Workshop on the Physics of Semiconductor Lasers (Leningrad, USSR, 1991)	91-58537	0-88318-936-4
No. 241	Scanned Probe Microscopy (Santa Barbara, CA, 1991)	91-76758	0-88318-816-3
No. 242	Strong, Weak, and Electromagnetic Interactions in Nuclei, Atoms, and Astrophysics: A Workshop in Honor of Stewart D. Bloom's Retirement (Livermore, CA, 1991)	91-76876	0-88318-943-7
No. 243	Intersections Between Particle and Nuclear Physics (Tucson, AZ, 1991)	91-77580	0-88318-950-X
No. 244	Radio Frequency Power in Plasmas (Charleston, SC, 1991)	91-77853	0-88318-937-2
No. 245	Basic Space Science (Bangalore, India, 1991)	91-78379	0-88318-951-8
No. 246	Space Nuclear Power Systems (Albuquerque, NM, 1992)	91-58793	1-56396-027-3 1-56396-026-5 (pbk.)
No. 247	Global Warming: Physics and Facts (Washington, DC, 1991)	91-78423	0-88318-932-1
No. 248	Computer-Aided Statistical Physics (Taipei, Taiwan, 1991)	91-78378	0-88318-942-9
No. 249	The Physics of Particle Accelerators (Upton, NY, 1989, 1990)	92-52843	0-88318-789-2
No. 250	Towards a Unified Picture of Nuclear Dynamics (Nikko, Japan, 1991)	92-70143	0-88318-951-8
No. 251	Superconductivity and its Applications (Buffalo, NY, 1991)	92-52726	1-56396-016-8
No. 252	Accelerator Instrumentation (Newport News, VA, 1991)	92-70356	0-88318-934-8
No. 253	High-Brightness Beams for Advanced Accelerator Applications (College Park, MD, 1991)	92-52705	0-88318-947-X
No. 254	Testing the AGN Paradigm (College Park, MD, 1991)	92-52780	1-56396-009-5
No. 255	Advanced Beam Dynamics Workshop on Effects of Errors in Accelerators, Their Diagnosis and Corrections (Corpus Christi, TX, 1991)	92-52842	1-56396-006-0

No. 256	Slow Dynamics in Condensed Matter (Fukuoka, Japan, 1991)	92-53120	0-88318-938-0
No. 257	Atomic Processes in Plasmas (Portland, ME, 1991)	91-08105	0-88318-939-9
No. 258	Synchrotron Radiation and Dynamic Phenomena (Grenoble, France, 1991)	92-53790	1-56396-008-7
No. 259	Future Directions in Nuclear Physics with 4π Gamma Detection Systems of the New Generation (Strasbourg, France, 1991)	92-53222	0-88318-952-6
No. 260	Computational Quantum Physics (Nashville, TN, 1991)	92-71777	0-88318-933-X
No. 261	Rare and Exclusive B&K Decays and Novel Flavor Factories (Santa Monica, CA, 1991)	92-71873	1-56396-055-9
No. 262	Molecular Electronics—Science and Technology (St. Thomas, Virgin Islands, 1991)	92-72210	1-56396-041-9
No. 263	Stress-Induced Phenomena in Metallization: First International Workshop (Ithaca, NY, 1991)	92-72292	1-56396-082-6
No. 264	Particle Acceleration in Cosmic Plasmas (Newark, DE, 1991)	92-73316	0-88318-948-8
No. 265	Gamma-Ray Bursts (Huntsville, AL, 1991)	92-73456	1-56396-018-4
No. 266	Group Theory in Physics (Cocoyoc, Morelos, Mexico, 1991)	92-73457	1-56396-101-6
No. 267	Electromechanical Coupling of the Solar Atmosphere (Capri, Italy, 1991)	92-82717	1-56396-110-5
No. 268	Photovoltaic Advanced Research & Development Project (Denver, CO, 1992)	92-74159	1-56396-056-7
No. 269	CEBAF 1992 Summer Workshop (Newport News, VA, 1992)	92-75403	1-56396-067-2
No. 270	Time Reversal—The Arthur Rich Memorial Symposium (Ann Arbor, MI, 1991)	92-83852	1-56396-105-9
No. 271	Tenth Symposium Space Nuclear Power and Propulsion (Vols. I–III) (Albuquerque, NM, 1993)	92-75162	1-56396-137-7 (set)

No. 272	Proceedings of the XXVI International Conference on High Energy Physics (Vols. I and II) (Dallas, TX, 1992)	93-70412	1-56396-127-X (set)
No. 273	Superconductivity and Its Applications (Buffalo, NY, 1992)	93-70502	1-56396-189-X
No. 274	VIth International Conference on the Physics of Highly Charged Ions (Manhattan, KS, 1992)	93-70577	1-56396-102-4
No. 275	Atomic Physics 13 (Munich, Germany, 1992)	93-70826	1-56396-057-5
No. 276	Very High Energy Cosmic-Ray Interactions: VIIth International Symposium (Ann Arbor, MI, 1992)	93-71342	1-56396-038-9
No. 277	The World at Risk: Natural Hazards and Climate Change (Cambridge, MA, 1992)	93-71333	1-56396-066-4
No. 278	Back to the Galaxy (College Park, MD, 1992)	93-71543	1-56396-227-6
No. 279	Advanced Accelerator Concepts (Port Jefferson, NY, 1992)	93-71773	1-56396-191-1
No. 280	Compton Gamma-Ray Observatory (St. Louis, MO, 1992)	93-71830	1-56396-104-0
No. 281	Accelerator Instrumentation Fourth Annual Workshop (Berkeley, CA, 1992)	93-072110	1-56396-190-3
No. 282	Quantum 1/f Noise & Other Low Frequency Fluctuations in Electronic Devices (St. Louis, MO, 1992)	93-072366	1-56396-252-7
No. 283	Earth and Space Science Information Systems (Pasadena, CA, 1992)	93-072360	1-56396-094-X
No. 284	US-Japan Workshop on Ion Temperature Gradient-Driven Turbulent Transport (Austin, TX, 1993)	93-72460	1-56396-221-7
No. 285	Noise in Physical Systems and 1/f Fluctuations (St. Louis, MO, 1993)	93-72575	1-56396-270-5